Collins and Lyne's
Microbiological Methods

Collins and Lyne's

Microbiological Methods

Seventh edition

C. H. Collins MBE, DSc, FRCPath, FIBiol, FIBMS
Senior Visiting Research Fellow, Department of Microbiology, National Heart and Lung Institute, University of London

Patricia M. Lyne CBiol, MIBiol
The Ashes, Hadlow, Kent

J. M. Grange MD, MSc
Director, Department of Microbiology, National Heart and Lung Institute, University of London and Honorary Consultant Microbiologist, Royal Brompton Hospital, London

BUTTERWORTH
HEINEMANN

Butterworth-Heinemann Ltd
Linacre House, Jordan Hill, Oxford OX2 8DP

A member of the Reed Elsevier plc group

OXFORD LONDON BOSTON
MUNICH NEW DELHI SINGAPORE SYDNEY
TOKYO TORONTO WELLINGTON

First published 1964
Reprinted 1966
Second edition 1967
Reprinted 1970
Third edition 1970
Fourth edition 1976
Reprinted with additions 1979
Fifth edition 1984
Revised reprint 1985
Reprinted 1987
Sixth edition 1989
Revised reprint 1991
Seventh edition 1995

British Library Cataloguing in Publication Data
Collins, C. H.
Collins and Lyne's Microbiological
Methods. – 7Rev.ed
I. Titles II. Lyne, Patricia M.
III. Grange, John M.
576.028

ISBN 0 7506 0653 3

Library of Congress Cataloguing in Publication Data
Collins, C. H. (Christopher Herbert)
Collins and Lyne's microbiological methods/C. H. Collins,
Patricia M. Lyne, J. M. Grange. – 7th ed.
p. cm.
Includes bibliographical references and index.
ISBN 0 7506 0653 3
1. Microbiology – Technique. 2. Microbiology – Laboratory manuals.
I. Lyne, Patricia M. II. Grange, John M. III. Title. IV. Title:
Microbiological methods.
QR65.C64
576′.026–dc20 94–21648
CIP

Printed in Great Britain

Contents

Contributors

L. Berg, BSc, MT, FIBMS, Clinical Laboratory, Encino-Tarzana Regional Medical Center, Tarzana, California, USA

J. S. Brazier, MSc, PhD, CBiol, MIBiol, FIBMS, Anaerobe Reference Laboratory, Public Health Laboratory, University Hospital of Wales, Heath Park, Cardiff.

E. Y. Bridson, MPhil, CBiol, MIBiol, FIBMS, 3 Bellever Hill, Camberley, Surrey

D. F. J. Brown, PhD, Public Health Laboratory, Addenbrooke's Hospital, Hills Road, Cambridge

C. K. Campbell, MSc, PhD, Mycology Reference Laboratory, Public Health Laboratory, Myrtle Road, Kingsdown, Bristol

C. H. Collins, MBE, DSc, FRCPath, CBiol, FIBiol, Department of Microbiology, National Heart and Lung Institute, Royal Brompton Hospital, Sydney Street, London

Janet E. L. Corry, BSc, MSc, PhD, CBiol, MIBiol, Department of Veterinary Medicine, University of Bristol, Langford, Bristol

D. A. B. Dance, MB, ChB, MSc, MRCPath, Department of Clinical Sciences, London School of Hygiene and Tropical Medicine, University of London, Keppel Street, London

T. J. Donovan, PhD, CBiol, MIBiol, FIBMS, Public Health Laboratory, William Harvey Hospital, Ashford, Kent

R. C. George, MB, BS, MSc, MRCPath, Respiratory and Systemic Infection Reference Laboratory, Central Public Health Laboratory, Colindale Avenue, London

J. M. Grange, MD, MSc, Department of Microbiology, National Heart and Lung Institute, Royal Brompton Hospital, Sydney Street, London

Patricia M. Lyne, CBiol, MIBiol, The Ashes, Hadlow, Kent

M.O. Moss, BSc, PhD, DIC, ARCS, Department of Microbiology, University of Surrey, Guildford, Surrey

S. F. Perry, BSc, CBiol, MIBiol, Quality Assurance Laboratory, Central Public Health Laboratory, Colindale Avenue, London

D. N. Petts, MSc, PhD, CBiol, MIBiol, FIBMS, Microbiology Department, Basildon Hospital, Basildon, Essex

P. Silley, BSc, PhD, CBiol, MIBiol, Don Whitley Scientific Ltd, Ottley Road, Shipley, West Yorkshire

P. Taylor, MSc, FIBMS, Microbiology Department, Royal Brompton Hospital, Sydney Street, London.

Preface to the seventh edition

The continued demand for this book and the rapid progress in microbial technology have encouraged the publishers and us to produce yet another edition.

We have followed the format and editorial procedures of earlier editions and asked our colleagues to revise existing text where necessary and to provide new material. More attention has been given to automation and new matter on quality assurance has been included. Some parts of the text have been radically rewritten, while basic techniques, so essential in the teaching of microbiology and in areas where modern technology is not yet available, have been retained.

As before, we have endeavoured to keep up to date with nomenclature, while also retaining well-known names. The revision of the tables of biochemical properties has been a major task, and where differences have been found in the publications consulted we have accepted the authority of the editors of *Cowan and Steel's Manual for the Identification of Medical Bacteria*, for which information we are indebted.

Again, we must thank our contributors for their assistance and advice, and also for permitting us to mould their material into the style which has been a feature of earlier editions.

C. H. Collins
Patricia M. Lyne
Hadlow

J. M. Grange
London

Chapter

1

Safety in microbiology

So far this century over 4000 laboratory workers have become infected with microorganisms in the course of their work and some have died. Unfortunately there is still an annual increment in these infections although most of them are clearly preventable.

To avoid becoming infected, therefore, laboratory workers should:

(1) Be aware of the potential hazards of the materials and organisms they handle, especially (a) those containing blood, and (b) those received from tropical and subtropical countries.
(2) Be aware of the routes by which pathogenic microorganisms can enter the human body and cause infections.
(3) Receive proper instruction and thereafter practise good microbiological technique (GMT) which effectively 'contains' microorganisms so that they do not have access to those routes.

Classification of microorganisms on the basis of risk

Microorganisms vary in their ability to infect individuals and cause disease. Some are harmless; some may be responsible for disease with mild symptoms; others can cause serious illnesses; a few have the potential for spreading in the community and causing serious epidemic disease. Experience and research into laboratory-acquired infections have enabled investigators to classify microorganisms and viruses into four categories (also known as groups or classes) numbered 1–4 on the basis of the hazards they present to laboratory workers and to the community should they 'escape' from the laboratory. These categories take into account the seriousness of the associated disease, the availability of prophylactic and therapeutic agents, the routes of infection and the history of laboratory infections caused by the organisms.

Three systems that are in current use are shown in Table 1.1, although similar systems have also been formulated by other states and organizations. All systems agree in principle, but there are slight differences in wording and emphasis. In this book the term 'Hazard Group' is used.

Lists of organisms in Hazard Groups 2, 3 and 4 have been compiled, e.g. by the US Public Health Service (USPHS, 1981), the UK Advisory Committee on Dangerous Pathogens (ACDP, 1990a), Standards Australia (AS, 1990), Association Française Normalization (AFNOR, 1990) and Berufsgenossenschaft der chemischen Industrie (BG Chemie, 1991, 1992). The Commission for the European Communities is currently developing lists for use in the European Union. Some variations exist between countries and these reflect the differences

Table 1.1 Summary of systems for classifying microorganisms on the basis of hazards to laboratory workers and the community

	Hazard			
	Low			*High*
USPHS[a] (1974)	Class 1 none or minimal	Class 2 ordinary potential	Class 3 special, to individual	Class 4 high, to individual
WHO[b] (1979)	Risk Group I low individual low community	Risk Group II moderate individual, low community	Risk Group III high individual low community	Risk Group IV high, to individual and community
ACDP[c] (1990[a])	Hazard Group I unlikely to cause human disease	Hazard Group 2 possibly to laboratory workers, unlikely to community	Hazard Group 3 some hazard to laboratory workers. May spread to community	Hazard Group 4 Serious hazard to laboratory workers. High risk to community

[a] United States Public Health Service
[b] World Health Organization (see WHO, 1993)
[c] United Kingdom Advisory Committee on Dangerous Pathogens

in the incidence of the microorganisms and the diseases concerned. Readers should use their national lists.

In this book the UK lists (ACDP, 1990a) are used. Most of the organisms are in Groups 1 and 2. Cautions, in italic type, are given for those in Group 3. No Group 4 agents (which are all viruses) are discussed in the text.

Routes of infection

There are at least four routes by which microorganisms may enter the human body.

Through the lungs

Aerosols, very small droplets (<0.5 μm) containing microorganisms, are frequently released during normal microbiological manipulations such as work with loops and syringes, pipetting, centrifuging, blending, mixing and homogenizing, pouring, opening culture tubes and dishes (see Collins, 1993). Inhalation of these has probably caused the largest number of laboratory-acquired infections.

Through the mouth

Organisms may be ingested during mouth pipetting and if they are transferred to the mouth by contaminated fingers and articles (food, cigarettes, pencils, etc.) that have been in contact with contaminated laboratory benches as a result of spillage, splashes and settled aerosol droplets.

Through the skin

The skin of hands, etc. is rarely free from (often minute) cuts and abrasions through which organisms can enter the blood stream. Infections may also arise as a result of accidental inoculation with infected hypodermic needles, broken glass and other sharp objects.

Into the eyes

Splashes of infected material into the eyes, and transfer of organisms on contaminated fingers to the eyes are not uncommon routes of infection.

Levels of containment

The Hazard Group number allocated to an organism indicates the appropriate Containment or Biosafety Level, i.e. the necessary accommodation, equipment, techniques and precautions. Thus, there are four of these levels (see Table 1.2). In the WHO classification the Basic Laboratory equates with Levels 1 and 2, the Containment Laboratory with Level 3 and the Maximum Containment Laboratory with Level 4.

The italic type 'Caution' at the head of sections dealing with Group 3 organisms indicates the appropriate containment level and any additional precautions.

Full details are given by ACDP (1990a), USPHS (1993), WHO (1993) and Collins (1993). Only a synopsis is presented here.

Table 1.2 Summary of Biosafety/Containment level requirements

Level	*Facilities*	*Laboratory practice*	*Safety equipment*
1	Basic, Level 1	GMT[a]	None. Work on open bench
2	Basic, Level 2	GMT plus protective clothing, biohazard signs	Open bench plus safety cabinet for aerosol potential
3	Containment, Level 3	Level 2 plus special clothing, controlled access	Safety cabinet for all activities
4	Maximum Containment, Level 4	Level 3 plus air lock entry, shower exit, special waste disposal	Class III safety cabinet, pressure gradient, double-ended autoclave

[a] GMT, good microbiological technique

Levels 1 and 2: the Basic laboratory

These are intended for work with organisms in Hazard Groups 1 and 2. Ample space should be provided. Walls, ceilings and floors should be smooth, non-absorbent, easy to clean and disinfect and resistant to the chemicals that are likely to be used. Floors should also be slip resistant. Lighting and heating should be adequate. Hand basins, other than laboratory sinks, are essential.

Bench tops should be wide, the correct height for work at a comfortable sitting position, smooth, easy to clean and disinfect and resistant to chemicals likely to be used. Adequate storage facilities should be provided.

Access should be restricted to authorized persons.

Level 3: the Containment laboratory

This is intended for work with organisms in Hazard Group 3. All the features of the Basic laboratory should be incorporated, plus the following.

The room should be physically separated from other rooms, with no communication (e.g. by pipe ducts, false ceilings) with other areas, apart from the door, which should be lockable, and transfer grilles for ventilation (see below).

Ventilation should be one way, achieved by having a lower pressure in the Containment laboratory than in other, adjacent, rooms and areas. Air should be removed to atmosphere (total dump, not recirculated to other parts of the building) by an exhaust system coupled to a microbiological safety cabinet (see p.24) so that during working hours air is continually extracted by the cabinet or directly from the room and airborne particles cannot be moved around the building. Replacement air enters through transfer grilles.

Access should be strictly regulated. The international biohazard warning sign, with appropriate wording, should be fixed to the door (Figure 1.1).

Figure 1.1 International biohazard sign

Level 4: the Maximum Containment laboratory

This is required for work with Hazard Group 4 materials and is outside the scope of this book. Construction and use generally require government licence or supervision.

Prevention of laboratory-acquired infections: a code of practice

The principles of containment involve good microbiological technique (GMT), with which this book is particularly concerned, and the provision of:

(1) primary barriers around the organisms to prevent their dispersal into the laboratory,
(2) secondary barriers around the worker to act as a safety net should the primary barriers be breached,
(3) tertiary barriers to prevent the escape into the community of any organisms which are not contained by the primary and secondary barriers.

This code of practice draws very largely on the requirements and recommendations published in other works (ACDP, 1990a; HSAC, 1991; Collins, 1993; WHO, 1993).

Primary barriers

These are techniques and equipment designed to contain microorganisms and prevent their direct access to the worker and their dispersal as aerosols.

(1) All body fluids and pathological material should be regarded as potentially infected. 'High Risk' specimens, i.e. those that may contain agents in Hazard Group 3 or originate from patients in the AIDS and hepatitis risk categories should be so labelled (see below).
(2) Mouth pipetting should be *banned in all circumstances*. Pipetting devices (p.35) should be provided.
(3) No article should be placed in or at the mouth. This includes mouthpieces of rubber tubing attached to pipettes, pens, pencils, labels, smoking materials, fingers, food and drink.
(4) The use of hypodermic needles should be restricted. Cannulas are safer.
(5) Sharp glass pasteur pipettes should be replaced by soft plastic varieties (p.35).
(6) Cracked and chipped glassware should be replaced.
(7) Centrifuge tubes should be filled only to within 2 cm of the lip. All Hazard Group 3 materials should be centrifuged in sealed centrifuge buckets (p.22).
(8) Bacteriological loops should be completely closed, with a diameter not more than 3 mm and a shank not more than 5 cm and be held in metal, not glass, handles
(9) Homogenizers should be inspected regularly for defects which might disperse aerosols (p.23). Only the safest models should be used. Glass Griffith's tubes and tissue homogenizers should be held in a pad of wadding in a gloved hand when operated.
(10) All Hazard Group 3 materials should be processed in a Containment laboratory (see Tertiary barriers below) and in a microbiological safety cabinet (p.24) unless exempted by agreed national regulations.
(11) A supply of suitable disinfectant (p.46) at use dilution should be available at every work station.
(12) Benches and work surfaces should be disinfected regularly and after spillage of potentially infected material.
(13) Discard jars for small objects and for re-usable pipettes should be provided at each work station. They should be emptied, decontaminated and refilled with freshly prepared disinfectant daily (p.53).
(14) Discard bins or bags supported in bins should be placed near to each work station. They should be removed and autoclaved daily (p.52).
(15) Broken culture vessels, e.g. after accidents, should be covered with a cloth. Suitable disinfectant should be poured over the cloth and the whole left for 30 min. The debris should then be cleared up into a suitable container (tray or dustpan) and autoclaved. The hands should be gloved and stiff cardboard used to clear up the debris.
(16) Infectious and pathological material to be sent through the post or by airmail should be packed in accordance with government, post office and airline regulations, obtainable from those authorities. Full instructions are also given elsewhere (p.7).

(17) Specimen containers should be robust and leak proof.
(18) Discarded infectious material should not leave the laboratory until it has been sterilized or otherwise made safe (See p.51).

Secondary barriers

These are intended to protect the worker should the primary barriers fail. They should, however, be observed as strictly as the latter.

(1) Proper overalls should be worn at all times and fastened. They should be kept apart from outdoor and other clothing. When agents in Hazard Group 3 and material containing or suspected of containing hepatitis B or the AIDS viruses are handled plastic aprons should be worn over the normal protective clothing. Surgeons' gloves should also be worn.
(2) Laboratory protective clothing should be removed when the worker leaves the laboratory and not worn in any other areas such as canteens and rest rooms.
(3) Hands should be washed after handling infectious material and always before leaving the laboratory.
(4) Any obvious cuts, scratches and abrasions on exposed parts of the worker's body should be covered with waterproof dressings.
(5) In laboratories where pathogens are handled there should be medical supervision.
(6) All illnesses should be reported to the medical or occupational health supervisors. They may be laboratory associated. Pregnancy should also be reported. It is inadvisable to work with certain organisms during pregnancy.
(7) Any member of the staff who is receiving steroids or immunosuppressive drugs should report this to the medical supervisor.
(8) Workers should be immunized where possible and practicable against likely infections (Wright, 1988; Collins, 1993; WHO 1993).
(9) In laboratories where tuberculous materials are handled the staff should have received BCG or have evidence of a positive skin reaction before starting work. They should have a pre-employment chest X-ray (Wright, 1988).

Tertiary barriers

These are intended to offer additional protection to the worker and to prevent the escape into the community of microorganisms that are under investigation in the laboratory.

They concern the provision of accommodation and facilities at the Biosafety/ Containment levels appropriate for the organisms handled, and of microbiological safety cabinets.

For more information on laboratory design see BOHS (1992) and Collins (1993).

Precautions against infection with hepatitis B (HBV) and the human immunodeficiency virus (HIV) and other blood-borne pathogens

Laboratory workers who may be exposed to these viruses in clinical material are advised to follow the special precautions described by one or more of the following: ACDP, 1990a, b; OSHA, 1991; WHO, 1991; Collins 1993; USPHS, 1993.

The UK control of substances hazardous to health regulations (COSHH) 1988 (under revision)

These apply to microorganisms as well as to chemicals. Assessments in relation to microorganisms should be made according to the Hazard Groups in which they are placed (ACDP, 1990a), the level of containment applied, and the techniques used. Hazard Group 3 organisms are identified as such in this book.

Some of the chemicals used in microbiology are known or thought to be hazardous to health. They are identified in the text by the word '*Caution*'.

The safe collection, receipt and transport of infectious materials

Specimen containers

These containers should be robust, preferably screw-capped, and unlikely to leak. Plastic bags should be provided to receive the container after the specimen has been collected. Request forms should not be placed in the same plastic bag as the specimen. Unless the source and laboratory are in the same premises, there should be robust outer containers provided with adequate absorbent packing material.

Receipt in the laboratory

Specimens from local sources, i.e. within state boundaries may be unpacked on trays by trained reception staff who wear laboratory protective clothing and gloves. The specimens should be left in their plastic bags pending examination.

Transfrontier specimens, which may contain exotic microorganisms, should be unpacked only by trained laboratory staff.

Mailing specimens

Most of these will be 'diagnostic specimens' which, according to international agreement, are human or animal material, including cultures of microorganisms being shipped for the purposes of diagnosis.

Within state boundaries the usual requirements are that the specimen is in a leak-proof primary container, securely cocooned in absorbent material and firmly packed in a robust outer container, which is labelled 'Pathological Specimen' or 'Diagnostic Specimen'.

Specimens for transfrontier shipment, packed as above, require an additional outer container and documentation (Figures 1.2 and 1.3). Full details, which

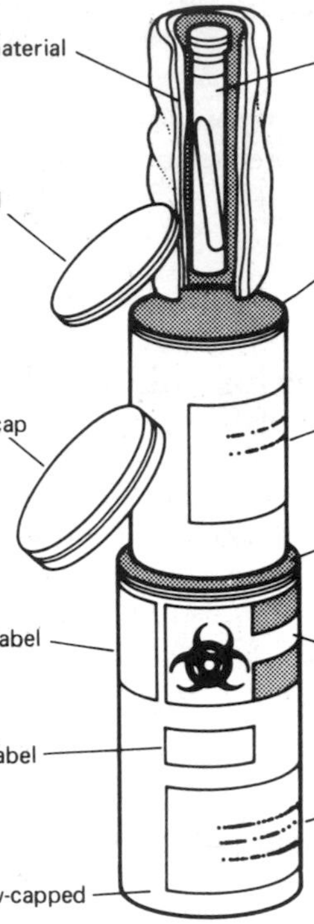

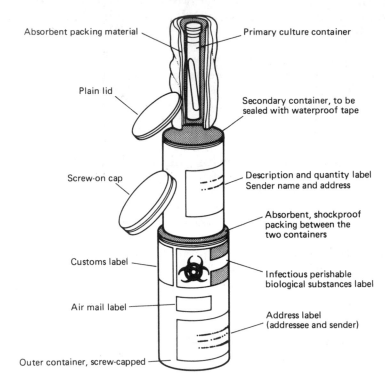

Figure 1.2 Packing infectious substances for the overseas post (airmail)

may vary from state to state, are published by national and international postal and civil aviation authorities (see also Collins, 1993; WHO, 1993).

The European Committee for Standardization (CEN) is currently preparing a standard: 'In-vitro Diagnostic Systems – Transport Packages for Medical and Biological Specimens . . .'

Instruction in safety

This instruction should be part of the general training given to all microbiology laboratory workers and should be included in what is known as good microbiological technique. In general, methods that protect cultures from contamination also protect workers from infection but this should not be taken for granted. Personal protection can be achieved only by good training and careful work. Programmes for training in safety in microbiology have been published (IAMLS, 1991, Collins, 1993, WHO, 1993).

For a review of the history, incidence, causes and prevention of laboratory-acquired infection see Collins (1993).

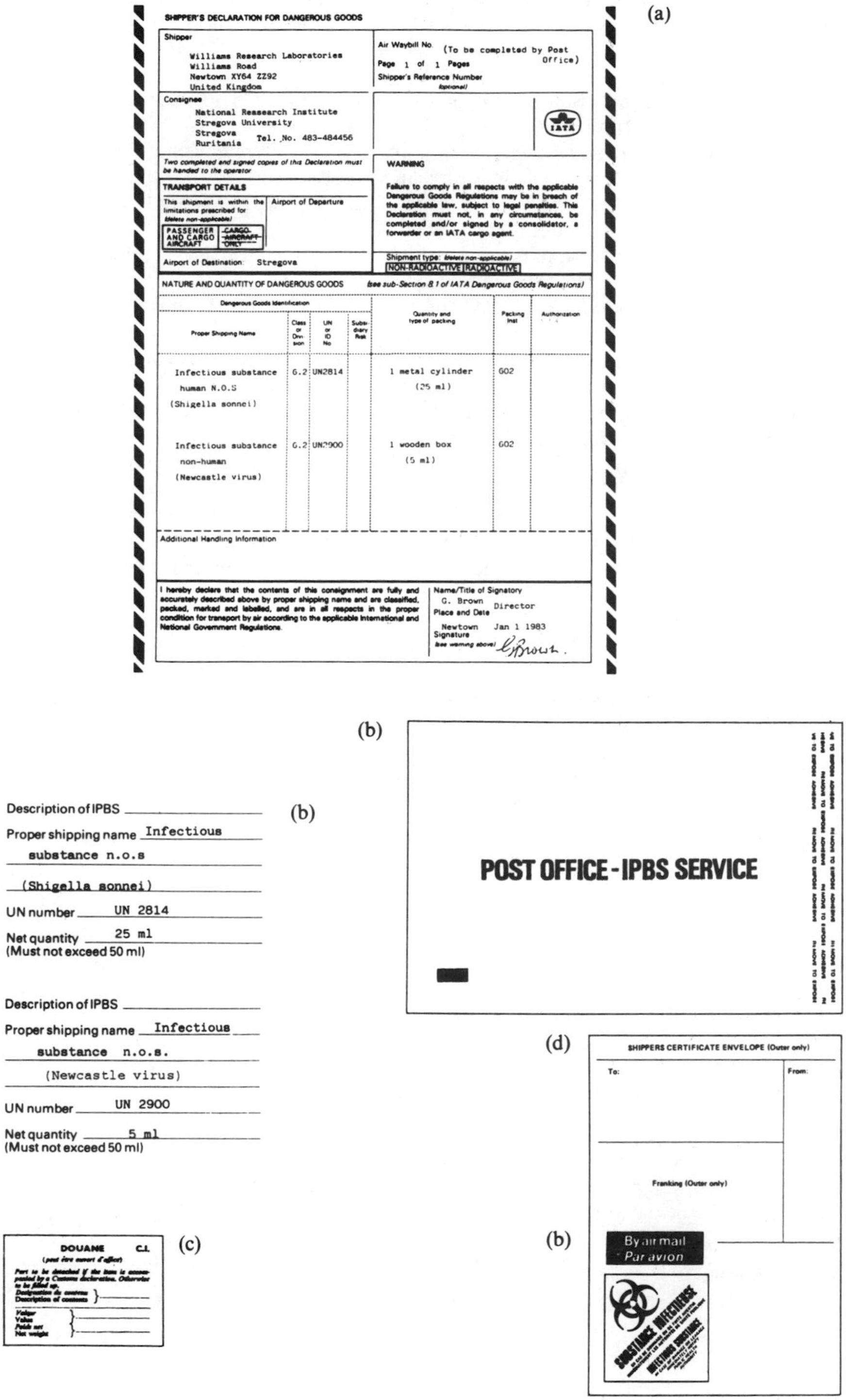

Figure 1.3 Documents for the air transport of infectious perishable biological substances. (a) Shippers' certificate for restricted articles. (b) Labels for contents, description and quantities. (c) Green customs label. (d) Address label

References

ACDP (1990a) *Categorization of Pathogens According to Hazard and Categories of Containment*, 2nd edn, Advisory Committee on Dangerous Pathogens, HMSO, London

ACDP (1990b) *HIV – the Causative Agent of AIDS and Related Conditions*, Advisory Committee on Dangerous Pathogens, HMSO, London

AS 2243 (1994) Australian Standard, *Safety in Laboratories. Part 3 Microbiology*, Office of Standards, Sydney, New South Wales

AFNOR (1990) *Contribution a l'establishment d'une liste de microorganismes pathogène pour l'homme.* Association Française de Normalization, Paris

BG Chemie (1991, 1992) *Sichere Biotechnologie: Pilze* (4/91), *Bakterien* (1/92), Berufusgenossenschaft der Chemische Industrie, Heidelberg.

BOHS (1992) Laboratory Design Issues. *British Occupational Hygiene Society Technical Guide No.10*, H & Scientific Consultants, Leeds

Collins, C. H. (1993) *Laboratory-acquired Infections*, 3rd edn, Butterworth–Heinemann, Oxford

HSAC (1991) *Safe Working and the Prevention of Infection in Clinical Laboratories*, Health Services Advisory Committee, HMSO, London

IAMLS (1991) A Curriculum for Instruction in Laboratory Safety. *Med Tec International*, 1, International Association of Medical Technology, Stockholm

OSHA (1991) Blood-borne Pathogens Standard. Occupational Safety and Health Act. *Federal Register*, **56**, 64175–64182

USPHS (1974) *Classification of Etiologic Agents on the Basis of Hazard*, Centers for Disease Control, Atlanta, Government Printing Office, Washington DC

USPHS (1981) Classification of etiologic agents on the basis of hazard, *Federal Register*, **46**, 59379–59380

USPHS (1993) *Biosafety in Microbiological and Biomedical Laboratories*, Centers for Disease Control and National Institutes of Health, 3rd edn, Government Printing Office, Washington DC

WHO (1991) Biosafety Guidelines for Diagnostic and Research Laboratories Working with HIV. *AIDS Series 9*, World Health Organization, Geneva

WHO (1993) *Laboratory Biosafety Manual*, 2nd edn, World Health Organization, Geneva

Wright, A. E. (1988) Health care in the laboratory. In *Safety in Clinical and Biomedical Laboratories* (ed. C. H. Collins), Chapman and Hall, London

Chapter

2

Quality assurance

Three terms require definition: quality control, quality assessment and quality assurance. Quality assurance is the total process that guarantees the quality of laboratory results and encompasses both quality control and quality assessment (Figure 2.1). The quality of a microbiological service is assured by quality control; i.e. a continual monitoring of working practices, equipment and reagents. Quality assessment is a system by which specimens of known but undisclosed content are introduced into the laboratory and examined by routine procedures. Quality control and quality assessment are therefore complementary activities and one should not be used as a substitute for the other (European Committee for Clinical Laboratory Standards, 1985).

The need for quality

There is increasing reliance on the quality of laboratory results and confidence in them is essential to the organization of all microbiological laboratories. Quality control of laboratory tests cannot be considered in isolation: the choice, collection, labelling and transport of specimens are all subject to error and require regular monitoring for quality and suitability.

Cost effectiveness is increasingly relevant in all areas of microbiology. In the health services, speed of response may affect the final outcome of health care

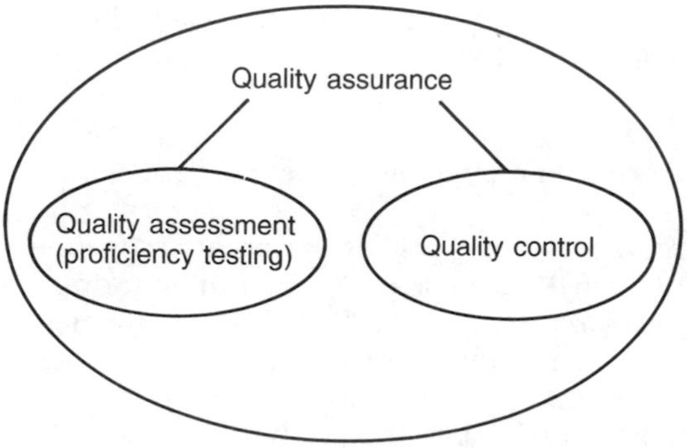

Figure 2.1 Relation between quality assurance, quality assessment and quality control

and the length and cost of a patient's stay in hospital and it is therefore as much an element of quality as the accuracy of the test result. There are also financial implications to the laboratory if poor quality of work results in time being spent in correcting errors and repeating tests as well in delays in issuing reports.

The most important contributory factors to high standards of quality are the competence and motivation of staff. These can be achieved and maintained by continued education and, above all, by communication within and between laboratories and with those using the services.

Practical quality control within the laboratory

Practical, realistic and economical quality control programmes should be applied to all procedures, irrespective of the size of the laboratory. Schedules should be based on the known stability of each item and its importance to the accuracy of the investigation as a whole. It is unnecessary to monitor all reagents and methods on every working day: the frequency of monitoring should be in direct proportion to the frequency with which errors are discovered.

Each laboratory must develop its own programme as the time spent on quality control will vary according to the size of the laboratory and the range of investigations. For its success, a senior member of staff should supervise the programme throughout the laboratory.

Internal quality control is divisible into four main areas: collection of specimens, preventive maintenance, manual procedures and monitoring of media and reagents.

Collection of specimens

In clinical microbiology, results of studies on specimens taken from sites containing indigenous or colonizing bacteria (e.g. the oropharynx, skin and gastrointestinal tract), or those likely to be contaminated during collection (e.g. sputum and urine specimens collected by patients with little or no supervision), are often difficult to interpret. Paradoxically, specimens containing the largest numbers of microbial species, with consequent high laboratory cost, are less likely to provide useful results than those yielding fewer types of organisms. Guidelines for the evaluation of the quality of specimens and the appropriateness of requests should be established by medical and laboratory staff (Bartlett *et al.*, 1978). The usefulness of rapid reporting of preliminary results should also be assessed (Bartlett, 1982).

Preventive monitoring and maintenance of equipment

This is becoming more important as many investigations are conducted with complex automated equipment. Regular monitoring and maintenance avoids costly repairs and long periods when the equipment is out of action ('down time') – particularly important when an uninterrupted 24 h service is required. Unique preventive maintenance systems, which are adaptable to the needs of an individual laboratory, have been described (Wilcox *et al.*, 1978). The following components belong to all preventive maintenance programmes.

Routine service tasks
The tasks needed to keep each item of equipment calibrated, running and clean are defined in the standard operating procedures manual (see below).

Interval of service
The frequency with which defined tasks are performed is assessed according to the type of equipment and its usage. While regularly used equipment will require more regular cleaning and servicing than that used infrequently, minimum standards of preventive maintenance must be defined for the latter.

Accountability
The concept of accountability must be introduced, and members of staff should be assigned specific responsibilities for maintenance tasks.

Staff
A list of trained persons who are readily available for trouble-shooting, verification tasks and repair should be compiled. While laboratory staff may perform uncomplicated maintenance work, advice with respect to complex equipment and procedures must be sought from the manufacturer's service department. The cost of a service contract must be weighed against the possibility of even more expensive repair costs and the consequences of prolonged 'down time'.

Implementation
Equipment maintenance is not a 'one-off' affair. Once tasks have been defined, and staff trained, the programme should continue regularly and uninterrupted.

Documentation
A system of recording is essential and may be adapted to the specific needs of a laboratory, either with inexpensive index cards or a computerized system.

Standard operating procedures manual

The standard operating procedures (SOP) manual is the most important document for the improvement and maintenance of the quality of the service. It is essential for accreditation of clinical microbiology laboratories. An out-of-date manual or one from another laboratory may be used as a guide to writing but, since each laboratory is unique, it must develop its own manual. If there are no written methods and new staff are taught by 'word of mouth', techniques will soon 'drift'. The headings standard to any SOP manual are listed and discussed by Fox *et al.* (1992). These are: principle of test and purpose of procedure; personnel and training requirements; specimen requirements; equipment; reagents; quality control; procedure and methodology; normal ranges and limitations of the procedure. To be of real benefit, the SOP manual should fulfil four criteria.

Authoritative
The manual must be written by a senior member of staff who has practical experience of the methods and who can be identified. The manual is authorized by the person who has responsibility for the procedures. Since the writing of the manual will entail a general review of working practices it is a good educational exercise for the person writing it.

Realistic
The manual should avoid theoretical ideals but should give a clear, concise and explicit description of the procedures in numbered steps. Unnecessarily complicated procedures will quickly lead to loss of respect and to the possibility of staff introducing unofficial short cuts. A well structured manual will permit a microbiologist from one laboratory to perform any of the tests described in another.

Up to date
The updating of manuals is vital. Review is a continuous process; the manual should be formally revised annually and amended when necessary. No new or modified procedure should be used until it has been checked by the supervisor and head of the laboratory and officially incorporated into the manual.

Availability
The manual must be freely available to all relevant personnel. It is the responsibility of senior staff to ensure that it is used and strictly adhered to. It need not be a bulky bound item; loose-leaf systems have the advantage that sections may be separated and kept on relevant benches. Each section should have a unique and clear code number for identification and for cross-referencing. The exclusive use of manufactures' kit inserts is not recommended, though their use as a source of information for the production of the manual is encouraged.

Culture media and reagents

Few laboratories in the UK now prepare media from basic materials. Those that do so must follow more comprehensive quality control procedures than those required for preparation of commercially produced media. Reference should be made to Chapter 5 and to other publications on culture media (Martin, 1991). Most laboratories now use dehydrated media which, if purchased from a reputable manufacturer, will have undergone extensive quality control testing before sale. Repetition of these tests is unnecessary but dehydrated media will have to be reconstituted, heated and possibly supplemented with additives during preparation and it is essential to control the effect of these processes. Media must fulfil four main parameters; sterility, support of bacterial growth, selective inhibition of growth, in some cases, and appropriate response of biochemical indicators.

All batches of media must be tested for sterility. Methods are usually a compromise as the sterility of every plate or tube of medium cannot be ensured without incubation of the whole batch with the concomitant risk of deterioration of its growth supporting ability. A representative portion of each new batch should be incubated overnight at 37°C to check for sterility before use. For batches of 100 or less, a 5% minimum sample should be tested; for larger batches, 10 plates or tubes taken randomly will suffice (Blazevic *et al.*, 1976). After overnight incubation media should be left at room temperature for a further 24 h as a check for contamination by psychrophiles. Low-level surface contamination of plates may not be detected on the controls but is usually recognizable during use, for instance, by the observation of growth part way through a streaked population.

Selective media, being inhibitory to many organisms, pose particular problems: even gross contamination may not be detected by the above methods but inoculation of a few ml of the final product into 10 times its volume of sterile broth (to dilute inhibitory substances) will permit detection of heavy contamination.

Tests for performance of media

The ability of each batch of new medium to support growth and exhibit the required differential and selective qualities should be determined before use. The organisms used should be stock strains from a recognized culture collection such as the NCTC or ATCC (Chapter 6). As characteristics of bacteria may alter when maintained by serial subculture, it is preferable to use stock strains stored at − 70°C, freeze-dried or in gelatin discs. Selected media are tested with organisms expected to grow and those expected to be inhibited. Biochemical indicator media are tested with two or more organisms to demonstrate positive and negative reactions. A practical approach based on a minimum number of strains has been described (Snell, 1991). The use of control organisms to monitor biochemical tests also serves to familiarize new and junior staff with the appearance of positive and negative reactions.

Reagents

Most pure chemicals and stains have greater stability than biological reagents. All reagents must be labelled as to contents, concentration, date of preparation and expiry date or shelf-life. They must be stored according to manufacturers' instructions and discarded when out of date. Freshly prepared reagents must be subjected to quality control before use; working solutions of stains are tested weekly and infrequently used reagents and stains are tested for their required purpose before use. Check chemical reagents for performance when new batches of media with which they are being used are tested (e.g. Kovacs reagent with indole detection medium).

Quality assessment

The concepts and statistical methods for the implementation of quality assessment were largely pioneered in clinical biochemistry laboratories. Schemes in microbiology laboratories are less easily organized because of the complexity, expense and labour intensity of the work. The diversity of microbiological investigations makes it difficult to produce control materials that cover all areas, and several technical difficulties have to be overcome in the preparation of quality assessment specimens (DeMello and Snell, 1985).

Quality assessment may be organized locally within one laboratory (internal quality assessment) or on a wider geographical basis (external quality assessment). An internal scheme may be organized by a senior member of staff within the laboratory. In such a scheme arrangements are made so that examinations of specimens or samples are repeated, thereby enabling direct comparisons with previously obtained results. An obvious advantage with this approach is that the specimens or samples are real, rather than simulated, ones. False identifications may be given to repeated specimens so that staff, being unaware that they are quality control specimens, do not give them special attention. Care should be taken with the use of stored materials as deterioration may prevent the reproduction of previous results. Quality assurance materials have been sources of laboratory-acquired infections (Collins, 1993).

External quality assessment is usually organized by a central laboratory, often on a national basis. Simulated samples of known but undisclosed content are dispatched to participants who examine them and report their findings to the organizing laboratory. The organizer compares the results from all participating laboratories and sends an analysis to each participant (Figure 2.2) thereby enabling them to gain an insight into their routine capabilities. If internal quality control procedures are functioning correctly, external assessment should reveal few unpleasant surprises.

Organizing laboratory		**Participants**
Prepare QA specimens	→	Examine specimens
Analyse results	←	Report results
Prepare report	→	Evaluate

Figure 2.2

Organization

Participation in the UK scheme is voluntary and strictly confidential. Full voluntary participation takes time to achieve, as participants can only be recruited through persistence, education and peer influence. Mandatory schemes, in use in some countries, ensure full and rapid participation but not necessarily full cooperation. Confidentiality is achieved by allocating each laboratory a unique identification number known only to the organizer and participant. This number is used in all routine transactions and individual participant's results are not revealed to other parties except under previously agreed and strictly defined circumstances.

Difficulties arise in the organization of external schemes in countries where the size and function of participant laboratories vary widely. There may also be differences in the type of laboratory report issued, particularly in the field of clinical microbiology. Thus some laboratories only report the results of tests while others edit and interpret the results. External quality assessment schemes must take account of these differences by offering basic and comprehensive options (Gavan, 1974). Other schemes reflect the philosophy that, although there may be differences in the types of laboratories, the same standards should be applied to their assessment. The organization of a comprehensive microbiology quality assessment scheme is costly. Thus it may be more cost-effective for countries with few laboratories to participate in another country's scheme. Although international schemes do exist, their aims differ from those of national schemes as they are principally used to assess selected national laboratories in developing countries as part of a wider support programme. International schemes may be used to compare the standards of major laboratories in different countries. These laboratories can then use the expertise and knowledge gained to initiate their own national schemes (Snell and Hawkins, 1992).

Specimens

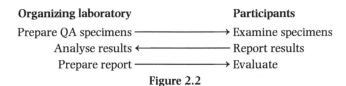

External quality assessment specimens should resemble, as closely as possible, materials usually examined by the participant laboratories. In some instances this is easily achieved, as with specimens for assay of antibodies in serum. In other cases, sufficient quantities of natural material may not be available and it is necessary to produce artificial specimens. Since external quality assessment should test a laboratory's routine capabilities, specimens must to a great extent reflect the routine workload. Thus, as the majority of specimens examined in a clinical laboratory do not yield pathogenic organisms, some (but not too many) of the simulated specimens should contain only non-pathogens. Commonly isolated organisms must be represented in quality assessment specimens but, as most laboratories experience little difficulty in isolating and identifying, e.g. *Staphylococcus aureus*, there is little to be learned from the inclusion of too many such specimens. Less frequently encountered organisms, such as *Corynebacterium diphtheriae* and chloramphenicol-resistant *Haemophilus influenzae*, must also be distributed as they play an important role in the education of new and junior members of staff.

To gain and maintain credibility, participants must have full confidence in the specimens being distributed. The homogeneity and stability of specimens must be established and controlled to ensure their comparability. Specimens designed to test isolation procedures must have the same number of organisms in each vial; this is most easily achieved by freeze-drying (DeMello and Snell, 1985). Specimens prepared for the detection of antigen or antibody must react in a normal and predictable manner in a full range of commonly used assays. If participants are to evaluate their performance, they need to know if their results are incorrect. Thus, in general, specimens should yield clear-cut results. In reality, this can be quite difficult. It is, for example, not easy to select strains of bacteria that are unequivocally susceptible or resistant to a range of relevant antibiotics. The use of strains giving 'borderline' results may, however, highlight the failure of laboratories to use optimal methods.

Benefits of quality assessment

The success of a quality assessment scheme is judged ultimately by its usefulness, the extent of which is primarily determined by the attitude of the participant. Quality assessment is a management tool and information gained from the scheme will be rendered ineffective if not used correctly. The whole philosophy of quality assessment is based on the assumption that the results obtained should reflect what is happening in day-to-day practice. For this to be true quality assessment specimens must be examined by the same staff and with the same methods and reagents as specimens in the normal workload.

Expenditure of extra effort on quality assessment specimens as a conscious attempt to 'cheat' or as a subconscious desire to excel in a challenge of ability (La Motte *et al.*, 1977) negates the usefulness of the assessment and fails to reveal any unsuspected problems. Pressures on laboratory staff to perform well in quality assessment schemes, especially if success is seen as contributing to economic viability of the laboratory by providing proof of excellence to customers, are understandable but should be resisted.

Accreditation

Accreditation is a valuable component of quality assurance in clinical microbiology and has been in use for some time in several countries, including the USA and Australia, but has only recently been introduced in the UK (Batstone, 1992). The purpose is simple: the audit by an external agency to see that laboratories conform to certain defined standards. An accreditation scheme should be comprehensive, covering all aspects of the laboratory, including organization and administration, staff development and education, facilities and equipment and policies and procedures. Participation in all relevant quality assessment schemes is a requirement for accreditation. The results of quality assessment should not, however, be given undue weight in the decision to grant or withhold accreditation as this could lead to increased effort in the examination of quality assessment specimens, thereby undermining the educational aspect of the scheme. The response to poor performance in quality assessment in terms of investigation and remedial action is a more useful measure of the quality of a laboratory.

References

Bartlett, R. C. (1982) Making optimum use in the microbiology laboratory. *Journal of the American Medical Association*, **247**, 857–859

Bartlett, R. C. *et al.* (1978) Quality assurance in clinical microbiology. In *Quality Assurance Practices for Health Laboratories* (ed. S. L. Horn), American Public Health Association, Washington, pp. 871–1005

Batstone, G. F. (1992) Medical audit in clinical pathology. *Journal of Clinical Pathology*, **45**, 284–287

Blazevic, D. J., Hall, C. T. and Wilson, M. E. (1976) Practical quality control procedures for the clinical microbiology laboratory (ed. A. Balows), *Cumitech*, **3**, 1–13, American Society for Microbiology, Washington DC

Collins, C. H. (1993) *Laboratory-Acquired Infections*, 3nd edn, Butterworth-Heinemann, Oxford

DeMello, J. V. and Snell, J. J. S. (1985) Preparation of simulated clinical material for bacteriological examination. *Journal of Applied Bacteriology*, **59**, 421–436

European Committee for Clinical Laboratory Standards (1985) *Standard for quality assurance. Part 1: Terminology and General Principles*, **2**, 1–13, Lund

Fox, M. C., Ward, K and Kirby, J. (1992) First steps to accreditation. *IMLS Gazette*, August, 412

Gavan, T. L. (1974) A summary of the bacteriology portion of the 1972 (CAP) quality evaluation program. *American Journal of Clinical Pathology*, **61**, 971–979

La Motte, L. C., *et al* (1977) Comparison of laboratory performance with blind and mail-distributed proficiency testing samples. *Public Health Report*, **99**, 554–560

Martin, R. J. (1991) Culture media. In *Quality Control Principles and Practice in the Microbiology Laboratory* (eds J. J. S. Snell, I. D. Farrell and C. Roberts), Public Health Laboratory Service, London, pp. 24–36

Snell, J. J. S. (1991) Bacteriological characterization tests. In *Quality Control Principles and Practice in the Microbiology Laboratory* (eds J. J. S. Snell, I. D. Farrell and C. Roberts), Public Health Laboratory Service, London, pp. 37–46

Snell, J. J. S. and Hawkins, J. M. (1992) Quality assurance – achievements and intended directions. *Reviews in Medical Microbiology*, **3**, 28–34

Wilcox, K. R. *et al.* (1978) Laboratory management. In *Quality Assurance Practices for Health Laboratories* (ed. S. L. Horn), American Public Health Association, Washington DC, p. 3–126

Chapter

3

Laboratory equipment

An extensive range of equipment is now available for laboratory work but not all of it is suitable for microbiology. Choice should be made solely on the principle of 'fitness for purpose'. In the UK the Medical Devices Directorate of the Department of Health arranges for the testing of equipment, investigates accidents involving equipment, and issues warning notices if defects are found. Reports are published in its journal *Health Equipment Information*. There is also a Directive of the European Union on medical devices.

Much new equipment is powered by electricity and electrical safety is therefore important. All current and new equipment should conform to the requirements of the International Electrotechnical Commission (IEC, 1990)

Before any new equipment is purchased it is also advisable to obtain the personal advice of microbiologists who have had experience in its use. It is bad policy to rely entirely on advertisements, catalogues, extravagant claims of representatives and the opinions of purchasing officers who are mainly concerned with balancing budgets. The best is not always the most expensive, but it is rarely the cheapest. Few microbiologists require very expensive research-type microscopes, but centrifuges should be the very best available. Laboratory autoclaves should be designed for laboratory, not pharmacy or hospital, use. Thermostatic equipment designed for chemical work is usually suboptimal for microbiological purposes.

In this chapter, we have indicated the types of equipment that we and our associates (and friends) have found adequate for use in medical, veterinary and food microbiology (except virology) work.

The operation of some microbiological equipment may result in the release of infectious aerosols (see Chapter 1 and Collins, 1993).

[Equipment for sterilization and filtration is considered in Chapter 4 and that for media preparation in Chapter 5.]

Microscopes

Although the microscope is an essential piece of microbiological apparatus, it is not used as often as might be expected and the purchase of elaborate and expensive (prestige) instruments is not justified. The medium priced instruments are adequate for most basic laboratory work.

Laboratory workers should be familiar with the general mechanical and optical principles, but a detailed knowledge of either is unnecessary and, apart from superficial cleaning, maintenance should be left to the manufacturers, who will arrange for periodic visits by their technicians. Most manufacturers publish handbooks containing useful explanations and information.

Wide-field compensating eyepieces should be fitted; they are less tiring than other eyepieces and are more convenient for spectacle wearers. For low power work, × 5 and × 10 objectives are most useful and for high power (oil immersion) microscopy, × 50 and × 100 fluorite objectives are desirable. These fluorite lenses are much better than achromatics and much less expensive than the hardly justifiable apochromats.

Careful attention to critical illumination, centring, and the position of diaphragms is essential for adequate microscopy.

Inverted high-power microscopes are useful in laboratories that do cell counts on urine specimens. Epifluorescence microscopes are essential for DEFT bacterial counting (Chapter 10).

Low-power microscopes

Low-power binocular dissecting microscopes are better than hand lenses for examining colonial morphology. Experience in their use may save much time and expense in pursuing unnecesary tests (see p.98). They also facilitate the subculture of small colonies from crowded petri dishes, thus avoiding mixed cultures.

Incubators and water-baths

Incubators

Incubators are available in various sizes. In general, it is best to obtain the largest possible model that can be accommodated, although this may create space problems in laboratories where several different incubation temperatures are employed. The small incubators suffer wider fluctuations in temperature when their doors are opened than do the larger models and most laboratory workers find that incubator space, like refrigerator space, is subject to 'Parkinson's law'.

Although incubators rarely develop faults, it is advisable, before choosing one, to ascertain that service facilities are available. The circuits are not complex but require expert technical knowledge to repair. Transporting incubators back to the manufacturers is most inconvenient.

In medical and veterinary laboratories, incubators are usually operated only at 35–37°C. The food or industrial laboratory usually requires machines operated at 15–20, 28–32 and 55°C. For temperatures above ambient there are no problems, but lower temperatures may need a cooling as well as a heating device.

Incubation in an atmosphere of 5–8% carbon dioxide is preferable for the cultivation of almost all bacteria of medical importance. Automatic control of carbon dioxide and humidity is now possible but as a back-up 'candle jars' should be available.

Water used to maintain humidity may become contaminated with fungi, especially aspergilli, causing problems with cultures that require prolonged incubation.

Cooled incubators
For incubation at temperatures below the ambient, incubators must be fitted with modified refrigeration systems with heating and cooling controls. These need to be correctly balanced.

Automatic temperature changes
These are necessary when cultures are to be incubated at different temperatures for varying lengths of time, as in the examination of water by membrane filtration, and when it is not convenient to move them from one incubator to another. Two thermostats are required, wired in parallel, and a time switch wired to control the thermostat set at the higher temperature. These are now built in to some models.

'Portable' incubators, which may be moved from room to room as desired, and also taken on ex-laboratory (e.g. environmental microbiology) investigations such as water examination (p.278) are now available.

Incubator rooms
Although advice on the in-house construction of these rooms was given in previous editions of this book it is now our opinion that they should be installed by specialist suppliers.

Metal shelving is preferable to wood: if the humidity in these rooms is high the growth of mould fungi may be encouraged by wood. Solid shelving should be avoided and space left at the rear of the shelves to allow for the circulation of air.

If, on the other hand, low humidity is a problem cultures may be prevented from drying up by placing them along with pieces of wet filter paper in plastic boxes such as those sold for the storage of food.

Water-baths

The contents of a test-tube placed in a water-bath are raised to the required temperature much more rapidly than in an incubator. These instruments are therefore used for short-term incubation. If the level of water in the bath comes one-half to two-thirds of the way up that of the column of liquid in the tube, convection currents are caused which keep the contents of the tube well mixed, and hasten reactions such as agglutination.

All modern water-baths are equipped with electrical stirrers and in some the heater, thermostat and stirrer are in one unit, easily detached from the bath for use elsewhere or for servicing. Water-baths must also be lagged to prevent heat loss through the walls. A bath that has not been lagged by the manufacturers can be insulated with slabs of expanded polystyrene.

Water-baths should be fitted with lids in order to prevent heat loss and evaporation. These lids must slope so that condensation water does not drip on the contents. To avoid chalky deposits on tubes and internal surfaces, only distilled water should be used, except, of course, in those baths which operate at or around ambient temperature, which are connected to the cold water supply and fitted with a constant level and overflow device. This kind of bath offers the same problems as the low-temperature incubators unless the water arrives at a much lower temperature than that required in the water-bath. In premises that have an indirect cold water supply from a tank in the roof, the water temperature in summer may be near or even above the required water-bath temperature, giving the thermostat no leeway to operate. In this case, unless it is possible to make direct connection with the rising main, the water supply to the bath must be passed through a refrigerating coil. Alternatively, the water-bath can be placed in a refrigerator and its electricity supply adapted from the refrigerator light or brought in through a purpose-made hole in the cabinet. The problem is best overcome by fitting laboratory refrigeration units, designed for this purpose.

Metal block heaters
These give better temperature control than water-baths and do not dry out, but they can be used only for tubes and thin glass bottles which must fit neatly into the holes in the steel or aluminium blocks.

Centrifuges

Safety, mechanical, electrical and microbiological, is an important consideration in the purchase and use of centrifuges. They should conform to national and international standards (e.g. BS 4402: 1982), IFCC (1990) and IEC (1990).

The ordinary laboratory centrifuge is capable of exerting a force of up to 3000 *g* and this is the force necessary to deposit bacteria within a reasonable time.

For general microbiology, a centrifuge capable of holding 15-ml and 50-ml buckets and working at a maximum speed of 4000 rev/min is adequate. The swing-out head is safer than the angle head as it is less likely to distribute aerosols (Collins, 1993).

For maximum microbiological safety, sealed buckets should be fitted. These confine aerosols if breakage occurs and are safer than sealed rotors. Sealed buckets ('safety cups') are supplied by several companies. Windshields offer no protection. There should be an electrically operated safety catch so that the lid cannot be opened when the rotors are spinning.

Centrifuge buckets are usually made of stainless steel and are fitted with rubber buffers. They are paired and their weight is engraved on them. They are always used in pairs, opposite to one another, and it is convenient to paint each pair with different coloured patches so as to facilitate recognition. If the buckets fit in the centrifuge head on trunnions, these are also paired.

Centrifuge tubes to fit the buckets are made of glass, plastic, nylon or spun aluminium. Conical and round-bottomed tubes are made. Although the conical tubes concentrate the deposit into a small button, they break much more readily than do the round-bottomed tubes. When the supernatant fluid, not the deposit, is required, round-bottomed tubes should be used.

Instructions for using the centrifuge

(1) Select two centrifuge tubes of identical lengths and thicknesses. Place liquid to be centrifuged in one tube and water in the other to within about 2 cm of the top.
(2) Place the tubes in paired centrifuge buckets and place the buckets on the pans of the centrifuge balance. This can be made by boring holes large enough to take the buckets in small blocks of hard wood and fitting them on the pans of a simple balance.
(3) Balance the tubes and buckets by adding 70% alcohol to the lightest bucket. Use a pasteur pipette and allow the alcohol to run down between the tube and the bucket.
(4) Place the paired buckets and tubes in diametrically opposite positions in the centrifuge head.
(5) Close the centrifuge lid and make sure that the speed control is at zero before switching on the current. (Many machines are fitted with a 'no-volt' release to prevent the machine starting unless this is done.)

(6) Move the speed control slowly until the speed indicator shows the required number of revolutions per minute.

Precautions

(1) Make sure that the rubber buffers are in the buckets, otherwise the tubes will break.
(2) Check the balancing carefully. Improperly balanced tubes will cause 'head wobble', spin-off accidents and wear out the bearings.
(3) Check that the balanced tubes are really opposite one another in multi-bucket machines.
(4) Never start or stop the machine with a jerk.
(5) Observe the manufacturer's instructions about the speed limits for the various loads.
(6) Open sealed centrifuge buckets in a microbiological safety cabinet.

Maintenance

The manufacturers will, by contract, arrange periodic visits for inspection and maintenance in the interests of safety and efficiency.

Blenders, tissue grinders and shakers

These range from instruments suitable for grinding and emulsifying large samples, e.g. of food, to glass devices for homogenizing small pieces of tissue.

Most electrically driven machines are efficient but suffer from the disadvantage of requiring a fresh, sterile cup for each sample. These cups are expensive and it takes time to clean and re-sterilize them.

Some of them may release aerosols during operation and when they are opened. A heavy Perspex or metal cover should be placed over them during use. This should be decontaminated afterwards. All blenders, etc. should be opened in a microbiological safety cabinet because the contents will be warm and under pressure. Aerosols will be released.

The Stomacher Lab-Blender overcomes these problems. Samples are emulsified in heavy-duty sterile plastic bags by the action of paddles. This is an efficient machine, capable of processing large numbers of samples in a short time without pauses for washing and re-sterilizing. There is little risk of aerosol dispersal. The bags are automatically sealed while the machine is working.

Tissue grinders (Griffith's tubes), essential for emulsifying small pieces of tissue for microbiological examination, are available in several sizes. A heavy glass tube is constricted near to its closed end and a pestle, usually made of glass covered with PTFE or of stainless steel, is ground into the constriction. With the pestle in place, the tissue and some fluid are put into the tube. The pestle is rotated by hand. The tissue is ground through the constriction and the emulsion collects in the bottom of the tube.

These grinders present some hazards. The tubes, even though made of borosilicate glass, may break and disperse infected material. They should be used inside a microbiological safety cabinet and be held in the gloved hand in a wad of absorbent material.

Small pieces of soft tissue, e.g. curettings, may be emulsified by shaking in a screw-capped bottle with a few glass beads and 1 or 2 ml of broth on a vortex mixer.

Vortex mixers are useful for mixing the contents of single bottles, e.g. emulsifying sputum, but should always be operated in a microbiological safety cabinet.

Shaking machines

Conventional shaking machines are sold by almost all laboratory suppliers. They are useful for mixing and shaking cultures and for serological tests, but should be fitted with racks, preferably of polypropylene, which hold the bottles or tubes firmly, and covered with a stout Perspex box when in use to prevent the dispersal of aerosols.

We believe that all bottles that contain infected material and which are shaken in any machine should be placed inside individual self-sealing or heat-sealed plastic bags in order to minimize the dispersal of infected airborne particles.

The better models allow bottles, flasks, etc. to be rotated at varying speeds as well as for shaking in two dimensions.

Microbiological safety cabinets

Microbiological safety cabinets are intended to capture and retain infected airborne particles released in the course of certain manipulations and to protect the laboratory worker from infection which may arise from inhaling them.

There are three kinds: Classes I, II and III. Class I and Class II cabinets are used in diagnostic, Level 2, and Containment, Level 3, laboratories for work with Hazard Group 3 organisms. Class III cabinets are used for Hazard Group 4 viruses.

A Class I cabinet is shown in Figure 3.1. The operator sits at the cabinet, works with the hands inside and sees what he/she is doing through the glass screen. Any aerosols released from the cultures are retained because a current of air passes in at the front of the cabinet. This sweeps the aerosols up through the filters which remove all or most of the organisms. The clean air then passes through the fan, which maintains the airflow, and is exhausted to atmosphere, where any particles or organisms that have not been retained on the filter are so diluted that they are no longer likely to cause infection if inhaled. An airflow between 0.7 and 1.0 m/s must be maintained through the front of the cabinet and modern cabinets have airflow indicators and warning devices. The filters must be changed when the airflow falls below this level.

The Class II cabinet (Figure 3.2) is more complicated and is sometimes called a laminar flow cabinet but as this term is also used for clean air cabinets which do not protect the worker it should be avoided (see p.31). In the Class II cabinet about 70% of the air is recirculated through filters so that the working area is bathed in clean (almost sterile) air. This entrains any aerosols produced in the course of the work and these are removed by the filters. Some of the air (about 30%) is exhausted to atmosphere and is replaced by a 'curtain' of room air which enters at the working face. This prevents the escape of any particles or aerosols released in the cabinet.

There are other types of Class II cabinets, with different air flows and

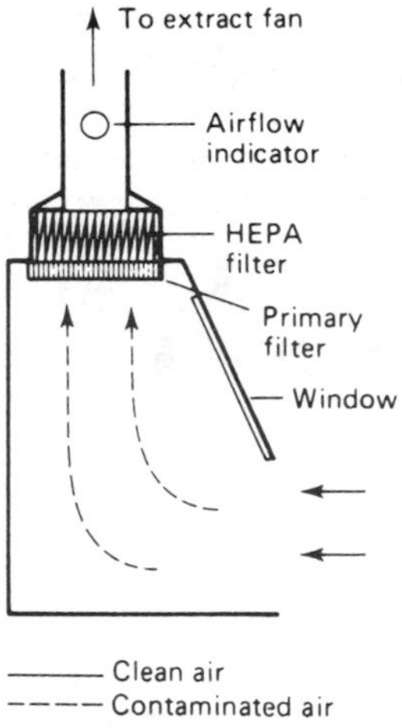

Figure 3.1 Class I microbiological safety cabinet

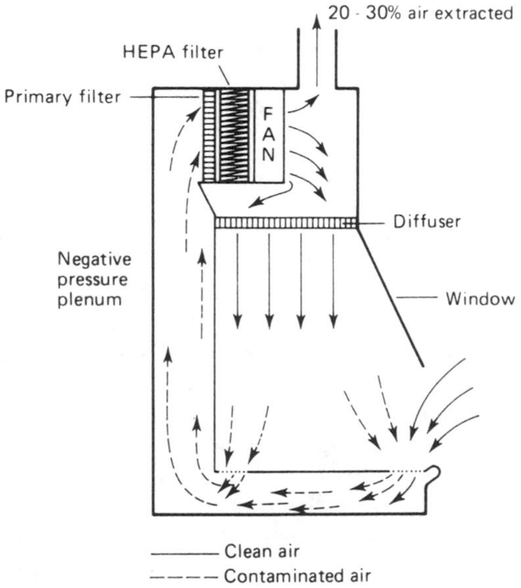

Figure 3.2 Class II microbiological safety cabinet

exhaust systems, which may also be used for toxic chemicals and volatile chemicals. Manufacturers should be consulted.

Class III cabinets are totally enclosed and are tested under pressure to ensure that no particles can leak from them into the room. The operator works with gloves which are an integral part of the cabinet. Air enters through a filter and is exhausted to atmosphere through one or two more filters (Figure 3.3).

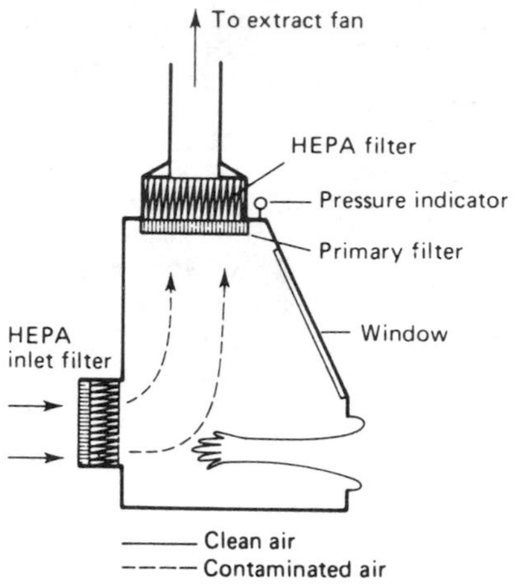

Figure 3.3 Class III microbiological safety cabinet

Purchasing

Safety cabinets should comply with national standards, e.g. British Standards Institution (BSI, 1992), Standards Association of Australia (SAA, 1983a, b), National Sanitation Foundation (USA) (NSF, 1992).

Use of safety cabinets

These cabinets are intended to protect the worker from airborne infection. They will not protect him from spillage and the consequences of poor techniques. The cabinet should not be loaded with unnecessary equipment or it will not do its job properly. Work should be done in the middle to the rear of the cabinet, not near the front and the worker should avoid bringing his hands and arms out of the cabinet while he is working. After each set of manipulations and before withdrawing his hands he should wait for 2–3 min to allow any aerosols to be swept into the filters. The hands and arms may be contaminated and should be washed immediately after ceasing work. Do not use bunsen burners, even micro-incinerators. They disturb the airflow. Use disposable plastic loops (p.37).

Siting

The efficiency of a safety cabinet depends also on correct siting and proper maintenance. Possible sites for cabinets in a room are shown in Figure 3.4. A is a poor site, as it is near to the door and airflow into the cabinet will be disturbed every time the door is opened and someone walks into the room and past the cabinet. Site B is not much better, as it is in almost a direct line between door and window, although no-one is likely to walk past it. Its left side is also close to the wall and airflow into it on that side may be affected by the 'skin effect', i.e. the slowing down of air when it passes parallel and close to a surface. Air passing across the window may be cooled, and will meet warm air from the rest of the room at the cabinet face, when turbulence may result.

Site C is better and site D is best of all. If two cabinets are required in the same room, sites C and D would be satisfactory, but they should not be too

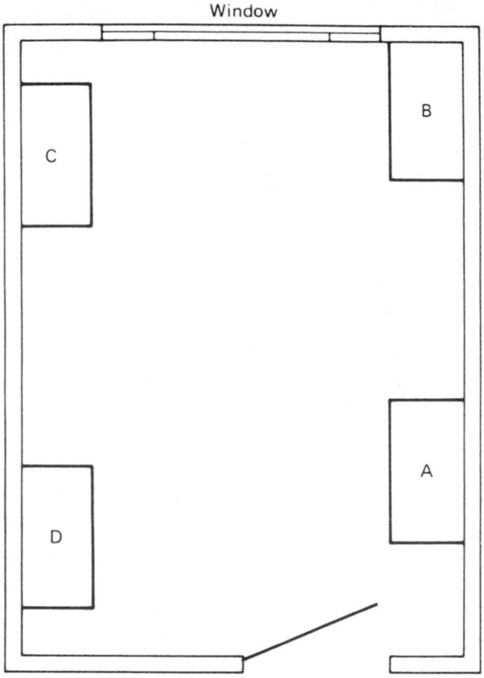

Figure 3.4 Possible sites for safety cabinets in relation to cross draughts from door and window and movement of staff. A is bad; B is poor; C is better; D is best

close together, or one may disturb the airflow of the other. We remember a laboratory where a Class I cabinet was placed next to a Class II cabinet. When the former was in use it extracted air from the latter and rendered that cabinet quite ineffective.

Care must also be taken in siting any other equipment that might generate air currents, e.g. fans and heaters. Mechanical room ventilation may be a problem if it is efficient, but this can be overcome by linking it to the electric circuits of the cabinets so that either, but not both, are extracting air from the room at any one time. Alternatively baffles may be fitted to air inlets and outlets to avoid conflicting air currents near to the cabinet. Tests with smoke generators (see below) will establish the directions of air currents in rooms.

None of these considerations need apply to Class III cabinets which are in any case usually operated in more controlled environments.

Testing airflows

The presence and direction of air currents and draughts are determined with 'smoke'. This may be generated from burning material in a device like that used by bee keepers, but is usually a chemical which produces a dense, visible vapour. Titanium tetrachloride is commonly used. If a cotton wool tipped stick, e.g. a throat swab, is dipped into this liquid and then waved in the air a white cloud is formed which responds to quite small air movements.

Commercial airflow testers are more convenient. They are small glass tubes, sealed at each end. Both ends are broken off with the gadget provided and a rubber bulb fitted to one end. Pressing the bulb to pass air through the tube

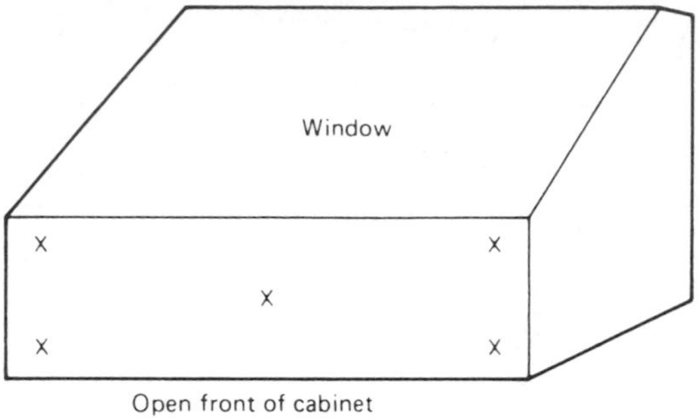

Figure 3.5 Testing airflow into a Class I safety cabinet. Anemometer readings should be taken at five places, marked with an X, in the plane of the working face with no-one working at the cabinet

causes it to emit white smoke. These methods are suitable for ascertaining air movements indoors.

To measure airflow an anemometer is required. Small, vane anemometers, timed by a stopwatch, are useful for occasional work but for serious activities electronic models, with a direct reading scale, are essential. The electronic vane type has a diameter of about 10 cm. It has a satisfactory time constant and responds rapidly enough to show the changes in velocity that are constantly occurring when air is passing into a Class I safety cabinet. Hot wire or thermistor anemometers may be used but they may show very rapid fluctuations and need damping. Both can be connected to recorders.

It is necessary to measure the airflow into a Class I cabinet in at least five places in the plane of the working face (Figure 3.5). At all points this should be between 0.7 and 1.0 m^3/s. No individual measurement should differ from the mean by more than 20%. If there is such a difference there will be turbulence within the cabinet. Usually some piece of equipment, inside or outside the cabinet, or the operator himself, is influencing the airflow.

Although airflows into Class I cabinets are usually much the same at all points on the working face this is not true for Class II cabinets, where the flow is greater at the bottom than at the top. The average inward flow can be calculated by measuring the velocity of air leaving the exhaust and the area of the exhaust vent. From this the volume per minute is found and this is also the amount entering the cabinet. Divided by the area of the working face it gives the average velocity. A rough and ready way, suitable for day-to-day use requires a sheet of plywood or metal which can be fitted over the working face. In the centre of this an aperture is made which is 2 × 2.5 cm. The inward velocity of air through this is measured and the average velocity over the whole face calculated from this figure and the area of the working face. The inflow at the working face should be not less than 0.4 m/s. A check should also be made with smoke to ensure that air is in fact entering the cabinet all the way round its perimeter and not just at the lower edge.

The downward velocity of air in a Class II cabinet should be measured with an anemometer at 18 points in the horizontal plane 10 cm above the top edge of the working face (Figure 3.6). The mean downflow should be between 0.25 and 0.5 m/s and no individual reading should differ from the mean by more than 20%.

It is usual, when testing airflows with an anemometer, to observe or record the readings at each position for several minutes, because of possible fluctuations.

To measure the airflow through a Class III cabinet the gloves should be

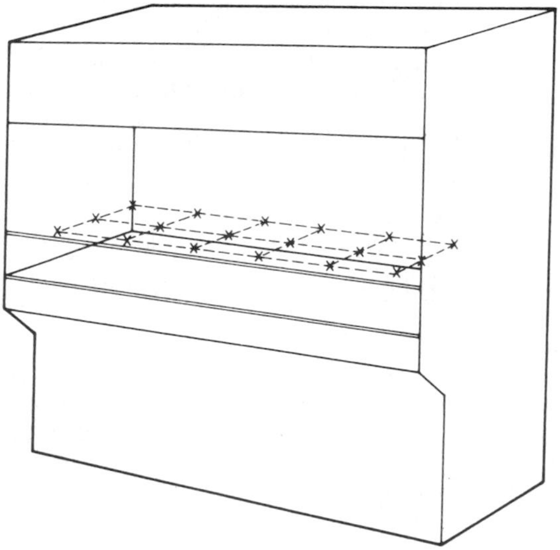

Figure 3.6 Testing the vertical airflow in a Class II safety cabinet. Anemometer readings should be taken at points marked X on an imaginary grid 6 inches within the cabinet walls and just above the level of the bottom of the glass window.

removed and readings taken at each glove port. Measurements should also be taken at the inlet filter face when the gloves are attached.

Decontamination

As safety cabinets are used to contain aerosols which may be released during work with Hazard Group 3 microorganisms the inside surfaces and the filters will become contaminated. The working surface and the walls may be decontaminated on a day-to-day basis by swabbing them with disinfectant. Glutaraldehyde is probably the best disinfectant for this purpose as phenolics may leave sticky residues and hypochlorites may, in time, corrode the metal. For thorough decontamination, however, and before filters are changed formaldehyde (*Caution* – see below) should be used and precautions should be taken before use. The installation should be checked to ensure that none of the gas can escape to the room or to other rooms. The front closure (night door) of the cabinet should seal properly onto the carcase, or masking tape should be available to seal it. Any service holes in the carcase should be sealed and the HEPA filter seating examined to ensure that there are no leaks. If the filters are to be changed and the primary or roughing filter is accessible from inside the cabinet it should be removed and left in the working area. The supply filter on a Class III cabinet should be sealed with plastic film.

Formaldehyde is toxic and explosive. In the UK the occupational exposure limit is 25 mg/m^3*. Mixtures of 7.75% (v/v) formaldehyde and dry air are explosive: the ignition point is 430°C.*

Formaldehyde is generated by:

1 boiling formalin, which is a 40% solution of the gas in water;
2 by heating paraformaldehyde, which is its solid polymer.

Boiling formalin

The volume used is important. Too little will be ineffective; too much leads to deposits of the polymer, which is persistent and which may contribute to the natural blocking of the filters. The British Standard 5726 BS (1992) specifies a concentration of 50 mg/m³. To achieve this 60 ml of formalin, mixed with 60 ml of water is required for each m³ of cabinet volume (this volume is specified by the manufacturer).

The formalin may be boiled in several ways. Some cabinets have built-in devices in which the liquid is dripped on to a hot-plate. Other manufacturers supply small boilers. A laboratory hot-plate, connected to a timing clock is adequate. The correct amount of formalin and water is placed in a flask.

Heating paraformaldehyde

Tablets (1 g) are available from chemical manufacturers. Three or four tablets are adequate for an average size cabinet. They may be heated on an electric frying pan connected to a time clock or placed in a dish and heated by an electric hair drier.

Exposure time

It is convenient to start decontamination in the late afternoon and let the gas act overnight. In the morning the fan should be switched on and the front closure 'cracked' open very slightly to allow air to enter and purge the cabinet. Some new cabinets have a hole in the front closure, fitted with a stopper which can be removed for this purpose. With Class III cabinets the plastic film is taken off the supply filter to allow air to enter the cabinet and purge it of formaldehyde. After several minutes the gloves may be removed. The fan is then allowed to run for about 30 min which should remove all formaldehyde.

Summary of decontamination procedure

The prefilter should be removed from inside the cabinet (where possible) and placed on the working surface. It is usually pushed or clipped into place and is easily removed, but as it is likely to be contaminated gloves should be worn. Formalin, 60 ml, plus water, 60 ml, per m³ of cabinet volume should be placed in its container on the heater in the cabinet or in the reservoir. The front closure is then put in place and sealed if necessary. Then the heater is switched on and the formalin boiled away. After switching off the heater the cabinet is left closed overnight. The next morning the cabinet fan is switched on and then the front closure is opened very slightly to allow air to pass in and purge the cabinet of formaldehyde. After several minutes the front closure is removed and the cabinet fan allowed to run for about 30 min. Any obvious moisture remaining on the cabinet walls and floor may then be wiped away and the filters changed.

Changing the filters and maintenance

The cabinet must be decontaminated before filters are changed and any work is done on the motor and fans. If these are to be done by an outside contractor, e.g. by the manufacturer's service engineer, the front closure should be sealed on again after the initial purging of the gas and a notice:

Cabinet decontaminated but not to be used placed on it awaiting the engineer's arrival. He will also require a certificate stating that the cabinet has been decontaminated.

The primary, or prefilters, should be changed when the cabinet airflow approaches its agreed local minimum. Used filters should be placed in plastic or tough paper bags, which are then sealed and burned. If the airflow is not restored to at least the middle of the range then arrangements should be made to replace the HEPA filter. This is usually done by the service engineer, but may be done by laboratory staff if they have received instruction. Unskilled operators often place the new filter in upside down or fail to set it securely and evenly in its place. Used filters should be placed in plastic bags, which are then sealed for disposal. They are not combustible. Some manufacturers accept used filters and recover the cases, but this is not usually a commercial practice – no refund is given. When manufacturers replace HEPA filters they may offer a testing service.

Testing; further information

This is beyond the scope of this book and the reader is referred to national standards and other publications (Clark, 1983; Collins, 1993).

In the UK the *Control of Substances Hazardous to Health Regulations* require that microbiological safety cabinets are tested to conform with the national standard at intervals of not less than 14 months.

Preparation room equipment

Sterilizers are discussed in Chapter 4.

Inspissators

In the preparation of slopes of medium consisting of egg or serum, the amount of heating required to coagulate the protein must be carefully controlled. A steamer heats the medium too rapidly and raises the temperature too high. The modern inspissator is thermostatically controlled at 75–85°C and fitted with a large internal circulating fan. The shelves on which the tubes are sloped are made of wire mesh so that circulation is not impeded. It is convenient to have wire mesh or aluminium racks made which hold tubes or bottles at the correct angle (5–10°) and which slide on the shelf brackets. These facilitate rapid loading and unloading while the instrument is hot.

An egg or serum medium is usually coagulated in 45–60 min at 80°C. Whether the inspissator is loaded hot or cold is a matter of personal choice, but better control, required for media that contain drugs, e.g. mycobacterial sensitivity tests, is obtained by raising the temperature first and then putting in a standard load for a constant time.

Laminar flow clean air work stations

These cabinets are designed to protect the work from the environment and are most useful for aseptic distribution of certain media and plate pouring. A stream of sterile (filtered) air is directed over the working area, either horizontally into the room or vertically downwards when it is usually re-circulated. They are particularly useful for preventing contamination when distributing sterile fluids.

Laminar flow clean air work stations are NOT microbiological safety cabinets and must not be used for manipulations with microorganisms or tissue culture cells. The effluent air, which may be contaminated, is blown into the face of the operator.

Glassware-washing machines

These are useful in large laboratories which use enough of any one size of tube or bottle to give an economic load.

Glassware drying cabinets

A busy wash-up room requires a drying cabinet with wire-rack shelves and a 3-kW electric heater in the base, operated through a three-heat control. An extractor fan on the top is a refinement, otherwise the sides near the top should be louvred.

Glassware, plastics and small equipment

For ordinary bacteriological work, soda-glass tubes and bottles are satisfactory. In assay work, the more expensive resistance glass might be justified. An important consideration is whether glassware should be washed or discarded. Purchasing cheaper glassware in bulk and using plastic disposable petri dishes and culture tubes may, in some circumstances, be more economical than employing labour to clean them.

Plastics fall into two categories:

1 disposable items, such as petri dishes, specimen containers and plastic loops, which are destroyed when autoclaved and cannot be sterilized by ordinary laboratory methods (but see Chapter 4), and
2 recoverable material, which must be sterilized in the autoclave.

Non-autoclavable plastics
These include polyethylene, styrene, acrylonitrile, polystyrene and rigid polyvinyl chloride.

Autoclavable plastics
These withstand a temperature of 121°C and include polypropylene, polycarbonate, nylon, PTFE (Teflon), polyallomer, TPX (methylpentene polymer), Viken and vinyl tubing.

It is best to purchase plastic apparatus from specialist firms who will give advice on the suitability of their products for specific purposes. Most recoverable plastics used in microbiology can be washed in the same way as glassware. Some plastics soften during autoclaving and may become distorted unless packed carefully.

Petri dishes

Disposable plastic petri dishes are used in most laboratories in developed countries. They are supplied already sterilized and packed in batches in polythene bags. They can be stored indefinitely, are cheap when purchased in bulk and are well made and easy to handle. Two kinds are available, vented and unvented. Vented dishes have one or two nibs that raise the top slightly from the bottom and are to be preferred for anaerobic and carbon dioxide cultures.

Glass petri dishes are, however, still popular in some areas. In general, two qualities are obtainable. The thin, blown dishes, usually made of borosilicate glass, are pleasant to handle, their tops and bottoms are flat and they stack safely. They do not become etched through continued use but are fragile, must be washed with care and are expensive. The thick pressed glass dishes are often convex, cannot always be stacked safely and are easily etched and scratched with use and washing. On the other hand, they are cheap and not fragile. Aluminium, stainless steel and plastic 'lids' are available which double the number of glass dishes.

Glass petri dishes are sterilized by hot air in aluminium boxes.

Test-tubes and bottles, plugs, caps and stoppers

Culture media can be tubed in either test-tubes or small bottles. Apart from personal choice, test-tubes are easier to handle in busy laboratories and take up less space in storage receptacles and incubators, but bottles are more convenient in the smaller workroom where media is kept for longer periods before use. Media in test-tubes may dry up during storage. Screw-capped test-tubes are available.

The most convenient sizes of test-tube are: 127 × 12.5 mm, holding 4 ml of medium: 152 × 16 mm, holding 5–10 ml; 152 × 19 mm, holding 10–15 ml; and 178 × 25 mm, holding 20 ml. Rimless test-tubes of heavy quality are made for bacteriological work. The lipped, thin glass chemical test-tubes are useless.

Cotton wool plugs have been used for many years to stopper test-tube cultures, but have largely been replaced by metal or plastic caps, or in some laboratories by soft, synthetic sponge bungs.

Aluminium test-tube caps were introduced some years ago but they have a limited tolerance and, in spite of alleged standard specifications of test-tubes, a laboratory very soon accumulates many tubes that will not fit the caps. Caps that are too loose are useless. These caps are cheap, last a long time, are available in many colours and save a great deal of time and labour. Those held in place by a small spring have a wider tolerance and fit most tubes. Polypropylene caps in several sizes and colours and which stand up to repeated autoclaving are obtainable.

Temporary closures can be made from kitchen aluminium foil.

Aluminium capped test-tubes are sterilized in the hot air oven in baskets. Polypropylene capped tubes must be autoclaved.

Rubber stoppers of the orthodox shape are useless as they are blown out of the tubes in the autoclave although synthetic sponge rubber stops remain *in situ*. The Astell rubber seals are designed to avoid this and to allow steam to penetrate into the tubes as readily as with cotton wool plugs or aluminium or polypropylene closures. These stoppers fit 16-mm tubes and also the Astell culture bottles used for roll-tube work, etc. They have a very long laboratory life.

Several sizes of small culture bottles are made for microbiological work. Some of these bottles are of strong construction and are intended to be reused. Others, although tough enough for safe handling, are disposable. The most useful sizes are: those holding 7, 14 and 28 ml; and the 'Universal Container', 28 ml, which has a larger neck than the others and is also used as a specimen container. These bottles usually have aluminium screw-caps with rubber liners. The liners should be made of black rubber; some red rubbers are thought to give off bactericidal substances. Polypropylene caps are also used; they need no liners but in our experience they may loosen spontaneously during long incubation or storage. The medium dries up. They are satisfactory in the short term. Astell culture bottles, holding about 20 ml and closed by Astell seals, are popular for roll-tube counting.

All of these bottles can be sterilized by autoclaving.

Media storage bottles

'Medical flats' or 'rounds' are made in sizes from 60 ml upwards. The most convenient sizes are 110 ml, holding 50–100 ml of medium, and 560 ml, for storing 250–500 ml. The flat bottles are easiest to handle and store but the round ones are more robust.

Specimen containers

The screw-capped glass or plastic 'Universal Containers' which hold about 28 ml is the most popular. There are many other containers, mostly made of plastic. There are too many different containers, many of plastic; some are satisfactory, others are not; some leak easily and others do not stand up to handling by patients. Only screw-capped containers with more than one and a half threads on their necks should be used. Those with 'push-in' or 'pop-up' stoppers are dangerous. They generate aerosols when opened. Waxed paper pots should not be used as they invariably leak.

For larger specimens, there is a variety of strong screw-capped jars.

Before any containers are purchased in bulk, it is advisable to test samples by filling them with coloured water and standing them upside down on blotting paper for several days after screwing the caps on moderately well. Patients and nurses may not screw caps on as tightly as possible. Similar bottles should be sent through the post, wrapped in absorbent material in accordance with postal regulations (see p. 7). Leakage will be evident by the staining of the blotting paper or wrapping. More severe tests include filling the containers with coloured water and centrifuging them upside down on a wad of blotting paper. Specimen container problems and tests are reviewed by Collins (1993).

There is a British Standard (BSI, 1975) for microbiological specimen containers.

Sample jars and containers

Containers for food samples, water, milk, etc., should conform to local or national requirements. In general, large screw-capped jars are suitable but are rather heavy if many samples are taken, and should not be used in food factories. Plastic containers may be used as leakage and spillage is not such a problem as with pathological material. Strong plastic bags are useful, particularly if they are of the self-sealing type. Otherwise 'quick-ties' may be used.

Pasteur pipettes

Glass pasteur pipettes are probably the most dangerous pieces of laboratory equipment in unskilled hands. They are used, with rubber teats, to transfer liquid cultures, serum dilutions, etc.

Very few laboratories make their own pasteur pipettes nowadays; they are obtainable, plugged and unplugged, from most laboratory suppliers and are best purchased in bulk. Long and short forms are available and most are made of 6–7 mm diameter glass tubing. Some are supplied sterile and ready for use.

After plugging, they can be sterilized in aluminium boxes made for this purpose.

New and safer pasteur pipettes with integral teats and made of low-density polypropylene are now available and are supplied ready sterilized.

Pasteur pipettes are used once only.

Graduated pipettes

Straight-sided blow-out pipettes, 1–10-ml capacity are used. They must be plugged with cotton wool at the suction end to prevent bacteria entering from the pipettor or teat and contaminating the material in the pipette. These plugs must be tight enough to stay where they are during pipetting but not so tight that they cannot be removed during cleaning. About 25 mm of non-absorbent cotton wool is pushed into the end with a piece of wire. The ends are then passed through a bunsen flame to tidy them. Wisps of cotton wool which get between the glass and the pipettor or teat may permit air to enter and the contents to leak.

Pipettes are sterilized in the hot air oven in square section aluminium containers similar to those used for pasteur pipettes (square-section containers which do not roll have supplanted the time-honoured round-section boxes). A wad of glass wool at the bottom of the container prevents damage to the tips.

Disposable 1 ml and 10 ml pipettes are available. Some firms supply them already plugged, wrapped and sterilized.

Pipetting aids

Rubber teats and pipetting devices provide an alternative to the highly dangerous practice of mouth pipetting.

Rubber teats

Choose teats with a capacity greater than that of the pipettes for which they are intended, i.e. a 1-ml teat for pasteur pipettes, a 2-ml teat for 1-ml pipette, otherwise the teat must be used fully compressed, which is tiring. Most beginners compress the teat completely, then suck up the liquid and try to hold it at the mark while transferring it. This is unsatisfactory and leads to spilling and inaccuracy. Compress the teat just enough to suck the liquid a little way past the mark on the pipette. Withdraw the pipette from the liquid, press the teat slightly to bring the fluid to the mark and then release it. The correct volume is now held in the pipette without tiring the thumb and without risking loss. To discharge the pipette, press the teat slowly and gently and then release it in the same way. Violent operation usually fails to eject all the liquid; bubbles are sucked back and aerosols are formed.

'Pipettors'
A large number of devices that are more sophisticated than simple rubber teats are now available. Broadly speaking there are four kinds of these:

1 rubber bulbs with valves that control suction and dispensing;
2 syringe-like machines that hold pipettes more rigidly than rubber bulbs and have a plunger operated by a rack and pinion or a lever;
3 electrically operated pumps fitted with flexible tubes in which pipettes can be inserted;
4 mechanical plunger devices which take small plastic pipette tips and are capable of repeatedly delivering very small volumes with great accuracy.

It is extremely difficult to give advice on the relative merits of the various devices. That which suits one operator, or is best for one purpose may not be suitable for others. Choice should therefore be made by the operators, not by managers or administrators who will not use them. None of those in categories 1, 2 and 4 above are expensive, and it should be possible for several different models to be available. What is necessary is some system of instruction in their use and in their maintenance. None should be expected to last forever.

Micro-slides and cover-glasses

Unless permanent preparations are required, the cheaper microscope slides are satisfactory. Slides with ground and polished edges are much more expensive. Most micro-slides are sold in boxes of 100 slides. They should not be washed and re-used but discarded.

Cover-glasses are sold in several thicknesses and sizes. Those of thickness grade No. 1, 16-mm square, are the most convenient. They are sold in boxes containing about 100 glasses. Plastic cover-glasses are available.

Durham's (fermentation) tubes

These are small glass tubes, usually 25–30 × 5–6 mm and closed at one end, which are placed inverted in tubes of culture media to detect gas formation. They are very cheap and are not worth washing.

Inoculating loops and wires

These are usually made of 25 SWG Nicrome wire, although this is more springy than platinum iridium. They should be short (not more than 5 cm long) in order to minimize vibration and therefore involuntary discharge of contents. Loops should be small (not more than 3 mm in diameter). Large loops are also inclined to empty spontaneously and scatter infected airborne particles. They should be completely closed, otherwise they will not hold fluid cultures. This can be carried out by twisting the end of the wire round the shank, or by taking a piece of wire 12 cm long, bending the centre round a nail or rod of appropriate diameter and twisting the ends together in a drill chuck. Ready-made loops of this kind are available.

Loops and wires should not be fused into glass rods. Aluminium holders are sold by most laboratory suppliers.

Plastic disposable loops are excellent. There are two sizes, 1 and 10 μl, and both are useful. They are sold sealed in packs of about 25, sterilized ready for use. Pointed plastic rods, for use in place of 'straight wire', for subculturing, are now available.

Spreaders

Cut a glass rod of 3–4 mm diameter into 180 mm lengths and round off the cut ends in a flame. Hold each length horizontally across a bat's wing flame so that it is heated and bends under its own weight approximately 36 mm from one end, and an L shape is obtained. Plastic spreaders are on the market, sterilized ready for use.

Racks and baskets

Test-tube and culture bottle racks should be made of polypropylene or of metal covered with polypropylene or nylon so that they can be autoclaved. These racks also minimize breakage, which is not uncommon when metal racks are used. Wooden racks are unhygienic.

Aluminium trays for holding from 10–100 bottles, according to size, are widely employed in the UK. They can be autoclaved, and are easily taken apart for cleaning.

The traditional wire baskets are unsafe for holding test-tubes. They contribute to breakage hazards and do not retain spilled fluids. For non-infective work, these baskets, in which the wire is covered with polypropylene or nylon, are satisfactory, but autoclavable plastic boxes of various sizes are safer for use with cultures.

Bunsen burners

The usual bunsen, with a bypass, is satisfactory for most work, but for material that may spatter or that is highly infectious a hooded bunsen should be used. There are several versions of these, but we recommend the Kampff and Bactiburner types which enclose the flame in a borosilicate tube.

Other bench equipment

A hand magnifier, forceps and a knife or scalpel should be provided, and also a supply of tissues for mopping up spilled material and general cleaning. Swab sticks, usually made of wood and about 6 in long, are useful for handling some specimens, and wooden throat spatulas are useful for food samples. These can be sterilized in large test-tubes or in the aluminium boxes used for pasteur pipettes.

Some form of bench 'tidy' or rack is desirable to keep loops and other small articles together.

Discard jars and disinfectant pots are considered in Chapter 4.

References

BSI (1975) BS 5213, *Medical Specimen Containers for Microbiology*, British Standards Institution, London

BSI (1982) BS 4402, *Specification for Safety Requirements for Laboratory Centrifuges*, British Standards Institution, London

BSI (1992) BS 5726, *Microbiological Safety Cabinets, Parts 1–4*, British Standard Institution, London

Clark, R. P. (1983) *The Performance, Testing and Limitations of Microbiological Safety Cabinets*, Science Reviews, Norwood

Collins, C.H. (1993) *Laboratory-acquired Infections*, 3rd edn, Butterworth-Heinemann, Oxford

IEC (1990) IEC 1010: Part 1, *Safety Requirements – Electrical Equipment for Measurement, Control and Laboratory Use*, International Electrical Commission, Geneva

IFCC (1990) *Guidelines for Selection of Safe Laboratory Centrifuges and their Safe Use*, International Federation of Clinical Chemistry, Copenhagen

NSF (1992) Standard No. 49. *Class II (Laminar Flow) Biohazard Cabinetry*, National Sanitation Foundation, Ann Arbor, MI, USA

SAA (1983a) AS 2252, *Biological Safety Cabinets*, Standards Association of Australia, Sydney

SAA (1983b) AS 2647, *Biological Safety Cabinets – Installation and Use*, Standards Association of Australia, Sydney

Chapter

4

Sterilization, disinfection and the treatment of infected material

'Sterilization' implies the complete destruction of all microorganisms, including spores.

'Disinfection' implies the destruction of vegetative organisms which might cause disease, or, in the context of the food industries, which might cause spoilage. Disinfection does not necessarily kill spores.

The two terms are not synonymous.

'Decontamination' is the preferred term in microbiological laboratories for rendering materials safe for use or disposal.

Sterilization

The methods commonly used in microbiological laboratories are:

(1) red heat (flaming);
(2) dry heat (hot air);
(3) steam under pressure (autoclaving);
(4) steam not under pressure (Tyndallization); and
(5) filtration.

Incineration is also a method of sterilization but as it is applied outside the laboratory for the ultimate disposal of laboratory waste it is considered separately (p. 51).

Red heat

Instruments such as inoculating wires and loops are sterilized by holding them in a bunsen flame until they are red hot. Hooded bunsens (p. 37) are recommended for sterilizing inoculating wires contaminated with highly infectious material (e.g. tuberculous sputum) to avoid the risk of spluttering contaminated particles over the surrounding areas.

Dry heat

This is applied in an electrically heated oven which is thermostatically controlled and is fitted with a large circulating fan to ensure even temperatures in all parts of the load. Modern equipment has electronic controls which can be set to raise the temperature to the required level, hold it there for a prearranged time and then switch off the current. A solenoid lock is incorporated in some models to prevent the oven being opened before the cycle is complete. This safeguards sterility and protects the staff from accidental burns.

Materials which can be sterilized by this method include glass petri dishes, flasks, pipettes and metal objects. Various metal canisters and cylinders which conveniently hold glassware during sterilization and keep it sterile during storage are available from laboratory suppliers. Some laboratory workers still prefer the time-honoured method of wrapping pipettes, etc. individually in brown (Kraft) paper.

Loading

Air is not a good conductor of heat so oven loads must be loosely arranged, with plenty of spaces to allow the hot air to circulate.

Holding times and temperatures

When calculating processing times for hot air sterilizing equipment, there are three time periods which must be considered.

1. *the heating-up period*, which is the time taken for the entire load to reach the sterilization temperature; this may take about 1 h;
2. *the holding periods* at different sterilization temperatures recommended by the UK Department of Health (DHSS, 1980) which are 160°C for 45 min, 170°C for 18 min, 180°C for $7\frac{1}{2}$ min and 190°C for $1\frac{1}{2}$ min;
3. *the cooling-down period*, which is carried out gradually to prevent glassware from cracking as a result of a too rapid fall in temperature; this period may take up to 2 h.

Control of hot air sterilizers

Tests for electrically operated fan ovens are described in British Standard 3421 (BSI, 1961) and in HTM10 (DHSS, 1980).

Hot air sterilization equipment should be calibrated with thermocouples when the apparatus is first installed and checked with thermocouples afterwards when necessary.

Ordinary routine control can be effected simply with commercial chemical indicator tubes.

Steam under pressure: the autoclave

The temperature of saturated steam at atmospheric pressure is approximately 100°C. Temperature increases with pressure, e.g. at 1 bar (*ca* 15 lb/in^2) it is 121°C. Bacteria are killed by autoclaving at this temperature for 15–20 min. Air has an important influence on the efficiency of autoclaving. The above relationship holds good only if no air is present. If about 50% of the air remains in the autoclave the temperature of the steam/air mixture will be only

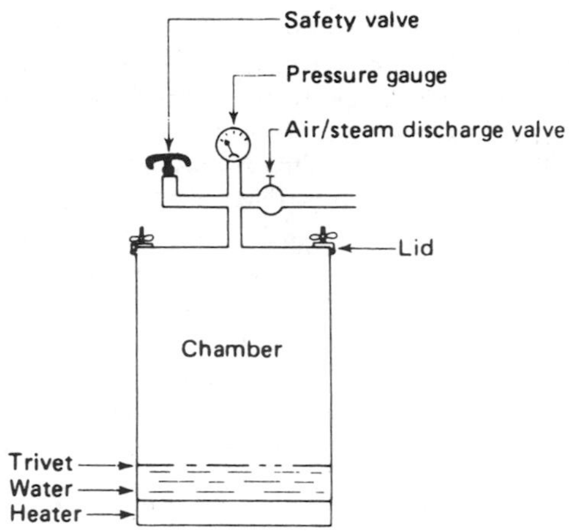

Figure 4.1 'Pressure cooker' laboratory autoclave

112°C. If air is present in the autoclave load it will also adversely affect penetration by the steam.

All the air that surrounds and permeates the load must first be removed before steam sterilization can commence.

Loads in autoclaves

As successful autoclaving depends on the removal of all the air from the chamber and the load, the materials to be sterilized should be packed loosely. 'Clean' articles may be placed in wire baskets, but contaminated material (e.g. discarded cultures) should be in solid bottomed containers not more than 20 cm deep (see Disposal of Infected Waste, below). Large air spaces should be left around each container and none should be covered.

Types of autoclave

Only autoclaves designed for laboratory work and capable of dealing with a 'mixed load' should be used. 'Porous load' and 'bottled fluid sterilizers' are rarely satisfactory for laboratory work. There are two varieties of laboratory autoclave:

1 pressure cooker types; and
2 gravity displacement models with automatic air and condensate discharge.

'Pressure cooker' laboratory autoclaves

These are still in use in many parts of the world. The most common type is a device for boiling water under pressure. It has a vertical metal chamber with a strong metal lid which can be fastened down and sealed with a rubber gasket. An air and steam discharge tap, pressure gauge and safety valve are fitted in the lid (Figure 4.1). Water in the bottom of the autoclave is heated by external gas burners, an electric immersion heater or a steam coil.

Operating instructions

There must be sufficient water inside the chamber. The autoclave is loaded and the lid is fastened down with the discharge tap open. The safety valve is then adjusted to the required temperature and the heat is turned on.

When the water boils, the steam will issue from the discharge tap and carry the air from the chamber with it. The steam and air should be allowed to escape freely until all of the air has been removed. This may be tested by attaching one end of a length of rubber tubing to the discharge tap and inserting the other end into a bucket or similar large container of water. Steam condenses in the water and the air rises as bubbles to the surface; when all of the air has been removed from the chamber, bubbling in the bucket will cease. When this stage has been reached, the air-steam discharge tap is closed and the rubber tubing removed. The steam pressure then rises in the chamber until the desired pressure and temperature are reached and steam issues from the safety valve.

When the load has reached the required temperature (see Testing autoclaves, below), the pressure is held for 15 min.

At the end of the sterilizing period, the heater is turned off and the autoclave allowed to cool.

The air and steam discharge tap is opened very slowly after the pressure gauge has reached zero (atmospheric pressure). If the tap is opened too soon, while the autoclave is still under pressure, any fluid inside (liquid media, etc.) will boil explosively and bottles containing liquids may even burst. The contents are allowed to cool. Depending on the nature of the materials being sterilized, the cooling (or 'run-down') period needed may be several hours for large bottles of agar to cool to 80°C, when they are safe to handle.

Gravity displacement autoclaves

These may be relatively simple in their construction and operation, as shown diagrammatically in Figure 4.2, or extremely sophisticated pieces of engineering in which air and, finally, steam are removed by vacuum pumps and in which the whole sterilization cycle may be programmed.

The jacket surrounding the autoclave consists of an outer wall enclosing a narrow space around the chamber, which is filled with steam under pressure to keep the chamber wall warm. The steam enters the jacket from the mains supply, which is at high pressure, through a valve that reduces this pressure to the working level. The working pressure is measured on a separate pressure gauge fitted to the jacket. This jacket also has a separate drain for air and condensate to pass through.

The steam enters the chamber from the same source which supplies steam to the jacket. It is introduced in such a way that it is deflected upwards and fills the chamber from the top downwards, thus forcing the air and condensate to flow out of the drain at the base of the chamber by gravity displacement. The drain is fitted with strainers to prevent blockage by debris. The drain discharges into a closed container (not shown in the diagram) so that there is a complete air break that prevents backflow. There is also a filter to ensure that aerosols are not released into the room.

The automatic steam trap or 'near-to-steam' trap is designed to ensure that only saturated steam is retained inside the chamber, and that air and condensate, which are at a lower temperature than saturated steam, are automatically discharged. It is called a 'near-to-steam' trap because it opens if the temperature falls to about 2°C below that of saturated steam and closes within 2°C or

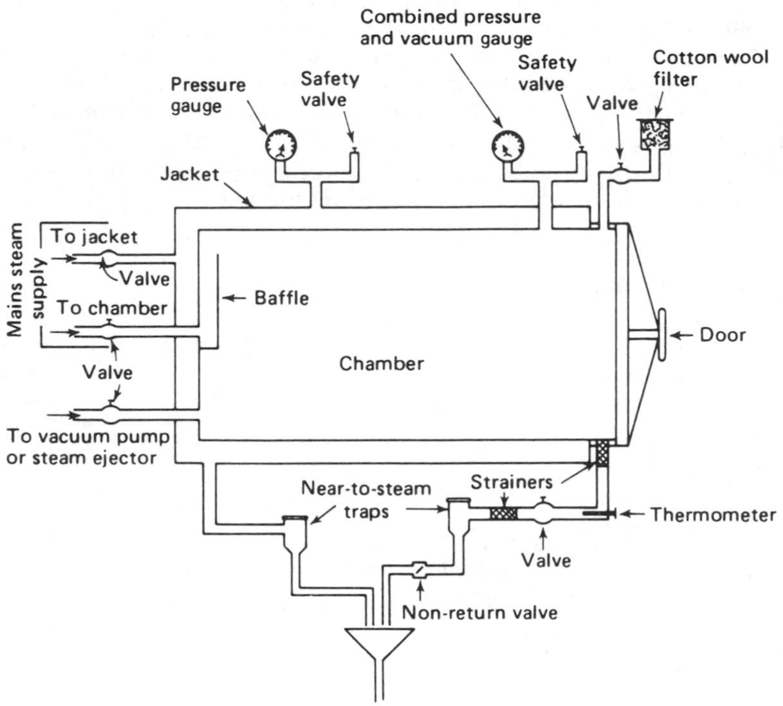

Figure 4.2 Gravity displacement autoclave

nearer to the saturated steam temperature. The trap operates by the expansion and contraction of a metal bellows which operates a valve.

There may be a thermometer probe in the drain but as this registers the temperature of steam at that point and not that in the load within the chamber this may be misleading. For example, the temperature in the drain may reach 121°C while that in the load is only 50°C (Collins, 1993).

In modern autoclaves flexible thermocouple probes are fitted in the chamber so that temperatures in various parts of the load may be recorded. In older models the thin thermocouple leads may be safely taken in through the door seals.

There are usually interlocking devices that prevent the opening of the door before the temperature in the chamber has fallen to 80°C. This does not imply that the temperature in the load has also fallen to a safe level. In large, sealed bottles it may still be over 100°C, when the contents will be at a high pressure. Sudden cooling may cause the bottles to explode. The autoclave should not be opened until the temperature in the load has fallen to 80°C or below. This may take a very long time, and in some autoclaves there are locks which permit the door to be opened only fractionally to cool the load further before it is finally released.

Operation of a gravity displacement autoclave

If the autoclave is jacketed, the jacket must first be brought to the operating temperature. The chamber is loaded, the door is closed and the steam-valve is opened, allowing steam to enter the top of the chamber. Air and condensate flow out through the drain at the bottom (Figure 4.2). When the drain thermometer reaches the required temperature a further period must be allowed for the load to reach the temperature. This should be determined initially and periodically for each autoclave as described below. Unless this is done the load is unlikely to be sterilized. The autoclave cycle is then continued for the holding time. When it is completed the steam valves are closed and the autoclave allowed to cool until the temperature dial reads less than 80°C. Not until then is the autoclave safe to open. It should first be 'cracked' or opened very slightly and left in that position for several minutes to allow steam to escape and the load to cool further (see Safety of the operator, below).

Time/temperature cycles

For most purposes the following time/temperature cycles will ensure sterilization of a properly-packed load (p. 41)

3 min at 134°C
10 min at 126°C
15 min at 121°C
25 min at 115°C

These are the holding times at temperature (HTAT), determined as described below. The usual HTAT in microbiology laboratories is 15 min at 121°C.

Material containing, or suspected of containing, the agents of spongiform encephalopathies should be autoclaved for 18 min or for six consecutive cycles of 3 min at 134°C (ACDP, 1990).

Testing autoclaves

The time/temperature cycles should be tested under 'worst load' conditions, e.g. a container filled with 5 ml screw-capped bottles. This should be placed in the centre of the chamber and, if space is available, other loaded containers may be placed around it. Thermocouple leads are placed in the middle of the load and at other places. The cycle is then started and timed.

There are three periods:

a *Warming up*, until the temperature in the middle of the load is 121°C.
b *Sterilization*, i.e. HTAT, in which the temperature in the load is maintained at 121°C for 15 min.
c *Cooling down*, after the steam valve has been closed and the temperature in the load has fallen to 80°C.

If the autoclave is to be operated manually these times should be noted and displayed. Automated autoclaves may be programmed.

Monitoring

During daily use or from time to time the recording thermometers may be supplemented with indicators. There are three types. Two give immediate results, the other is retrospective:

1 *Bowie-Dick autoclave tape* (*DHSS, 1980*). This is tape impregnated with a chemical and which is placed in the load. The colour changes if there is adequate steam penetration. These tapes should be reserved for vacuum-type autoclaves.

2 *Chemical indicators*. These are usually in sealed tubes or sachets, and change colour if the correct time and temperature combination has been achieved.
3 *Biological indicators*. Spores of two organisms are used: *Bacillus stearothermophilus* (NCTC 1007, ATCC 7935) and *Clostridium sporogenes* (NTCC 8596, ATCC 7955). They are used as suspensions or absorbed on carriers such as filter paper strips. Laboratory preparations may be unreliable as the heat resistance of the spores depends on the culture media used. It is best to purchase commercial products. Strips are placed at various locations in the load and, after autoclaving, are added to tubes of broth which are then incubated. Turbidity indicates unsuccessful processing.

There is a British Standard for the design, installation, maintenance and testing of laboratory autoclaves (BSI, 1988). In the UK autoclaves are 'pressure vessels' under the *Pressure Systems and Transportable Gas Container Regulations 1989*.

For further information on autoclaving see Kennedy (1988), Gardner and Peel (1991) and Russell *et al.* (1992).

Protection of the operator

Serious accidents, including burns and scalds to the face and hands have occurred when autoclaves have been opened, even when the temperature gauge read below 80°C and the door lock has allowed the door to be opened. Liquids in bottles may still be over 100°C and under considerable pressure. The bottles may explode on contact with air at room temperature.

When autoclaves are being unloaded operators should wear full-face visors of the kind that cover the skin under the chin and throat. They should also wear thermal-protective gloves (see Kennedy, 1988).

Steam at 100°C ('Tyndallization')

This process, named after the Irish physicist cum bacteriologist John Tyndall, employs a Koch or Arnold steamer, which is a metal box in the bottom of which water is boiled by a gas burner, electric heater or steam coil. The articles to be processed rest on a perforated rack just above the water level. The lid is conical so that the condensation runs down the sides instead of dripping on the contents. A small hole in the top of the lid allows air and steam to escape. This method is used to sterilize culture media that might be spoiled by exposure to higher temperatures, e.g. media containing easily hydrolysed carbohydrates or gelatin. These are steamed for 30–45 min on each of three successive days. On the first occasion, vegetative bacteria are killed; any spores that survive will germinate in the nutrient medium overnight, producing vegetative forms that are killed by the second or third steaming.

Filtration

Bacteria can be removed from liquids by passing them through filters with very small pores that trap bacteria but, in general, not mycoplasmas or viruses. The method is used for sterilizing serum for laboratory use, antibiotic solutions and special culture media that would be damaged by heat. It is also used for separating the soluble products of bacterial growth (e.g. toxins) in fluid culture media.

Filters of historic interest, now rarely used, include the Berkefeld (made of Kieselguhr), the Chamberland (unglazed porcelain), the Seitz (asbestos) and sintered glass.

Membrane filters

These filters are made from cellulose esters (cellulose acetate, cellulose nitrate, collodion, etc.). A range of pore sizes is available. Bacterial filters have a pore size of less than 0.75 μm. The membranes can be sterilized by autoclaving.

For use, the membrane is mounted on a perforated platform, usually made of stainless steel, which is sealed together between the upper and lower funnels. Filtration is achieved by applying either a positive pressure to the entrance side of the filter or a negative pressure to the exit side.

Small filter units for filtering small volumes of fluid (e.g. 1–5 ml) are available. The fluid passes through the filter by gravitational force in a centrifuge or is forced through a small filter from a syringe.

Membrane filters and filter holders to suit different purposes are obtainable and several manufacturers publish useful booklets or leaflets about their products.

Chemical disinfection

Some disinfectants present health hazards (see Table 4.1). It is advisable to wear eye and hand protection when making dilutions.

Many different chemicals may be used and they are collectively described as disinfectants or biocides. The former term is used in this book. Some are ordinary reagents, others are special formulations, marketed under trade names. Disinfectants should not be used to 'sterilize' materials or when physical methods are available. In some circumstances, e.g. in food establishments, cleaning with detergents is better. The effects of time, temperature, pH, and the chemical and physical nature of the article to be disinfected and of the organic matter present are often not fully appreciated.

Types and laboratory uses of disinfectants

There is an approximate spectrum of susceptibility of microorganisms to disinfectants. The most susceptible are vegetative bacteria, fungi and lipid-containing viruses. Mycobacteria and non lipid-containing viruses are less susceptible and spores are generally resistant.

Some consideration should be given to the disinfectant's toxicity and any harmful effects that they may have on the skin, eyes and respiratory tract.

Only those disinfectants which have a laboratory application are considered here. For further information and other applications see Ayliffe *et al.* (1984), Gardner and Peel (1991) and Russell *et al.* (1992).

The most commonly used disinfectants in laboratory work are clear phenolics and hypochlorites. Aldehydes have a more limited application, and alcohol and alcohol mixtures are less popular but deserve greater attention. Iodophors and quaternary ammonium compounds (QAC) are more popular in the USA than in the UK, while mercurial compounds are the least used. The properties of these disinfectants are summarized below and in Table 3.1. Other substances, such as ethylene oxide and propiolactone, are used commercially in the preparation

Table 4.1 Properties of some disinfectants

	Active against							Inactivated by					Toxicity		
	Fungi	Bacteria G+	G−	Myco-bacteria	Spores	Lipid viruses	Non lipid viruses	Protein	Natural materials	Man-made materials	Hard water	Deter-gent	Skin	Eyes	Lungs
Phenolics	+ + +	+ + +	+ + +	+ +	−	+	v	+	+ +	+ +	+	C	+	+	−
Hypochlorites	+	+ + +	+ + +	+ +	+ +	+	+	+ + +	+	+	+	C	+	+	+
Alcohols	−	+ + +	+ + +	+ + +	−	+	v	+	+	+	+	−	−	+	−
Formaldehyde	+ + +	+ + +	+ + +	+ + +	+ + +[a]	+	+	+	+	+	+	−	+	+	+
Glutaraldehyde	+ + +	+ + +	+ + +	+ + +	+ + +[b]	+	+	NA	+	+	+	−	+	+	+
Iodophors	+ + +	+ + +	+ + +	+ + +	+	+	+	+ + +	+	+	+	A	+	+	−
QAC	+	+ + +	+ +	−	−	−	−	+ + +	+ +	+ + +	+ + +	A(C)	+	+	−

From Collins (1993)
+ + +, good
+ +, fair
+, slight
−, nil
v depends on virus
[a] above 40°C
[b] above 20°C

C, cationic
A, anionic
NA, not applicable

of sterile equipment for hospital and laboratory use. They are not normally used in laboratories but are used in some hospitals for medical equipment. Commercial equipment is available for this purpose.

Clear phenolics
These compounds are effective against vegetative bacteria (including mycobacteria) and fungi. They are inactive against spores and non lipid-containing viruses. Most phenolics are active in the presence of considerable amounts of protein but are inactivated to some extent by rubber, wood and plastics. They are not compatible with cationic detergents. Laboratory uses include discard jars and disinfection of surfaces. Clear phenolics should be used at the highest concentration recommended by the manufacturers for 'dirty situations', i.e. where they will encounter relatively large amounts of organic matter. This is usually 2–5%, as opposed to 1% for 'clean' situations where they will not encounter much protein. Dilutions should be prepared daily and diluted phenolics should not be stored for laboratory use for more than 24 h, although many diluted clear phenolics may be effective for more than 7 days.

Skin and eyes should be protected.

Hypochlorites
The activity is due to chlorine, which is very effective against vegetative bacteria (including mycobacteria), spores and fungi. Hypochlorites are considerably inactivated by protein and to some extent by natural non-protein material and plastics and they are not compatible with cationic detergents. Their uses include discard jars and surface disinfection but as they corrode some metals care is necessary. They should not be used on the metal parts of centrifuges and other machines which are subjected to stress when in use.

The hypochlorites sold for industrial and laboratory use in the UK contain 100 000 ppm available chlorine. They should be diluted as follows:

Reasonably clean surfaces	1:100	giving	1 000 ppm
Pipette and discard jars	1:40		2 500
Blood spillage	1:10		10 000

Some household hypochlorites (e.g. those used for babies' feeding bottles) contain 10 000 ppm and should be diluted accordingly. Household 'bleaches' in the UK and USA contain 50 000 ppm available chlorine and dilutions of 1:20 and 1:5 are appropriate.

Hypochlorites decay rapidly in use, although the products as supplied are stable. Diluted solutions should be replaced after 24 h. The colouring matter added to some commercial hypochlorites is intended to identify them: it is not an indicator of activity.

Hypochlorites may cause irritation of skin, eyes and lungs.

Sodium dichloroisocyanate (NADCC) is a solid chlorine-releasing agent. The tablets are useful for preparing bench discard jars and the powder for dealing with spillages, especially of blood.

Aldehydes
Formaldehyde (gas) and glutaraldehyde (liquid) are good disinfectants. They are active against vegetative bacteria (including mycobacteria), spores and fungi. They are active in the presence of protein and are not very much inactivated by natural or man-made materials, or detergents.

Formaldehyde is not very active at temperatures below 20°C and requires a relative humidity of at least 70%. It is not supplied as a gas, but as a solid polymer, paraformaldehyde, and a liquid, formalin, which contains 37–40% of formaldehyde. Both forms are heated to liberate the gas, which is used for disinfecting enclosed spaces such as safety cabinets and rooms. Formalin, diluted 1:10 to give a solution containing 4% formaldehyde, is used for disinfecting surfaces and, in some circumstances, cultures. Solid, formaldehyde-releasing compounds are now on the market and these may have laboratory applications but they have not yet been evaluated for this purpose. Formaldehyde is used mainly for decontaminating safety cabinets (p.29) and rooms.

Some glutaraldehyde formulations need an activator, which is supplied with the bulk liquid. Most activators contain a dye so that the user can be sure that the disinfectant has been activated. Effectiveness and stability after activation varies with product and the manufacturers' literature should be consulted.

Aldehydes are toxic. Formaldehyde is particularly unpleasant as it affects the eyes and causes respiratory distress. Special precautions are required (see below).
Glutaraldehyde is moderately toxic and is also an irritant, especially to eyes, skin and the upper respiratory tract.

Alcohol and alcohol mixtures

Ethanol and propanol, at concentrations of about 70–80% in water, are effective, albeit slowly, against vegetative bacteria. They are not effective against spores or fungi. They are not especially inactivated by protein and other material or detergents.

Effectiveness is enhanced by the addition of formaldehyde, e.g. a mixture of 10% formalin in 70% alcohol, or hypochlorite to give 2000 ppm of available chlorine

Alcohols and alcohol mixtures are useful for disinfecting surfaces and, alcohol-hypochlorite mixtures excepted, for balancing centrifuge buckets.

They are relatively harmless to skin but may cause eye irritation.

Quaternary ammonium compounds

These are cationic detergents known as QACs or quats, and are effective against vegetative bacteria and some fungi but not against mycobacteria or spores. They are inactivated by protein and by a variety of natural and plastic materials and by anionic detergents and soap. Their laboratory uses are therefore limited but they have the distinct advantages of being stable and of not corroding metals. They are usually employed at 1–2% dilution for cleaning surfaces and are very popular in food hygiene laboratories because of their detergent nature.

QACs are not toxic and are harmless to the skin and eyes.

Iodophors

Like chlorine compounds these iodines are effective against vegetative bacteria (including mycobacteria), spores, fungi, and both lipid-containing and non lipid-containing viruses. They are rapidly inactivated by protein, and to a certain extent by natural and plastic substances and are not compatible with anionic detergents. For use in discard jars and for disinfecting surfaces they should be diluted to give 75–150 ppm iodine, but for hand-washing or as a sporicide, diluted in 50% alcohol to give 1600 ppm iodine. As sold, iodophors usually contain a detergent and they have a built-in indicator: they are active as long as they remain brown or yellow. They stain the skin and surfaces but stains may be removed with sodium thiosulphate solution.

Iodophors are relatively harmless to skin but some eye irritation may be experienced.

Mercurial compounds
Activity against vegetative bacteria is poor and mercurials are not effective against spores. They do have an action on viruses at concentrations of 1:500 to 1:1000 and a limited use, as saturated solutions, for safely making microscopic preparations of mycobacteria.

Their limited usefulness and highly poisonous nature make mercurials unsuitable for general laboratory use.

Precautions in the use of disinfectants
As indicated above, some disinfectants have undesirable effects on the skin, eyes and respiratory tract. Disposable gloves and safety spectacles, goggles or a visor should be worn by anyone who is handling strong disinfectants, e.g. when pouring from stock and preparing dilutions for use.

Testing disinfectants

Several countries and organizations have 'official' tests for disinfectants (e.g. DGHM, 1982, 1984; AOAC, 1984; BSI, 1985, 1986; AFNOR, 1989). Specifications for these are detailed. They are used by manufacturers and have no place in clinical or public health laboratories which might use them only occcasionally, as reproducibility is then a problem. Some, e.g. the Rideal-Walker and the Chick-Martin tests, have fallen into disrepute because they have been used to compare unlikes. Those that use *Salmonella typhi*, even strains that are claimed to be avirulent, should be abandoned altogether. Currently, the European Committee for Standardization (CEN) is preparing a European standard.

Precise details of current 'official' tests will not, therefore, be given here – only the principle.

Suspensions containing known numbers of colony forming units (cfu)/ml of *Staphylococcus aureus* and/or *Pseudomonas aeruginosa* are prepared and added to various dilutions of the test product. After predetermined contact times a known volume of the mixture is removed to a neutralizing fluid that deactivates the disinfectant. The number of surviving colony forming units is then counted. A satisfactory disinfectant might be expected to reduce the number of colony forming units by 5 log units within an hour.

Laboratory tests

Although the standard or official tests are best left to manufacturers there is one, the 'in use' test, that is simple and useful.

The 'in-use' test
Samples of liquid disinfectants are taken from such sources as laboratory discard jars, floor-mop buckets, mop wringings, disinfectant liquids in which cleaning materials or lavatory brushes are stored, disinfectants in Central Sterile Supply Departments, used instrument containers and stock solutions of diluted disinfectants. The object is to determine whether the fluids contain living bacteria, and in what numbers. The test is described here in detail for use in any laboratory, because meaningful results can be obtained only in the light of local circumstances.

Method (as described by Maurer, 1972)
Stage 1 A 1-ml volume of disinfectant solution in use is taken from each pot or bucket with a separate sterile pipette.
Stage 2 The 1-ml sample is added to 9 ml of diluent in a sterile universal

container or a 25-ml screw-capped bottle. The diluent should be selected according to the group to which the disinfectant belongs (Table 4.2).
Stage 3 The bottle of diluent is returned to the laboratory within 1 h of the addition of the disinfectant. A separate sterile pasteur pipette or 50-drop pipette is used to withdraw a small volume of the disinfectant/diluent and to place ten drops, separately, on the surface of each of two well-dried nutrient agar plates.
Stage 4 The two plates are incubated as follows:

1. One plate is incubated for 3 days at 32 or 37°C, whichever is most convenient. The optimum temperature for most pathogenic bacteria is 37°C but those which have been damaged by disinfectants often recover more readily at 32°C.
2. The second plate is incubated for 7 days at room temperature.

Stage 5 After incubation, the plates are examined and bacterial growth is recorded.

'In-use' test results
The growth of bacterial colonies on one or both of a pair of plates is evidence of the survival of bacteria in the particular pot from which the sample is taken. One or two colonies on a plate may be ignored. A disinfectant is not a sterilant and the presence of a few live bacteria in a pot is to be expected.

However, the growth of five or more colonies on one plate should arouse suspicion that all is not well. The relationship between the number of colonies on the plate and the number of live bacteria in the pot can easily be calculated, as the disinfectant sample is diluted 1 in 10 and the 50-drop pipette delivers 50 drops/ml. When five colonies are grown from ten drops of disinfectant/diluent, then five live bacteria were present in one drop of disinfectant and 250 live bacteria were present in 1 ml of disinfectant.

The use of disinfectants in hospitals

It is advisable that hospitals work out a detailed code of practice on the use of disinfectants and antiseptics which applies to their own particular circumstances. Guides for this purpose have been published by the Public Health Laboratory Service (Ayliffe *et al.*, 1984).

For further information about disinfection and sterilization see the books by Gardner and Peel (1991) and Russell *et al.* (1992).

Table 4.2 'In-use' test neutralizing diluents

Diluent	*Disinfectant group*
Nutrient broth	Alcohols Aldehydes Hypochlorites Phenolics
Nutrient broth + Tween 80, 3% w/v	Hypochlorites + detergent Iodophors Phenolics + detergent QACs

Treatment and disposal of infected materials

It is a cardinal rule that no infected material shall leave the laboratory. (Collins *et al.*, 1974)

Laboratory waste that contains microorganisms is technically clinical waste (HSAC, 1993) and should be autoclaved on site.

In the context of the treatment and disposal of contaminated laboratory waste (including that contained in re-usable articles), this simple precept offers no problems to properly-equipped and well-managed laboratories.

It is clearly the responsibility of the laboratory management to ensure that no waste containing viable microorganisms leaves the laboratory premises. The only possible exception would be when it is to be incinerated under the direct supervision of a member of the laboratory staff. Unfortunately, placing this responsibility firmly on the laboratory management may not solve the problem. In some establishments there are very sketchy ideas on freeing material from living organisms. There is a touching faith in the ability of disinfectants at varying concentrations and indefinite ages to kill microbes submerged in or even placed near to them.

We have always believed that disinfection is a first line defence, and for discarded bench equipment it is a temporary measure, to be followed as soon as possible by autoclaving or incineration. Disinfectants should not be used as the sole method of treating bacterial cultures, even if they are completely submerged and all air bubbles are removed.

Table 4.3, adapted from Collins and Kennedy (1993), lists items that should be regarded as infectious and therefore should be autoclaved. (If incineration is contemplated see p. 57.)

Table 4.3 Proposed classification of clinical and biomedical laboratory waste

Disposables other than sharps
Specimens or their remains (in their containers) submitted for tests: containing blood, faeces, sputum, urine, secretions, exudates, transudates, other normal or morbid fluids but not tissues
All cultures made from these specimens, directly or indirectly
All other stocks of microorganisms that are no longer required
Used diagnostic kits (which may contain glass, plastics, chemicals and biologicals)
Used disposable transfer loops, rods, plastic pasteur pipettes
Disposable cuvettes and containers used in chemical analyses
Biologicals, standards and quality control materials
Food samples submitted for examination in outbreaks of food poisoning
Paper towels and tissues used to wipe benches and equipment and to dry hands
Disposable gloves and gowns
Sharps
Hypodermic needles (with syringes attached if custom so requires)
Disposable knives, scalpels, blades, scissors, forceps, probes
Glass pasteur pipettes; slides and cover glasses
Broken glass, ampoules and vials
Tissues and animal carcases
Bedding from animal cages

(Adapted from Collins and Kennedy, 1993)

Containers for discarded infected material

In the laboratory there should be five important types of receptacles for discarded infected materials:

1 colour-coded discard bins or plastic bags for specimens and cultures;
2 discard jars to receive slides, pasteur pipettes and small disposable items;
3 pipette jars for graduated (recoverable) pipettes; and
4 colour-coded plastic bags for combustibles such as specimen boxes and wrappers which might be contaminated;
5 Colour-coded 'sharps' containers for hypodermic needles and syringes.

Recommended colours in the UK (HSAC, 1992) are:
Yellow For incineration.
Light blue or transparent with blue inscription For autoclaving (but may subsequently be incinerated).
Black Normal household waste: local authority refuse collection.
White or clear plastic Soiled linen (e.g. laboratory overalls).

Discard bins and bags

Discard bins should have solid bottoms which should not leak; otherwise contaminated materials may escape. To overcome steam penetration problems these containers should be shallow, not more than 20 cm deep and as wide as will fit loosely into the autoclave. They should never be completely filled. Suitable plastic (polypropylene) containers are available commercially although not specifically designed for this purpose.

Plastic bags are popular but only those made for this purpose should be used. They should be supported in buckets or discard bins. Even the most reliable may burst if roughly handled.

For safe transmission to the preparation room the bins may require lids and the bags may be fastened with wire ties.

Discard bins and bags should be colour coded, i.e. marked in a distinctive way so that all workers recognize them as containing infected material.

Discard jars

The jars or pots of disinfectant that sit on the laboratory bench and into which used slides, pasteur pipettes and other rubbish are dumped have a long history of neglect and abuse. In all too many laboratories these jars are filled infrequently with unknown dilutions of disinfectants, are overloaded with protein and articles that float and are infrequently emptied. The contents are rarely disinfected.

Choice of container
Old jam jars and instant coffee jars are not suitable. Glass jars are easily broken and broken glass is an unnecessary laboratory hazard especially if it is likely to be contaminated. Discard jars should be robust and autoclavable and the most serviceable articles are 1-litre polypropylene beakers or screw-capped polypropylene jars. These are deep enough to hold submerged most of the things that are likely to be discarded, are quite unbreakable, and survive many autoclave cycles. They go dark brown in time, but this does not affect their use. Screw-capped polypropylene jars are better because they can be capped after use,

inverted to ensure that the contents are all wetted by the disinfectants and air bubbles which might protect objects from the fluid are removed.

Correct dilution
A 1-litre discard jar should hold 750 ml of diluted disinfectant and leave space for displacement without overflow or the risk of spillage when it is moved. A mark should be made at 750 ml on each jar, preferably with paint (grease pencil and felt-pen marks are less permanent). The correct volume of neat disinfectant to be added to water to make up this volume for 'dirty situations' can be calculated from the manufacturers' instructions. This volume is then marked on a small measuring jug, e.g. of enamelled iron, or a plastic dispenser is locked to deliver it from a bulk container. The disinfectant is added to the beaker and water added to the 750-ml mark.

Sensible use
Laboratory supervisors should ensure that inappropriate articles are not placed in discard jars. There is a reasonable limit to the amount of paper or tissues that such a jar will hold, and articles that float are unsuitable for disinfectant jars, unless these can be capped and inverted from time to time to wet all the contents.

Large volumes of liquids should never be added to dilute disinfectants. Discard jars, containing the usual volume of neat disinfectant, can be provided for fluids such as centrifuge supernatants, which should be poured in through a funnel that fits into the top of the beaker. This prevents splashing and aerosol dispersal. At the end of the day water can be added to the 750 ml mark and the mixture left overnight. Material containing large amounts of protein should not be added to disinfectants but should be autoclaved or incinerated.

Regular emptying
No material should be left in disinfectant in discard jars for more than 24 h, or surviving bacteria may grow. All discard jars should therefore be emptied once daily, but whether this is at the end of the day or the following morning is a matter for local choice. Even jars that have received little or nothing during that time should be emptied.

'Dry discard jars'

Instead of jars containing disinfectants there is a place in some laboratories for the plastic containers used in hospitals for discarded disposable syringes and their needles. These will accommodate pasteur pipettes, slides, etc. and can be autoclaved or incinerated.

Pipette jars

Jars for recoverable pipettes should be made of polypropylene or rubber. These are safer than glass. The jars should be tall enough to allow pipettes to be completely submerged without the disinfectant overflowing. A compatible detergent should be added to the disinfectant to facilitate cleaning the pipettes at a later stage. Tall jars are inconvenient for short people, who tend to place them on the floor. This is hazardous. The square based rubber jars may be inclined in a box or on a rack which is convenient and safer.

Plastic bags for combustibles

These should be colour coded (yellow in the UK, see p.52).

Treatment and disposal procedures

There are three practical methods for the treatment of contaminated, discarded laboratory materials and waste:

1 autoclaving,
2 chemical disinfection,
3 incineration.

The choice is determined by the nature of the material: if it is disposable, or recoverable, and if the latter if it is affected by heat. With certain exceptions none of the methods excludes the others. It will be seen from Figure 4.3 that incineration alone is advised only if the incinerator is under the control of the laboratory staff. Disinfection alone is advisable only for graduated, recoverable pipettes.

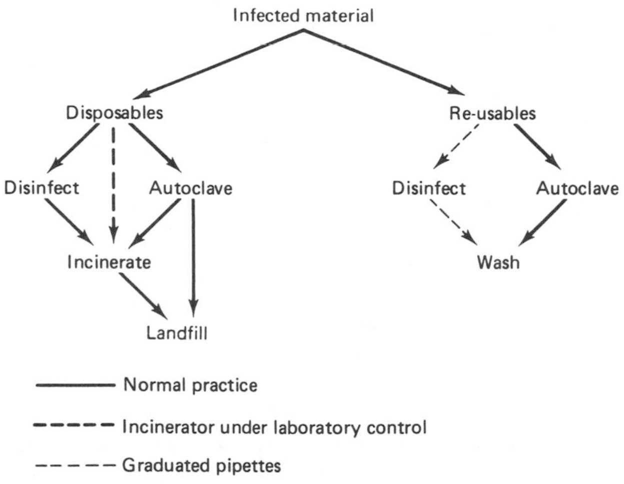

Figure 4.3 Flow chart for the treatment of infected material

Organization of treatment

The design features of a preparation room for dealing with discarded laboratory materials should include autoclaves, a sluice, a waste disposal unit plumbed to the public sewer, deep sinks, glassware washing machines, drying ovens, sterilizing ovens and large benches.

These should be arranged to preclude any possible mixing of contaminated and decontaminated materials. The designers should therefore work to a flow, or critical pathway chart provided by a professional microbiologist.

Such a chart is shown in Figure 4.4. The contaminated materials arrive in coloured coded containers onto a bench or into an area designated and used

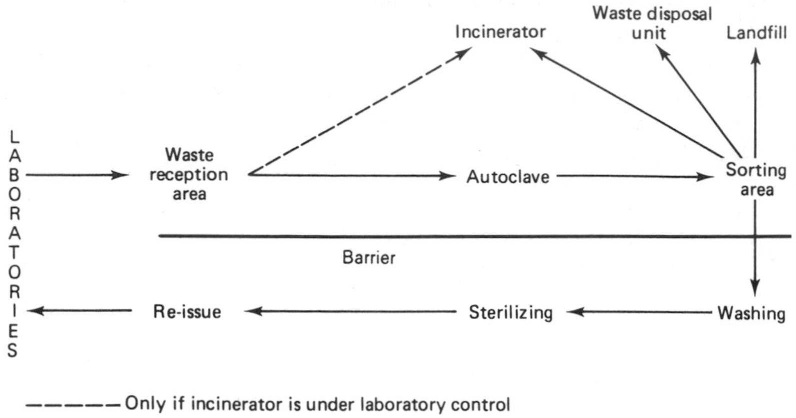

Figure 4.4 Design of preparation (utility) rooms. Flow chart for the disposal of infected laboratory waste and re-usable materials

for that purpose only. They are then sorted according to their colour codes and despatched to the incinerator or loaded into the autoclave. Nothing bypasses this area. After autoclaving, the containers are taken to the sorting bench where the contents are separated into:

(1) waste for incineration, which is put into colour coded bags,
(2) waste for the rubbish tip, which is also put into different colour coded bags,
(3) waste suitable for the sluice or waste disposal unit,
(4) recoverable material which is passed to the next section or room for washing and re-sterilizing.

This room should have a separate autoclave. Contaminated waste and materials for re-use or re-issue should not be processed in the same autoclave.

Procedures for various items

Before being autoclaved the lids of discard bins should be removed and then included in the autoclave load in such a way that they do not interfere with steam penetration. Plastic bags should have the ties removed and the bags opened fully in the bins or buckets that support them.

Contaminated glassware
After autoclaving, culture media may be poured away or scraped out and the tubes and bottles, etc. washed by hand or mechanically with a suitable detergent. The washing liquid or powder used will depend on the hardness of the water supply and the method of washing. The advice of several laboratory detergent manufacturers should be sought.

Busy laboratories require glassware washing machines. Before purchasing one of these machines it is best to consider several and to ask other laboratories which models they have found satisfactory. A prerequisite is a good supply of distilled or deionized water.

If hand-washing is practised, double sinks, for washing and then rinsing, are necessary, plus plastic or stainless steel bowls for final rinsing in distilled or deionized water. Distilled water from stripper stills, off the steam line, is rarely satisfactory for bacteriological work.

Rubber liners should be removed from screw-caps and the liners and caps washed separately and re-assembled. Colanders or sieves made of polypropylene are useful for this procedure.

New glassware, except that made of borosilicate or similar material, may require neutralization. When fluids are autoclaved in new soda-glass tubes or bottles, alkali may be released and alter the pH. Soaking for several hours in 2–3% hydrochloric acid is usually sufficient, but it is advisable to test a sample by filling with neutral water plus a few drops of suitable indicator and autoclaving.

Discard jars

After standing overnight to allow the disinfectant to act, the contents of the jars should be poured carefully through a polypropylene colander and flushed down the sluice sink. The colander and its contents are then placed in a discard bin and autoclaved. Rubber gloves should be worn for these operations.

The empty discard jars should be autoclaved before they are returned to the laboratory for further use. There may be residual contamination.

Re-usable pipettes

After total immersion in disinfectant (e.g. hypochlorite, 2500 ppm available chlorine: an anionic detergent may be added) overnight the pipettes should be removed with gloved hands.

Before the pipettes are washed, the cotton wool plugs must be removed. This can be carried out by inserting the tip into a piece of rubber tubing attached to a water tap. Difficult plugs can be removed with a small crochet hook. Several excellent pipette washing machines are manufactured that rely on water pressure and/or a siphoning action, but the final rinse should be in distilled or deionized water.

24-h urines

Although rare in microbiology laboratories other pathology departments may send them for disposal on the grounds that they may contain pathogens. Ideally they should be processed in the department concerned as follows.

Sufficient disinfectant, e.g. hypochlorite, should be added to the urine to give the use-dilution. After standing overnight the urine should be poured carefully down the sink or sluice to join similar material in the public sewer. The containers, which are usually plastic, may then be placed in colour coded bags for incineration.

Incineration

The problems with this method of disposal of infected waste that has not been autoclaved are in ensuring that the waste actually reaches the incinerator, and that if it does it is effectively sterilized; and that none escapes, either as unburned material or up the flue. Incinerators are rarely under the control of laboratory staff. Sometimes they are not even under the control of the staff of the hospital or institution, but are some distance away and contaminated and infectious material has to be sent to them on the public highway. Some older incinerators are inefficient.

Unburned material may be found among the ash, and from its appearance it may be deduced that it may not have been heated enough to kill microorganisms. We have recovered unconsumed animal debris, including fur and feathers and entrails, from a laboratory incinerator. The up-draught of air may carry microorganisms up the flue and into the atmosphere if the load is too large or badly distributed.

In the UK there are strict regulations about the incineration of clinical waste, which includes infected laboratory waste (see Collins and Kennedy, 1993) and inefficient incinerators are being phased out. The ACDP (1990) states that all waste must be made safe before disposal or removal to the incinerator but the HSAC (1991) permits it to be transported to an incinerator provided that 'it is securely packaged' and the incinerator is 'under the supervision of an operator who is fully conversant with appropriate safe working procedures for handling contaminated waste'. As infected laboratory waste, especially discarded cultures, is the most hazardous of all clinical waste, and as all reputable microbiological laboratories possess autoclaves, this practice should be resisted and the material made safe by autoclaving it before it leaves the laboratory (Collins *et al.*, 1974, ACDP, 1990; Collins and Kennedy, 1993; Collins, 1993, 1994).

There is a British Standard (BSI, 1987) for the incineration of hospital waste.

References

ACDP (1990) *Categorization of Pathogens according to Hazard and Categories of Containment*, Advisory Committee on Dangerous Pathogens, HMSO, London.

AFNOR (1989) *Recuil de Normes française. Antiseptiques et Désinfectants*, Association Française de Normalisation, Paris.

AOAC (1984) *Official Methods of Analysis of the Association of Analytical Chemists*, Washington DC

Ayliffe, G. A. J., Coates, D. and Hoffman, P. N. (1984) *Chemical Disinfectants in Hospitals*, Public Health Laboratory Service, London

BSI (1961) BS 3421 *Specification for the Performance of Electrically Heated Sterilizing Ovens*, British Standards Institution, London

BSI (1985) BS 541: *Determination of the Rideal-Walker Coefficient of Disinfectants*, British Standards Institution, London

BSI (1986) BS 808: *Assessing the Efficiency of Disinfectants by the Modified Chick-Martin Test*, British Standards Institution, London

BSI (1987) BS 3316: *Specification for Standard Performance Requirements for Incineration Plant for the Destruction of Hospital Waste*, British Standards Institution, London

BSI (1988) BS 2426: *Autoclaves for Sterilization in Laboratories*, British Standard Institution, London

Collins, C. H. (1993) *Laboratory-acquired Infections*, 3rd edn, Butterworth-Heinemann, Oxford

Collins, C. H. (1994) Infected Laboratory waste. *Letters in Applied Microbiology*, **19**, 61–62

Collins, C. H. and Kennedy, D. A. (1993) *Treatment and Disposal of Clinical Waste*, Science Reviews, Leeds

Collins, C. H., Hartley, E. G. and Pilsworth, R. (1974) *The Prevention of Laboratory-Acquired Infections*, PHLS Monograph No. 6, HMSO, London

DGHM (1982) Prüfung und Bewertung chemische Desinfectionsverfahren – Anforderungen für die Aufnahme in die VII Liste. Deutsche Gesellschaft für Hygiene un Mikrobiologie. *Hygiene und Medizin*, **9**, 453–455

DGHM (1984) Prüfung und Bewertung chemische Desifectionsverfahren – Anforderungen für die Aufnahme in die VII Liste. Deutsche Gesellschaft für Hygiene un Mikrobiologie. *Hygiene und Medizin*, **9**, 41–46

DHSS (1980) *Sterilizers*, Health Technical Memorandum No. 10, HMSO, London
Gardner, J. F. and Peel, M. M. (1991) *Introduction to Sterilization and Disinfection*, 2nd ed. Churchill Livingstone, London.
HSAC (1991) *Safe Working and the Prevention of Infection in Clinical Laboratories*, Health Services Advisory Committee, HMSO, London
HSAC (1992) *Safe Disposal of Clinical Waste*, Health Services Advisory Committee, HMSO, London
Kennedy, D. A. (1988) Equipment-related hazards. In *Safety in Clinical and Biomedical Laboratories* (ed. C. H. Collins), Chapman & Hall, London, pp. 11–46
Maurer, I. M. (1972) The management of laboratory discard jars. In *Safety in Microbiology* (eds D. A. Shapton and R. G. Board), Society for Applied Bacteriology Technical Series No. 6, Academic Press, London, pp. 53–59
Russell, A. D., Hugo, W. B. and Ayliffe, G. A. J. (1992) *Principles and Practice of Disinfection, Preservation and Sterilization*, Blackwell Scientific, Oxford

Chapter

5

Culture media

Most laboratories now purchase their culture media in dehydrated form or ready for use.

Commercially prepared media, which are subjected to strict internal quality control, may be purchased from large companies that have international supply organizations, or from smaller companies that may supply certain regions only. Many culture media have generic names and are supplied by the majority of companies, but it should not be assumed that they are identical in performance because this may be influenced by the unique characteristics of the individual manufacturer's ingredients.

Some culture media have proprietary names and are therefore obtainable only from particular companies. Although similarly formulated media may be purchased elsewhere it must be stressed that such 'equivalent' products should not be assumed to be identical.

Only two commercially available media are rigorously controlled to meet rigorous external regulatory specifications: Standard Plate Agar (American Public Health Association; APHA 1985) and Mueller-Hinton agar (National Committee for Clinical Laboratory Standards, 1990). Culture media meeting these standards should give a uniform performance irrespective of manufacturer or batch number.

The information given in this chapter is intended to guide users in their choice of media for particular organisms and purposes. For more details the reader is referred to MacFaddin (1985), Baird *et al.* (1987), Barrow and Feltham (1993), Bridson (1994) and the current editions of the Oxoid, Difco, BBL and LABM Manuals and catalogues.

Formulas are given for those media which are not commercially available or for workers who prefer their own products. Further details of culture media may be found in the technical publications of the various manufacturers, e.g. current editions of the Oxoid, BBL and Difco Manuals and the LABM catalogue.

Ingredients of some culture media and reagents (e.g. azides, barbitones, cyclohexamide, selenites, tellurites) may be hazardous to health. It is advisable to wear dust masks (e.g. 3M) when handling these. See manufacturers' handbooks and data sheets.

General purpose media

These are non-selective media that are designed to support the growth of a wide spectrum of heterotrophic organisms, i.e. those that require organic carbon. These media contain enzyme or acid digests of animal and/or plant proteins as the main sources of organic carbon and nitrogen. They are usually supplemented with meat or yeast extracts which provide essential metal ions, minerals and vitamins. There are broth formulations for enrichment and agar formulations for isolation. The growth of the more fastidious organisms is encouraged by growth-promoting supplements.

Growth promoting supplements

These include:

Horse or sheep blood	Horse or rabbit serum
Fildes' extract	Yeast autolysate
Menadione/vitamin K	Sodium pyruvate
Ferric/ferrous salts	Vitamins

Selective isolation media

These media are designed for the isolation of specific organisms and contain selective chemicals which suppress the growth of others. Although some chemical agents, e.g bile and selenium salts, have been used for many years the recent use of antibiotics has increased the isolation rate of many pathogens and has permitted the growth of 'new' or unsuspected organisms.

Selective broth media restrict the growth of unwanted species, particularly in the early hours of incubation, and allow the slower growing and often less numerous pathogens to be enriched in numbers. These media usually contain selective chemical agents but some also contain antibiotics. Table 5.1 gives the basal medium and selective agents for a variety of bacterial pathogens.

Media for the selective isolation of bacteria

Not all of the media mentioned in this chapter are referred to in other parts of the book. This is because the editors and contributors have no experience of them. They are included to indicate that readers have a wider choice.

We have followed the convention of using capital initials for the names of proprietary media, most of which are described in the *Oxoid Manual* (1990).

Aeromonas
Blood agar plus ampicillin; Aeromonas medium.

Bacillus cereus
B. Cereus Selective agar.

Bacteroides
Blood agar plus menadione-antibiotic supplements.

Bordetella
Bordet Gengou or Charcoal Agar Base plus antibiotic supplements.

Brucella
Blood agar or proprietary Brucella media plus antibiotic supplements.

Table 5.1 Selective agents for the isolation of pathogenic bacteria

Organisms	*Basal medium*	*Selective agents*
Actinomyces	Brain-heart infusion agar	Sodium nalidixate/metronidazole
Aeromonas	Blood/nutrient agar	Ampicillin
Bacillus cereus	Special base	Polymyxin
Bacteroides	Blood agar or Wilkins Chalgren agar*	Kanamycin/vancomycin Sodium nalidixate/vancomycin
Bordetella	Charcoal agar or Bordet-Gengou agar	Cephalexin
Brucella	Blood agar or Brucella base	Polymyxin/bacitracin/cycloheximide/sodium nalidixate/nystatin/vancomycin
Campylobacter	Blood agar or Columbia agar or special bases*	Vancomycin/polymyxin/trimethoprim/cycloheximide/colistin/amphotericin or cefoperazone
Clostridia	Blood agar or Wilkins Chalgren agar*	Neomycin/kanamycin
C. difficile	Special base*	Cycloserine/cefoxitin
Coliforms	Broth bases	Brilliant green, bile salts, lauryl sulphate
	Agar bases	Bile salts, deoxycholate, eosin/methylene blue
Corynebacteria	Blood agar*	Tellurite
Enterococci	Barnes, Edwards, Mead, Slanetz & Bartley, KF	Sodium azide, ethyl violet, kanamycin, bile salts
Fusobacteria	Blood agar or Wilkins-Chalgren agar	Neomycin/vancomycin, sodium nalidixate/vancomycin
Gardnerella	Columbia agar*	Gentamicin/sodium nalidixate/amphotericin B
Haemophilus	Blood agar*	Sodium oleate
Lactobacilli	Glucose/tomato juice, minerals, agar	Acid pH, sorbitan, cycloheximide/phenylethanol

* These media also require growth supplements.

Campylobacter
Campylobacter agar plus supplements, Campylobacter Medium.

Clostridia
Wilkins-Chalgren agar plus antibiotics; Reinforced Clostridial agar; Tryptose-Sulphite Cycloserine agar, TSC; Oleandomycin Polymyxin Sulphadiazine Perfringens agar, OPSP; Iron Sulphite agar; Clostrisel; Clostridium Difficile agar; Perfringens Selective Agar; Clostridium Selective Medium; Willis and Hobbs medium (p.87).

Coliform bacilli
Broths Brilliant Green Bile; Brilliant Green Lactose Bile Broth (BGLBB); lauryl sulphate; MacConkey; Minerals Modified Glutamate; Lactose Broth; EC broth; EE broth. *Agars* Cystine Lactose Electrolyte-Deficient (CLED); deoxycholate; Endo; Eosin Methylene Blue (EMB); Violet Red Bile (VRB); Brilliant Green.

Corynebacteria
Hoyle; other commercial tellurites; Loeffler.

Enterobacteria
See under Coliform bacilli, Salmonella, Shigella.

Table 5.1—Continued

Organisms	*Basal medium*	*Selective agents*
Legionella	Blood agar or special CYE bases*	Cefamandole/polymyxin/anisomycin/vancomycin/colistin/trimethoprim
Listeria	Oxford, PALCAM, USDA-UVM bases*, Fraser broth	Polymyxin/acriflavine/sodium nalidixate/colistin/cefotetan/fosfomycin/cycloheximide
Mycobacteria	Lowenstein-Jensen, Kirchner, Middlebrook	Acid pH/malachite green/penicillin, sorbitan mono-oleate
Mycoplasmas	Mycoplasma base	Thallous acetate, penicillin, enrichments
Neisseria	GC base, New York City medium*	Vancomycin/colistin/trimethoprim/nystatin/amphotericin
Pediococci	Raka-Ray, Wort agar	Sorbitan mono-oleate/cycloheximide/phenylethanol
Pseudomonas	Nutrient agar, King's medium, special base	Cetrimide/fucidin/cephaloridine/sodium nalidixate
Salmonella	Broth enrichment	Selenite, tetrathionate brilliant or malachite green
	Agar isolation	Bismuth sulphite, deoxycholate citrate, brilliant green, crystal violet
Shigella	Agar isolation	Deoxycholate citrate, bile salts
Staphylococci	Baird-Parker, Kranep Mannitol-salt, Giolitti-Cantoni, phenolphthalein-polymyxin agar	Tellurite, polymyxin, lithium, azide, salt, cycloheximide, glycine
Streptococci	Blood agar*, Islam, COBA*, CNA	Colistin/oxolinic acid/sodium nalidixate
Vibrio	Alkaline peptone water, TCBS, glucose-salt-lauryl sulphate, salt-colistin, Monsur	Bile salt, sodium chloride, colistin, tellurite
Yersinia enterocolitica	CIN medium	Cefsulodin/irgasan/novobiocin

* These media also require growth supplements.

Enterococci
See under Streptococci (p.64)

Gardnerella
Gardnerella agar and supplements.

Haemophilus
Blood agar plus Fildes or commercial supplements.

Lactobacilli
L-S Differential; MRS; Raka-Ray; Tomato Juice; Rogosa.

Legionella
Blood agar plus supplements; Legionella Selective Media.

Leptospira
EMJH medium and EMJH selective media (p.78).

Listeria
MacBride's; Listeria Selective Agar (Oxford); Listeria Enrichment Broth; Frazer Medium; Palcam Agar

Mycobacteria
Lowenstein-Jensen (p.81); Acid Egg, Kirchner; Middlebrook 7H9, 7H10, 7H11; Mycoplasma agar, broth and supplements

Neisseria
GC base plus supplements; New York City.

Pediococci
Raka-Ray; Wort agar.

Pseudomonas
Commercial selective media plus cetrimide and/or antibiotics; King's medium A (p.80) Pseudomonas Selective Medium.

Salmonellas
Enrichment Selenite; Mueller-Kauffman Tetrathionate; Brilliant Green; Rappaport-Vassiliadis (RV).

Plating Bismuth sulphite; Deoxycholate-citrate (many formulations: SS, DCA, XLD, etc.); Hektoen; Mannitol Lysine Crystal Violet Brilliant Green Agar; brilliant green agars (various formulations).

Shigellas
Deoxycholate – citrate (many formulations) as for salmonella; Hektoen.

Staphylococci
Baird-Parker; Mannitol-Salt; milk salt agar (p.75); phenolphthalein polymyxin phosphate (PPPA) (p.84); Tellurite Polymyxin Egg Yolk (TPEY); Kranep; Staphylococcus 110; Giolotti-Cantoni.

Streptococci and Enterococci
Blood agar plus selective supplements; Azide media; BAGG; Barnes'; Edwards'; Mead's; Slanetz and Bartley's; KF Streptococcus; COBA; Islam's (p.80). Kanamycin Aesculin Azide Agar; Colistin-oxolinic acid agar (p.360)

Vibrios
Alkaline peptone water (p.71); Thiosulphate Citrate Bile Sucrose (TCBS); Salt colistin

Yersinia enterocolitica
Yersinia Selective medium plus supplements.

Media for dermatophytes and pathogenic yeasts

General media
Czapek-Dox agar; wort agar; Sabouraud agar (various formulations).

Dermatophytes
Littman ox gall agar; malt agar with and without chloramphenicol and cycloheximide. Dermasel.

Yeasts
BiGGY (Nickerson) medium; Yeast and Mould agar; corn meal agar.

Transport media

Semisolid media with and without charcoal and antibiotics, intended to keep delicate microbes alive during transit to the laboratory are sold under the following names:

Amies; Cary-Blair; Stuart; Transgrow; Transport; Pertussis.

Media for food microbiology

Most general and selective media are useful, especially glucose (dextrose) tryptone media.

Bacterial counts
Standard (plate) count agar; Standard methods agar; milk agar; yeast extract milk agar. Tryptose Glucose Extract Agar.

Yeasts and moulds
Buffered yeast agar (p.74). OGYE agar; WL agar; Aspergillus AFPA; Lysine agar; glucose salt agar; potato dextrose agar; Malt extract agar; Rose Bengal chloramphenicol agar; Dichloran Glycerol agar; Dichloran Rose Bengal chloramphenicol agar.

Bacillus cereus
B. cereus Selective Medium

Acidophilic yeasts
Malt agar.

Beer spoilage
Universal Beer agar; Raka-Ray agar.

Brochothrix
Gardner's STAA (p.79)

Coliform bacilli
See p.62.

Clostridia
See p.62.

Enterobacteria
See coliform bacilli.

Enterococci
See p.64.

Flat sour organisms
Glucose tryptone agar.

Hydrogen swell organisms
Glucose tryptone agar.

Lactobacilli
See p.63.

Lipolytic organisms
Tributyrin agar.

Listeria
See p.64.

Moulds and yeasts in dairy products and utensils
Malt agar plus chloramphenicol and cycloheximide; Buffered yeast agar (p.74); Dichloran Glycerol (DG 18) agar; Dichloran Rose-Bengal (DRBC) agar; Rose-Bengal Selective agar; Czapek yeast agar (p.76); Synthetischer nährstoffarmer agar (SNA) (p.86); Potato dextrose agar.

Moulds on meat
Potato glucose agar.

Staphylococci
See p.64.

Streptococci
See p.64.

Media for water and environmental bacteriology

Bacterial counts
Standard methods agar; standard plate count agar.

Coliform bacilli
See p.62.

Clostridium perfringens
See p.62.

Enterobacteria
See coliform bacilli, p.62.

Legionella
See p.63.

Salmonellas
See p.64.

Membrane filtration media

Companies that sell these media place a suffix before the names of their products: BBL, *M*-; Difco, m-; Gibco, MF-; Merck, Membrane Filtration; Oxoid, M-. Membrane filtration media at present available include the following.

Counting bacteria
Glucose tryptone broths; standard methods broth; tryptone soya broths.

Coliform bacilli
Brilliant green broths; Endo media; eosin methylene blue broth; enrichment broth; lauryl sulphate media; MacConkey broths; resuscitation broths.

Flat sour organisms
Glucose tryptone broths.

Salmonellas, including S.typhi
Bismuth sulphite broth; tetrathionate broth.

Staphylococci
Staphylococcus 110 broth.

Streptococci
Azide broths; enterococcus media.

Moulds
Czapek-Dox broth.

Antibiotic sensitivity test (AST) media

Mueller Hinton is probably the most widely used AST medium. For the Kirby-Bauer test the medium must conform to the requirement of the National Committee for Clinical Laboratory Standards (NCCLS, 1990). Other AST media, such as Isosensitest and Direct Sensitivity Test (DST) agars, are used in various antibiotic dilution and disc diffusion tests (see Chapter 12).

Sterility test media

For products not containing preservatives – any non-selective broth medium. For products containing preservatives – Clausen medium; thioglycollate media (various formulations).

Identification media

These are used for various 'biochemical' tests. Most are available commercially; page numbers are given for those that are not. See also 'Kit' bacteriology (p.106).

Aesculin hydrolysis
Aesculin medium, Edwards' medium.

Arginine hydrolysis
Arginine broth, combined arginine-ornithine media.

Casein hydrolysis
Skim milk agar (p.86).

Carbohydrate utilization (most available commercially)
These are usually a peptone broth base containing an indicator (phenol red or Andrade) to which the fermentable substance is added at 0.5–2% (1% for most bacteria). A Durham's tube is included if detection of gas is desired.

These simple media are not suitable for all purposes, however, and the following modifications are advisable for special purposes.

Baird-Parker carbohydrate media
For staphylococci and micrococci (p.75).

Basal synthetic medium
Add carbohydrate for aerobic spore bearers and pseudomonads (p.72).

GC serum-free sugar medium
For fastidious organisms (p.71).

Gillies medium
Contains glucose, mannitol, sucrose and salicin for screening enterobacteria.

Kohn two-tube medium
This is a modification of Gillies medium.

Krumwiede double sugar agar
Contains lactose and sucrose.

MRS medium
Plus carbohydrates for lactobacilli (p.83).

Robinson's serum water sugars
Preferred for pathogenic corynebacteria (p.75).

Russell's double sugar agar
Contains glucose and lactose.

Triple sugar iron agar (TSI)
Contains glucose, lactose and sucrose and also indicator for H_2S production.

Citrate utilization
Koser; Simmons.

Decarboxylase tests
Falkow and Moeller lysine media, basal media to which arginine, lysine and ornithine must be added (see also p.76).

Esculin hydrolysis
See Aesculin hydrolysis.

DNase
DNase agars with or without indicator.

Gelatin liquefaction
Nutrient gelatin, gelatin agar, various formulations (p.84). Charcoal gelatin discs for use with nutrient broths.

Gluconate utilization
Gluconate broth (p.79).

Hippurate hydrolysis
Hippurate broth (p.80).

Hugh and Leifson test
Add glucose to commercial base (p.80).

Hydrogen sulphide production
Various formulations of iron agar, iron plus carbohydrate and iron plus lysine medium. Kligler Agar; Kohn Agar.

Indole production
Any commercial peptone or tryptone water medium.

Lecithinase
Egg yolk agar and broth (p.77). Willis and Hobbs medium (p.87). Egg yolk salt agar (p.78).

Malonate utilization
Malonate broth (p.81).

Milk reactions
Litmus milk, bromocresol purple milk, Crossley milk medium.

Methyl red (MR) and Voges Proskauer (VP)
Glucose phosphate broth, MRVP media.

Motility
Semisolid media. See also Craigie method (p.86).

Nitratase test
Nitrate broths, indole-nitrate broths and agars (p.84).

Nagler reaction
Egg yolk agar (p.77). Willis and Hobbs medium (p.87).

OF (Ox-ferm: Hugh and Leifson) test
Add glucose to commercial base (see also p.80).

ONPG test
ONPG broth (p.84).

Organic acid utilization
Add 1–2% organic acid to commercial organic acid base medium (see also p.72).

Oxygen preference
Use sloppy agar (p.86).

PPA (phenylalanine deamination)
Phenylalanine agar.

Phosphatase test
Phenolphthalein phosphate agar (p.84).

Pyocyanin production
King's medium A (p.80).

Starch hydrolysis
Starch agar (p.86).

Sulphatase test
Phenolphthalein sulphate agar and broth (p.85).

Urease test
Usually sold as a base, with sterile (filtered) urea separately.

Tyrosine decomposition
Tyrosine agar (p.87).

Xanthine decomposition
Xanthine agar (p.87).

Laboratory-prepared culture media

Keep a manual or set of standard operational procedures (SOPs) that contains lists of suppliers and sources of raw materials, formulas, details of preparation and performance of quality control tests. Also, keep a 'day book' to record details of media prepared.

Raw materials

Purchase raw materials from reliable specialist companies who will provide quality specifications (e.g. to BPC, AR or other national standards). Before making bulk purchases obtain a sample, with a specific batch number, and test it (p.90).

Use only glass-distilled or deionized water unless otherwise specified in the original formula. Do not store it for any length of time. Stored water, and inadequately maintained deionizers may support the growth of microorganisms, e.g. pseudomonads.

Apparatus

Use only stainless steel and glass vessels and carefully and thoroughly clean them after each use.

Agar preparators, which save much time and labour, are self-contained, electrically-operated, variable-temperature pressure cookers which can be preset to dissolve agar and then cool and maintain the medium at 50°C. Ports are provided so that blood and sterile fluids may be added. Models vary in their capacity from 2 to 12 litres.

Laboratory prepared media

Acid broth

Used for high acid thermally processed foods

Proteose peptone	5 g
Yeast extract	5 g
Glucose	5 g
Dipotassium phosphate K_2HPO_4	4 g
Distilled water to	1000 ml

Dissolve and dispense in 12–15 ml amounts, sterilize at 121°C for 15 min. Final pH should be 5.0.

Alkaline peptone water

This is used for isolating *V. cholerae*. Adjust the pH of peptone water to pH 8.6 with N NaOH.

Antifungal drug sensitivity testing

1 Glucose monohydrate	20 g
Casamino acids	20 g
Sodium glycerophosphate	5 g
Yeast extract 5%	2 ml
Distilled water	1000 ml

Heat at 50–60°C to dissolve and distribute in 100-ml lots.

2 Agar	3 g
Distilled water	1000 ml

Dissolve by autoclaving and distribute in 100-ml lots.

For sensitivity to amphotericin and imidazole add 2 ml of 1% cytosine to medium (1). For the assay of 5-fluorocytosine mix 100 ml of (1) and (2) and distribute in 10-ml lots for tests on slopes or 20-ml lots for tests in petri dishes.

Antifungal drug assay medium

Yeast Nitrogen Base	3.5 g
Glucose	10.0 g
KH_2PO_4	1.5 g
$Na_2HPO_4.2H_2O$	1.0 g
Distilled water	1000 ml

Dissolve, dispense in 110-ml lots. Add 1.2 g of agar to each bottle and autoclave.

Ammonium salt sugars

See under Carbohydrate media (p.74).

Arginine broth

This is useful in identifying some streptococci and Gram-negative rods.

Tryptone	5 g
Yeast extract	5 g
Dipotassium hydrogen phosphate	2 g
L-Arginine monohydrochloride	3 g
Glucose	0.5 g
Water	1000 ml

Dissolve by heating, adjust to pH 7.0; dispense in 5–10-ml amounts and autoclave at 115°C for 10 min.

For arginine breakdown by lactobacilli, use MRS broth in which the ammonium citrate is replaced with 0.3% arginine hydrochloride.

Baird-Parker's sugar media

See under Carbohydrate media (p.75).

Basal synthetic media

These are used when investigations are being made into the ability of bacteria to use various carbon sources or, with the addition of a suitable carbon source, of growth factor or vitamin requirements.

(a) *For carbohydrate utilization*

Ammonium dihydrogen phosphate	1.0 g
Potassium chloride	0.2 g
Magnesium sulphate	0.2 g
Agar	10.0 g
Water	1000 ml

Dissolve by heating, add 4 ml of 0.2% bromothymol blue and a final 1% of sterile (filtered) carbohydrate solution (0.1% aesculin; 0.2% starch; these are sterilized by steaming).

To investigate amino acid sources, add 1% glucose and 0.1 g of amino acid.

(b) *For organic acid utilization*

Magnesium sulphate	1.0 g
Sodium chloride	1.0 g
Diammonium hydrogen phosphate	1.0 g
Potassium dihydrogen phosphate	0.5 g
Agar	12.0 g
Water	1000 ml

Dissolve salts by heating in 200 ml of water. Dissolve agar in the remaining water. Mix and adjust to pH 6.8. Bottle in 100-ml amounts. Melt, add 0.2 g of organic acid (sodium salts of acetic, benzoic, citric, oxalic, propionic, pyruvic, succinic, tartaric acids; calcium salt of malic acid; mucic acid as free acid), re-adjust to pH 6.8 if necessary and add 0.4 ml of 0.2% phenol red. Dispense and give a final steaming to sterilize.

To investigate amino acid sources, add 1% glucose and the indicator and 0.1 g of the amino acid.

Blood agar media

Blood agar
This is nutrient agar or one of the commercial dehydrated Blood Agar Bases to which sterile horse blood (sheep blood in the USA and some other countries) is added. It is used to detect haemolytic organisms and to encourage the growth of organisms that grow poorly or not at all on nutrient agar. Some specially enriched commercial blood agar bases are available for diagnostic and sensitivity tests.

Melt 100 ml of nutrient agar or prepare 100 ml of Blood Agar Base. Cool to 48°C in a water-bath. Add 10 ml of sterile defibrinated or oxalated horse blood and mix gently so as to avoid bubbles. Pour into petri dishes by raising the lid only enough to introduce the neck of the bottle. This amount should give seven poured plates.

In pathology laboratories, to conserve blood and to see more easily the clear zones around colonies, a very thin layer of blood agar is poured on top of a thin layer of ordinary nutrient agar. Sometimes saline agar, a 1% agar containing 0.9% sodium chloride, at pH 7.4 is used.

Chloral hydrate agar
Used in the purification of cultures of anthrax bacilli. Add 5% chloral hydrate to blood agar while still fluid.

Chocolate agar
Some fastidious bacteria grow best on a blood agar in which the cells have been disrupted and the haemoglobin altered by heat (horse blood is better than sheep blood).

Make blood agar as described above but replace the bottle in the water-bath, raise the temperature of the water-bath slowly to 80°C and leave for 5 min. Cool slightly and pour into petri dishes. The medium should be homogeneous and chocolate coloured, not flaky.

Chocolate cystine agar
For the cultivation of *Neisseria* and *Haemophilus*.

Solution A	
Cystine	0.1 g
Nutrient broth	50 ml
Solution B	
p-Aminobenzoic acid	0.1 g
Nutrient broth	50 ml

Make up when required and steam for 1 h.

To 950 ml of melted blood agar base add 25-ml amounts of each of solutions *A* and *B*. Bottle and sterilize at 115°C for 10 min. For use, melt, cool to 50°C, add 10% horse blood and heat at 70–80°C until reddish brown in colour. Pour plates and use as soon as possible.

Colistin-nalidixic acid agar
To isolate Gram-positive cocci from material heavily contaminated with *Pseudomonas, Proteus, Klebsiella*, etc., add 10 μg/ml of colistin and 15 μg/ml of nalidixic acid to blood agar base before adding the blood (Petts, 1984).

Crystal violet blood agar
The dye inhibits many organisms but not streptococci. Add 1 ml of a 0.001% solution of crystal violet in water to 500 ml of blood agar while still melted.

Gentamicin blood agar
Add 500 μg of gentamicin to 100 ml of blood agar.

Lysed blood agar
This is used for sulphonamide sensitivity tests. Lysed horse blood (not sheep) is obtainable commercially (see above) or made by alternate freezing and thawing of whole blood. The medium is made as described under Blood Agar.

Neomycin blood agar
Add 75 μg of neomycin sulphate to 100 ml of blood agar.

PABA blood agar
To neutralize sulphonamides in an inoculum, add 10 mg of *p*-aminobenzoic acid to 100 ml of the medium before sterilizing it.

Buffered yeast agar

This is useful for the detection of and counting yeasts and moulds in the food industry, utensils, washes, etc.

Yeast extract	5.0 g
Glucose	20.0 g
Ammonium sulphate	0.72 g
Ammonium dihydrogen phosphate	0.26 g
Agar	12.0 g
Water	1000 ml

Dissolve by heating and adjust to pH 5.5, bottle and sterilize at 115°C for 10 min. If required for bottle counts, increase agar to 20 g/litre.

To adjust this medium to a more acidic pH, cool to 50°C and add 1% citric acid monohydrate and 10% lactic acid to each 100 ml as follows (Davis, 1931, 1982)

pH	*1% citric acid (ml)*	*10% lactic acid (ml)*
4.75	1.26	0.125
4.50	2.24	0.20
4.25	3.92	0.30
4.00	6.16	0.45
3.75	9.52	0.70
3.50	14.56	1.17

Carbohydrate fermentation and oxidation tests

Formerly known as 'peptone water sugars', some of these media are now more nutritious and do not require the addition of serum except for very fastidious organisms. Liquid, semisolid and solid basal media are available commercially and contain a suitable indicator, e.g. Andrade (pH 5–8), phenol red (pH 6.8–8.4) or bromothymol blue (pH 5.2–6.8).

Some sterile carbohydrate solutions (glucose, maltose, sucrose, lactose, dulcitol, mannitol, salicin) are available commercially. Prepare others as a 10% solution in water and sterilize by filtration. Add 10 ml to each 100 ml of the reconstituted sterile basal medium and tube aseptically. Do not heat carbohydrate media. Durham's (fermentation) tubes need to be added to the glucose tubes only; gas from other substrates is not diagnostically significant.

These media are not suitable for some organisms; variations are given below.

Ammonium salt 'sugars'

Pseudomonads and spore bearers produce alkali from peptone water and the indicator may not change colour. Use this basal synthetic medium instead.

Ammonium dihydrogen phosphate	1.0 g
Potassium chloride	0.2 g
Magnesium sulphate	0.2 g
Agar	10.0 g
Water	1000 ml

Dissolve by heating, add 4 ml of 0.2% bromothymol blue and a final 1% of sterile (filtered) carbohydrate solution (0.1% aesculin; 0.2% starch; these are sterilized by steaming).

'Anaerobic sugars'

Peptone water or serum peptone water sugar media need to have a low oxygen tension for testing the reactions of anaerobes, even under anaerobic conditions. Add a clean wire nail to each tube before sterilization. Do not add indicator until after growth is seen.

Baird-Parker's carbohydrate medium

For the sugar reactions of staphylococci and micrococci use Baird-Parker (1966) formula.

Yeast extract	1 g
Ammonium dihydrogen phosphate	1 g
Potassium chloride	0.2 g
Magnesium sulphate	0.2 g
Agar	12 g
Water	1000 ml

Steam to dissolve and adjust to pH 7.0. Add 20 ml of 2% bromocresol purple and bottle in 95-ml amounts. Sterilize at 115°C for 10 min. For use, melt and add 5 ml of 10% sterile carbohydrate.

Lactobacilli fermentation medium

For sugar reactions of lactobacilli make MRS base (p.83) without Lab-Lemco and glucose and adjust to pH 6.2–6.5. Bottle in 100-ml amounts and autoclave at 115°C for 15 min. Melt 100 ml, add 10 ml of 10% sterile (filtered) carbohydrate solution and 2 ml of 0.2% chlorophenol red. Dispense aseptically.

Robinsons's serum water sugars

Some organisms will not grow in peptone water sugars. Robinson's serum water medium is superior to that of Hiss.

Peptone	5 g
Disodium hydrogen phosphate	1 g
Water	1000 ml

Steam for 15 min, adjust to pH 7.4 and add 250 ml of horse serum. Steam for 20 min, add 10 ml of Andrade indicator and 1% of appropriate sugar (0.4% starch).

Unheated serum contains diastase and may contain a small amount of fermentable carbohydrate, so a buffer is desirable, particularly in starch fermentation tests.

Carbon assimilation media for yeasts

Ammonium sulphate	5 g
Magnesium sulphate	0.5 g
Potassium dihydrogen phosphate	1.0 g
Agar	1.5 g
Water	1000 ml

Dissolve by heating. Autoclave in 15-ml amounts at 115°C for 15 min. Melt and add sterile carbon source to 3%.

Czapek yeast autolysate agar (CYA)

For yeasts and moulds

Sodium nitrate	3.0 g
Dipotassium hydrogen phosphate	1.0 g
Magnesium sulphate (7 H_2O)	0.5 g
Ferrous sulphate (7 H_2O)	0.01 g
Yeast extract	5.0 g
Agar	15.0 g
Distilled water	1000 ml

Dissolve by heat. Distribute and autoclave at 115°C for 15 min.

Decarboxylase medium

(Moeller, 1955)

Peptone (Evans)	5 g
Lab-Lemco	5 g
Pyridoxine HCl	5 mg
Glucose	0.5 g
Distilled water	990 ml

Dissolve and adjust to pH 6.0. Add 5 ml of 1:500 bromocresol purple and 2.5 ml of 1:500 cresol red. Make up to 1000 ml and divide into three batches. To one add 2 g% DL-lysine and adjust to pH 6.0 with 0.1 N NaOH. To another batch add 2 g% DL-ornithine and adjust to pH 6.0 with 0.1 N NaOH. The third is the control medium. Dispense in 1.1-ml amounts in narrow tubes so that the column of medium is about 2 cm. Seal with a few millimetres of liquid paraffin, plug and steam for 1 h.

Medium to demonstrate arginine decarboxylase or dihydrolase is made in the same way.

Decarboxylase media

(Falkow, 1958)

This formula is preferred for those organisms which require a less anaerobic medium than Moeller's medium.

Peptone	5 g
Yeast extract	3 g
Glucose	1 g
Water	1000 ml

Dissolve by heating, adjust to pH 6.7 and add 10 ml of 0.2% bromocresol purple. Bottle in 100-ml amounts and autoclave at 115°C for 10 min.

To 100 ml add 0.5 g of the appropriate amino acid (lysine, ornithine, arginine) and tube. Ordinary-sized tubes or bottles can be used. Dispense also one batch without amino acid as control. Sterilize by steaming.

GC serum-free sugars

(Flynn and Waitkins, 1972)

Oxoid GC base	33 g
Distilled water	900 ml
Phenol red (0.2%)	9 ml
C-H supplement	18 ml

Bring to boil to dissolve. Adjust to pH 7.6 with N NaOH, distribute in 90 ml lots and autoclave at 115°C for 10 min. Store in cold room.

C-H supplement

L-glutamine	1 g
Ferric nitrate	0.05 g
Sterile distilled water	100 ml

For use melt four bottles of base, cool to 56°C and add 10 ml of each filter-sterilized (10%) appropriate sugar solutions and add sterile indicator solution before autoclaving.

Donovan's (1966) medium

Originally devised to differentiate klebsiellas, this is useful for screening lactose fermenting Gram-negative rods.

Tryptone	10 g
Sodium chloride	5 g
Triphenyltetrazolium chloride	0.5 g
Ferrous ammonium sulphate	0.2 g
1% aqueous bromothymol blue	3 ml
Inositol	10 g
Water	1000 ml

Dissolve by heating and adjust to pH 7.2. Distribute in 4-ml amounts and autoclave at 115°C for 15 min. Allow to set as butts.

Dorset's egg medium

Wash six fresh eggs in soap and water and break into a sterilized basin. Beat with a sterile fork and strain through sterile cotton gauze into a sterile measuring cylinder. To 3 parts of egg add 1 part of nutrient broth. Mix, tube or bottle and inspissate in a sloped position for 45 min at 80–85°C. Glycerol (5%) may be added if desired. (Griffith's egg medium uses saline instead of broth and is useful for storing stock cultures.)

Egg yolk agar

(Lowbury and Lilly, 1955)

This is selective for *Cl. perfringens* and shows the Nagler and pearly layer reactions.

Nutrient agar or blood agar, base, melted	100 ml
Fildes' extract	5 ml

Steam for 20 min to remove chloroform from Fildes' extract, cool to 55°C, and add 10 ml of egg yolk suspension, 100–125 μg/ml of neomycin and pour plates.

Egg yolk salt broth and agar

For lecithinase tests

Nutrient broth or blood agar base	10 ml
Egg yolk emulsion	1 ml
NaCl (5%)	0.2 ml

Elek's medium (modified)

(Davies, 1974)
Used for plate toxigenicity test on diphtheria bacilli.

Solution A

Difco proteose peptone	20 g
Maltose	3 g
Lactic acid	0.7 g
Distilled water	500 ml

Heat to dissolve. Add 3.5 ml of 10 N NaOH. Mix well, heat to boiling, filter, adjust to pH 7–8.

Solution B

Agar	15 g
Sodium chloride	5 g
Distilled water	500 ml

Heat to dissolve. Adjust to pH 7.8.

Mix solutions A and B and bottle in 15-ml amounts. Autoclave at 115°C for 10 min.

EMJ medium for leptospires

The formula is given by Waitkins (1985) but the complete medium and enrichment is obtainable from Difco.

Selective EMJH/5FU medium

(Waitkins, 1985)
Dissolve 1 g of 5-fluorouracil in approx. 50 ml of distilled water containing 1–2 ml of N NaOH by gentle heat (do not exceed 56°C). Adjust to pH 7.4–7.6 with N HCl, make up to 100 ml with distilled water and sterilize through a membrane filter (0.45 μm). Dispense in 1-ml lots and store at − 4°C.

Thaw 1 ml at 56°C and add it to 100 ml of the EMJH medium (final concentration of 5FU is 100 μg/ml).

Ellner's (1956) medium

This is used to persuade *Cl. perfringens* to form spores.

Peptone	10 g
Yeast extract	3 g
Starch, soluble	3 g
Magnesium sulphate ($MgSO_4.7H_2O$)	0.1 g
Disodium hydrogen phosphate ($Na_2HPO_4.7H_2O$)	50 g
Potassium dihydrogen phosphate	1.5 g
Water	1000 ml

Dissolve by heating, adjust to pH 7.8 and tube. Before use, steam to drive off oxygen, cool rapidly and inoculate heavily at the bottom of the tube with a pasteur pipette without introducing air.

Gardner's (1966) STAA medium

For the isolation of *Brochothrix* from meat.

Peptone	20 g
Yeast extract (Oxoid)	2 g
Glycerol	15 g
K_2HPO_4	1 g
$MgSO_4.7H_2O$	1 g
Agar	13 g
Distilled water	1000 ml

Heat to dissolve, adjust to pH 6.0 and sterilize at 121°C for 15 min. Tube in suitable amounts. Before use melt, cool to 50°C and add these inhibitory agents to give the final concentrations stated: streptomycin sulphate to 500 µg/ml; cycloheximide to 50 µg/ml; thallous acetate to 50 µg/ml.

Gelatin agar

Melt nutrient agar and add 5% bacteriological gelatin.

Gluconate broth

For differentiation among enterobacteria.

Yeast extract	1.0 g
Peptone	1.5 g
Dipotassium hydrogen phosphate	1.0 g
Potassium gluconate	40.0 g
(or sodium gluconate)	37.25 g
Water	1000 ml

Dissolve by heating, adjust to pH 7.0, filter if necessary, dispense in 5–10–ml amounts and autoclave at 115°C for 10 min.

Glucose salt Teepol broth

For the isolation of *Vibrio parahaemolyticus* from water and foods.

Beef extract	3 g
Tryptone	10 g
NaCl	30 g
Glucose	5 g
Methyl violet	0.002 g
Teepol 610	4 ml
Distilled water	1000 ml

Dissolve by heat, adjust to pH 9.4 and autoclave at 115°C for 10 min.

Glucose phosphate medium

For MR and VP tests. Add 0.5% each of glucose and K_2HPO_4 to commercial peptone broth.

Griffith's egg medium

For storing stock cultures, see Dorset's egg medium.

Hippurate broth

Dissolve 10 g of sodium hippurate in 1000 ml of nutrient broth. Steam to dissolve and dispense in small amounts.

Hugh and Leifson's medium

Peptone	2.0 g
Sodium chloride	5.0 g
Dipotassium hydrogen phosphate	0.3 g
Agar	3.0 g
Water	1000 ml

Heat to dissolve, adjust to pH 7.1 and add 15 ml of 0.2% bromothymol blue and sterile (filtered) glucose to give a final 1% concentration. Dispense aseptically in narrow (less than 1 cm) tubes in 8–10 ml amounts.

Baird-Parker (1966) modification for staphylococci and micrococci:

Tryptone	10 g
Yeast extract	1 g
Glucose	10 g
Agar	2 g
Water	1000 ml

Steam to dissolve, adjust to pH 7.2. Add 20 ml of 0.2% bromocresol purple, tube in 10-ml amounts in narrow (12 mm) tubes and sterilize at 115°C for 10 min.

Islam's (1977) medium

For Group B streptococci

Proteose peptone No. 3 (Difco)	23 g
Soluble starch	5 g
$NaH_2PO_4.2H_2O$	1.5 g
Na_2HPO_4	5.75 g
Agar	10 g
Water	1000 ml

Steam to dissolve, adjust to pH 7.4 and sterilize at 115°C for 10 min. For use melt, cool to 55°C and add 5% inactivated horse serum.

King's A broth (modified)

(Drake, 1966)

Peptone	20 g
Ethanol	25 ml

Potassium sulphate	10 ml
Magnesium chloride	1.4 g
Cetrimide	0.5 g
Distilled water	1000 ml

Steam to dissolve, distribute in screw-capped bottles and sterilize at 115°C for 10 min.

Kirchner's medium

For the culture and differentiation of mycobacteria.

$Na_2HPO_4.12H_2O$	19.0 g
KH_2PO_4	2.5 g
$MgSO_4.7H_2O$	0.6 g
Trisodium citrate	2.5 g
Asparagine	5.0 g
Glycerol	20.0 ml
Phenol red 0.4% aq.	3.0 ml
Distilled water	1000 ml

Steam to dissolve. The pH should be 7.4–7.6. Bottle in 9-ml amounts and autoclave at 115°C for 10 min. To each bottle 1 ml of horse serum and antibiotics as desired.

Loeffler medium: inspissated serum

Mix 3 parts of horse serum with 1 part of glucose broth, tube and inspissate in slopes.

Lowenstein-Jensen medium

This is used for the isolation of mycobacteria.

KH_2PO_4 (AnalaR)	4.0 g
$MgSO_4.7H_2O$ (AnalaR)	0.4 g
Magnesium citrate	1.0 g
Asparagine	6.0 g
Glycerol (AnalaR)	20.0 ml
Distilled water to	1000 ml

Dissolve in this order, steam for 2 h.

Clean fresh eggs with soap and water and break them into a sterile graduated cylinder. Take 1600 ml of egg fluid, mix in a screw-capped jar or polypropylene container with some large glass beads to break the yolks, filter through gauze and add to 1 litre of salt mixture. Add 50 ml of 1% aqueous malachite green, dispense and inspissate.

A 4-ml volume of pyruvic acid (neutralized with 3 N NaOH solution) or 12 g sodium pyruvate can replace glycerol.

Malonate broth

$(NH_4)_2SO_4$	2 g
K_2HPO_4	0.6 g
KH_2PO_4	0.4 g
Sodium malonate	3 g

Yeast extract	1 g
Distilled water	1000 ml
Bromothymol blue (0.2%)	12.5 ml

Heat to dissolve, add indicator and autoclave at 115°C for 10 min.

Malt agar modifications

To isolate acidophilic yeasts, cool the sterilized medium to 50°C, add 1 ml of sterile 10% lactic acid and pour plates at once.

For dermatophytes or pathogenic yeasts add 10 ml of 5% cycloheximide in acetone and 10 ml of 0.5% chloramphenicol in alcohol. Dispense in 10-ml amounts and autoclave at 115°C for 10 min.

To suppress bacterial growth and isolate moulds, add only the chloramphenicol.

Mead's (1963) medium

This medium allows *Enterococcus* spp. to be recognized by their ability to ferment sorbitol, decompose tyrosine and reduce triphenyltetrazolium chloride. Cultures are incubated at 45°C.

Peptone	10 g
Yeastrel	1 g
Sorbitol	2 g
Tyrosine	5 g
Agar	12 g
Water	1000 ml

Dissolve by autoclaving at 115°C for 10 min and adjust to pH 6.2. Add 0.1 g of triphenyltetrazolium chloride and 1 g of thallous acetate. When these have dissolved, add a further 4 g of tyrosine to give a suspension. Cool to 50°C, mix to give an even suspension of tyrosine and pour plates. Layered plates, the lowest layer of the same medium without tyrosine, are best.

Milk salt agar

(Gilbert *et al.*, 1969)

This is used for the recovery of *S. aureus* from food samples.

Peptone	5 g
Lab Lemco	3 g
Sodium chloride	6.5 g
Agar	15 g
Water	1000 ml

Sterilize at 121°C for 15 min. Add 10% skim milk and pour plates.

MRS medium

(De Man *et al.*, 1960)

Lab Lemco	10 g
Yeast extract	5 g
Peptone	10 g
Dipotassium hydrogen phosphate	2 g
Sodium acetate	5 g
Magnesium sulphate ($MgSO_4.7H_2O$)	0.2 g

Manganese sulphate ($MnSO_4.4H_2O$)	0.05 g
Tween 80	1 ml
Triammonium citrate	2 g
Water	1000 ml

Mycoplasma media

As these media vary in their ability to support the growth of mycoplasmas it is best to purchase the commercial base which may be known as Mycoplasma or PPLO (pleuropneumonia-like organisms) agar and broth. Some supplements are also available commercially.

Yeast extract
Freshly-prepared extract is best. Suspend 20 g of fresh bakers' yeast in 1000 ml of deionized water. Autoclave at 115°C for 15 min, allow to cool and centrifuge. Dispense the supernatant fluid in 20 ml amounts. Autoclave again and store at −20°C.

Additives
Substrates Prepare separate 10% solutions of glucose, urea and arginine in deionized water and sterilize by filtration. Store at 4°C.
Indicators Dissolve 1 g of phenol red in 2.5 ml of N NaOH and make up to 100 ml with deionized water. Filter through filter paper and sterile by filtration.
Selective agents Prepare penicillin solution, 100000 i.u./ml and amphotericin B, 0.5 mg/ml, in sterile deionized water. Store at −20°C.

Prepare a 10% solution of thallous acetate (*Caution*) in deionized water. Sterilize by filtration and store at 4°C.

Mycoplasma agar and broth
Add the following to 70 ml of commercial agar (melted and cooled to 54°C) or broth:

Horse serum	20 ml
Yeast extract	10 ml
Substrate	10 ml
Phenol red stock	0.2 ml
Thallous acetate stock	0.2 ml
Penicillin stock	0.2 ml
Amphotericin B stock	0.1 ml

Pour agar media into Petri dishes (small dishes are economical as the base is expensive). Dispense broth in 5 ml amounts in small screw-capped bottles. Store at 4°C for not more than 4 weeks.

Selective diphasic medium
Dispense 1.5 ml amounts of mycoplasma agar in 7 ml screw-capped bottles. Prepare overlay broth:

Mycoplasma broth	30 ml
Horse serum	30 ml
Yeast extract	15 ml
Glucose stock	15 ml
Methylene blue (0.1% in water)	1.5 ml
Phenol red stock	0.6 ml
Thallous acetate stock	0.4 ml
Penicillin stock	0.65 ml

Add NaOH until the medium is grey or purple and then add 2.5 ml to each bottle of mycoplasma agar.

Dissolve by heating, bottle and autoclave. For use, add 2% sterile (filtered) glucose or other carbohydrate solution.

N medium

(Collins, 1962)
This is used in the differentiation of mycobacteria.

NaCl	1.0 g
$MgSO_4.7H_2O$	0.2 g
KH_2PO_4	0.5 g
$Na_2HPO_4.12H_2O$	3.0 g
$(NH_4)_2SO_4$	10.0 g
Glucose	10.0 g
Water	1000 ml

Dissolve in that order, adjust to pH 6.8–7.0 and distribute. Autoclave at 115°C for 10 min. This medium can be modified by the addition of 1.2% agar and 0.04% bromocresol purple.

Nitrate broth

Dissolve 1 g of potassium nitrate in 1000 ml of nutrient broth. Steam to dissolve and bottle in small amounts.

Nutrient gelatin

Choose a high-quality gelatin and add 12–15% to a nutrient, Lemco or yeast extract broth. Heat to dissolve at 115°C for 10 min.

ONPG broth

o-Nitrophenyl-D-galactopyranoside	6 g
0.001 M Na_2HPO_4	1000 ml

Tube in 2-ml amounts. Do not heat.

Phenolphthalein phosphate agar

Phenolphthalein phosphate agar for the phosphatase test is made by adding 1 ml of a 1% solution of phenolphthalein phosphate to 100 ml of nutrient agar or commercial blood agar base and pouring plates.

Phenolphthalein phosphate polymyxin agar (PPPA)

For isolating *Staphylococcus aureus* from foods.

Add 125 units of polymyxin to 1 litre of phenolphthalein phosphate agar.

Phenolphthalein sulphate agar

This medium for the arylsulphatase test is made by dissolving 0.64 g of potassium phenolphthalein sulphate in 100 ml of water (0.01 M) Dispense one of the Middlebrook Broth media in 2.7-ml amounts and add 0.3 ml of the phenolphthalein sulphate solution to each tube.

Phenylalanine agar

DL-phenylalanine	2 g
Yeast extract	3 g
Na_2HPO	1 g
NaCl	5 g
Agar	20 g
Distilled water	1000 ml

Dissolve by heat; sterilize at 115°C for 10 min and tube as slopes.

Robertson's cooked meat medium

This is a useful general purpose medium and is used for the cultivation of anaerobes.

Mince 500 g of fresh lean beef and simmer for 1 h in 500 ml of boiling water containing 2 ml of 1 N NaOH. This neutralizes the acid in the medium; filter, dry the meat in a cloth and when dry put it into 28-ml screw-capped bottles so that there is about 25 mm of meat in each. Add 10–12 ml of infusion or Lemco broth to each tube, cap and autoclave at 115°C for 10 min.

Purple milk

Supersedes litmus milk

Skim milk	1000 ml
Bromocresol purple, 0.2%	10 ml

Dispense in 5 or 10-ml lots and autoclave at 110°C for 10 min.

Salt meat broth

Advantage is taken of the salt-tolerance (halophily) of *S.aureus* in investigating outbreaks of food poisoning due to *S.aureus*. Many other organisms are inhibited by high salt concentrations.

Prepare Robertson's cooked meat medium as above but add 10% sodium chloride to the liquid before tubing.

Semisolid oxygen preference medium

For identifying mycobacteria. Add 0.1% agar to one of the fluid Middlebrook liquid media. Autoclave at 115°C for 10 min. Dispense in 12-ml amounts in narrow bottles. Before use add 0.5 ml of oleic acid albumin dextrose catalase (OADC) supplement.

Skim milk agar

Reconstitute skim milk powder and mix equal volumes of this and double-strength nutrient agar melted and at 55°C. Mix and pour plates.

'Sloppy agar' Craigie tube

Broth containing 0.1–0.2% agar will permit the migration of motile organisms but will not allow convection currents. This permits motile organisms to be separated from non-motile bacteria in the Craigie tube.

Dispense sloppy agar in 12-ml amounts in screw-capped bottles. To each bottle add a piece of glass tubing of dimensions 50 × 5 or 6 mm. There must be sufficient clearance between the top of the tube and the surface of the medium so that a meniscus bridge does not form and the only connection between the fluid and the inner tube is through the bottom.

Starch agar

Prepare a 10% solution of soluble starch in water and steam for 1 h. Add 20 ml of this solution to 100 ml of melted nutrient agar and pour plates.

Sucrose agar

To demonstrate levan or dextran formation, add 5% sucrose to 100 ml of melted nutrient agar. Steam for 30 min, cool to 55°C, add 5 ml of sterile horse serum and pour plates.

Synthetic nutrient-poor medium

For yeasts and moulds.

Potassium dihydrogen phosphate	1.0 g
Potassium nitrate	1.0 g
Magnesium sulphate ($7H_2O$)	0.5 g
Potassium chloride	0.5 g
Glucose	0.2 g
Sucrose	0.2 g
Agar	20.0 g
Distilled water	1000 ml

Dissolve by heat.

Tellurite blood agar media for corynebacteria

The addition of 0.04% potassium tellurite (*caution*) to blood agar inhibits many other organisms and permits differentiation of *gravis, mitis* and *intermedius* types of *C. diphtheriae* and other corynebacteria. The base medium should be an infusion, not a digest medium, and Hoyle's base gives good results. For recognition and differentiation at 18–24 h, Brain-Veal Infusion agar is excellent. Some workers prefer lysed to fresh blood. Horse blood may be lysed by alternate freezing and thawing or by the addition of 5 ml of 2% saponin solution to 100 ml of blood.

Potassium tellurite is stored as a 2% solution.

Tributyrin agar

Fat-splitting organisms cause spoilage in butter, etc. Most of these bacteria split glyceryl tributyrate (tributyrin).

Peptone	5 g
Yeast extract	3 g
Tributyrin	10 g
Agar	12 g
Water	1000 ml

Heat to dissolve, adjust to pH 7.5, dispense and autoclave at 115°C for 15 min. Colonies of lipolytic organisms clear the medium.

Tyrosine agar: xanthine agar

To 100 ml of melted nutrient agar, add 5 g of tyrosine or 4 g of xanthine and steam for 30 min. Mix well to suspend amino acid and pour plates. See also Mead's medium.

Willis and Hobbs (1959) medium

This excellent medium for isolating and identifying clostridia sometimes needs minor modification to give the best results for particular species. Nagler reaction, 'pearly layer' proteolysis and lactose fermentation can be observed.

Nutrient broth	1000 ml
Lactose	12 g
Agar	12 g
Water	1000 ml

Heat to dissolve, adjust to pH 7.0, add 3.5 ml of 1% neutral red and 1 g of sodium thioglycollate, bottle in 100-ml portions and autoclave at 115°C for 10 min.

To each 100 ml of melted base add 4 ml of egg yolk emulsion and 15 ml of sterile whole milk. Pour plates.

Neomycin may be added to inhibit non-spore-bearing anaerobes and aerobic spore bearers but should be used with caution. It is desirable to experiment with varying concentrations up to 100 μg/ml.

Yeast extract milk agar

This is the same as commercial Milk agar.

Diluents

Saline and Ringer solutions

Physiological saline, which has the same osmotic pressure as microorganisms, is a 0.85% solution of sodium chloride in water. Ringer solution, which is ionically balanced so that the toxic effects of the anions neutralize one another, is better than the saline for making bacterial suspensions. It is used at one-quarter the original strength.

Sodium chloride (AnalaR)	2.15 g
Potassium chloride (AnalaR)	0.075 g
Calcium chloride, anhydrous (AnalaR)	0.12 g
Sodium thiosulphate pentahydrate	0.5 g
Distilled water	1000 ml

The pH should be 6.6.

Tablets of sodium chloride and Ringer are useful.

Calgon Ringer

Alginate wool, used in surface swab counts, dissolves in this.

Sodium chloride	2.15 g
Potassium chloride	0.075 g
Calcium chloride	0.12 g
Sodium hydrogen carbonate	0.05 g
Sodium hexametaphosphate	10.00 g

The pH is 7.0.

It is conveniently purchased as tablets.

Thiosulphate Ringer

This may be preferred for dilutions of rinses where there may be residual chlorine. Conveniently manufactured in tablet form.

Peptone water diluent (maximum recovery diluent)

The safest and least lethal of diluents is 0.1% peptone water.

Phosphate buffered saline

This is a useful general diluent at pH 7.3.

Sodium chloride	8.0 g
Potassium dihydrogen phosphate	0.34 g
Dipotassium hydrogen phosphate	1.21 g
Water	1000 ml

Indicators

Andrade

Dissolve 0.5 g of acid fuchsin in 100 ml of water. Add 1 N NaOH solution until the colour changes to yellow. The indicator is colourless at pH 7.2, pink in acidic solutions and yellow in alkaline solutions.

Bromocresol purple

Dibromo-*o*-cresolsulphonphthalein. Dissolve 0.1 g in 9.2 ml of 0.02 N NaOH and dilute to 250 ml with water. The indicator (0.4% solution) is yellow at pH 5.8 and blue at 6.8.

Bromothymol blue

Dibromothymolsulphonphthalein. Dissolve 0.1 g in 8 ml of 0.02 N NaOH and make up to 250 ml with water. This indicator (0.4% solution) is yellow at pH 6.0 and blue at pH 7.6.

Congo red

This indicator (0.1% solution in water) is blue at pH 3.0 and red at pH 5.2.

Methyl red

4′-Dimethylaminoazobenzene-4-sulphonate. Dissolve 0.1 g in 300 ml of ethanol and add 200 ml of water. This solution is red at pH 4.4 and yellow at pH 6.2.

Phenolphthalein

Dissolve 1 g in 100 ml of 95% ethanol. The solution is colourless at pH 8.3 and red at pH 10.

Phenol red

Phenolsulphonphthalein. Dissolve 0.1 g in 14.1 ml of 0.02 N NaOH and dilute to 250 ml with water. This indicator (0.4% solution) is yellow at pH 6.8 and red at pH 8.4.

Thymol blue

Thymolsulphonphthalein. Dissolve 0.1 g in 10.75 ml of 0.02 N NaOH and dilute to 250 ml with water. It has an acid range, red at pH 1.2 and yellow at 2.8, and an alkaline range, yellow at pH 8.0 and blue at pH 9.6.

Adjustment of pH

The pH of reconstituted and laboratory-prepared media should be checked with the meter, but for practical purposes devices such as the BDH Lovibond Comparator are good enough.

Pipette 10 ml of the medium into each of two 152 × 16 mm test-tubes. To one add 0.5 ml of 0.04% phenol red solution. Place the tubes in a Lovibond Comparator with the blank tube (without indicator) behind the phenol red colour disc, which must be used with the appropriate screen. Rotate the disc until the colours seen through the apertures match. The pH of the medium can be read in the scale aperture. Rotate the disc until the required pH figure is seen, add 0.05 N sodium hydroxide solution or hydrochloric acid to the medium plus indicator tube and mix until the required pH is obtained.

From the amount added, calculate the volume of 1 or 5 N alkali or acid that must be added to the bulk of the medium. After adding, check the pH again.

Example 10 ml of medium at pH 6.4 requires 0.6 ml of 0.05 N sodium hydroxide solution to give the required pH of 7.2.

Then the bulk medium will require

$$\frac{0.6 \times 100}{20} \text{ ml of 1 N NaOH/litre} = 3.0 \text{ ml}$$

Discs of other pH ranges are available.

All final readings of pH must be made with the medium at room temperature because hot medium will give a false reaction with some indicators. Note the volume of acid and alkali used and compare it with records of previous batches. Major differences in pH or buffering capacity may indicate an error in calculation or changes in the content of raw materials.

Dispensing culture media

Liquid media can be dispensed with the aid of a large funnel held in a retort stand and fitted with a short piece of rubber tubing and a spring clip. The automatic commercial fillers are more convenient. Some distributors may be connected to agar preparators.

To make agar slopes, dispense the medium in 3–7 ml amounts according to the size of the tube required. After sterilization and before the medium sets, slope the tubes individually on the bench by leaning them against a length of glass or metal rod 6 or 7 mm in diameter. When making large batches, slope the tubes or bottles on wire racks, e.g. old refrigerator shelves, sloped at a convenient angle.

When pouring plates, raise the lid only far enough to permit the mouth of the tube or bottle to enter. Pour about 12–15 ml in each plate. Dry the plates slightly open in an incubator, and store them medium side-up in a refrigerator, protected from moisture loss.

General precautions

Ensure that all equipment, including autoclaves, preparators and distributors, is regularly serviced and maintained in good condition.

Do not overheat media or quality will be impaired. Do not autoclave concentrates before they are dissolved by prior heating or they will set at the bottom of the vessel. Do not hold partly processed media overnight.

Screw caps, aluminium or propylene closures are better than cotton wool plugs.

Contamination

This is a major cause of media loss. It may be caused by inadequate heat processing, or failure to sterilize the distributing apparatus. The appearance of a few colonies on plated media after storage may indicate environmental fall-out. This may be reduced or even eliminated by using automated plate pouring machines and/or by pouring media in laminar-flow clean air cabinets.

Quality control and performance tests

Commercial manufacturers of dehydrated or prepared media carry out extensive quality control tests on the raw materials and final products before release for sale. These tests need not be repeated in laboratories that purchase such media, but some tests are desirable after reconstitution. This is because there may be variations within laboratories in manipulations, such as heat processing and the addition of supplements.

Detailed information about quality control testing of microbiological culture media and reagents is given by Snell *et al.* (1991). See also Chapter 2.

Two simple procedures, one for isolation media and the other for identification media, are given below.

Isolation and selective media: efficiency of plating technique (EOP)

New batches of culture media may vary considerably and should be tested in the following way.

Prepare serial tenfold dilutions, for example. 10^{-2} to 10^{-7} of cultures of various organisms which will grow on or be inhibited by the medium. For example, when testing DCA use *S. sonnei, S. typhi, S. typhimurium*, several other salmonellas and *E. coli*. Use at least four well-dried plates of the test medium for each organism and at least two plates each of a known satisfactory control medium, and of a non-selective medium, e.g. nutrient agar. Do Miles and Misra drop counts with the serial dilutions of the organisms on these plates so that each plate is used for several dilutions (Figure 5.1) (see p.153).

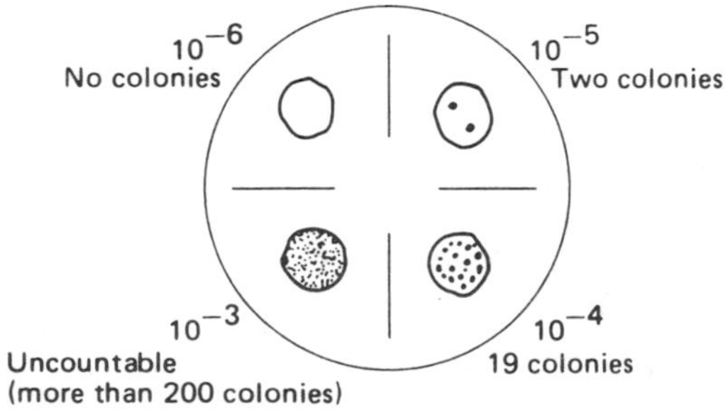

Figure 5.1 Miles and Misra count

Count the colonies, tabulate the results and compare the performances of the various media. These may suggest that in a new batch of a special medium it is necessary to alter the proportion of certain ingredients.

Stock cultures, however, will frequently grow on media which will not support the growth of 'damaged' organisms from natural materials. It is best, therefore, to use dilutions or suspensions of such material.

With some culture media, these drop counts do not give satisfactory results because of reduced surface tension. An alternative procedure is recommended.

Make cardboard masks to fit over the tops of petri dishes and cut a square 25 × 25 mm in the centre of each. Place a mask over a test plate and drop one drop of the suspension through the square on the medium. Spread the drop over the area limited by the mask. This method has the additional advantage of making colony counts easier to perform but more plates must, of course, be tested.

In addition to these EOP tests, ordinary plating methods should be used to compare colony size and appearance.

Identification media

Maintain stock cultures (p.99) of organisms known to give positive or negative reactions with the identification tests in routine use. The same organism may often be used for a number of tests. When a new batch of any identification medium is anticipated, subculture the appropriate stock strain so that a young, active culture is available. Test the new medium with this culture.

Keep stock cultures of the organisms likely to be encountered, preferably from a type culture collection and maintained freeze dried until required. Subculture them to test new batches of culture media.

Caution: Bacteria in Hazard Group 3 should not be used unless the tests can be conducted in Level 3 laboratories.

References

APHA (1985) *Standard Methods for the Examination of Dairy Products*, 14th edn, American Public Health Association, Washington DC

Baird, R.M., Corry, J.E.L. and Curtis, G.D.W. (eds) (1987) Pharmacopoeia of culture media for food microbiology. *International Journal of Food Microbiology*, **5**, 187–229

Baird-Parker, A.C. (1966) Methods for classifying staphylococci and micrococci. In *Identification Methods for Microbiologists* Part A (eds B.M. Gibbs and F.A. Skinner), Society for Applied Bacteriology Technical Series No. 1, Academic Press, London, pp. 59–64

Barrow, G.I. and Feltham, R.K.A. (eds) (1993) *Cowan and Steel's Manual for the Identification of Medical Bacteria*, 3rd edn, Cambridge University Press, Cambridge

Bridson, E.Y. (1994) *The Development, Manufacture and Control of Microbiological Culture Media*, Unipath, Basingstoke

Collins, C.H. (1962) The classification of 'anonymous' acid fast bacilli from human source. *Tubercle*, **43**, 293–298

Davis, J.G. (1931) Standardisation of media in the acid ranges with special reference to the use of citric acid and buffer mixtures for yeast and mould media. *Journal of Dairy Research*, **3**, 133–144

Davis, J.G. (1982) Unhopped beer wort as a medium for yeast and mould counts. *Laboratory Practice*, **31**, 219

Davies, J.R. (1974) Elek's test for the toxigenicity of *C. diphtheriae*. In *Laboratory Methods, 1*, Public Health Laboratory Service Monograph No. 5, HMSO, London

De Man, J.C., Rogosa, M. and Sharpe, M.E. (1960) A medium for the cultivation of lactobacilli. *Journal of Applied Bacteriology*, **23**, 130–133

Donovan, T.J. (1966) A *Klebsiella* screening medium. *Journal of Medical Laboratory Technology*, **23**, 194–196

Drake, C.H. (1966) Evaluation of culture media for the isolation and enumeration of *Pseudomonas aeruginosa. Health Laboratory Service*, **3**, 10–14

Ellner, P.D. (1956) A medium promoting rapid quantitative sporulation in *Clostridium perfringens. Journal of Bacteriology*, **71**, 495–496

Falkow, S. (1958) Activity of lysine decarboxylase as an aid in the identification of salmonellae and shigellae. *American Journal of Clinical Pathology*, **29**, 598–560

Flynn, J. and Waitkins, S.A. (1972) A serum-free medium for testing the fermentation reactions in *Neisseria. Journal of Clinical Pathology*, **25**, 525–527

Gardner, G.A. (1966) A selective medium for the enumeration of *Microbacterium thermosphactum* in meat and meat products. *Journal of Applied Bacteriology*, **29**, 455–460

Gilbert, R.J., Kendall, M. and Hobbs, B.C. (1969) Media for the isolation and enumeration of coagulase positive staphylococci. In *Isolation Methods for Microbiologists* (eds D.A. Shapton and G.W. Gould), Society for Applied Bacteriology Technical Series No. 3, Academic Press, London, pp. 9–14

Islam, A.K.M.S. (1977) Rapid recognition of Group B streptococci. *Lancet*, **i**, 356–357

Lowbury, G.J.L. and Lilly, H.A. (1955) A selective plate medium for *Cl. welchii*. *Journal of Pathology and Bacteriology*, **70**, 105–107

MacFaddin, J.F. (1985) *Media for the Isolation-Cultivation-Identification-Maintenance of Medical Bacteria*, Vol. 1, Williams and Wilkins, Baltimore

Mead, G.C. (1963) A medium for the isolation of *Streptococcus faecalis sensu stricto*. *Nature (London)*, **197**, 1323–1324

Moeller, V. (1955) Simplified tests for some amino acid decarboxylases and for the arginine dihydrolase system. *Acta Pathologica et Microbiologica Scandinavica*, **36**, 158–172

National Committee for Clinical Laboratory Standards (1990) Approved Standard ASM-2. Performance standards for antimicrobic disc susceptibility tests, 4th edn, Villanova, PA, USA

Oxoid Manual (1990) Unipath, Basingstoke

Petts, D.N. (1984) Colistin-oxolinic acid-blood agar: a new selective medium for streptococci. *Journal of Clinical Microbiology*, **19**, 4–7

Snell, J.J.S., Farrell, I.D. and Roberts, C. (eds) (1991) *Quality Control: Principles and Practice in the Microbiology Laboratory*, Public Health Laboratory Service, London

Waitkins, S. (1985) Leptospiras and Leptospirosis. In *Isolation and Identification of Microorganisms of Medical and Veterinary Importance* (eds C.H. Collins and J.M. Grange), Society for Applied Bacteriology Technical Series No. 21, Academic Press, London, pp. 251–296

Willis, A.T. and Hobbs, G. (1959) A medium for the identification of clostridia producing opalescence in egg yolk emulsions. *Journal of Pathology and Bacteriology*, **77**, 299–300

Chapter

6

Cultural methods

General bacterial flora

Samples and specimens for microbiological examination may contain mixtures of many different organisms. To observe the general bacterial flora of a sample of any material, use non-selective media. No one medium nor any one temperature will support the growth of all possible organisms. It is best to use several different media, each favouring a different group of organisms, and to incubate at various temperatures aerobically and anaerobically. The pH of the medium used should approximate that of the material under examination.

For bacteria, examples are nutrient agar, glucose tryptone agar and blood agar, and for moulds and yeasts, malt agar and Sabouraud agar. These and other suitable general purpose media are described in Chapter 5.

Methods for estimating bacterial numbers are given in Chapter 10, and the dilution methods employed for that purpose can be used to make inocula for the following techniques. Three methods are given here for plating material. They depend on separating clumps or aggregates of bacteria so that each will grow into a separate colony. These colonies can be subcultured or picked for further examination.

Pour plate method

Melt several tubes each containing 15 ml of medium. Place in water-bath at 45–50°C to cool. Emulsify the material to be examined in 0.1% peptone water and prepare 1:10, 1:100 and 1:1000 dilutions in the same solution (see p.152). Add 1 ml of each dilution to 15 ml of the melted agar, mix by rotating the tube between the palms of the hands and pour into a petri dish. Make replicate pour plates of each dilution for incubation at suitable temperatures. Allow the medium to set, invert the plates and incubate. This method gives a better distribution of colonies than that used for the plate count (p.152), but cannot be used for that purpose because some of the inoculum, diluted in the agar, remains in the test-tube.

Spreader method

Dry plates of a suitable medium. Make dilutions of the emulsified material as described above. Place about 0.05 ml (1 drop) of dilution in the centre of a plate and spread it over the medium by pushing the glass spreader backward and forward while rotating the plate. Replace the petri dish lid and leave for 1–2 h to dry before inverting and incubating as above.

Looping-out method

Make dilutions of material as described above. Usually the neat emulsions and at 1 : 10 dilution will suffice. Place one loopful of material on the medium near the rim of the plate and spread it over the segment (Figure 6.1). Flame the loop and spread from area *A* over area *B* with parallel streaks, taking care not to let the streaks overlap. Flame the loop and repeat with area *C*, and so on. Each looping-out dilutes the inoculum. Invert the plates and incubate.

When looping-out methods are used, two loops are useful; one is cooling while the other is being used.

These surface plate methods are said to give better isolation of anaerobes than pour plates.

Spiral or rotary plating

This mechanical method is invaluable for plating large numbers of specimens or cultures and for doing sensitivity tests by Stokes method (p.183). Hand-operated (with electrically-driven turntable) and fully-automated models are available. With the latter, exact volumes may be plated. A laser colony counter is on the market.

To use the hand-operated machine place the open plate on the turntable. Touch the rotating plate with the charged loop at the centre of the medium and draw it slowly towards the circumference. A spiral of bacterial growth will be obtained after incubation which will show discrete colonies (Figure 6.1).

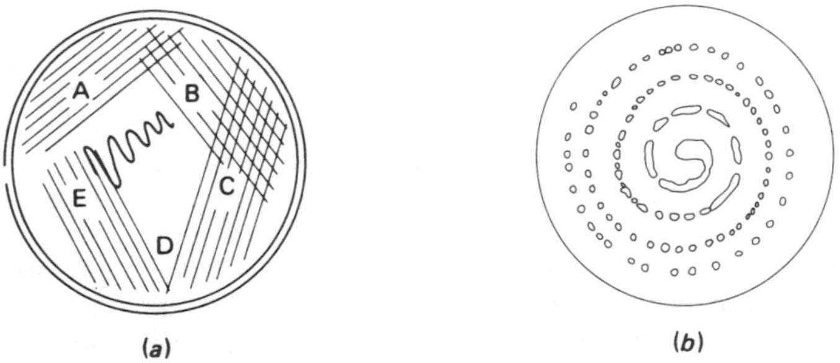

Figure 6.1 (a) 'Spreading' or 'looping-out' in a plate; (b) rotary plating

Multipoint inoculators

These are convenient tools when many replicate cultures are needed. They are fully or semi-automatic and spot-inoculate large numbers of petri dish cultures. A number of pins take up the inoculum from a single reservoir or different inocula from the wells in, e.g., a microtitre plate and spot-inoculate the medium on one or a succession of petri dishes. Multipoint inoculators have heads for up to 36 pins for 90-mm dishes and 96 for larger dishes. A variety of these instruments is available commercially. They are particularly useful for combined antibiotic susceptibility, minimum inhibitory concentration tests, the quality control of culture media, the assessment of bacterial loads of large numbers of samples (e.g. urines, foods) and in some identification procedures, e.g. of staphylococci and enterobacteria.

The UK Public Health Laboratory Service has published an excellent guide to multipoint inoculation methods (Faiers *et al.*, 1991).

Shake tube cultures

These are useful for observing colony formation in deep agar cultures, especially of anaerobic or microaerophilic organisms.

Dispense media, glucose agar or thioglycollate agar, in 15–20-ml amounts in bottles or tubes 20–25 mm in diameter. Melt and cool to approximately 45°C. Add about 0.1 ml of inoculum to one tube, mix by rotating between the palms, remove one loopful to inoculate a second tube, and so on. Allow to set and incubate. Submerged colonies will develop and will be distributed as follows: obligate aerobes grow only at the top of the medium, obligate anaerobes only near the bottom; microaerophiles grow near but not at the top; facultative organisms grow uniformly throughout the medium.

Stab cultures

These can be used to observe motility, gas production and gelatin liquefaction. The medium is dispensed in 5-ml amounts in tubes or bottles of 12.5 mm diameter and inoculated by stabbing the wire down the centre of the agar.

Subcultures

For further examination, pure cultures are prepared from the various types of colonies in mixed cultures.

Place the plate culture under a low-power binocular microscope and select the colony to be examined. With a flamed straight wire (*not* a loop) touch the colony. There is no need to dig or scrape; too vigorous handling may result in contamination with adjacent colonies or with those beneath the selected colony.

With the charged wire, touch the medium on another plate, flame the wire and then use a loop to spread the culture to obtain individual colonies. This should give a pure culture but the procedure may need repeating.

To inoculate tubes or slopes, follow the same procedure. Hold the charged wire in one hand, the tube in the other in a nearly horizontal position. Remove the cap or plug of the tube by grasping it with the little finger of the hand holding the wire. Pass the mouth of the tube through a bunsen flame to kill any organism that might fall into the culture and inoculate the medium by drawing the wire along the surface of a slope or touching the surface of a liquid. Replace the cap or plug and flame the wire. Several tubes can be inoculated in succession without recharging the wire.

Anaerobic culture

Anaerobic jars

Modern anaerobic jars are made of metal or transparent polycarbonate, are vented by Schrader valves and use 'cold' catalysts. They are therefore safer than earlier models; internal temperatures and pressures are lower. More catalyst is used and the escape of small particles of catalyst, which can ignite hydrogen–air mixtures is prevented. Nevertheless the catalyst should be kept dry, dried after use and replaced frequently.

For ordinary clinical laboratory work the commercial sachets are a convenient source of hydrogen and hydrogen–carbon dioxide mixtures. A recent development is a sachet that achieves reliable anaerobic conditions without generating hydrogen. The final atmosphere contains less than 1% oxygen, supplemented with carbon dioxide.

Place the plates 'upside down', i.e. with the media at the bottom of the vessel. If they are incubated the right way up, the medium sometimes falls into the lid when the pressure is reduced. Dry the catalyst capsule by flaming and replace the lid.

Some workers prefer to use pure hydrogen or a mixture of 90% hydrogen and 10% carbon dioxide. A non-explosive mixture of 10% hydrogen, 10% carbon carbon dioxide and 80% nitrogen is safer (see Kennedy, 1988) but may be difficult to obtain in some countries. There is a common misconception about storing these cylinders horizontally: they may safely be stored vertically.

For the evacuation-replacement jar technique with a mixture containing 90% or more hydrogen attach a vacuum pump to the jar across a manometer and remove air to about −300 mmHg. Replace the vacuum with the gas mixture, disconnect both tubes and allow the secondary vacuum to develop as the catalyst does its work. Leave for 10 min and reconnect the manometer to check that there is a secondary vacuum of about −100 mmHg. This indicates that the seals of the jar are gas tight and that the catalyst is working properly. Replace the partial vacuum with gas to atmospheric pressure.

NB If the gas mixture contains only 10% hydrogen the jar must be evacuated to at least −610 mmHg so that enough oxygen is removed for the catalyst and hydrogen to remove the remainder.

Indicators of anaerobiosis

'Redox' indicators based on methylene blue and resazurin are available commercially. There is a simple test to demonstrate that anaerobiosis is achieved during incubation: inoculate a blood agar plate with *Clostridium perfringens* and place on it a 5 μg metronidazole disc. This bacterium is not very exacting in its requirement for anaerobiosis and will grow in suboptimal conditions. A zone of inhibition around the growth indicates that conditions were sufficiently reduced for the drug to work and that anaerobiosis is adequate. There is little to commend the use of *Pseudomonas aeruginosa* as a 'negative' control. Even if a 'positive' control such as *Cl. tetani* fails to grow it does not necessarily suggest inadequate anaerobic conditions.

Anaerobic cabinets

These are now commonly used in diagnostic laboratories. They may be used simply as anaerobic incubators with the added advantage that cultures do not have to be removed for examination, or as anaerobic work stations for inoculating pre-reduced media. Modern cabinets are very sophisticated, allowing 'bare hand' working, humidity control and atmospheric detoxification.

Culture under carbon dioxide

Some microaerophiles grow best when the oxygen tension is reduced. Capnophiles require carbon dioxide. For some purposes, candle jars are adequate. Place the cultures in a can or jar with a lighted candle or night-light. Replace the lid. The light will go out when most of the oxygen has been removed and replaced with carbon dioxide.

Alternatively, place 25 ml of 0.1 N HCl in a 100-ml beaker in the container and add a few marble chips before replacing the lid.

Commercial carbon dioxide sachets, hydrogen–carbon dioxide sachets or carbon dioxide from a cylinder may be used with an anaerobic jar. Remove air from the jar down to −76 mmHg and replace with carbon dioxide from a football bladder. This is the best method for capnophiles, which require 10–20% carbon dioxide to initiate growth, but if extensive work of this nature is carried out, a carbon dioxide incubator is invaluable. A nitrogen–carbon dioxide mixture can be used instead of pure carbon dioxide.

Incubation of cultures

Incubators are discussed in Chapter 2. Instead of using an incubator for culturing at 20–22°C, plates and tubes may be left on the bench, but a cupboard is better as draughts can be avoided. A maximum and minimum thermometer is necessary. Ambient temperatures often fluctuate widely when windows are open or heating is switched off at night.

Psychrophiles are usually incubated at 4–7°C. A domestic refrigerator can be used provided that it is not opened too often. Temperatures of 20–25°C are widely used for food bacteria and 35–37°C for medical bacteria. Thermophiles require 55–60°C.

Petri dish cultures are incubated upside down, i.e. medium uppermost (except in anaerobic vessels), otherwise condensation water collects on the surface of the medium and prevents the formation of isolated colonies.

If it is necessary to incubate petri dish cultures for several days, they must be sealed in order to prevent the medium drying up. Plastic bags or plastic food containers can be used.

Colony and cultural appearances

Much time and labour may be saved in the identification of bacteria if workers familiarize themselves with colonial appearances on specific media. These appearances and the terms used to describe them are illustrated in Figure 6.2.

Examine colonies and growths on slopes with a hand-lens or, better, plate microscope. Observe pigment formation both on top of and under the surface of the colony and note if pigment diffuses into the medium. In broth cultures, surface growth may occur in the form of a pellicle. Some cultures develop characteristic odours.

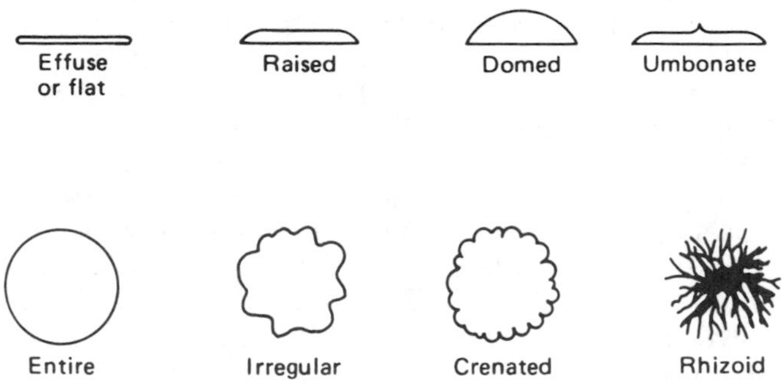

Figure 6.2 Description of colonies

Auxanograms

These are used to investigate the nutritional requirements of bacteria and fungi, for example, which amino acid or vitamin substance is required by 'exacting strains' or which carbohydrates the organisms can utilize as carbon sources.

Make a minimal inoculum suspension of the organism to be tested in phosphate-buffered saline. Centrifuge, re-suspend in fresh phosphate-buffered saline. Wash again in the same way and add 1 ml to several 15-ml amounts of basal medium (p.72) with indicator. Pour into plates. On each plate place a filter-paper disc of 5-mm diameter and on the filter-paper place one loopful of a 5% solution of the appropriate carbohydrate. Incubate and observe growth. Commercially-prepared discs are available.

When testing for amino acid or vitamin requirements in this way, use the basal medium plus glucose. Several discs, each containing a different amino acid, can be used on a single plate.

Auxanogram kits are available commercially.

Stock cultures

Most laboratories need stock cultures of 'standard strains' of microorganisms for testing culture media, controlling certain tests and so that workers may become familiar with cultural appearances and other properties. These must be subcultured periodically to keep them alive, with the consequent risk of contamination, and more importantly in laboratories engaged on assay work, the involuntary selection of mutants. It is because of mutations in bacterial cultures that the so-called 'standard strains' or 'type cultures' so often vary in behaviour from one laboratory to another.

To overcome these difficulties, Type Culture Collections are maintained by government laboratories. In these establishments, the cultures are preserved in a freeze-dried state, and in the course of maintenance of the stocks they are tested to ensure that they conform to the official descriptions of the standard or type strains.

Cultures of microorganisms may be obtained from: the National Collection of Type Cultures (NCTC), Central Public Health Laboratory, Colindale Avenue, London NW9 5HT; the National Collection of Industrial and Marine Bacteria

(NCIMB), 23 Machar Drive, Aberdeen AB2 1RY, the American Type Culture Collection (ATCC), 12301 Parklawn Drive, Rockville, Maryland 20852; the National Collection of Dairy Organisms (NCDO), National Institute for Research in Dairying, Shinfield near Reading, Berkshire; the National Collection of Pathogenic Fungi (NCPF) Public Health Laboratory, Myrtle Road, Kingsdown, Bristol BS2 8EL; National Collection of Plant Pathogenic Bacteria (NCPPB), Harpenden Laboratories, Hatching Green, Harpenden, Herts AL5 2BD, the Commonwealth Mycological Collection (CMC), Ferry Lane, Kew, Surrey; the National Collection of Yeast Cultures (NCYC), Food Research Institute, Colney Lane, Norwich, Norfolk NR4 7UA. A fee is usually charged.

Further information and lists of collections may be obtained from the United Kingdom Federation of Culture Collections, Biochemistry Group, Research and Development, ICI Agricultural Division, Billingham, Cleveland TS23 1LB.

The Comité Europeén de Normalization (CEN) is currently preparing a standard for strain conservation.

Strains that are to be used occasionally, or rarely, should be obtained when required and not maintained in the laboratory. Strains that are in constant use may be kept for short periods by serial subculture, for longer periods by one of the drying methods and indefinitely by freeze-drying. When stocks of dried or freeze-dried cultures need renewing, however, it is advisable to start again with a culture from one of the National Collection rather than perpetuate one's own stock.

Serial subculture

The principle is to subculture as infrequently as possible in order to keep the culture alive and to arrest growth as early as possible in the logarithmic phase so as to avoid the appearance and possible selection of mutants. Useful media for keeping stock cultures of bacteria are Griffith's egg, fastidious anaerobe medium, Robertson's cooked meat and litmus milk. Fungi can be maintained on Sabouraud medium. These media must be in screw-capped bottles. It is difficult to keep cultures of some of the very fastidious organisms, for example gonococci, and it is advisable to store them above liquid nitrogen (see below).

Inoculate two tubes, incubate until growth is just obvious and active, for example, 12–18 h for enterobacteria, other Gram-negative rods and micrococci, 2 or 3 days for lactobacilli. Cool the cultures and store them in a refrigerator for 1–2 months.

Keep one of these cultures, *A*, as the stock, and use the other, *B*, for laboratory purposes.

Stock $A_1 \rightarrow A_2 \rightarrow A_3 \rightarrow A_4$

Use   use *B*   use *B*   use *B*

Preservation of cultures

Drying cultures

Two methods are satisfactory. To dry by Stamp's method, make a small amount of 10% gelatin in nutrient broth and add 0.25% of ascorbic acid. Add a thick suspension of the organisms before the medium gels (to obtain a final density of about 10^{10} organisms/ml). Immerse paper sheets in wax, drain and allow to set. Drop the suspension on the waxed surface. Dry in a vacuum desiccator over phosphorous pentoxide. Store the discs that form in screw-capped bottles in a refrigerator. Petri dishes smeared with a 20% solution of silicone in light petroleum and dried in a hot-air oven can be used instead of waxed paper.

Rayson's method is slightly different. Use 10% serum broth or gelatin broth and drop on discs of about 10 mm diameter cut from Cellophane and sterilized in petri dishes. Dry in a vacuum desiccator over phosphorous pentoxide. Store in screw-capped bottles in a refrigerator.

To reconstitute, remove one disc with sterile forceps into a tube of broth and incubate.

Freezing

Make thick suspensions of bacteria in 0.5 ml of sterile tap water or skim milk in screw-capped bottles and place in a deep freeze at −40°C or, better, −70°C. Many organisms remain viable for long periods in this way.

Make heavy suspensions in 30% glycerol in 1% peptone water (preserving medium) and distribute in small vials. Freeze at −70°C (−20°C may suffice). There are commercial systems (e.g. Prolab, UK) in which suspensions are added to vials containing prepared beads. The excess fluid is removed and the vial frozen. One bead is removed to culture medium as required.

Suspensions may also be stored in the vapour phase above liquid nitrogen.

Freeze-drying

Suspensions of bacteria in a mixture of 1 part of 30% glucose in nutrient broth and 3 parts of commercial horse serum are frozen and then dried by evaporation under vacuum. This preserves the bacteria indefinitely and is known as lyophilization. Yeasts, cryptococci, nocardias and streptomycetes can be lyophilized by making a thick suspension in sterile skimmed milk fortified with 5% sucrose. This simple protective medium can be sterilized by autoclaving in convenient portions, e.g. 3 ml in bijou bottles.

Several machines are available. The simplest consists of a metal chamber in which solid carbon dioxide and ethanol are mixed. The bacterial suspension is placed in 0.2-ml amounts in ampoules or small tubes together with a label and these are held in the freezing mixture, which is at about −78°C, until they freeze. They are then attached to the manifold of a vacuum pump capable of reducing the pressure to less than 0.01 mmHg. Between the pump and the manifold there is a metal container, usually part of the freezing container, in which a mixture of solid carbon dioxide and phosphorous pentoxide condenses the moisture withdrawn from the tubes. The tubes or ampoules are then sealed with a small blowpipe.

Simultaneous freezing and drying is carried out in centrifugal freeze-driers in which bubbling of the suspension is prevented by centrifugal action.

References

Faiers, M., George, R., Jolly, J. and Wheat, P (1991) *Multipoint Methods in the Clinical Laboratory*, Public Health Laboratory Service, London

Kennedy, D. A. (1988) Equipment-related hazards. In *Safety in Clinical and Biomedical Laboratories*, (ed. C. H. Collins), Chapman & Hall, London, pp. 11–46

Chapter

7

Identification methods

Pure cultures are essential for the identification of microorganisms. Methods for obtaining them are described in Chapter 6 but it may be necessary to plate out subcultures repeatedly to ensure that only one kind of organism is present. General-purpose media should be used, providing that they support the growth of the organism in question. Colonies taken from selective media may contain organisms whose growth has been suppressed.

Cultures for identification should first be examined microscopically. Stained films may reveal mixed cultures.

Preparation of films for microscopy

The morphology of bacteria may be observed in wet, unstained preparations with light, dark-field or phase contrast microscopy but these are not permanent. It is usual to stain thin films of organisms prepared as follows.

Place a very small drop of saline or water in the centre of a 76 × 25 mm glass microslide. Remove a small amount of bacterial growth with an inoculating wire or loop, emulsify the organisms in the liquid, spreading it to occupy about 1 or 2 cm^2. Allow to dry in air or by waving high over a bunsen burner, taking care that the slide becomes no warmer than can be borne on the back of the hand. Pass the slide, film side down, once only through the bunsen flame to 'fix' it. This coagulates bacterial protein and makes the film less likely to float off during staining. It may not kill all the organisms, however, and the film should still be regarded as a source of infection. After staining, drain, blot with fresh, clean filter paper and dry by gentle heat over a bunsen.

Most laboratories now buy stains in solution ready for use, but some formulas are given below.

In the USA only stains certified by the Biological Stains Commission should be used. In the UK there is no comparable system and stains prepared by one of the specialist organizations should be used or purchased in solution ready for use.

A large number of stains, mostly basic aniline dyes, have been used and described. For practical purposes one or two simple stains, the Gram method and an acid-fast stain, are all that are required, plus one stain for fungi. A few other stains are described here, including methods for staining capsules and spores.

Methylene blue stain

Stain for 1 min with a 0.5% aqueous solution of methylene blue or Loeffler's methylene blue: mix 30 ml of saturated aqueous methylene blue solution with 100 ml of 0.1% potassium hydroxide. Prolonged storage, with occasional shaking, yields 'polychrome' methylene blue for McFadyean's reaction for *B. anthracis* in blood films.

Fuchsin stain

Stain for 30 s with the following: dissolve 1 g basic fuchsin in 100 ml 95% ethanol. Stand for 24 h. Filter and add 900 ml water. *Or* use carbol fuchsin (see 'Acid-fast stain') diluted 1:10 with phenol saline.

Gram stain

There are many modifications of this. This is Jensen's version:

1 Dissolve 0.5 g methyl violet in 100 ml distilled water.
2 Dissolve 2 g potassium iodide in 20 ml distilled water. Add 1 g of finely ground iodine and stand overnight. Make up to 300 ml when dissolved.
3 Dissolve 1 g of safranin or 1 g of neutral red in 100 ml of distilled water.

Stain with methyl violet solution for 20 s. Wash off and replace with iodine solution. Leave for 1 min. Wash off iodine solution with 95% alcohol or acetone, leaving on for a few seconds only. Wash with water. Counterstain with fuchsin or safranin for 30 s.

Some practice is required with this stain to achieve the correct degree of decolorization. Acetone decolorizes much more rapidly than alcohol.

The Hucker method is commonly used in US laboratories.

1 Dissolve 2 g of crystal violet in 20 ml of 95% ethanol. Dissolve 0.8 g of ammonium oxalate in 80 ml of distilled water. Mix these two solutions, stand for 24 h and then filter.
2 Dissolve 2 g of potassium iodide and 1 g of iodine in 300 ml of distilled water using the method described for Jensen's Gram stain.
3 Grind 0.25 g of safranin in a mortar with 10 ml of 95% ethanol. Wash into a flask and make up to 100 ml with distilled water.

Stain with the crystal violet solution for 1 min. Wash with tap water. Stain with the iodine solution for 1 min. Decolorize with 95% ethanol until no more stain comes away. Wash with tap water. Stain with the safranin solution for 2 min.

Acid-fast stain

The Ziehl–Neelsen method employs carbol fuchsin, acid–alcohol and a blue or green counterstain. Colour-blind workers should use picric acid counterstain.

(1)	Basic fuchsin	5 g
	Crystalline phenol (*caution*)	25 g
	95% alcohol	50 ml
	Distilled water	500 ml

Dissolve the fuchsin and phenol in the alcohol over a warm water-bath, then add the water. Filter before use.

(2)	95% ethyl alcohol	970 ml
	Conc. hydrochloric acid	30 ml

(3) 0.5% methylene blue or malachite green, or 0.75% picric acid in distilled water.

Pour carbol-fuchsin on the slide and heat carefully until steam rises. Stain for 3–5 min but do not allow to dry. Wash well with water, decolorize with acid–alcohol for 10–20 s, changing twice and counterstain for a few minutes with methylene blue or malachite green.

Some workers prefer the original decolorizing procedure with 20% sulphuric acid followed by 95% alcohol.

When staining nocardias or leprosy bacilli use 1% sulphuric acid and no alcohol.

Cold staining methods are popular with some workers. For the Muller–Chermack method add 1 drop of Tergitol 7 (sodium heptadecyl sulphate) to 25 ml of carbol fuchsin and proceed as for Ziehl–Neelsen stain.

For Kinyoun's method make the carbol-fuchsin as follows:

Basic fuchsin	4 g
Melted phenol (*caution*)	8 g
95% ethanol	20 ml
Distilled water	100 ml

Dissolve the fuchsin in the alcohol. Shake gently while adding the water. Then add the phenol. Proceed as for Ziehl–Neelsen stain but do not heat. Decolorize with 1% sulphuric acid in water.

Fluorescent stain for acid-fast bacilli

There are several of these. We have found this auramine–phenol stain adequate.

(1)	Phenol crystals (*caution*)	3 g
	Auramine	0.3 g
	Distilled water	100 ml
(2)	Conc. HCl	0.5 ml
	NaCl	0.5 g
	Ethanol	75 ml

(3) Potassium permanganate 0.1% in distilled water.

Prepare and fix films in the usual way. Stain with auramine-phenol for 4 min. Wash in water. Decolorize with acid alcohol for 4 min. Wash with potassium permanganate solution.

Staining corynebacteria

To show the barred or beaded appearance and the metachromatic granules of these organisms Laybourn's modification of Albert's stain may be used.

(1) Dissolve 0.2 g malachite green and 0.15 g toluidine blue in a mixture of 100 ml water, 1 ml glacial acetic acid and 2 ml 95% alcohol.

(2) Dissolve 3 g of potassium iodide in 50–100 ml of distilled water. Add 2 g of finely ground iodine and leave overnight. Make up to 300 ml.

Stain with Solution (1) for 4 min. Wash, blot and dry and stain with Solution

(2) for 1 min. The granules stain black and the barred cytoplasm light and dark green.

Spore staining

Make a thick film, stain for 3 min with hot carbol-fuchsin (Ziehl–Neelsen). Wash, flood with 30% aqueous ferric chloride for 2 min. Decolorize with 5% sodium sulphite solution. Wash and counterstain with 1% malachite green.

Flemming's technique substitutes nigrosin for the counterstain. Spread nigrosin over the film with the edge of another slide.

Spores are stained red. Cells are green, or with Flemming's method are transparent on a grey background.

Capsule staining

Place a small drop of india ink on a slide. Mix into it a small loopful of bacterial culture or suspension. Place a cover-glass over the drop avoiding air bubbles and press firmly between blotting paper. Examine with high-power lens. (Dispose of blotting paper in disinfectant.)

For dry preparations mix one loopful of indian ink with one loopful of a suspension of organisms in 5% glucose solution at one end of a slide. Spread the mixture with the end of another slide, allow to dry and pour a few drops of methyl alcohol over the film to fix. Stain for a few seconds with methyl violet (Jensen's Gram solution No. 1). The organisms appear stained blue with capsules showing as haloes.

Giemsa stain may be used. Fix the films in absolute methanol for 3 min. Pour on Giemsa stain and leave until it has almost dried up. Wash rapidly in water and then in phosphate buffer (0.001 mol/l, pH 7.0). Blot and dry. Capsules are stained pink, bacterial cells blue.

Flagella stains

This is a modification of Ryu's method (Kodaka *et al.*, 1982).

(1)	Tannic acid powder	10 g
	Phenol 5% in water (*caution*)	50 ml
	Aluminium potassium sulphate ($12\,H_2O$), saturated solution	50 ml
(2)	Crystal violet	12 g
	Ethanol	100 ml

Mix 10 parts of (1) with 1 part of (2) and stand for 3 days before use.

Use a young culture and dilute it with water until it is barely turbid. Place a small drop on a very clean, flamed slide. Allow it to spread and dry in air. Stain for 15 min, wash and dry.

Diene's stain for mycoplasmas

Methylene blue	2.5 g
Azure II	1.25 g
Sodium bicarbonate	0.15 g
Maltose	10 g
Benzoic acid	0.2 g

Dissolve in 100 ml of distilled water.

Identification procedures

There are three approaches to the identification of bacteria: 'traditional', 'kit' and 'automated'. As many of the traditional methods have now been replaced in most clinical laboratories by paper discs, strips and batteries of tests in kits, these will be considered first. Automated methods for identification are described in Chapter 8.

Controls

As indicated in Chapter 2, quality control is important. Stock cultures of known organisms should therefore be included in test procedures. Suitable organisms, with their NCTC and ATCC catalogue numbers, are listed in Table 7.1. They may be stored and resuscitated for use as described on p.100.

Kits for bacteriology

These methods offer certain advantages, e.g. the saving of time and labour but before an arbitrary choice is made between conventional tests and the newer methods, and between the several products available, the following factors should be considered.

(1) There are no 'Universal Kits'. Some organisms cannot be identified by these methods.
(2) Caution is required in interpretation. A kit result may suggest unlikely organisms, e.g. *Yersinia pestis, Brucella melitensis, Pseudomonas mallei* and ill-considered reporting of these could cause havoc.
(3) Conventional, paper strip and disc methods enable the user to make his own choice of tests. Kit methods do not; they give 10–20 test results whether the user needs them or not.
(4) Identification of organisms by conventional, paper strip or disc methods, usually requires the judgement of the user. With kit methods the results are often interpreted by numeric charts or computers.
(5) If final identification depends on serology then a few screening tests are usually adequate. If identification depends on biochemical tests then the more tests used the more reliable will be the results. Kits may then be the methods of choice, especially with unusual organisms and in epidemiological investigations when the kit manufacturers' services are particularly useful.
(6) Some techniques may be more hazardous than others. Those that involve pipetting increase the risk of dispersing aerosols and of environmental contamination. Those employing syringe and needle work increase the risk of self-inoculation.
(7) Some methods are much more expensive per identification than others.
(8) Kits devised for identifying bacteria of medical importance may not give the correct answers if they are used to identify organisms of industrial significance.

It follows that choices should be made according to the nature of the investigations, the professional knowledge and skill of the workers and the size of the

Table 7.1 Bacterial strains for use in control tests in identification procedures

Strain	*NCTC*	*ATCC*
Acinetobacter		
calcoaceticus	7 844	15 308
lwoffi	5 866	15 309
Aeromonas hydrophila	8 049	7 966
Alkaligenes faecalis	415	19 018
Bacillus		
cereus	10 876	7 464
subtilis	6 633	10 400
Clostridium histolyticum	19 401	503
perfringens	13 124	8 237
Edwardsiella tarda	10 396	19 547
Enterobacter		
aerogenes	13 048	10 006
cloacae	10 005	—
Enterococcus faecalis	29 212	—
Escherichia coli	25 922	10 418
Mycobacterium		
fortuitum	10 349	6 841
kansasii	10 268	14 471
phlei	8 151	19 249
terrae	10 856	15 755
Nocardia		
braziliensis	11 274	19 296
otitidiscaviarum	1 934	14 629
Proteus		
mirabilis	10 975	—
rettgeri	7 475	—
Pseudomonas aeruginosa	27 853	10 662
Serratia marcescens	13 880	10 218
Staphylococcus		
epidermidis	12 228	—
aureus	25 923	6571
Streptococcus		
agalactiae	13 813	8 181
milleri	10 708	—
pneumoniae	6 303	—
salivarius	8 618	7 073

laboratory budget. The relative merits of the individual products mentioned below are not assessed here but assistance may be obtained from the survey of an Advisory Group (Bennett and Joynson, 1986) in which 10 commercial kits were tested with over 1000 bacterial strains. Kit manufacturers will supply other references on request.

A summary of kits known to the authors and editors is given in Table 7.2.

Table 7.2 Some identification kits, strips and discs[a]

Test	*Kit*	*Manufacturer*[b]
Aflatoxin	Aflatext M	15
Anaerobes	API	3
	Minitek	1
Bacillus cereus	BCET RPLA	12
Bacillary diarrhoea	TEKRA BDE	3
Bacitracin discs	—	12
β-lactamase	DrySlide	5
Borrelia (Lyme disease)	Lyme	20
Campylobacter	API	3
	Microscreen	18
Candida albicans	Rapidex albicans	3
	—	19
Clostridium		
perfringens	CET RPLA	12
difficile	CD Tox	13
	Toxin A	10
	—	19
	Microscreen	18
Cholera	VET RPLA	12
Cryptococcus	Slidex Crypto	3
Coryneforms	API	3
Diarrhoeal enterotoxin	TECRA BDE	3
E. coli 0157	Microscreen	18
Enterobacteria	API	3
	Enterotubes	16
	Enteroset	6
	Micro ID	7
Enterococci	API	3
	—	20
Faecal pathogens	API	3
Gram reaction	Bactident aminopeptidase	8
Haemophilus	API	3
	Slidex Meningite	3
	—	19
Helicobacter	API	3
	—	13
Indole	—	9
Legionella	Microcheck	4
Listeria	API	3
	—	19
Meningococci	Slidex Meningite	3
Moraxella/*M. catarrhalis*	—	19
Non-enteric Gram negatives	API	3
ONPG discs	—	12
Optochin discs	—	12
Oxidase	DrySlide	5
PPA	—	9, 12
Shigella	Bactigen	2
Salmonella	API	3
	Rapid test	12
	Salmonella TEK	11
	TECRA Salmonella	3
	—	19
	Microscreen	18

Table 7.2—continued

Test	*Kit*	*Manufacturer*[b]
Staphylococci	API	3
	Readychek Staph	13
	Minitek	1
	Slidex Staph	3
	Staphylase	12
	Microscreen	18
Streptococci	API	3
	Slidex Meningite	3
	Slidex Strept B	3
	Strep grouping	12, 14
	—	19
Toxins, various	—	12, 14
Treponema	Reagin Card	12
	RPR Card	17
	—	13
Urease	—	9
Urinary pathogens	API	3
Yeasts	API	3
	Minitek	1
X and V factor discs	—	12

[a] This list is not inclusive. There are other products. Some manufacturers, e.g. Biomerieux (API), market more than one kit for any one group of organisms. As it is not possible to detail them all here the manufacturers' catalogues should be consulted. Advertisements carrying the addresses of these companies are to be found in various technical and scientific journals.

[b] *Key to manufacturers:* 1, Becton Dickinson; 2, BioConnections; 3, BioMerieux; 4, Boots; 5, Difco; 6, Fisher; 7, General Diagnostics; 8, Merck; 9, Medical Wire; 10, Meridian; 11, Organon; 12, Oxoid (Unipath); 13, Porton Cambridge; 14, Prolab; 15, Rhône Poulenc; 16, Roche; 17, Seradyn; 18, Mercia; 19, Lab M; 20, Daro.

Conventional tests

Aesculin hydrolysis

Inoculate aesculin medium or Edwards' medium and incubate overnight. Organisms that hydrolyse aesculin blacken the medium.

Controls: Positive, *S. marcescens*; negative. *E. tarda*.

Ammonia test

Incubate culture in nutrient or peptone broth for 5 days. Wet a small piece of filter-paper with Nessler reagent and place it in the upper part of the culture tube. Warm the tube in a water-bath at 50–60°C. The filter-paper turns brown or black if ammonia is present.

Arginine hydrolysis

Incubate the culture in arginine broth for 24–48 h and add a few drops of Nessler reagent. A brown colour indicates hydrolysis. (But for lactobacilli, see p.72)

Controls: Positive, *E. cloacae*; negative, *P. rettgeri*.

Carbohydrate fermentation and oxidation

Liquid, semisolid and solid media are described on p.74. They contain a fermentable carbohydrate, alcohol or glucoside and an indicator to show the production of acid. To demonstrate gas production, a small inverted tube (Durham's tube; gas tube) is placed in the fluid media. In solid media (stab tube p.96), gas production is obvious from the bubbles and disruption of the medium. Normally, gas formation is recorded only in the glucose tube and any that occurs in tubes of other carbohydrates results from the fermentation of glucose formed during the first part of the reaction.

For most bacteria, use peptone or broth-based media.

For *Lactobacillus* spp., use MRS base (p.83) without glucose or meat extract and adjust to pH 6.2–6.5. Add chlorphenol red indicator.

For *Bacillus* spp. and *Pseudomonas* spp., use ammonium salt 'sugar' media (p.75). These organisms produce ammonia from peptones and this may mask acid production.

For *Neisseria* spp. and other fastidious organisms use solid or semisolid media. Neisserias do not like liquid media. Enrich with Fildes' extract (5%) or with rabbit serum (10%) or use the GC serum-free medium of Flynn and Waitkins. Horse serum may give false results as it contains fermentable carbohydrates. Robinson's buffered serum sugar medium (p.75) overcomes this problem and is the medium of choice for *Corynebacterium* spp.

Treat all anaerobes as fastidious organisms. The indicator may be decolorized during incubation so add more after incubation. A sterile iron nail added to liquid media may improve the anaerobic conditions.

Tests may be carried out on solid medium containing indicator. Inoculate the medium heavily, spreading it all over the surface and place carbohydrate discs on the surface. Acid production is indicated by a change of colour in the medium around the disc. We do not recommend placing more than four discs on one plate.

Casein hydrolysis

Use skim milk agar (p.86) and observe clearing around colonies of casein-hydrolysing organisms. The milk should be dialysed so that acid production by lactose fermenters does not interfere. To detect false clearing due to this, pour a 10% solution of mercuric chloride (*caution*) in 20% hydrochloric acid over this medium. If the cleared area disappears, casein was not hydrolysed.

Controls: Positive, *N. braziliensis*; negative, *N. otitidiscaviarum*.

Catalase test

1 Emulsify some of the culture in 0.5 ml of a 1% solution of Tween 80 in a screw-capped bottle. Add 0.5 ml of 20-vol hydrogen peroxide (*caution*) and replace the cap. Effervescence indicates the presence of catalase. Do not do this test on an uncovered slide as the effervescence creates aerosols. Cultures on low-carbohydrated medium give the most reliable results.
2 Add a mixture of equal volumes of 1% Tween 80 and 20-vol hydrogen peroxide to the growth on an agar slope. Observe effervescence after 5 min.
3 Test mycobacteria by Wayne's method (p.421).
4 Test minute colonies which grow on nutrient agar by the blue slide test.

Place one drop of a mixture of equal parts of methylene blue stain and 20-vol hydrogen peroxide on a slide. Place a cover-slip over the colonies to be tested and press down firmly to make an impression smear. Remove the cover-slip and place it on the methylene blue-peroxide mixture. Clear bubbles, appearing within 30 s, indicate catalase activity.

Controls: Positive, *S. epidermidis*; negative, *E. faecalis*.

Citrate utilization

Inoculate solid medium (Simmons) or fluid medium (Koser) with a straight wire. Heavy inocula may give false-positive results. Incubate at 30–35°C and observe growth.

Controls: Positive, *P. rettgeri*; negative, *S. epidermidis*.

Coagulase test

Staphylococcus aureus and a few other organisms coagulate plasma (see p.354).

Controls: Positive, *S. aureus*; negative, *S. epidermidis*.

Decarboxylase tests

Falkow's method can be used for most Gram-negative rods but Moeller's method gives the best results with *Klebsiella* spp. and *Enterobacter* spp. Commercial Falkow media allow only lysine decarboxylase tests but commercial Moeller media permit lysine, ornithine and arginine tests to be used.

Inoculate Falkow medium (p.76) and incubate for 24 h at 37°C. The indicator (bromocresol purple) changes from blue to yellow due to fermentation of glucose. If it remains yellow, the test is negative; if it then changes to purple, the test is positive.

When using Moeller medium (p.76), include a control tube that contains no amino acid. After inoculation, seal with liquid paraffin to ensure anaerobic conditions and incubate at 37°C for 3–5 days. There are two indicators, bromothymol blue and cresol red. The colour changes to yellow if glucose is fermented. Decarboxylation is indicated by a purple colour. The control tube should remain yellow.

Controls, arginine: Positive, *E. cloacae*; negative, *E. aerogenes*.
lysine: Positive, *S. marcescens*; negative, *P. rettgeri*.
ornithine: Positive, *S. marcescens*; negative, *P. rettgeri*.

DNase test

Streak or spot the organisms heavily on DNase agar. Incubate overnight and flood the plate with 1 N hydrochloric acid, which precipitates unchanged nucleic acid. A clear halo around the inoculum indicates a positive reaction.

Controls: Positive, *S. marcescens*; negative, *E. aerogenes*.

Eijkman test

Inoculate suspected *E. coli* into brilliant green broth with a gas tube. Incubate at 44 ± 0.2°C for 24 h. *E. coli* is one of the few organisms that produce gas at this temperature.

Controls: Positive, *E. coli*; negative, *E. cloacae*.

Gelatin liquefaction

1 Inoculate nutrient gelatin stabs, incubate at room temperature for 7 days and observe digestion. For organisms that grow only at temperatures when gelatin is fluid, include an uninoculated control and after incubation place both tubes in a refrigerator overnight. The control tube should solidify. This method is not entirely reliable.
2 Inoculate nutrient broth. Place a denatured gelatin charcoal disc in the medium and incubate. Gelatin liquefaction is indicated by the release of charcoal granules, which fall to the bottom of the tube.
3 For a rapid test, use 1 ml of calcium chloride 0.01 mol/l in saline. Inoculate heavily (whole growth from slope or petri dish). Add a gelatin charcoal disc and place in a water-bath at 37°C. Examine at 15-min intervals for 3 h.
4 Inoculate gelatin agar medium. Incubate overnight at 37°C and then flood the plate with saturated ammonium sulphate solution. Haloes appear around colonies of organisms producing gelatinase.

Controls: Positive, *A. hydrophila*; negative, *E. coli*.

Gluconate oxidation

Inoculate gluconate broth (p. 79) and incubate for 48 h. Add an equal volume of Benedict's reagent and place in a boiling water-bath for 10 min. An orange or brown precipitate indicates gluconate oxidation.

Controls: Positive, *E. cloacae*; negative, *P. rettgeri*.

Hippurate hydrolysis

Inoculate hippurate broth, incubate overnight and add excess of 5% ferric chloride. A brown precipitate indicates hydrolysis.

Controls: Positive, *S. agalactiae*; negative, *S. salivarius*.

Hugh and Leifson (HL) test. Oxidation fermentation test

This is also known as the 'oxferm' test. Some organisms metabolize glucose oxidatively, i.e. oxygen is the ultimate hydrogen acceptor and culture must therefore be aerobic. Others ferment glucose, when the hydrogen acceptor is another substance: this is independent of oxygen and cultures may be aerobic or anaerobic.

Heat two tubes of medium (p. 80) in boiling water for 10 min to drive off oxygen, cool and inoculate; incubate one aerobically and the other either anaerobically or seal the surface of the medium with 2 cm of melted Vaseline or agar to give anaerobic conditions.

Oxidative metabolism: acid in aerobic tube only.

Fermentative metabolism: acid in both tubes.

For testing staphylococci and micrococci, use the Baird-Parker modification (p. 80).

Controls: Oxidation, *A. calcoaceticus*; fermentation, *E. coli;* no action, *A. faecalis*.

Hydrogen sulphide production

1 Inoculate a tube of nutrient broth. Place a strip of filter-paper impregnated with a lead acetate indicator in the top of the tube, holding it in place with a cotton wool plug. Incubate and examine for blackening of the paper.
2 Inoculate one of the iron or lead acetate media. Incubate and observe blackening.

TSI (Triple Sugar Iron Agar) and similar media do not give satisfactory results with sucrose-fermenting organisms. The indicator paper (lead acetate) is the most sensitive method and the ferrous chloride media the least sensitive method, but the latter is probably the method of choice for identifying salmonellas and for differentiation in other groups where the amount of hydrogen sulphide produced varies with the species.

Controls: Positive, *E. tarda*; negative, *P. rettgeri*.

Indole formation

Grow the organisms for 2–5 days in peptone or tryptone broth.

1 *Ehrlich's method* Dissolve 4 g of *p*-dimethylaminobenzaldehyde in a mixture of 80 ml of concentrated hydrochloric acid and 380 ml of ethanol (do not use industrial spirit; this gives a brown instead of a yellow solution). To the broth culture add a few drops of xylene and shake gently. Add a few drops of the reagent. A rose pink colour indicates indole.
2 *Kovac's method* Dissolve 5 g of *p*-dimethylaminobenzaldehyde in a mixture of 75 ml of amyl alcohol and 25 ml of concentrated sulphuric acid. Add a few drops of this reagent to the broth culture. A rose pink colour indicates indole.
3 *Spot test* This is described by Miller and Wright (1982). Dissolve 1 g of *p*-dimethylamino cinnamaldehyde (DMAC) (*caution*) in 100 ml of 10% hydrochloric acid. Moisten a filter paper with it and smear on a colony from blood or nutrient agar. A blue green colour is positive, pink negative. This test may be applied directly to colonies but is unreliable in the presence of carbohydrates, e.g. on MacConkey or CLED media.

Controls: Positive, *P. rettgeri*; negative, *S. marcescens*.

Lecithinase activity

1 Inoculate egg yolk agar (p.77) or Willis and Hobbs medium (p.87) and incubate for 3 days. Lecithinase-producing colonies are surrounded by zones of opacity.
2 Inoculate egg yolk salt broth and incubate for 3 days. Lecithinase producers make the broth opalescent. Some organisms (e.g. *B. cereus*) give a thick turbidity in 13 h.

Controls: Positive, *B. cereus*; negative, *B. subtilis*.

Levan production

Inoculate nutrient agar containing 5% of sucrose. Levan producers give large mucoid colonies after incubation for 24–48 h.

Controls: Positive, *S. salivarius*; negative, *S. milleri*.

Lipolytic activity

Inoculate tributyrin agar. Incubate for 48 h at 25–30°C. A clear zone develops around colonies of fat-splitting organisms. These media can be used for counting lipolytic bacteria in dairy products.

Controls: Positive, *S. epidermidis*; negative, *P. mirabilis*.

Malonate test

Inoculate one of the malonate broth media and incubate overnight. Growth and a deep blue colour indicate malonate utilization.

Controls: Positive, *E. cloacae*; negative, *P. rettgeri*.

Methyl red test

Inoculate glucose phosphate broth (p.80) and incubate for 5 days at 30°C. Add a few drops of methyl red solution, prepared by dissolving 0.1 g of the dye in 300 ml of ethanol and making up to 500 ml with distilled water. A red colour indicates that the pH has been reduced to 4.5 or less. A yellow colour indicates a negative reaction.

Controls: Positive, *P. rettgeri*; negative, *S. marcescens*.

Motility

Hanging drop method

Place a very small drop of liquid bacterial culture in the centre of a 16-mm square No. 1 cover-glass, with the aid of a small (2-mm) inoculating loop. Place a small drop of water at each corner of the cover-glass. Invert over the cover-glass a microslide with a central depression – a 'well-slide'. The cover-glass will adhere to the slide and when the slide is inverted the hanging drop is suspended in the well.

Instead of using 'well-slides' a ring of Vaseline may be made on a slide. This is supplied in collapsible tubes or squeezed from a hypodermic syringe.

Bring the edge of the hanging drop, or the air bubbles in the water seal into focus with the 16-mm lens before turning to the high-power dry objective to observe motility.

Bacterial motility must be distinguished from Brownian movement. There is usually little difficulty with actively motile organisms, but feebly motile bacteria may require prolonged observation of individual cells.

Careful examination of hanging drops may indicate whether a motile organism has polar flagellae – a darting zig-zag movement – or peritrichate flagellae – a less vigorous and more vibratory movement.

To examine anaerobic organisms for motility grow them in a suitable liquid medium. Touch the culture with a capillary tube 60–70-mm long and about 0.5–1-mm bore. Some culture will enter the tube. Seal both ends of the tube in the bunsen flame and mount on Plasticine on the microscope stage. Examine as for hanging drops.

Craigie tube method
Craigie tubes (p.86) can be used instead of hanging drop cultures. Inoculate the medium in the inner tube. Incubate and subculture daily from the outer tube. Only motile organisms can grow through the sloppy agar.

TTC method
Donovan's medium detects motility. This motility agar is semisolid and contains triphenyl tetrazolium chloride (TTC). Motility is shown by a diffuse red line in stab culture. Compartment petri dishes may be used.

Controls: motile, *S. marcescens*; non-motile, *A. lwoffi*

Nagler test

This tests for lecithinase activity (see p.113) but the word has come to mean the half antitoxin plate test for *Cl. perfringens*. Egg yolk agar containing Fildes enrichment or Willis and Hobbs' medium are satisfactory.

Control: *C. perfringens* – no activity in presence of its antitoxin.

Nitratase test (nitrate reduction)

(1) Inoculate nitrate broth medium and incubate overnight.
(2) Make a dense suspension of the test organisms in sodium nitrate 0.01 mol/l in M/45 phosphate buffer of pH 7.0 and incubate at 37°C for 24 h
(3) Grow the organisms in a suitable broth. Add a few drops of 1% sodium nitrate solution and incubate for 4 h.

Acidify with a few drops of 1 N hydrochloric acid and add 0.5 ml each of a 0.2% solution of sulphanilamide and 0.1% *N*-naphthylethylenediamine hydrochloride (*caution*) (these two reagents should be kept in a refrigerator and freshly prepared monthly). A pink colour denotes nitratase activity. But as somc organisms further reduce nitrite, if no colour is produced add a very small amount of zinc dust. Any nitrate present will be reduced to nitrite and produce a pink colour, i.e. a pink colour in this part of the test indicates no nitratase activity and no colour indicates that nitrates have been completely reduced.

Controls: Positive, *S. marcescens*; negative *A. lwoffii*.

ONPG test

Lactose is fermented only when β-galactosidase and permease are present. Deficiency of the latter gives late fermentation. True non-lactose fermenters do not possess β-galactosidase.

Inoculate ONPG broth (p.84) and incubate overnight. If β-galactosidase is present a yellow colour due to *o*-nitrophenol is formed.

Controls: Positive, *S. marcescens*; negative, *P. rettgeri*.

Optochin test

Pneumococci are sensitive but streptococci are resistant to optochin (ethylhydrocupreine hydrochloride).

Streak the organisms on blood agar and place an optochin disc on the surface. Incubate overnight and examine the zone around the disc.

Controls: Positive, *S. pneumoniae*; negative *S. milleri*.

Oxidase test (cytochrome oxidase test)

(1) Soak small pieces of filter-paper in 1% aqueous tetramethyl-*p*-phenylenediamine dihydrochloride or oxalate (which keeps better). Some filter-papers give a blue colour and these must not be used. Dry or use wet. Scrape some of a fresh young culture with a clean platinum wire or a glass rod (dirty or Nichrome wire gives false positives) and rub on the filter-paper. A blue colour within 10 s is a positive oxidase test. Old cultures are unreliable. Tellurite inhibits oxidase as do fermentable carbohydrates. Organisms which have produced acid from a carbohydrate should be subcultured to a sugar-free medium.

(2) Incubate cultures on nutrient agar slopes for 24–48 h at the optimum temperature for the strain concerned. Add a few drops each of freshly prepared 1% aqueous *p*-aminodimethylaniline oxalate and 1% α-naphthol in ethanol. Allow the mixture to run over the growth. A deep blue colour is a positive reaction.

Controls: Positive, *P. aeruginosa*; negative *A. lwoffii*.

Oxidative or fermentative metabolism of glucose

See Hugh and Leifson test above.

Phenylalanine test (PPA or PPD test)

Inoculate phenylalanine agar and incubate overnight. Pour a few drops of 10% ferric chloride solution over the growth. A green colour indicates deamination of phenylalanine to phenylpyruvic acid. Among the enterobacteria only proteus and providencia strains have this property.

Controls: Positive, *P. rettgeri*; negative, *E. cloacae*.

Phosphatase test

Some bacteria, e.g. *S. aureus*, can split ester phosphates. Inoculate a plate of phenolphthalein phosphate agar (p. 84) and incubate overnight. Expose the culture to ammonia vapour (*caution*). Colonies of phosphatase producers turn pink.

Controls: Positive, *S. aureus*; negative, *S. epidermidis*.

Proteolysis

Inoculate cooked meat medium (p.85) and incubate for 7–10 days. Proteolysis is indicated by blackening of the meat or digestion (volume diminishes). Tyrosine crystals may appear.

Controls; Positive, *C. histolyticum*; negative, *C. perfringens*.

Starch hydrolysis

Inoculate starch agar (p.86). Incubate for 3–5 days then flood with dilute iodine solution. Hydrolysis is indicated by clear zones around the growth. Unchanged starch gives a blue colour.

Controls: Positive, *B. subtilis*; negative, *E. coli*.

Sulphatase test

Some organisms (e.g. certain species of mycobacteria) can split ester sulphates. Grow the organisms in media (Middlebrook 7H9 for mycobacteria) containing potassium phenolphthalein disulphate 0.001 mol/l for 14 days. Add a few drops of ammonia solution. A pink colour indicates the presence of free phenolphthalein.

Controls: Positive, *M. fortuitum*; negative, *M. phlei* (3 day test only).

Tellurite reduction

Some mycobacteria reduce tellurite to tellurium metal (p.421).

Controls: Positive, *M. fortuitum*; negative, *M. terrae*.

Tween hydrolysis

Some mycobacteria can hydrolyse Tween (polysorbate) 80, releasing fatty acids that change the colour of an indicator. This test is used mostly with mycobacteria (p.421).

Controls: Positive, *M. kansasii*; negative, *M. fortuitum*.

Tyrosine decomposition

Inoculate parallel streaks on a plate of tyrosine agar (p.87) and incubate for 3–4 weeks. Examine periodically under a low-power microscope for the disappearance of crystals around the bacterial growth.

Controls: Positive, *N. braziliensis*; negative, *N. otitidiscaviarum*.

Urease test

Inoculate one of the urea media heavily and incubate for 3–12 h. The fluid media give more rapid results if incubated in a water-bath. If urease is present, the urea is split to form ammonia, which changes the colour of the indicator from yellow to pink.

Controls: Positive, *P. rettgeri*; negative, *S. marcescens*.

Voges Proskauer reaction (VP test)

This tests for the formation of acetyl methyl carbinol (acetoin) from glucose.

This is oxidized by the reagent to diacetyl, which produces a red colour with guanidine residues in the media.

Inoculate glucose phosphate broth (p.80) and incubate for 5 days at 30°C. A very heavy inoculum and 6 h may suffice. Test by one of the following methods.

(1) Add 3 ml of 5% alcoholic α-naphthol solution and 3 ml of 40% potassium hydroxide solution (Barritt's method).
(2) Add a 'knife point' of creatinine and 5 ml of 40% potassium hydroxide solution (O'Meara's method).
(3) Add 5 ml of a mixture of 1 g of copper sulphate (blue) dissolved in 40 ml of saturated sodium hydroxide solution plus 960 ml of 10% potassium hydroxide solution (APHA method).

A bright pink or eosin red colour appearing in 5 min is a positive reaction. For testing *Bacillus* spp., add 1% sodium chloride to the medium.

Controls: Positive, *S. marcescens*; negative, *P. rettgeri*.

Xanthine decomposition

Inoculate parallel streaks on a plate of xanthine agar (p.87) and incubate for 3–4 weeks. Examine periodically under a low-power microscope for the disappearance of xanthine crystals around the bacterial growth.

Controls: *N. otitidiscaviarum*; negative, *N. braziliensis*.

Agglutination tests

These tests are performed in small test-tubes (75 × 9 mm). Dilutions are made in physiological saline with graduated pipettes controlled by a rubber teat or (for a single-row test) with pasteur pipettes marked with a grease pencil at approximately 0.5 ml. Automatic pipettes or pipettors are useful for doing large numbers of tests.

Standard antigen suspensions and agglutinating sera can be obtained commercially. Antigen suspensions are used mainly in the serological diagnosis of enteric fever, which may be caused by several related organisms, and of brucellosis.

Standard agglutinating sera are used to identify unknown organisms.

Testing unknown sera against standard H and O antigen suspensions

To test a single suspension

Prepare a 1:10 dilution of serum by adding 0.2 ml to 1.8 ml of saline. Set up a row of seven small tubes. Add 0.5 ml of saline to tubes 2–7 and 0.5 ml of 1:10 serum to tubes 1 and 2. Rinse the pipette by sucking in and blowing out saline several times. Mix the contents of tube 2 and transfer 0.5 ml to tube 3. Rinse the pipette. Continue with doubling dilutions but discard 0.5 ml from tube 6 instead of adding it to tube 7. Rinse the pipette between each dilution. The dilutions are now:

Tube no.	1	2	3	4	5	6	7
Dilution	1:10	1:20	1:40	1:80	1:160	1:320	0

Add 0.5 ml of standard suspension to each tube. The last tube, containing no serum, tests the stability of the suspension. The dilutions are now:

Tube no.	1	2	3	4	5	6	7
Dilution	1:20	1:40	1:80	1:160	1:320	1:640	0

To test with several suspensions
In some investigations, e.g. in enteric fevers, a number of suspensions will be used. Set up six large tubes (152 × 16 mm). To tube 1 add 9 ml of saline, to tubes 2–6 add 5 ml of saline and to tube 1 add 1 ml of serum. Mix and transfer 5 ml from tube 1 to tube 2, and continue in this way, rinsing the pipette between each dilution.

Set up a row of seven small tubes (75 × 9 mm) for each suspension to be tested. Transfer 0.5 ml from each large tube to the corresponding small tube. Work from right to left, i.e. weakest to strongest dilution to avoid unnecessary rinsing of the pipette. The final dilutions are now 1:20 to 1:640.

Incubation temperature and times
Incubate tests with *O* suspensions in a water-bath at 37°C for 4 h, then allow to stand overnight in refrigerator.

Incubate *H* tests for 2 h in a water-bath at 50–52°C (*not* 55–56°C, as the antibody may be partly destroyed). The level of water in the bath should be adjusted so that only about half of the liquid in the tubes is below the surface of the water. This encourages convection in the tubes, mixing the contents.

Brucella agglutinations
To avoid false negative results due to the prozone phenomenon, double the dilutions for at least three more tubes, i.e. use 1:20 to 1:5120.

Reading agglutinations
Examine each tube separately from right to left, i.e. beginning with the negative control. Wipe dry and use a hand-lens. If the tubes are scratched, dip them in xylene. The titre of the serum is that dilution in which agglutination is easily visible with a low-power magnifier.

Testing unknown organisms against known sera

Preparation of O suspensions
The organism must be in the smooth phase. Grow on agar slopes for 24 h. Wash off in phenol–saline and allow lumps to settle. Remove the suspension and dilute it so that there are approximately 1×10^9 bacteria/ml by the opacity tube method (p.151). Heat at 60°C for 1 h. If this antigen is to be stored, add 0.25% chloroform.

If a K antigen is suspected, heat the suspension at 100°C in a water-bath for 1 h (but the B antigen is thermostable).

Preparation of H suspensions
Check that the organism is motile and grow it in nutrient broth for 18 h, or in glucose broth for 4–6 h. Do not use glucose broth for the overnight culture as the bacteria grow too rapidly. Add formalin to give a final concentration of 1.0% and leave for 30 min to kill the organisms. Heat at 50–55°C for 30 min (this step may be omitted if the suspension is to be used at once). Dilute to approximately 1×10^9 organisms/ml by the opacity tube method.

Agglutination tests
Use the same technique as that described under 'Testing unknown sera . . .' but

as most sera are issued with a titre of at least 1:250 (and labelled accordingly) it is not necessary to dilute beyond 1:640. In practice, as standard sera are highly specific and an organism must be tested against several sera, it is usually convenient to screen by adding 1 drop of serum to about 1 ml of suspension in a 75 × 9 mm tube. Only those sera which give agglutination are then taken to titre.

Slide agglutination
This is the normal procedure for screening with *O* sera.

Place a loopful of saline on a slide and next to it a loopful of serum. With a straight wire, pick a colony and emulsify in the saline. If it is sticky or granular, or autoagglutinates, the test cannot be done. If the suspension is smooth, mix the serum in with the wire. If agglutination occurs it will be rapid and obvious. Dubious slide agglutination should be discounted.

Only *O* sera should be used. Growth on solid medium is used for slide agglutination and this is not optimum for the formation of flagella. False-negative results may be obtained with *H* sera unless there is fluid on the slope.

Suspensions of organisms in the R phase are agglutinated by 1:500 aqueous acriflavine solution.

Fluorescent antibody techniques

The principle of fluorescent antibody techniques is that proteins, including serum antibodies, may be labelled with fluorescent dyes by chemical combination without alteration or interference with the biological or immunological properties of the proteins. These proteins may then be seen in microscopical preparations by fluorescence microscopy.

The preparation is illuminated by ultraviolet or ultraviolet blue light. Any fluorescence emitted by the specimen passes through a barrier filter above the object. This filter transmits only the visible fluorescence emission. Microscopes suitable for this purpose are now readily available, and it is a simple matter to convert standard microscopes for fluorescence work. Fluorescent dyes are used instead of ordinary dyes as they are detectable in much smaller concentrations. They are available in a form which simplifies the conjugation procedure considerably. The fluorochromes in frequent current use are fluorescein isothiocyanate (FITC) and Lissamine rhodamine B (RB200). Of these, FITC is the most commonly used. It gives an apple-green fluorescence.

Fluorescence labelling is often used in microbiology and immunology along with or instead of traditional serological tests. Immune serum globulin conjugated with fluorochrome is usually employed to locate the corresponding antigen in microbiological investigations.

Reagents, accompanied by technical instructions, are available commercially.

References

Bennett, C. H. N. and Joynson, D. H. M. (1986) Kit systems for identifying Gram-negative aerobic bacilli: report of the Welsh Standing Specialist Advisory Working Party in Microbiology. *Journal of Clinical Pathology*, **39**, 666–671

Kodaka, H., Armfield, A. Y., Lombard, G. and Dowell, V. R. (1982) Practical procedure for demonstrating bacterial flagella. *Journal of Clinical Microbiology*, **16**, 948

Miller, J. M. and Wright, J. W. (1982) Spot indole test: evaluation of four methods. *Journal of Clinical Pathology*, **15**, 589–592

Chapter

8

Automated methods

In medical and industrial microbiology there is now a need for 'rapid methods' that will shorten the time between receipt of a specimen or sample and the issue of a report. This has led to the development of instrumentation, usually known as automation, which has the additional advantage that it does not require constant attention, can proceed overnight, and has therefore earned the title of 'walk-away'. Such automation may be described as the operation of an instrument by mechanical, electrical or electronic procedures alone or in combination, thereby removing the need for direct human action.

The equipment described here requires the minimum of instrumentation and is widely available. Clearly, not all such equipment can be included. Nor can details be given of use and operation; these may be found in the manufacturers' instruction manuals.

Identification and antimicrobial suceptibility

A pure culture is a prerequisite. Thereafter the user has a wide choice. Most systems incorporate antimicrobial agent testing options.

Vitek

Initially introduced as the Automated System in 1976 by BioMerieux (France), this is based on the detection of microbial growth in microwells within plastic cards. The original cards were designed for the identification of urine isolates and used the most probable number (MPN) system to determine significant bacteriuria. Additional cards became available for the identification of pure cultures and for antimicrobial susceptibility testing (several antibiotics per card). Results are available in 4–8 h and customized cards may be purchased.

Results from the identification cards are interpreted automatically, although some cards, including those for anaerobes, *Neisseria* spp., *Haemophilus* spp. and the Enteric Pathogens kit, require off-line incubation and manual entry into the Vitek computer.

For antimicrobial susceptibility testing over 40 antimicrobial agents are available and the results from each test include an interpolated minimum inhibitory concentration (MIC) as well as the National Committee of Clinical Laboratory Standards of susceptible, moderately susceptible, intermediate and resistant (NCCLS 1988).

The Vitek is an integrated modular system consisting of a filler-sealer unit, reader-incubator, computer and printer and may be interfaced with other laboratory information management systems (LIMS). The inoculum is prepared from a predetermined medium and after adjustment to either MacFarlane 1 or 2 standard the suspension is automatically transferred to the test card during the vacuum cycle of the filling module. The cards are then incubated in trays at 35°C. The optical density is monitored hourly; the first reading establishes a base line, after which the light reduction caused by growth or biochemical reaction is recorded.

Although the Vitek may be considered to be a comprehensive system several workers (see Stager and Davis, 1992) suggest that further improvement is needed for the identification of some *Staphylococcus* and *Streptococcus* spp. It was also suggested that it would be desirable to shorten the time taken for the identification of non-glucose fermenting Gram-negative bacilli, Gram-positive bacteria and yeasts.

API/ATB

The API system (BioMerieux) is widely used and while it is not as sophisticated as the Vitek, particularly with sample inoculation, it has been semi-automated as the ATB system with an accompanying change from 20 to 32 biochemical test options per strip. (Lapage *et al.* (1973) designed a computer program which showed that an average of 29–32 tests were required for reliable identification of aberrant strains of Enterobacteriaceae and non-fermenting Gram-negative rods and that additional tests did not improve the accuracy.)

The system consists of a densitometer, inoculator, reader and data-handler and allows a choice of rapid or overnight incubation, depending on the test type. Like the Vitek it has both identification and antimicrobial susceptibility testing options. A particularly useful output feature is the typicality index which provides the user with culture identification and indicates whether the isolate is typical or atypical, compared with the base profile for that species.

Sensititre

The ARIS (Automated Reading and Incubation System) is based on the Sensititre (Radiometer Instruments, UK) and provides walk-away automation and plate reading. It is a modular system which currently identifies Gram-negative bacteria only in either 5 or 18 h and provides comprehensive break-point or MIC results for Gram-negative and Gram-positive organisms against up to 54 antimicrobial agents.

The major difference between the Sensititre and other systems is that the former uses fluorescence-based technology. Substrates are linked to a fluorophore such as 4-methylumbelliferone (4MU) or 7-aminomethylcoumarin (7AMC). The substrate fluorophore is normally non-fluorescent but in the presence of specific enzymes the substrates are cleaved from the fluorophore, when the unbound 4MU or 7AMC fluoresces under UV light.

The test media contain specifically designed probes for a number of reactions. Carbohydrate utilization is detected by monitoring the pH shift with a fluorescent indicator. When the pH in the test well is alkaline the indicator is fluorescent but as the carbohydrate is oxidized or fermented, producing acid, the fluorescence decreases, indicating a positive reaction. Carbon-source utilization, decarboxylase and urease reactions also depend on monitoring pH changes with fluorescence indicators. Bacterial enzyme tests detect the production of various enzymes including peptidase, pyranosidase, phosphatase and glucuronidase. Enzymes are determined by their ability to cleave a quenched fluorophore, resulting in fluorescence. Aesculin, a fluorescent glycoside, breaks down to the non-fluorescent aescletin and glucose; non-fluorescence therefore indicates a positive result. Tryptophan deamination results in a coupled reaction in which the formation of a dark colour suppresses the fluorescent signal and thus indicates a positive result.

The real benefit of a fluorescence endpoint is that emission may be detected with greater sensitivity and a reaction can therefore be observed much earlier than is possible with colour changes or turbidity readings. Fluorescence also overcomes some of the problems associated with conventional approaches such as loss of resolution caused by pigmentation, scanty or variable growth, and the use of opaque or pigmented supplements.

The system is based on the standard 96-well microtitre tray and consists of an auto-inoculator, plate reader and data-handler. Inocula are prepared, with the aid of the built-in nephelometer, by transferring colonies from the isolation media to sterile distilled water to equal a 0.5 McFarlane standard. The inoculator dispenses 50 μl into each test well. The test plate is read after incubation for 5–16 h. The light source is a broad-band xenon lamp (360 m) that generates microsecond pulses of high-peak-power light. This passes through interference filters and a beam-splitting cube with wavelength-selective coatings to lenses that focus the light on a test well and the detector. The detector is a photomultiplier tube that transmits raw fluorescence data to the DEC PRO 380 computer. Approximately 30 s are required to read a plate and the biocode generated is matched to the Sensititre data base. There may be additional test prompts and a reincubate prompt also occurs at 5 h if there is no, or a low, probability of identification. The result of each test is printed and the quality of each identification is determined on the basis of calculated probability values.

Stager and Davis (1992) reviewed the performance reports of the Sensititre and stated that the limitations were the requirement for off-line incubation and the necessity for reincubation of some panels. The development of systems for the rapid identification of Gram-positive bacteria would extend its usefulness.

Biolog

This system uses 96-well microtitre plates and depends on the ability of the test organism to reduce tetrazolium violet, incorporated in the test cells, to the purple formazan. When a carbon source is not used the microwell remains colourless, as does the control well. The resulting pattern produces what is called a 'metabolic fingerprint' which is matched to the Biolog database.

The system consists of a manual 8-channel repeating pipettor, a turbidometer, a MicroPlate reader, a MicroLog Program Disk, and any DOS-based IBM-compatible PC, including XT and AT 286, 386 and 486 machines. The computer must have a hard drive of at least 20 Mbyte and one floppy drive. The operator can use the Biolog Gram-positive or Gram-negative databases or compile a user-defined file. An unknown biochemical profile may then be compared with either data bases or a combination of the two. Other features of the software include on-line information about any species in the library, cluster analysis programmes in the form of dendrograms and two- or three-dimensional plots to demonstrate the relatedness of strains or species, and the separation of the Gram-negative database into clinical and environmental sectors.

It is the last of these that gives some distinction to the Biolog as it identified a niche in the market that other systems had failed to fill and it includes non-clinical isolates as a feature of its initial database, thereby addressing the particular needs of environmental and research workers. This is best illustrated by considering the ES Microplate that is designed for characterizing and/or identifying different *Escherichia coli* and *Salmonella* strains, for characterizing mutant strains and for quality control tests on *E. coli* and *S. typhimurium* strains carrying recombinant plasmids.

Cobas Bact

The Cobas Bact was designed for the clinical laboratory and initially targeted at antimicrobial susceptibility testing and identification. It uses turbidity and pH indicator changes but is of particular interest in its mode of operation. A sample is taken from a discrete colony, diluted to a 1.0 MacFarlane standard and placed in the centre of a designated rotor. The rotor is then inserted in the instrument and taken up by the transport system into a cassette in which it is held during the incubation period. After incubation for 20 min the rotor reaches the first measuring station and then distributes a fixed volume of sample to each test compartment. The rotor continues to move through the incubator so that it reaches the measuring station every 20 min. All measurements are made at four different wavelengths (605, 564, 546 and 430 nm) and five replicate readings are taken with a xenon lamp. Samples may be introduced at any time so the system is suitable for same-day, walk-away or overnight testing. The operation is fully automatic and after completion of a 5-h cycle the rotor is ejected automatically into the waste compartment and the results are printed.

Geiss and Geiss (1992) showed that the system is very accurate for the identification of most clinically-important Gram-negative rods but had some shortcomings with non-fermenting bacteria.

Midi microbial identification system

The Midi (Microbial ID Inc., USA) is a fully automated, computerized gas chromatography system that can analyse more than 300 C9 to C20 fatty acid methyl esters. The value of this type of analysis is that the fatty acid composition is a stable genetic trait that is highly conserved within a taxonomic group.

The system software includes operational procedures, automatic peak naming, data storage and comparison of unknown profiles with the database, using pattern recognition algorithms. The database contains more than 100 000 strain profiles, including representatives of Enterobacteriaceae and *Pseudomonas, Staphylococcus* and *Bacillus* spp. It also includes mycobacteria, anaerobes and yeasts.

Representative isolates must be subcultured in defined media before preparation of the methyl ester, after which the whole process is automated. The system is calibrated by a mixture of straight-chain fatty acids. This is necessary after analysis of 10 samples and corrects for any changes in sample injection volume and variability in gas flow rates. Calibration samples can be set up automatically. The autosampler will take up to 2 days' analytical capacity and may therefore be considered as truly walk-away.

GC analysis is a well-established tool in the anaerobe laboratory and extensive work has been done in this field, notably at the Virginia Polytechnic Institute. McAllister *et al.* (1991) reported that the Midi system correctly identified 97% of the anaerobes tested.

Mastascan

The Mastascan Colour (Mast Laboratories, UK) was the first commercially-available instrument that read and automatically recorded results from multipoint-inoculated agar dilution antibiotic susceptibility tests as well as biochemical identification tests (see the review by Limb and Wheat, 1991).

The system consists of an image analysis scanner module, an 84 Mbyte PC with tape-streamer, colour monitor and printer. The principle is that the colour video camera measures reflected light levels from bacterial growth on agar media containing defined levels of antibiotic. The data are then digitally compared with the appropriate controls and the results interpreted on the basis of user-defined parameters. The camera can also analyse colour changes in the growth media and may thus be used for biochemical identification tests. Results generated by the camera can be manually overridden. Data from both the susceptibility tests and identification programmes can be combined for ease of reporting.

Other instruments

The systems decribed below are included because of their widespread use in Europe or as examples of particular modes of operation. The list is not exhaustive and could have included, for example, the Sceptor system (Becton Dickinson, USA) and the WalkAway instruments (Baxter Diagnostics, USA). The latter systems incubate Microtitre panels and automatically interpret biochemical or susceptibility results with either a photometric or fluorogenic reader.

Stager and Davis (1992) reviewed five studies in which the accuracy of either two or three of the systems were compared: percentage accuracy varied from as low as 35% to as high as 99.2%. This type of study is of somewhat limited value, however, as the companies are continually improving their systems, particularly with respect to the quality of the databases, which are, of course, fundamental to correct identification. This is best seen in the studies of Kelly *et al.* (1984), Stevens *et al.* (1984) and Truant *et al.* (1989), who all presented evidence that variation in biotypes of individual species from different geographical areas may in part be responsible for performance variability. This needs to be considered both by the manufacturers and the users, especially by the latter when they are considering a purchase.

Blood culture instrumentation

Bactec

The Bactec series of instruments (Becton Dickinson, USA) are generally associated with blood cultures but have other uses. Samples of blood, spinal, synovial, pleural and other normally sterile body fluids are injected into the Bactec vials. Growth of organisms is detected by the production of CO_2 from culture media.

Bactec 460

This early Bactec system used the makers' culture media containing ^{14}C-labelled substrate. The instrument takes 60 vials which are inoculated with

5 ml of the test fluid. It then automatically measures the radioactivity in the head space gas for CO_2 each hour and prints the result. If the activity exceeds a preset threshold the blood culture is considered to be positive.

Bactec NR-660, NR730 and NR-860

The NR- series replaced the radioactive substrate by employing infrared (IR) spectrometry to detect the CO_2. Infrared light is absorbed by the CO_2 in the test cell and the amount passing through to the detector is registered. The amplifier converts this measurement of conductivity to voltage which is converted to a read-out value. There is an inverse relation between the amount of CO_2 generated and the amount of IR light detected. This is accounted for in the calculations that generate the growth value (GV).

During the test a pair of needles penetrates the vial septum and the headspace gas is drawn through one of them into the detector. The other needle is connected to an external gas cylinder so that the appropriate gas for aerobiosis or anaerobiosis can be replaced. The various models in the NR-series use different IR systems. In the NR-860 a positive culture is flagged if the GV, or the difference between two consecutive measurements, exceeds a predetermined threshold. The vial headspace is also measured and when that exceeds a default setting the sample is flagged as potentially positive. Vial testing is automatic and an automated tray transport mechanism shuttles the trays between the incubator and the IR sensor unit. A bar code scanner reads the sample label. The NR-860 can store 480 samples in eight drawers. The two bottom drawers, containing the most recent aerobic cultures, are mounted on an orbital shaker. The working capacity is 48 new cultures per day with a 5-day test protocol. When positive samples are identified they are removed for further investigation.

Bactec 9240

This latest instrument is a fluorescence-based walk-away detection system that reputedly offers increased sensitivity and, as it is non-invasive, it offers a high level of built-in microbiological safety. As no venting is required the samples may be collected in Vacutainer blood collecting sets. A dye in the sensor reacts with CO_2, thereby modulating the amount of light that is absorbed by a fluorescent material in the sensor. The photodetectors measure the fluorescence, which is related to the amount of CO_2 released in the culture. Upon initiation of a test the machine carries out a diagnostic routine after which it commences continuous and automated on-line testing. A test cycle of all racks is completed every 10 min and positive cultures are flagged by a light on the front of the instrument and displayed on the monitor.

The Bactec 9240 can monitor 240 culture vials, arranged in six racks. The vials are incubated at 35°C with agitation. The working capacity is 24 new culture sets per day with a 5-day test protocol.

Bactec 460/TB for mycobacteria

The Bactec 460 instrument was modified to the 460/TB for the detection and susceptibility testing of mycobacteria. In this the cover was replaced by a hood with a forced, recirculating air supply and HEPA filters to avoid dispersion of any aerosols into the laboratory atmosphere. This instrument, subsequently modified to detect CO_2 by infrared spectrometry, instead of radioactivity, has proved very useful in the early detection of tubercle bacilli in sputum cultures and of antimicrobial resistance of mycobacteria.

BacT/Alert

The BacT/Alert system (Organon Teknika, USA) was introduced in 1990/91 as a fully automated system and therefore as an alternative to semi-automated radiometric and infrared systems. Like the Bactec it utilizes the production of CO_2 and employs a novel colorimetric sensor in the base of the aerobic and anaerobic blood culture bottles. The sensor changes colour as CO_2 is produced; the rate of change is detected by a reflectometer and the data are passed to a computer whose algorithm can distinguish between the constant CO_2 production by blood cells and the accelerating CO_2 production from a positive blood culture. The sensor is covered by an ion exclusion membrane which is permeable to CO_2 but not to free hydrogen ions, media components or whole blood. Indicator molecules in the water-impregnated sensor are dark green in their alkaline state, changing progressively to yellow as the pH decreases. CO_2 from a positive culture passes through the membrane, giving the following reaction:

$$CO_2 + H_2O \longrightarrow HCO_3^- + H_2$$

The free hydrogen ions react with the indicator molecules causing the sensor to change colour. This is detected by a red light-emitting diode and measured by a solid-state detector.

Aerobic and anaerobic culture bottles are available and are maintained at negative pressure for ease of inoculation. The recommended adult sample is 5–10 ml per bottle. It is necessary to vent the aerobic bottles occasionally. Cultures taken outside normal laboratory hours may be incubated externally and entered into the system on the following day. The system holds up to 240 bottles, each with its own detector. Up to four data-detection units may be interfaced with the same computer, giving a maximum throughput for one system of 960 bottles at any one time. A single detection unit has 10 blocks of 24 wells, each block operating independently, thereby allowing easy removal and routine servicing. The only maintenance required is cleaning the back-up tapes fortnightly and quality controlling each cell once every 30 days as prompted by the system. Organon Teknika supplies a Reflectance Standard Kit with each system which allows the user to calibrate it and test for quality control. The IBM-compatible system analyses all data, interpreting both positive and negative growth curves. It also allows complete patient files to be kept and will interface with existing in-house systems. The system contains a 40 Mbyte hard disk which can store the results from up to 50 000 cultures. A back-up is made automatically three times a day to ensure that there is no loss of data should the computer fail.

The system has been evaluated against the radiometric Bactec 460 and shown to give comparable results (Thorpe *et al.* 1990). Wilson *et al.* (1992) also concluded that the BacT/Alert and Bactec 660/730 non-radiometric systems were comparable for recovering clinically significant microorganisms from adult patients with bacteraemia or fungaemia. The BacT/Alert, however, detected microbial growth earlier than the Bactec and gave significantly fewer false-positive results. As yet, there do not appear to be any reports of comparisons between the BacT/Alert and the Bactec 9240.

BioArgos

BioArgos (Sanofi Diagnostics Pasteur, France) is another fully-automated blood culture system that detects CO_2 by infrared spectroscopy through a glass bottle. It is run from an IBM PS/2 computer and is 'hands-off' in that no gas is injected into the vials during processing. It can handle a maximum of 720 samples in its incubator. Courcol *et al.* (1992) compared BioArgos with the Bactec NR-60 and observed no significant differences between the two systems. Anaerobes were detected earlier with BioArgos but the detection of some organisms that had a strict requirement for oxygen was slightly delayed.

Sentinel

Sentinel (Difco, USA and UK) is a novel system. It consists of a base plinth on which there are 1–5 drawers each holding up to 80 blood culture bottles. Each bottle contains two electrodes, one gold plated, the other of an aluminium alloy. These are concealed and when the culture bottle is placed in the incubator drawer they pierce a thin membrane and enter the culture medium. The bottle then behaves as a simple battery and microbial growth is detected by measuring the relative changes in voltage generated by the aluminium anode which slowly dissolves, liberating aluminium ions. The resulting electrons are transferred to the gold cathode and, in turn, to a reducible chemical species in the culture medium. In the aerobic bottle the major electron acceptor is dissolved oxygen but in the anaerobic bottle the acceptors are compounds included in the culture medium. Microbial growth results in fall in voltage between the two electrodes. This is detected electronically and an algorithm function in the computer software determines when the fall in voltage is sufficient to record a positive culture. Voltage changes are monitored every 15 s and transmitted to the computer every 15 min.

Stevens *et al.* (1992) compared the performance of Sentinel with the Bactec 460, NR-660 and NR-730 in a multicentre evaluation commissioned by the UK health departments. Blood culture sets (2180) consisting of aerobic and anaerobic yielded 218 (10%) clinically important isolates of which both systems detected 155 (71%). Sentinel alone detected 35 (16%) and Bactec alone 28 (13%). False positive results were 130 (3%) for Sentinel and 67 (1.5%) for Bactec. The results were passed to the manufacturers so that the algorithms could be redesigned to detect positives at a minimum of 2.5 h. It was considered that Sentinel was easy to use and that the menu-based software was user-friendly. The report mentioned that the 24 V electricity supply to the incubator drawers is a good safety feature but that the venting needle in the aerobic bottle is a potential hazard.

Urine examination

Questor

Questor (Difco, USA) reports counts of microorganisms as well as of red, white and epithelial cells. Eight sample trays are loaded in a stacker which holds up to 112 samples. Patient details are entered into the computer and the process is then fully automated. Four ml of saline are added to the urine in each well and the mixture is drawn through the counting orifice into which a movable, cone-shaped probe is gradually inserted. During the 1-min counting cycle the probe moves inwards, narrowing the aperture to an annulus. The voltage across this is monitored, allowing separate monitoring of the variously sized particles cells. Up to 50 samples per hour can be processed.

Blockages have caused problems with earlier instruments of this type, but Difco claims that the flow effects at the probe tip divert large cells away when the orifice is closing, obviating blockages.

Impedance instrumentation

Three systems are generally available: Bactometer (bioMerieux, France), Malthus (Radiometer, Denmark), RABIT (Don Whitley, UK). A fourth, Bactrac (Sy-Lab, Austria) is a recent addition. All operate on the same principle.

Impedance may be defined as resistance to flow of an alternating current as it passes through a conductor (see Eden and Eden (1984) and Kell and Davey (1990) for detailed theory). When two metal electrodes are immersed in a conducting fluid the system behaves either as a resistor and capacitator in series or as a conductor and capicitator in parallel (Kell and Davey, 1990). When it is a series combination the application of an alternating sinusoidal potential will produce a current which is dependent on the impedance, Z, of the system which, in turn, is a function of its resistance, R, capacitance, C, and applied frequency, F; thus

$$Z = \sqrt{R^2 + (\tfrac{1}{2}\pi FC)^2}$$

Any increase in conductance, defined here as the reciprocal of resistance, or capacitance results in a decrease in impedance and in increase in current. The AC equivalent of conductance is admittance, defined as the reciprocal and the units of measurement are Siemens (S).

Microbial metabolism usually results in an increase in both conductance and capacitance, causing a decrease in impedance and a consequent increase in admittance. The concepts of impedance, conductance, capacitance and resistance are only different ways of monitoring the test system and are interrelated. An important factor, however, is that all are dependent upon the frequency of the alternating current and all are complex quantities in that they contain both real and imaginary parts (Kell and Davey, 1990). In practical terms the user must be aware that the electrical response is frequency-dependent, has a conductive and capacitative component and is temperature-dependent. Temperature control in any impedance system is of critical importance. A temperature increase of 1°C will result in an average increase of 0.9% in capacitance and 1.8% in conductance (Eden and Eden, 1984) and from this it can be calculated that a less than a 100 millidegree temperature drift can result in a false detection. It is interesting to note that three of the four systems mentioned above have adopted a different method of temperature control: Bactometer, a hot-air oven;, Malthus, a water-bath; and RABIT a solid block heating system.

These systems provide a measurement of net changes in conductivity of the culture medium. Tests are monitored continually and when the rate of change in conductivity exceeds the user-defined pre-set criterion the system reports growth. The time required to reach the point of detection is the 'time to detection' (TTD) and is a function of the size of the initial microbial population, its growth kinetics and the properties of the culture medium. For a given protocol the TTD is inversely proportional to the initial microbial load of the sample. At the point of detection it is generally considered that there will be approximately 10^6 cfu/ml of the test organism present in the sample. This will vary according to the type of organism, growth media, etc., but will be constant for any one organism growing under defined test conditions.

The attributes of the three principal systems are outlined in Table 8.1 which shows that there are differences in the measuring modes. The Bactometer has the option of measuring either conductance or capacitance; the Malthus measures conductance; and the principal signal component in the RABIT is conductance, although there is a also a capacitance component. The benefits of capacitance are best exemplified in attempts to detect organisms in which normal metabolism results in no net increase in medium conductivity. This does not preclude the Malthus and RABIT from such applications, however, as both can utilize the indirect technique (Owens *et al.*, 1989) in which an electrical signal is generated by CO_2 production. The technique also has the advantage of being fully compatible with samples and test media that have a high salt content, which is not the case with the Bactometer. The indirect technique also allows the user a greater choice of culture media as these do not have to be optimized for electrical response but simply for growth of the target organism.

Table 8.1 Comparison of the three principal impedance systems/Bactometer, Malthus and RABIT

	Bactometer	*Malthus 2000*	*RABIT*
Test capacity	512	480	512
Modular system	Yes	No	Yes
Temperature options per system	8	1	16 max
Cell volume	2 ml max	2 and 5 ml	2–10 ml
Cell format	Block of 16	Single	Single
Reusable/disposable	Disposable	Reusable and disposable	Reusable
Media availability	Yes	Yes	Yes
Measuring mode	Conductance or capacitance	Conductance	Impedance
Indirect mode	No	Yes	Yes
Full access to cell throughout test period	No	No	Yes
Temperature control	Hot-air oven	Water-bath	Dry heating block

Fluorescence-based technology

The mechanism of bioluminescence assay, mediated by the luciferin/luciferase reaction, for adenosine triphosphate (ATP) has been reviewed by Stewart and Williams (1992). The luciferin/luciferase complex reacts specifically with the ATP in living cells and emits a light signal proportional to the amount of ATP present. There is a linear relationship between the ATP and the plate count (Stannard and Wood, 1983). The ATP in the sample originates from both living cells and other sources (the somatic ATP) and the two must be differentiated if the technique is to be used in microbiological testing (although it is not necessary for rapid hygiene monitoring, as in the food industry). The two commonly-used methods are separation of the organisms from the sample before extracting the ATP, and prior destruction of the somatic ATP.

Bactofoss

The Bactofoss (Foss Electric, Denmark) is a unique, fully automated instrument that combined the principles mentioned above to remove sample 'noise'. All functions are monitored by the central processing unit and messages are presented on the screen. The Bactofoss is a simple push-button instrument which provides a result within 5 min.

The test sample is taken up automatically and deposited in a temperature-controlled funnel where several filtrations and pretreatments remove somatic ATP. The microorganisms are left on a filter paper which is then positioned in an extraction chamber. Extraction reagent is added to extract microbial ATP, the amount of which is determined by measurement of luminescence after the addition of luciferin and luciferase. All reagents, etc. are supplied by the manufacturer and the instrument can perform 20 analyses per hour. Limond and Griffiths (1991) have shown that it is applicable to meat and milk samples, with detection sensitivities of $3 \times 10^4 - 3 \times 10^8$ cfu/g and $1 \times 10^4 - 1 \times 10^8$ respectively.

Other instruments

Several other instruments use bioluminescence, e.g. Sonco (Batley, UK), Biotrace (Bridgend, UK) and Amersham International (UK). The principles are all the same but the degree of automation varies.

Direct epifluorescence techniques

The direct epifluorescence technique (DEFT) was originally developed for counting bacteria in raw milk (Pettipher *et al.*, 1980, 1989). It takes less than 30 min and uses membrane filtration and epifluorescence microscopy. Pre-treatment of samples may be necessary to facilitate rapid filtration and distribution of bacteria on the membrane. A suitable fluorescent stain shows organisms that are easily distinguished from debris by an epifluorescence microscope which is linked to an automatic counting system.

DEFT has been used in manual or semi-automatic modes for several years but the new Cobra system (France) is fully automated.

DEFT is particularly useful for liquid samples such as milk and beverages but could be considered for screening urine in clinical laboratories.

Bactoscan

The Bactoscan (Foss Electric, Denmark) is fully automated, employs the DEFT principle but is dedicated to the quality controlling of milk by making direct bacterial counts.

A 2.5-ml sample of milk is treated with lysing fluid to dissolve protein and somatic cells. The bacteria are then separated from the milk by gradient centrifugation. The bacterial suspension is mixed with a protease enzyme to dissolve particulate protein and is then stained with acridine orange. The stained organisms are counted by a continuously-functioning epifluorescence microscope and the results are displayed on screen. Eighty samples per hour can be screened and the system is flushed with rinsing fluid between each sample. When testing is finished the machine is automatically cleaned.

The enzyme solution must be prepared daily but the other reagents need be made up only every other day. The start-up procedure takes about 30 min and involves a calibration routine, lens checking and setting the discrimination level of the counting module. Milk samples are placed in special racks, heated to 40°C and loaded on the instrument. The sample identification can be keyed in manually or entered by a bar code reader. Results are displayed on a screen. Bactoscan counts correlate well with standard plate counts and may therefore be converted to equivalent cfu/ml.

Flow cytometry

Flow cytometry, widely use in research, has now been developed for routine industrial use. The technique allows cell by cell analysis of the test samples and, coupled with fluorescent labels, provides a rapid and automated method for detecting microbial flora and examining its metabolic state. The sample is injected into a 'sheath' fluid which passes under the objective lens via a hydrodynamic focusing flow cell. The sheath fluid passes continuously through the flow cell, thereby focusing the sample stream into a narrow, linear flow. The sample then passes through a light beam which causes suitably-labelled cells to emit fluorescent pulses. Each pulse is detected and subsequent analysis allows pulses to be recognized as separate counts and graded in terms of fluorescence intensities.

Systems currently available include the ChemFlow AutoSystem (Chemunex, France) which can detect up to 400 cells/s at sample flow rates of approximately 0.4 ml/min. The results may be printed as hard copy or transferred to a computer. The labelling and counting takes less than 30 min per sample. The Epics Elite System (Coulter Electronics, UK) is currently used by a large water undertaking to screen for *Cryptosporidium*.

The ability of flow cytometry to analyse and sort cells into defined populations on the basis of cell size, density and discriminatory labelling is a powerful new microbiological tool.

Pyrolysis mass spectrometry

Pyrolysis mass spectrometry (PMS) has considerable potential for the identification, classification and typing of bacteria (Magee *et al.*, 1989; Freeman *et al.*, 1990; Sisson *et al.*, 1992). In the RAPyD-400 benchtop automated PMS (Horizon Instruments, UK) samples are spread on V-shaped Ni-Fe pyrolysis foils held in pyrolysis tubes. Samples are then heated by Curie-point techniques in a vacuum which causes pyrolysis in a controlled and reproducible manner. The gas produced is passed through a molecular beam and analysed by a rapid scanning quadruple mass spectrometer to produce a fingerprint of the original sample over a mass range of 12–400 Da. Data are analysed by onboard multivariate statistical routines, providing principal component analysis, discriminant function analysis, cluster analysis and factor spectra. The system is entirely automated, with sample loading, extraction, indexing to next sample and data collection controlled by a computer. Routine servicing and maintenance are minimal. Routine analysis takes approximately 90 s per sample with a batch of 300 samples.

It is important that the organisms are originally cultured on media that do not impose stress and consequent alteration of phenotype expression which may obscure the relatedness of isolates. It may be desirable to facilitate expression of certain phenotypes before analysis, thereby allowing the technique to differentiate between unrelated strains, e.g. differentiation between toxigenic and non-toxigenic strains (Sisson *et al.*, 1992). Experience in UK public health laboratories suggests that speed, low running costs and versatility of PMS makes it suitable for the initial screening in outbreaks of infection.

Further instrumentation

Although it is not possible to review here all automated instrumentation it is worthwhile mentioning some others that might be of interest.

The Micro-Oxymax (Columbus Instruments, USA) is a 'closed' multi-channel indirect calorimeter which functions as a computerized respirometer. Other microcalorimetry systems are available; the heat produced during analysis flows through thermoelectric couples resulting in an electrical current flow that is directly proportional to the heat produced. The signal may then be amplified and sent to a suitable recording device.

The Omnispec 4000 Monitor (Westcor, USA) uses reflectance colorimetry by which microbial activities are monitored by mediation of dye pigmentation changes (see Manninen and Fung, 1992). Results with food samples correlate well with standard plate counts. The Bioscreen C Microbiological Growth Analyzer (Labsystems, Finland), an automated optical density system, can produce up to 200 growth curves simultaneously. For the automation of conventional plate counting there are the Spiral Plating Systems (Don Whitley, UK), and the image analysis systems dedicated for counting both conventional and spiral plates of Protos (Don Whitley, UK), Seescan (Seescan, UK) and Domino (Perceptive Instruments, UK).

Although immunoassays are outside the scope of this chapter there is scope for their automation by microtitre plate technology (Dynatech, UK) combined with robotic sample processors (Quatro Biosystems, UK). Another example of a technique suitable for automation is that which uses enzyme-linked amperometric sensors (Brooks *et al.*, 1992).

Molecular microbiology

Advances in molecular microbiology have added to diagnostic techniques, mainly with DNA probes and the polymerase chain reaction (PCR).

DNA probes

A DNA probe is a piece of single-stranded DNA which can recognize and consequently hybridize with a complementary DNA sequence. The probe also carries a label which 'lights up' after hybridization. Probe specificity depends on whether it is made from total cellular DNA, short-chain oligonucleotides or by cloning DNA fragments which gives greater specificity. There are many assay formats, although all follow the same principle whereby the DNA double helix of the target organism is separated, usually by heating, and immobilized on a membrane. Immobilization prevents the complementary strands re-hybrizing and allows access for the probe.

After challenge with the probe the membrane is washed to remove any unbound probe and the resulting hybridization is visualized. The dot-blot assay immobilizes target DNA on a membrane and visualizes hybrization as a coloured spot. Colony-blot assays take up cells from a plate on to a sterile membrane, thereby forming a replica of that plate. The cells are then lysed to release the DNA which is immobilized and challenged by the probe. Liquid hybridization formats use a probe that carries a chemifluorescent label which is added directly to lysed cells. Non-hybridized probe is broken down by adding a proprietary reagent. The amount of fluorescence measured is directly proportional to the number of target organisms. Paddle and bead formats carry the probe on their solid surfaces. After hybrization the unbound probe is washed off and detection is visualized. Several commercial kits are available for this type of examination.

DNA fingerprinting

Otherwise known as restriction fragment length polymorphism (RFLP) analysis, this combines probe and restriction enzyme technologies. The assay format is generally known as Southern blotting. Restriction enzymes cut DNA at sites characterized by short sequences of certain bases, resulting in a number of fragments of different lengths. The number and sizes of the fragments vary considerably from one organism to another, even within a species. The fragments can be separated according to size and transferred directly to a membrane by electrophoretic techniques. The immobilized fragments are then probed, resulting in a DNA fingerprint that is characteristic of the source DNA.

PCR technology

The best established DNA amplification method is the polymerase chain reaction (PCR). This is based on the repetitive cycling of three simple reactions: denaturation of double-stranded DNA, annealing of single-stranded complementary oligonucleotides and extension of oligonucleotides to form a DNA copy. The conditions for these reactions vary only in incubation temperatures, all occurring in the same tube in a cascade manner. The repetitive cycling is therefore self-contained and can be automated in a programmable thermocycler. The first step is the heat denaturation of native DNA which melts as the hydrogen bonds of the double helix break. The single strands are then available to re-anneal with complementary sequence DNA. The second step takes place at a reduced temperature. Two short DNA primers are annealed to complementary sequences on opposite strands of a template, thereby flanking the region to be amplified. They act as a starting point for DNA polymerase and consequently define the DNA region to be amplified. In the third step synthesis of complementary new DNA occurs through the extension of each annealed primer by the action of *Taq* polymerase in the presence of excess deoxyribonuclease triphosphates. The new strand formed consists of the primer at its 5′ end trailed by a string of linked nucleotides that are complementary to those of the corresponding template. An essential feature of the amplification process is that all previously synthesized products in the previous cycle act as templates for the subsequent cycle, resulting in geometric amplification. Twenty cycles, which take as little as 2 h, result in a million-fold increase in the amount of target DNA which can be visualized by electrophoresis and staining with ethidium bromide, or by using a DNA probe.

Although this technique is currently being used much development work is still necessary. Many samples carry substances that inhibit *Taq* DNA polymerase and partial processing may therefore be necessary. Cross contamination can be a problem. Complete automation of this technology may not be far away and would almost certainly increase it attractiveness to the routine laboratory.

For further details of the principles of PCR, DNA probes and RFLP analysis see Grange *et al.* (1991).

References

Brooks, J. L., Mirhabibollahi, B. and Kroll, R. G. (1992) Experimental enzyme-linked amperometric immunosensors for the detection of salmonellas in foods. *Journal of Applied Bacteriology*, **73**, 189–196

Courcol, R. J., Duhamel, M., Decoster, A., Lemaire, V. M., Rastorgoueff, M. L., Ochin, D. and Martin, G. R. (1992) A fully automated blood culture system. *Journal of Clinical Microbiology*, **30**, 1995–1998

Eden, R. and Eden, G. (1984) *Impedance Microbiology*, Research Studies Press Ltd, Herts

Freeman, R., Goodfellow, M., Gould, F. K., Hudson, S. J. and Lightfoot, N. F. (1990) Pyrolysis mass spectrometry (Py-MS) for the rapid epidemiological typing of clinically significant bacterial pathogens. *Journal of Medical Microbiology*, **23**, 283–286

Geiss, H. K. and Geiss, M. (1992) Evaluation of a new commercial system for the identification of Enterobacteriaceae and non-fermentative bacteria. *European Journal of Clinical Microbiology*, **11**, 610–616

Grange, J. M., Fox, A. and Morgan, N. L. (eds) (1991) *Genetic Manipulation: Techniques and Applications*, Society for Applied Bacteriology Technical Series No. 28; Blackwell, Oxford

Kell, D. B. and Davey, C. L. (1990) Conductimetric and impedimetric devices. In *Biosensors, A Practical Approach* (ed. A. E. G. Cass), Oxford University Press, Oxford, pp. 125–154

Kelly, M. T., Matsen, J. M., Morello, J. A., Smith, P. B. and Tilton, R. C. (1984) Collaborative clinical evaluation of the Autobac IDX system for identification of gram-negative bacilli. *Journal of Clinical Microbiology*, **19**, 529–533

Lapage, S. P., Bascomb, S., Willcox, W. R. and Curtis, M. A. (1973) Identification of bacteria by computer: general aspects and perspectives. *Journal of General Microbiology*, **77**, 273–290

Limb, D. I. and Wheat, P. F. (1991) Evaluation of a commercial automated system for the identification of Gram-negative enteric bacilli. *European Journal of Clinical Microbiology and Infectious Diseases*, **10**, 749–752

Limond, A. and Griffiths, M. W. (1991) The use of the Bactofoss instrument to determine the microbial quality of raw milks and pasteurized products. *International Dairy Journal*, **1**, 167–182

Magee, J. T., Hindmarsh, J. M., Bennett, K. W., Duerden, B. I. and Aries, R. E. (1989) A pyrolysis mass spectrometry study of fusobacteria. *Journal of Medical Microbiology*, **28**, 227–236

Manninen, M. T. and Fung, D. Y. C. (1992) Estimation of microbial numbers from pure bacterial cultures and from minced beef samples by reflectance colorimetry with Omnispec 4000. *Journal of Rapid Methods and Automation in Microbiology*, **1**, 41–58

McAllister, J. M., Master, R. and Poupard, J. A. (1991) Comparison of the Microbial Identification System and the Rapid ANA II system for the identification of anaerobic bacteria. *Abstract 91st General Meeting of the American Society for Microbiology 1991*, American Society for Microbiology, Washington, DC.

NCCLS (1988) *Performance Standards for Antimicrobial Disc Sensitivity Tests*, Document M2-T4, Vol. 8 No. 7, 4th edn, National Committee for Clinical Laboratory Standards, Villanova, PA, USA.

Owens, J. D., Thompson, D. S. and Timmerman, A. W. (1989) Indirect conductimetry; a novel approach to the conductimetric enumeration of microbial populations. *Letters in Applied Microbiology*, **9**, 245–249

Pettipher, G. L., Mansell, R., McKinnon, C. H. and Cousins, C. M. (1980) Rapid membrane filtration epifluorescent microscopy technique for direct enumeration of bacteria in raw milk. *Applied and Environmental Microbiology*, **39**, 423–429

Pettipher, G. L., Kroll, R. G., Farr, L. J. and Betts, R. P. (1989) DEFT: Recent developments for food and beverages. In *Rapid Microbiological Methods for Foods, Beverages and Pharmaceuticals* (eds C. J. Stannard, S. B. Pettit and F. A. Skinner), Society for Applied Bacteriology Technical Series No. 25, Blackwell, Oxford, pp. 33–46

Sisson, P. R., Freeman, R., Magee, J. G. and Lightfoot, N. F. (1992) Rapid differentiation of *Mycobacterium xenopi* from mycobacteria of the *Mycobacterium avium-intracellulare* complex by pyrolysis mass spectrometry. *Journal of Clinical Pathology*, **45**, 355–370

Stager, C. E. and Davis, J. R. (1992) Automated systems for identification of microorganisms. *Clinical Microbiology Reviews*, **5**, 302–327

Stannard, C. J. and Wood, J. M. (1983) The rapid estimation of microbial contamination of raw beef meat by measurement of adenosine triphosphate (ATP). *Journal of Applied Bacteriology*, **55**, 429–438

Stevens, M., Feltham, R. K. A., Schneider, F., Grasmick, C., Schaak, F. and Roos, P. (1984) A collaborative evaluation of a rapid automated bacterial identification system: the Autobac IDX. *European Journal of Clinical Microbiology*, **3**, 419–423.

Stevens, M., Parel, H., Walters, A., Burch, K., Jay, A., Dowling, N., Mitchell, C. J., Swann, R. A., Willis, A. T., Shanson, D. C. and MacDonald, C. A. (1992) Comparison of Sentinel and Bactec blood culture systems. *Journal of Clinical Microbiology*, **45**, 815–818

Stewart, G. S. A. B. and Williams, P. (1992) *Lux* genes and the applications of bacterial bioluminescence. *Journal of General Microbiology*, **138**, 1289–1300

Thorpe, T. C., Wilson, M. L., Turner, J. E., DiGuiseppi, J. L., Willert, M., Mirrett, S. and Reller, L. B. (1990) BacT/Alert: an automated colorimetric microbial detection system. *Journal of Clinical Microbiology*, **28**, 1608–1612

Truant, A. L., Starr, E., Nevel, C. A., Tsolakis, M. and Fiss, E. F. (1989) Comparison of AMS-Vitek, MicroScan, and Autobac Series II for the identification of gram-negative bacilli. *Diagnostic Microbiology and Infectious Diseases*, **12**, 211–215

Wilson, M. L. Weinstein, M. P., Reimer, L. G., Mirrett, S. and Reller, L. B. (1992) Controlled comparison of the BacT/Alert and Bactec 660/730 Nonradiometric blood culture systems. *Journal of Clinical Microbiology*, **30**, 323–329

Chapter

9

Mycological methods

Direct examination

Arrange hair, skin or nail fragments on a slide in a drop of mounting fluid (see below). Apply a cover-glass and then leave for a few minutes till the preparation 'clears'. Wet films of pus may be made with an equal volume of this mountant or an equal volume of glycerol if a more permanent preparation is required. Mix cerebrospinal fluid deposit with an equal volume of 2.5% nigrosin in 50% glycerol. Glycerol prevents mould growth in the solution, delineates capsules more sharply and prevents preparations, which should be thin, from drying out.

Mounting fluid and stain

Dimethyl sulphoxide–potassium hydroxide (DMS/KOH)

This is used to 'clear' specimens so that mycelium may be more easily seen.

Add 40 ml of dimethyl sulphoxide (*caution*) to 60 ml of water and dissolve 40 g of potassium hydroxide in the mixture.

Lactophenol–cotton blue

Phenol, melted (*caution*)	20 g
Lactic acid	20 ml
Glycerol	40 ml
Water	20 ml

Warm the water and add the melted phenol, followed by the lactic acid and glycerol.

Add 0.05% cotton blue (methyl blue) to lactophenol.

Mount small portions of moulds from foods in water, or if the material is water-repellent, in lactophenol with or without cotton blue.

Cultural examination for moulds and yeasts

Primary isolation

Inoculate slopes or plates of chloramphenicol malt agar. If dermatophytes are sought, use a medium containing cycloheximide, either malt agar or one of the proprietary media containing an indicator. Push pieces of hair or skin into the medium at several points. Scrape nail clippings to a powder with a scalpel; do not put them on in lumps. Pieces of skin, hair or nail powder can be picked up more easily if the heated inoculating wire is first pushed into the medium to coat it with agar.

Spread pus, CSF deposit or biopsy material from cases of the deeper mycoses on slopes of glucose peptone agar with chloramphenicol. Cut tissues into small pieces or grind them in a Griffith's tube. Place in the bottom of a petri dish and pour on chloramphenicol malt agar (moulds) or glucose peptone agar (actinomycetes), melted and at 45°C.

Wash mycetoma grains several times with sterile saline to remove most contaminants before culturing.

Incubate cultures at 28°C. Most yeasts grow in 2–3 days, though cryptococci are rather slower. Common moulds, too, will grow in a few days while cultures for dermatophytes should be kept for 2 weeks though most are identifiable after 1 week. It is useful, then, to examine all mould cultures once or twice a week and ensure adequate aeration.

Mould growths on and from foods, etc.

Examine with a lens or low-power microscope *in situ* – to see the arrangement of spores, etc. before disturbing the growth. Remove a small piece of mycelium with a wire or needle, the end of which has been bent at right angles to give a short, sharp hook. Use the point of the needle to cut out a piece of the growth near the edge where sporulation is just beginning. Transfer it to a drop of lactophenol cotton blue or PVA mountant on a slide, apply a cover-slip then examine after 30 min when the stain has penetrated the hyphae. If bubbles are present warm gently to expel them.

Cultures from food, soil, plant material, etc.

Inoculate plates of chloramphenicol or other selective media (see pp.65, 66) or use dichloran glycerol (DG18) medium.

Dip slides, coated with appropriate media, are useful for liquid samples or homogenates and are available commercially. Zygomycetes, if present, will outgrow everything else but are markedly restrained by Czapek–Dox. To grow moulds from physiologically dry materials such as jam increase the sucrose content of the Czapek–Dox to 20%. Some yeasts can tolerate low pH and can be isolated on malt agar to which 1% lactic acid has been added after melting and cooling to 50°C. Incubate cultures at room temperature and at 30°C.

Counting moulds in foods, etc.

Make a 10% suspension of the material in sterile water, Ringer or peptone water diluent. Homogenise in a Stomacher (other blenders may raise the temperature and injure the moulds). Make ten-fold dilutions and surface plate on selective media (e.g. AFPA, OGYE or (for xerophilic moulds), DG18). Incubate and count colonies on plates that have between 50 and 100. Calculate the numbers of mould propagules per gram.

Methods for individual foods are given in Chapter 16. Useful sources of information are Pitt and Hocking (1985), King *et al.* (1986), Krogh (1987), Sampson and van Reenan-Hoekstra (1988).

Sampling air for mould spores

A cheap, simple device for this purpose is the Porton impinger (p.265) in which particles from a measured volume of air are trapped in a liquid medium on which viable counts are done. Slit or SAS samplers (p.264) are also useful. More information is gained by using the Andersen sampler, which sizes airborne particles using the principle that the higher the air speed the smaller the particle which will escape the air stream and be impacted on an agar surface. Plastic petri dishes give lower counts than glass ones but they make the whole machine much easier to handle.

The medium and time of incubation will depend on the organism sought. For general use one run with Czapek–Dox and one with chloramphenicol malt agar will serve; for thermophilic actinomycetes use glucose peptone agar without antibiotics; incubate one set at 40°C, the other at 50°C.

Identification of moulds

Only well-sporing mould colonies can be readily identified. If spores are not seen reincubate and/or subculture to other media. Sporulation may sometimes be stimulated by incubation under near UV ('black') light.

The usual methods of examining sporing structures are needle mounts (small fragments of mycelium taken from the colony with a mounted needle) and tape mounts as follows.

Cut a strip about 1.5 cm wide from a roll of 2.5 cm Sellotape, then attach by a narrow end to a mounted needle, forming a flag. Blot the surface of the colony with this near the growing edge, then mount the tape, sticky side up, in a drop of mountant, add another drop of mountant and overlay this with a cover glass, and examine at once.

The centre of a well-sporing colony may show nothing but spores. If so make a second preparation nearer to the edge of the colony. Conversely, a poorly-sporing colony may shown spores only in its centre.

Slide culture

This permits a more critical examination. Cut small blocks of medium (0.5 by 0.5 cm) from a poured plate and transfer them to the centres of sterile slides. Inoculate the edges of each block and apply a cover-slip. Incubate in a petri dish containing moist filter paper and glass rods to support the slide (wet chamber). Watch the progress of growth using the low power of the microscope when visible growth has developed. When typical structures are seen, prepare a slide and cover-slip each with a drop of mountant. Lift the cover-slip off the slide culture and mount on the prepared slide, then lift off and discard the agar block: the slide is now mounted on the prepared cover-slip. Examine both preparations after 30 min.

Cultural examination of yeasts

Morphology

To study yeast morphology use corn meal agar: streak the organism on a plate of the medium, cover with a sterile cover-glass and incubate at 30°C. Examine after 3–4 days through the cover glass with a ×10 and ×40 objective. Look for cell shape, mycelium and spore formation.

To encourage ascospore formation inoculate sodium acetate agar made by adjusting the pH of 0.5% sodium acetate to 6.5 with acetic acid, then solidifying it with agar. Stain by the malachite green-safranin method: prepare a smear in a drop of water and dry in air. Fix by heat and cover with 1% malachite green. Heat until steam rises, wash with tap-water and counterstain with safranin. Wash, blot and dry. Ascospores are stained green, vegetative cells red.

Fermentation tests

These differ from those used in bacteriology in that the sugar concentration is 3%. Durham tubes are always used and a 10-ml volume of medium is preferable to encourage fermentation rather than oxidation.

Inoculate these 'sugar' media (with Durham's tubes): glucose, maltose, lactose, raffinose and galactose. Incubate cultures at 25–30°C for at least 7 days. Gas production indicates fermentation (see p.110).

Auxanograms

Fermentation tests alone are insufficient to identify species of yeasts. Make a thick suspension by adding about 5 ml of sterile water to a malt agar slope: it is not necessary to wash the organisms. Melt tubes of auxanogram carbon-free base (for carbon assimilation) and nitrogen-free base (for nitrogen assimilation) and cool to 45°C. Add about 0.25 ml of the yeast suspension to each 20 ml of auxanogram medium and immediately prepare pour plates in 9 cm petri dishes. When set place up to 5 discs of carbon or nitrogen sources well apart on the surface of the appropriate medium and examine the plates daily for up to 7 days for a halo of growth around the substrate. Carbon sources commonly used are glucose (control), maltose, sucrose, lactose, inositol, galactose, raffinose, mannitol and cellobiose. Nitrogen sources include sodium nitrate, asparagine and ethylamine hydrochloride (see p.99).

Mycotoxins and seed-borne moulds

Mycotoxins are metabolites of fungi which may be produced during mould growth on foods and animal feeds. The most seriously contaminated commodities are cereals and oilseeds. The myotoxigenic moulds most frequently encountered belong to the genera *Penicillium*, *Aspergillus* and *Fusarium* (Moss, 1989).

Surface-sterilize seeds by immersion for 2 min in sodium hypochlorite, 0.4% available chlorine. Place up to 10 small or 5 large seeds on DG-18 agar in a petri dish and incubate at room temperature.

Subculture from colonies to oatmeal agar (*Aspergillus*), Czapek yeast autolysate agar (*Penicillium*) or synthetic nutrient-poor agar (*Fusarium*). These media are described on pp. 64 – 66.

'Kit' tests

The commercial kits are satisfactory for the identification of most strains encountered in clinical laboratories. Some use methods based on traditional criteria; others rely on novel approaches. In all of them, however, the data bases are from a restricted list of species and occasional misidentifications may occur.

Serological methods

Antigens

To detect circulating antibodies a crude antigen is made. The fungus is grown in a well-aerated broth culture. It is harvested by filtration and stored overnight at − 20°C before mechanical disruption, extraction and lyophilization. Alternatively, antigens may be made by concentrating the culture filtrate.

The freeze-dried antigens are reconstituted in phosphate buffered saline at a strength indicated by previous titration against positive antisera. Several commercial kits are now available and these obviate the difficulties of standardizing reagents.

Antibody detection

These tests are useful in histoplasmosis and coccidioidomycosis. Both yeast and mycelial phase antigens of histoplasma are used. Those who have facilities and expertise for growing and handling dangerous pathogens can prepare histoplasma mycelial antigen as described above. The cultures must be inoculated in a cabinet and killed by exposure for 3 days to 1 : 5000 thiomersal before processing. A yeast phase produced on chocolate cystine agar can be used to inoculate glucose peptone broth fortified with 0.1% of cystine. The culture is stirred or shaken at 37°C for 3–4 days and killed with thiomersal. The washed yeasts are suspended in saline with 1 : 5000 thiomersal and this whole yeast cell suspension constitutes the antigen.

The preparation of *Coccidioides immitis* antigen is best left to the experts; laboratory infections are almost always fatal.

Immunodiffusion

Agar base

This is made by dissolving 2 g of agar (Oxoid No. 1) in 100 ml of water by autoclaving. 100 ml of buffer is heated to 50°C and mixed with the agar. The medium is kept in a water-bath at 50°C and discarded if not used within 48 h of preparation.

Buffer

Boric acid (H_2BO_3)	10 g
Powdered borax ($Na_2B_4O_7.10H_2O$)	20 g
EDTA, disodium salt	10 g
Water	to 1000 ml

The pH should be 8.2. This buffer is, of course, double strength; it is diluted with agar for plates and slides or with an equal volume of water for use in electrophoresis tanks.

Plates

A 30-ml volume of agar is measured into a plastic petri dish and a Perspex jig with metal pegs is fitted in place of the lid to produce the pattern of wells shown in Figure 9.1. The large wells are 6 mm in diameter and the smaller wells 2 mm. The distance between the central and peripheral wells is also 6 mm. The test serum occupies the central well, antigens the pairs of large and small wells (shown shaded) and appropriate control sera the top and bottom wells. Two antigens can thus be tested and a single solution suffices as the relative volumes (60 and 6 μl in the two holes) give high and low concentration gradients in the agar. This arrangement also allows reactions of identity to be obtained between test and control sera, thus eliminating some anomalous reactions, particularly among the aspergilli.

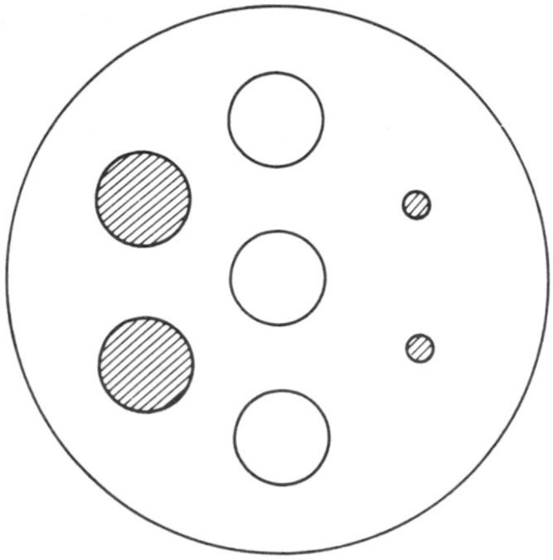

Figure 9.1 Double diffusion tests

Counter immunoelectrophoresis (CIE)

Reagents

Veronal buffer	Barbitone (*caution*)	6.88 g
	Sodium barbitone (*caution*)	15.14 g
	Sodium azide (*caution*)	1.00 g
	Distilled water	1000 ml
Agarose gel	1% agarose	50 ml
	Veronal buffer	50 ml
Coomassie blue	Coomassie brilliant blue	2 g
	Trichloroacetic acid, 10%	50 ml
	Distilled water	1000 ml

Destain solution	Methanol	300 ml
	Acetic acid	200 ml
	Distilled water	500 ml

Method
Pre-coat clean glass microscope slides with a very thin layer of 1% agarose gel and heat to 50°C for 30 min.

Pour 5.5 ml of the same gel on the slides and allow to set. Cut wells, 4.5 mm diameter (to hold 20 μl) for sera and 2.5 mm (to hold 8 μl) for the antigens. Allow 5 mm between the neatest edges of respective wells.

Add sera and antigens and include suitable controls.

Electrophorese with antigens at the cathode side of a CIE tank, using veronal buffer diluted 1:1 with water at 4 V/cm for 90 min.

If rapid results are required wash the wet gel for 1 h and examine by dark-ground illumination. This gives a preliminary result only and must be confirmed after washing and staining:

(1) Wash with 50% trisodium phosphate for 30 min.
(2) Wash with two changes of saline, 40 min each.
(3) Place in saline overnight.
(4) Wash in distilled water for 30 min.
(5) Cover with blotting paper and dry at 60°C for 1 h.
(6) Stain in Coomassie brilliant blue for 5 min.
(7) Wash in distilled water for 5 min.
(8) Place in destaining solution for 5 min.

Antigen detection

Commercial kits are now available for the early diagnosis of infections by *Cryptococcus, Candida* and *Aspergillus*. An example of the latex agglutination technique often employed in kits is given below for the detection of *Cryptococcus* antigen in serum or CSF.

Antigen
Polystyrene latex, coated with rabbit anti-cryptococcus globulin, is agglutinated by cryptococcal capsular polysaccharide. Because of extensive cross-reaction, type A globulin acts as a 'polyvalent' reagent.

Preparation of globulin
Rabbits are immunized with a small capsule strain of cryptococcus until their serum has an agglutination titre of 1 in 500 or better. The animals are then bled and the globulin fraction of their serum isolated using ammonium sulphate or caprylic acid. After dialysis, it is freeze-dried and reconstituted at a concentration of 4% for use.

Sensitization of latex
The stock latex solution is made from Dow or Difco latex 0.81. A 1:20 dilution is normally satisfactory. When this is further diluted 1:100 in a round 12-mm tube, it should have an absorbance of 0.29–0.31 at 650 nm. To this stock an equal volume of the maximally reactive dilution of globulin in glycine–saline is added. This dilution is determined by a chess-board titration using latex sensitized with 1:100, 1:200, 1:400, 1:600 and 1:800 globulin against a

known positive human serum or CSF. Alternatively, diluted washings from cryptococcal cells can be used.

The routine reagent can usually be made by mixing equal volumes of 1:250 globulin and 1:10 latex and leaving the mixture at 4°C overnight.

Diluent for cryptococcal latex agglutination

Glycine	7.31 g
Sodium chloride	10 g
1 N sodium hydroxide solution	3.5 ml
Water	to 1000 ml

Leave the solution overnight to equilibrate then check the pH, which should be 8.2.

Use this buffer as it is to dilute the latex; for diluting test and control sera, add 0.1% of bovine albumin. (This buffer is an excellent culture medium; keep the stock frozen at −20°C.)

Screening test

Eight drops of diluent are placed in each of two wells in a WHO plastic tray for each specimen. Three areas are marked off on a microscope slide with a felt pen. Two drops of serum or CSF are added to the left-hand area of the slide and two drops to the first well. After mixing, two drops of this 1:5 dilution are added to the centre area of the slide and two to the second well, making this 1:25. Two drops of the 1:25 dilution from the second well are added to the right-hand area of the slide. One drop of sensitized latex is added to each area on the slide and the slide rocked for 3 min. A slight granularity always appears: this should be disregarded. Definite agglutination is considered positive.

When the screening test is positive or if cryptococci have been isolated, doubling dilutions are tested to obtain an end-point.

Antifungal drugs

At least one kit (API ATBFUNGUS) is available for testing susceptibility to antifungal agents (Philpot and Charles, 1993). The method given below for fluorocytosine may be adapted for use with other agents.

Methods of susceptibility testing and assay

Medium

5-Fluorocytosine is sulphonamide-like in that it requires a medium free of antagonists. This excludes the use of peptone and meat extract. Some organisms, particularly cryptococci, are reluctant to grow on a medium whose only source of nitrogen is inorganic. Growth is much improved by using casamino acids, a nitrogen source free of nucleic acids and their components, supplemented by sufficient yeast extract to give good growth without antagonizing 5-fluorocytosine. Difco or Oxoid yeast extract is suitable for this; Marmite is not. A small amount of glycerophosphate is added as a buffer. This does not precipitate on autoclaving, and the sterilized medium remains clear. The other antifungal drugs have no special requirements but it is obviously simpler to use one medium for all. For preparation of the medium see p.71.

Drug solutions
5-Fluorocytosine is freely soluble in water. A stock solution containing 1000 µg/ml will keep indefinitely if frozen at –20°C. When this solution is thawed for use, a few crystals of 5-fluorocytosine may remain undissolved. Complete solution is assured by placing it in the 56°C water-bath for 10 min before use. Amphotericin B is available as Fungizone, a complex with sodium deoxycholate in 50-mg quantities. Dissolve contents of one bottle in 10 ml of sterile water (avoid saline at all stages, it precipitates the amphotericin). This stock solution will keep for 6 months if frozen at –20°C. For a working solution make two tenfold dilutions in 0.4% sodium deoxycholate solution, to give a concentration of 50 µg/ml. This keeps for several months if frozen at –20°C. Pimaricin must be dissolved in dimethyl formamide; Nystatin and the imidazole compounds in equal volumes of acetone and water. Intermediate dilutions of these drugs are opalescent, and should be prepared just before use; if kept for more than about 30 min there is an appreciable loss through precipitation on the glass of the bottle. Solutions of the imidazole compounds – 20 mg of solid weighed into a sterile bottle then taken up in 10 ml of water plus 10 ml of acetone – have been found to keep for 4 months in the refrigerator. There is no point in keeping them at –20°C since the acetone prevents them from freezing.

Working solutions for sensitivity testing
5-Fluorocytosine: 10 µg and 100 µg/ml
Miconazole, ketoconazole, pimaricin and nystatin (the last in units): 50 µg and 5 µg/ml.
Amphotericin B: 10 µg and 1 µg/ml.

Sensitivity testing
Melt the measured volumes of agar, then cool them to 48°C. Use 10-ml lots for slopes, adding drug solutions as follows:

5-Fluorocytosine
(The drug may be added to measured volumes of agar before autoclaving.)

100 µg/ml				0.1	0.2	0.4	0.8 ml
10 µg/ml	0.1	0.2	0.5				ml
Concentration	0.1	0.2	0.5	1.0	2.0	4.0	8.0 µg/ml

Miconazole, ketoconazole, pimaricin and nystatin

50 µg/ml				0.1	0.2	0.4	0.8 ml
5 µg/ml	0.1	0.2	0.4				ml
Concentration	0.05	0.1	0.2	0.5	1.0	2.0	4.0 µg/ml

Amphotericin B

10 µg/ml				0.1	0.2	0.4	0.8 ml
1.0 µg/ml	0.1	0.2	0.5				
Concentration	0.01	0.02	0.05	0.1	0.2	0.4	0.8 µg/ml

The volume of antibiotic added has been taken into consideration. The maximum error introduced is 8%, too small a difference to affect the result. Divide each 10-ml lot into two slopes, and include a pair of control slopes without drugs. For plates, 20-ml lots of medium are used and the above concentrations are of course doubled.

The inoculum

Grow the organism on the assay medium at 30°C. Results are less satisfactory if this step is omitted. Yeasts will grow well enough after 24 h, and the growth should be suspended in sterile water, then this suspension added, drop by drop, to 10 ml of sterile water till a faint but distinct opalescence results. One drop of this diluted suspension is run down each slope, and incubation continued at 30°C until growth on the control slopes is adequate. *Candida* species reach this stage in 24 h: cryptococci need a further 24 h incubation. Filamentous organisms—aspergilli, histoplasmas and the like, are grown on the same medium again at 30°C. The young growth is stripped off the surface of the medium after 2–3 days before spore formation begins and macerated in a small macerator with 3 ml of sterile water, using 15-s bursts until enough of the growth is disintegrated to give an opalescent suspension. The macerator bottle is allowed to stand for 10 min, so that aerosols within may settle. The macerating bead is then removed and the cap of a sterile bottle used to cap the bottle. The contents of the bottle are now mixed gently, large lumps are allowed to settle, then the supernatant suspension is taken off and treated exactly like the yeast suspension, i.e. it is diluted until just opalescent. One drop is used as inoculum and the tubes are incubated as before until the growth of the controls is adequate.

If the sensitivities of several yeast strains are required at the same time, up to four organisms can be accommodated by using duplicate drops of suspension on plates marked out in sectors as in viable counting. As the drops are concentrated on a small area, a 1/25 dilution of the suspension described above is necessary to obtain comparable results. In all cases the aim should be to produce discrete colonies, not a confluent growth.

When several strains are to be tested it is convenient to prepare the slopes or plates containing 5-fluorocytosine, miconazole or ketoconazole in the afternoon, then keep them in the refrigerator overnight. If plates are used, they can be dried next morning while the amphotericin series is being prepared. 5-Fluorocytosine or miconazole incorporated in agar slopes will keep for 3 weeks in the refrigerator at 4°C with no detectable loss of potency. Where many tests are done, production in quantity is feasible – and easier.

Short method for yeast sensitivity tests

In practice, yeast sensitivities generally fall within the four lowest drug concentrations given above. Using a square petri dish with 25 compartments, four drug concentrations and a control can be put in the five columns across one plate using volumes of 1.2 ml. If the centre row is left empty, two organisms can be tested in duplicate. Plates containing 5-fluorocytosine or ketoconazole will keep for a week in the refrigerator.

Disc test sensitivity methods

Disc methods may be used for yeasts but are not satisfactory for moulds because of the difficulty in preparing suitable seeding suspensions. Spore suspensions may give misleading results because the MICs of germinating spores may differ from those of vegetative forms.

Methods of assay

Amphotericin B, miconazole and ketoconazole assay

Standard solutions
Amphotericin B, 2.0 μg/ml; miconazole, 4.0 μg/ml or ketoconazole, 4.0 μg/ml is diluted 1/10 in horse serum. The appropriate standard is diluted in duplicate and in parallel with duplicate dilutions of the patient's serum in one of the following ways:

1 Four rows of eleven 75 × 12 mm test-tubes with aluminium caps (Oxoid) are sterilized in a rack, 0.5 ml of the test organism suspension is added to each; 0.5 ml of standard to the first tube of rows 1 and 2 and 0.5 ml of the test serum to rows 3 and 4. Each one is then serially diluted to tube 10, leaving tube 11 as a growth control.
2 A sterile Microtitre plate (M24AR) with lid (M42R) is used with 0.1-ml volumes of diluent and sera. The whole operation is made much easier by using a four-place multichannel automatic pipette with autoclaved tips for dispensing and diluting. The suspension may be dispensed from a 5-cm plastic petri dish or a sterile 50-ml beaker.

In either case, the results of amphotericin B assay can be read after overnight incubation. Assays of imidazoles should be left for 24 h, then the end point taken where growth is sharply reduced by comparison with the controls. Ignore traces of growth.

5-Fluorocytosine assay

5-Fluorocytosine diffuses readily in agar while amphotericin B, miconazole and ketoconazole, the other drugs commonly encountered, diffuse little if at all. Fortunately, the expected levels of these latter drugs are low and it is possible to estimate serum levels by dilution in broth in parallel with a standard prepared in horse serum.

Media
The liquid medium for dilution tests is given on p.71.

Organisms
The test organism is *Saccharomyces cerevisiae* NCPF 3178. It is the least granular when grown on glucose peptone agar and safer to use than a potential pathogen. A distinctly cloudy suspension is made in 10 ml of sterile distilled water; 0.1 ml is used in broth for the liquid assay and the rest added to 110 ml of 5-fluorocytosine assay medium melted and cooled to 45°C.

Pour the inoculated medium into a bioassay plate 250 mm square.

Standard and test sera
Prepare standards containing 10, 15, 25, 40 and 60 μg/ml of 5-fluorocytosine. Keep frozen (will last for 6 months). Dilute patient's serum as follows (any amphotericin B which is present will be neutralized within one hour):

Horse serum	0.4 ml
Patient's serum	0.5 ml
Ergosterol 1% in acetone	0.1 ml

If the patient is being treated with an imidazole drug, use 0.1 ml of 10% Tween 80 in place of ergosterol.

The test

Lay the dish on a sheet of paper (template) having a pattern of 25 positions to which test or standard numbers are assigned at random. Punch holes in the medium with a 4-mm cork borer and remove the agar. Add 15 μl volumes of test and standard solutions in duplicate in the pattern shown on the template. Incubate at 30°C overnight.

Measure the diameters of the zones of inhibition. Plot the standards on one decade semilogarithmic paper and obtain the concentration of drug in the test serum by interpolation.

For more details of technical methods see Mackenzie *et al.* (1980), Mackenzie and Philpot (1981) and Campbell *et al.* (1985).

References

Campbell, C. K., Davis, C. and Mackenzie, D. W. R. (1985) Detection and isolation of pathogenic fungi. In *Isolation and Identification of Micro-organisms of Medical and Veterinary Importance* (eds C. H. Collins and J. M. Grange), Society for Applied Bacteriology Technical Series, No. 21, Academic Press, London, pp. 329–343

King, A. D., Pitt, J. I., Beauchat, L. R. and Corry, J. E. L. (1986) *Methods for the Mycological Examination of Food*, Plenum, New York

Krogh, P. (1987) *Mycotoxins in Food*, Academic Press, London

MacKenzie, D. W. R., Philpot, C. M. and Proctor, A. G. J. (1980) *Basic Serodiagnosis Methods for Diseases caused by Fungi and Actinomyces*, Public Health Laboratory Service Monograph No. 12, HMSO, London

Mackenzie, D. W. R. and Philpot, C. M. (1981) *Isolation and Identification of Ringworm Fungi*, Public Health Laboratory Service Monograph, No. 15, HMSO, London

Moss, M. O. (1989) Mycotoxins of *Aspergillus* and other filamentous fungi. In *Filamentous Fungi in Foods and Feeds*, Society for Applied Bacteriology Symposium Series No. 18, *Journal of Applied Bacteriology*, **87** (Supp.) pp. 69S–82S

Philpot, C. M. and Charles, D. (1993) Determination of sensitivity to antifungal drugs: evaluation of an API kit. *British Journal of Biomedical Science*, **50**, 27–30

Pitt, J. J. and Hocking, A. D. (1985) *Fungi and Food Spoilage*, Academic Press, Sydney

Samson, R. A. and van Reenan-Hoekstra, E. S. (1988) *Introduction to Food Borne Fungi*, Centraalbureau voor Schimmelcultures, Baarn

Chapter

10

Counting methods

It is often necessary to report on the size of the bacterial population in a sample. Unfortunately, industries and health authorities have been allowed to attach more importance to these 'bacterial counts' than is permitted by their technical or statistical accuracy (see also p.213).

If a *total (microscopic) count* is required, many of the organisms counted may be dead or indistinguishable from other particulate matter. A *viable count* assumes that a visible colony will develop from each organism. Bacteria are, however, rarely separated entirely from their fellows and are often clumped together in large numbers particularly if they are actively reproducing. A single colony may therefore develop from one organism or from hundreds or even thousands of organisms. Each colony develops from one *viable unit.* Because any agitation, as in the preparation of dilutions, will either break up or induce the formation of clumps, it is obviously difficult to obtain reproducible results. Bacteria are seldom distributed evenly throughout a sample and as only small samples are usually examined very large errors can be introduced. Viable counts are usually given as numbers of *colony-forming units* (cfu).

Many of the bacteria present in a sample may not grow on the medium used, at the pH or incubation temperatures or gaseous atmosphere employed, or in the time allowed.

Accuracy is often demanded where it is not needed. If it is decided that a certain product should contain less than, say, 10 viable bacteria/g this suggests that, on average, of 10 tubes each inoculated with 0.1 g seven would show growth and three would not, and out of 10 tubes each inoculated with 0.01 g only one or two would show growth. If all the 0.1-g tubes or five of the 0.01-g tubes showed growth, there would be more than 10 organisms/g. It does not matter whether there are 20 or 10 000: there are too many. There is no need to employ elaborate counting techniques.

In viable count methods, it is recognized that large errors are inevitable even if numbers of replicate plates are used. Some of these errors are, as indicated above, inherent in the material, others in the technique. Errors of ±90% in counts of the order of 10 000 to 100 000/ml are not unusual even with the best possible technique. It is, therefore, necessary to combine the maximum of care in technique with a liberal interpretation of results. The figures obtained from a single test are valueless. They can be interpreted only if the product is regularly tested and the normal range is known.

Physical methods are used to estimate total populations, i.e. dead and living organisms. They include direct counting and measurements of turbidity. Biological methods are used for estimating the numbers of viable units. These include the plate count, roll-tube count, drop count, surface colony count, dip slide count, contact plate, membrane filter count, most probable number estimations, and automated methods (Chapter 8).

Direct counts

Counting chamber method

The Helber counting chamber is a slide 2–3 mm thick with an area in the centre called the platform and surrounded by a ditch. The platform is 0.02 mm lower than the remainder of the slide (Figure 10.1). The top of the slide is ground so that when an optically plane cover-glass is placed over the centre depression, the depth is uniform. On the platform an area of 1 mm^2 is ruled so that there are 400 small squares each 0.0025 mm^2 in area. The volume over each small square is 0.02×0.0025 mm^3, i.e. 0.000 05 ml.

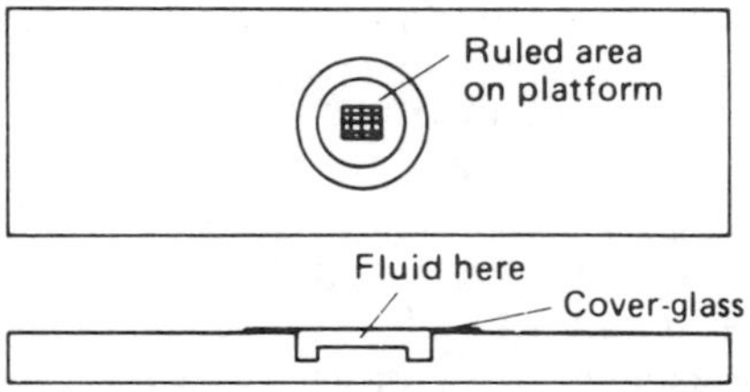

Figure 10.1 Bacterial counting chamber (**depth is exaggerated**)

Add a few drops of formalin to the well mixed suspension to be counted. Dilute the suspension so that when the counting chamber is filled there will be about five or 10 organisms per small square. This requires initial trial and error. The best diluent is 0.1% peptone water containing 0.1% lauryl sulphate and (unless phase contrast or dark field is used for counting) 0.1% methylene blue. Always filter before use.

Place a loopful of suspension on the ruled area and apply the cover-glass, which must be clean and polished. The amount of suspension must be such that the space between the platform and the cover-glass is just filled and no fluid runs into the ditch; this requires practice. If the cover-glass is applied properly, Newton's rings will be seen. Allow 5 min for the bacteria to settle.

Examine with a 4-mm lens with reduced light, dark field or phase contrast if available. Count the bacteria in 50–100 squares selected at random so that the total count is about 500. Divide the count by number of squares counted. Multiply by 20 000 and by the original dilution factor to obtain the total number of bacteria/ml. Repeat twice more and take the average of the three counts. Clumps of bacteria, streptococci, etc., can be counted as units or each cell counted as one organism.

With experience, reasonably accurate counts can be obtained but the chamber and cover-glass must be scrupulously cleaned and examined microscopically to make sure bacteria are not left adhering to either.

The Breed count

Named after its originator, this is a rough but useful technique. Breed slides have areas of 1 cm^2 marked on them. Into a square, place 0.01 ml of the fluid to be counted, for example milk, with a commercially available microsyringe or loop. Allow to dry and stain with methylene blue. Examine with an oil immersion lens giving a known field diameter, which is usually about 0.16 mm. The area seen is thus πr^2 or 3.14×0.08^2, i.e. 0.02 mm^2, and as the total area is 100 mm^2, one field represents 1/5000 part of the whole area, or 1/5000 part of the original 0.01 ml. One organism per field equals

5000/0.01 ml or 500 000/ml. Count the number of organisms in several fields in different parts of the ruled area and use the equation

$$\text{Count/ml} = \frac{N \times 4 \times 10^4}{\pi d^2}$$

where N is the number per field and d is the diameter of the field.

Opacity tube method

International reference opacity tubes, containing glass powder or barium sulphate are available and are known as Brown's or McFarland opacity tubes. These are numbered narrow glass tubes of increasing opacity and a table is provided equating the opacity of each tube with the number of organisms/ml. The unknown suspension is matched against the standards in a glass tube of the same bore. It may be necessary to dilute the unknown suspension. These physical methods should not be used with Hazard Group 3 organisms.

Colony counts

In these techniques, the material containing the bacteria is serially diluted and some of each dilution is placed in or on suitable culture media. Each colony developing is assumed to have grown from one viable unit, which, as indicated earlier, may be one organism or a group of many.

Diluents

Some diluents, e.g. saline or distilled water, may be lethal for some organisms. Diluents must not be used direct from a refrigerator as cold-shock may prevent organisms from reproducing. Peptone-water diluent (p.88), generally known as 'maximum recovery diluent', is most commonly used. If residues of disinfectants are present add suitable quenching agents, e.g Tween 80 or lecithin for quaternary ammonium compounds and sodium thiosulphate for chlorine and iodine. See Chapter 4 and Bloomfield (1991).

Pipettes

The 10-ml and 1-ml straight-sided blow-out pipettes are commonly used and are specified for some statutory tests. Disposable pipettes are labour saving. Mouth pipetting must be expressly forbidden, regardless of the nature of the material under test. A method of controlling teats is given on p.35. Automatic pipettors or pipette pumps, described on p.36, should be used. Some of these can be pre-calibrated and used with disposable polypropylene pipette tips.

Pipettes used in making dilutions must be very clean, otherwise bacteria will adhere to their inner surfaces and may be washed out into another dilution. Siliconed pipettes may be used. Fast-running pipettes and vigorous blowing-out should be avoided: they generate aerosols. As much as 0.1 ml may remain in pipettes if they are improperly used.

Preparation of dilutions

Dispense 9-ml volumes of diluent in screw-capped bottles. If tubes with loose caps are used then sterilize them first and dispense the diluent aseptically otherwise a significant amount of diluent may be lost during autoclaving.

When diluting liquids, for example milk for bacterial counts, proceed as follows.

Mix the sample by shaking. With a straight-sided pipette dipped in half an inch only, remove 1 ml. Deliver into the first dilution blank, about half an inch above the level of the liquid. Wait 3 s, then blow out carefully to avoid aerosol formation. Discard the pipette. With a fresh pipette, dip half an inch into the liquid, suck up and down 10 times to mix, but do not blow bubbles. Raise the pipette and blow out. Remove 1 ml and transfer to the next dilution blank. Discard the pipette. Continue for the required number of dilutions, and remember to discard the pipette after delivering its contents, otherwise the liquid on the outside will contribute to a cumulative error. The dilutions will be

Tube no.	*1*	*2*	*3*	*4*	*5*
Dilution	1:10	1:100	1:1000	1:10 000	1:100 000
Vol. of original fluid/ml	0.1 (or 10^{-1})	0.01 (10^{-2})	0.001 (10^{-3})	0.0001 (10^{-4})	0.00001 (10^{-5})

When counting bacteria in solid or semisolid material, weigh 10 g and place in a Stomacher or blender. Add 90 ml of diluent and homogenize.

Alternatively, cut into small pieces with a sterile scalpel, mix 10 g with 90 ml of the diluent and shake well. Allow to settle. Assume that the bacteria are now evenly distributed between the solid and liquid.

Both of these represent dilutions of 1:10, i.e. 1 ml contains or represents 0.1 g, and further dilutions are prepared as above. The dilutions will be:

Tube	*1*	*2*	*3*	*4*
Dilution	1:100	1:1000	1:10 000	1:100000
Weight of original material	0.01 (or 10^{-2})	0.001 (10^{-3})	0.0001 (10^{-4})	0.00001 g (10^{-5} g)

Mechanical aids for preparing the initial 1:10 dilutions are available commercially ('Gravimetric diluters').

Plate count

Melt nutrient agar or other suitable media in bottles or tubes of appropriate volumes (usually 10 or 20 ml) amounts. Cool to 45°C in a water-bath.

Set out petri dishes, two or more per dilution to be tested and label with the dilution number. Pipette 1 ml of each dilution into the centre of the appropriate dishes, using a fresh pipette for each dilution. Do not leave the dish uncovered for longer than is absolutely necessary. Add the contents of one agar tube to each dish in turn and mix as follows. Move the dish gently six times in a clockwise circle of diameter about 150 mm. Repeat counter-clockwise. Move the dish back-and-forth six times with an excursion of about 150 mm. Repeat with to-and-fro movements. Allow the medium to set, invert and incubate for 24–48 h.

Economy in pipettes

In practice, the pipette used to transfer 1 ml of a dilution to the next tube may be used to pipette 1 ml of that dilution into the petri dish. Alternatively, after the dilutions have been prepared a single pipette may be used to plate them out, starting at the highest dilution to avoid significant carry-over.

Surface count method

This is often used instead of the pour-plate method in the examination of foods, particularly because the growth of some organisms (e.g. pseudomonads) may be impaired if agar is used at too high a temperature. Also, it it is easier to subculture colonies from surface plates than from pour plates.

Place 0.1 ml or another suitable amount of the sample measured with a standard (commercial) loop or a micropipette in the centre of a well-dried plate of suitable medium and spread it with a loop or spreader all over the surface (p.95). Incubate and count colonies.

When counting colonies, difficulties arise with large 'smears' of small colonies caused when a large viable unit is broken up but not dispersed during manipulation. Treat these as single units. Usually, other colonies can be seen and counted through 'smears'. There may be problems with swarming organisms, such as some *Proteus* spp. In this case pour melted agar over the inoculated plates.

Counting colonies on pour or surface plates

To count on a simple colony counter select plates with between 30 and 300 colonies. Place the open dish, glass side up, over the illuminated screen. Count the colonies using a 75-mm magnifier and a hand-held counter. Mark the glass above each colony with a felt tip pen. Calculate the colony or viable count/ml by multiplying the average number of colonies per countable plate by the reciprocal of the dilution. Preferably count the colonies from at least two dilutions and calculate the cfu in the original sample by the method of weighted means (BSI, 1991). Report as 'colony forming units/ml' (cfu) or as 'viable count/ml' not as 'bacteria/g or /ml'.

If all plates contain more than 300 colonies rule sectors, e.g. of one-quarter or one-eighth of the plate, count the colonies in these and include the sector value in the calculations.

For large work loads semi- or fully automatic counters are essential. In the former the pen used to mark the glass above the colonies is connected to an electronic counter which displays the numbers counted on a small screen. In the fully automatic models a TV camera or laser beam scans the plate and the results are displayed or recorded on a screen or read-out device.

Roll-tube count

Instead of using plates, tubes or bottles containing media are inoculated with diluted material and rotated horizontally until the medium sets. After incubation, colonies are counted.

Dispense the medium in 2–4-ml amounts in 25-ml screw-capped or Astell bottles. The medium should contain 0.5–1.0% *more* agar than is usual. Melt and cool to 45°C in a water-bath. Add 0.1 ml of each dilution and rotate horizontally in cold water until the agar is set in a uniform film around the walls of the bottle. This requires some practice. An alternative method employs a slab of ice taken from the ice tray of a refrigerator. Turn it upside down on a cloth and make a groove in it by rotating horizontally on it a bottle similar to that used for the counts but containing warm water. The count tubes are then rolled in this groove. If the roll-tube method is to be used often, the Astell Roll Tube apparatus, which includes a water-bath, saves much time and labour.

Incubate roll-tube cultures inverted so that condensation water collects in the neck and does not smear colonies growing on the agar surface. To count, draw a line parallel to the long axis of the bottle and rotate the bottle, counting colonies under a low-power magnifier.

The roll-tube method is popular in the dairy and food industries. Machines for speeding up and taking the tedium out of this method are available.

A roll-tube method for counting organisms in bottles (container sanitation) is given on p. 267

Drop count method

In this method, introduced by Miles and Misra (1938) and usually referred to by their names, small drops of the material are placed on agar plates. Colonies are counted in the inoculated areas after incubation.

Pipettes or standard loops that deliver known volumes are required. Disposable cotton wool-plugged Pasteur pipettes are available commercially and will deliver drops of 0.02 ml. These are sufficiently accurate for the purpose provided that the tips are checked immediately before use to ensure that they are undamaged. It is advisable, however, to check several pipettes in each batch by counting the number of drops of distilled water that yield 1 g.

Dry plates of suitable medium very well before use. Drop at least five drops from each dilution of sample from a height of not more than 2 cm (to avoid splashing) on each plate. Replace the lid but do not invert until the drops have dried. After incubation, select plates showing discrete colonies in drop areas, preferably one which gives less than 40 colonies per drop (10–20 is ideal). Count the colonies in each drop, using a hand magnifier. Divide the total count by the number of drops counted, multiply by 50 to convert to 1 ml and by the dilution used.

This method lends itself to arbitrary standards; for example, if there are less than 10 colonies between five drops of sample, this represents less than 100 colonies of that material. If these drops are uncountable (more than 40 colonies), then the colony count/ml is greater than 2000.

Spread-drop plate count

This is a modification of the Miles and Misra method, often used in the examination of foods.

Spread standard volume drops (e.g. 0.02 ml) on quadrants marked out on the bottom of the dish. Inoculate two dishes, each with drops from different decimal dilutions. Do counts on plates (usually two) from dilutions that yield ≤ 50 colonies. Ideally, calculate the cfu/g of test product by the weighted means method (BSI, 1991).

The spiral plate method

The apparatus is described on p.95. The results are said to compare favourably with those obtained by the surface spread plate method. A commercial imaged-based colony counting system is available for the rapid and accurate counting of colonies on spiral plates.

Membrane filter counts

These offer a more accurate alternative to the most probable number (MPN) estimations (p. 157). Liquid containing bacteria is passed through a filter that will retain the organisms. The filter is then laid on a a sterile absorbent pad containing liquid culture medium or on the surface of an agar medium and incubated, when the colonies which develop can be counted.

The filter-carrying apparatus is made of metal, glass or plastic and consists of a lower funnel, which carries a fritted glass platform surrounded by a silicone rubber ring. The filter disc rests on the platform and is clamped by its periphery between the rubber ring and the flange of the upper funnel. The upper and lower funnels are held together by a clamp.

The filters are thin, porous cellulose ester discs, varying in diameter and about 120 μm thick. The pores in the upper layers are 0.5–1.0 μm diameter enlarging to 3–5 μm diameter at the bottom. Bacteria are thus held back on top, but culture medium can easily rise to them by capillary action. Filters are stored interleaved between absorbent pads in metal containers. A grid to facilitate counting is ruled on the upper surface of each filter.

Membrane filter apparatus may be re-usable or sold as kits by the companies that make membranes.

Shallow metal incubating boxes with close-fitting lids, containing absorbent pads to hold the culture media, are also required.

The most convenient size of apparatus for counting colonies takes filters 47 mm in diameter, but various sizes larger or smaller are available.

Sterilization

Assemble the filter carrier with a membrane filter resting on the fritted glass platform, which in turn should be flush on top of the rubber gasket. Screw up the clamping ring, but not to its full extent. Wrap in oven-resistant plastic film or foil and autoclave for 15 min at 121°C. Alternatively, loosely assemble the filter carrier, wrap and autoclave; sterilize the filters separately interleaved between absorbent pads in their metal container also by autoclaving. This is the most convenient method when a number of consecutive samples are to be tested; in this case, sterilize also in the same way the appropriate number of spare upper funnels.

Sterilize the incubation boxes, each containing its absorbent pad, by autoclaving.

Culture media

Ordinary or standard culture media do not give optimum results with membrane filters. It has been found necessary to vary the proportions of some of the ingredients. Membrane filter versions of standard media are noted on p. 67. A 'Resuscitation' medium to revive injured bacteria (e.g. damaged by chlorine) is also desirable for some purposes.

Method

Erect the filter carrier over a filter flask connected by a non-return valve to a pump giving a suction of 25–50 mmHg. Check that the filter is in place and tighten the clamping ring. Pour a known volume, neat or diluted, of the fluid to be examined into the upper funnel and apply suction.

Pipette about 2.5–3 ml of medium on a Whatman 5 cm, No. 17, absorbent pad in an incubation container or petri dish. This should wet the pad to the edges but not overflow into the container.

When filtration is complete, carefully restore the pressure. Unscrew the clamping ring and remove the filter with sterile forceps. Apply it to the surface of the wet pad in the incubation box so that no air bubbles are trapped. Put the lid on the container and incubate. A fresh filter and upper funnel can be placed on the filter apparatus for the next sample.

For total aerobic counts, use tryptone soya membrane medium. For counting coliform bacilli, either 'presumptive' or *E. coli*, see Water examination, Chapter 20 For counting clostridia in water see p.274.

For anaerobic counts, use the same medium incubated anaerobically, or roll the filter and submerge in a 25-mm diameter tube of melted thioglycollate agar at 45°C and allow to set. To count anaerobes causing sulphide spoilage, roll the filter and submerge in a tube of melted iron sulphite medium at 42°C and allow to set. For yeast and mould counts, use the appropriate Sabouraud-type or Czapek-type media.

Counting

Count in oblique light under low-power magnification. If it is necessary to stain colonies to see them, remove the filter from the pad and dip in a 0.01% aqueous solution of methylene blue for half a minute and then apply to a pad saturated with water. Colonies are stained deeper than the filter. The grid facilitates counting. Report as membrane colony count per standard volume (100 ml, 1 ml, etc.).

Count anaerobic colonies in thioglycollate tubes by rotating the tube under strong illumination. Count black colonies in iron sulphite medium.

Bacteria in air

Membrane filters may be used with an impinger. Methods are given on p.264.

Dip slides

This useful method, commonly used in clinical laboratories for the examination of urine, now has a more general application, especially in food and drink industries for viable coliform and yeast counts. The slides are made of plastic and are attached to the caps of screw-capped bottles. There are two kinds: one is a single- or double-sided tray containing agar culture media; the other consists of a membrane filter bonded to an absorbent pad containing dehydrated culture media. Both have ruled grids, either on the plastic or the filter, which facilitate counting.

The slides are dipped into the samples, drained, replaced in their containers and incubated. Colonies are then counted and the bacterial load estimated.

Direct epifluorescence filtration technique (DEFT)

This is a rapid, sensitive and economical method for estimating the bacterial content of milk and beverages. It may also be used for foods if they are first treated with proteolytic enzymes and/or surfactants.

A small volume (e.g. 2 ml) of the product is passed through a 24-mm polycarbonate membrane which is then stained by acridine orange and examined with an epifluorescence microscope. Kits are available from several companies. For detailed information see Chapter 8.

Rapid automated methods

Conventional viable counts are expensive in time and labour. Holding products for 24 h or more adds to the cost of production. Rapid methods are therefore desirable, but they must be reliable, sensitive and specific. Initial and running costs must be balanced against those of conventional techniques and storage.

Rapid automated methods use one or other of the following: electronic particle counting; bioluminescence (as measured by bacterial ATP); changes in pH and Eh by bacterial growth; changes in optical properties; detection of ^{14}C in CO_2 evolved from a substrate; microcalorimetry; changes in impedance or conductivity; flow cytometry. See Chapter 8.

Most probable number (MPN) estimates

These are based on the assumption that bacteria are *normally* distributed in liquid media, that is, repeated samples of the same size from one source are expected to contain the same number of organisms *on average*: some samples will obviously contain a few more, some a few less. The average number is the *most probable number*. If the number of organisms is large, the differences between samples will be small; all the individual results will be nearer to the average. If the number is small the difference will be relatively larger.

If a liquid contains 100 organisms/100 ml, then 10-ml samples will contain, on average, 10 organisms each. Some will contain more, perhaps one or two samples will contain as many as 20; some will contain less, but a sample containing none is most unlikely. If a number of such samples is inoculated into suitable medium, every sample would be expected to show growth.

Similarly, 1-ml samples will contain, on average, one organism each. Some may contain two or three and others will contain none. A number of tubes of culture media inoculated with 1-ml samples would therefore yield a proportion showing no growth.

Samples of 0.1 ml, however, could be expected to contain only one organism per 10 samples and most tubes inoculated would be negative.

It is possible to calculate the most probable number of organisms/100 ml for any combination of results from such sample series. Tables have been prepared (Tables 10.1–10.4) for samples of 10 ml, 1 ml and 0.1 ml using five tubes or three tubes of each sample size and, for water testing, using one 50-ml, five 10-ml and five 1-ml samples.

This technique is used mainly for estimating coliform bacilli, but it can be used for almost any organisms in liquid samples if growth can be easily observed, e.g. by turbidity, acid production. Examples are yeasts and moulds in fruit juices and beverages, clostridia in food emulsions and rope-spores in flour suspensions.

'Black tube' MPN counts for anaerobes are described on p.407.

Mix the sample by shaking and inverting vigorously. Pipette 10-ml amounts into each of five tubes (or three) of 10 ml of double-strength medium, 1-ml amounts into each of five (or three) tubes of 5 ml of single-strength medium and 0.1-ml amounts of (or 1 ml of a 1:10 dilution) into each of five (or three) tubes of 5 ml of single-strength medium. For testing water, also add 50 ml of water to 50 ml of double-strength broth.

Double-strength broth is used for the larger volumes because the medium would otherwise be too dilute.

Incubate for 24–48 h and observe growth, or acid and gas, etc. Tabulate the numbers of positive tubes in each set of five (or three) and consult the appropriate table.

Table 10.1[a] MPN/100 ml, using one tube of 50 ml, and five tubes of 10 ml

50-ml tubes positive	*10-ml tubes positive*	*MPN/100 ml*
0	0	0
0	1	1
0	2	2
0	3	4
0	4	5
0	5	7
1	0	2
1	1	3
1	2	6
1	3	9
1	4	16
1	5	18 +

[a] Tables 10.1 to 10.3 indicate the estimated number of bacteria of the coliform group present in 100 ml of water, corresponding to various combinations of positive and negative results in the amounts used for the tests. The tables are basically those originally computed by McGrady (1918) with certain amendments due to more precise calculations by Swaroop (1951). A few values have also been added to the tables from other sources, corresponding to further combinations of positive and negative results which are likely to occur in practice. Swaroop has tabulated limits within which the real density of coliform organisms is likely to fall, and his paper should be consulted by those who need to know the precision of these estimates.

These tables and the accompanying information are reproduced from *The Bacteriological Examination of Drinking Water Supplies*, Reports on Public Health and Medical Subjects, No. 71 (1985) by permission of the Controller of Her Majesty's Stationery Office, London.

Table 10.2 MPN/100 ml, using one tube of 50 ml, five tubes of 10 ml and five tubes of 1 ml

50-ml tubes positive	*10-ml tubes positive*	*1-ml tubes positive*	*MPN/100 ml*
0	0	0	0
0	0	1	1
0	0	2	2
0	1	0	1
0	1	1	2
0	1	2	3
0	2	0	2
0	2	1	3
0	2	2	4
0	3	0	3
0	3	1	5
0	4	0	5
1	0	0	1
1	0	1	3
1	0	2	4
1	0	3	6
1	1	0	3
1	1	1	5
1	1	2	7
1	1	3	9
1	2	0	5
1	2	1	7
1	2	2	10
1	2	3	12
1	3	0	8
1	3	1	11
1	3	2	14
1	3	3	18
1	3	4	20
1	4	0	13
1	4	1	17
1	4	2	20
1	4	3	30
1	4	4	35
1	4	5	40
1	5	0	25
1	5	1	35
1	5	2	50
1	5	3	90
1	5	4	160
1	5	5	180+

Table 10.3 MPN/100 ml, using five tubes of 10 ml, five tubes of 1 ml and five tubes of 0.1 ml

10-ml tubes positive	*1-ml tubes positive*	*0.1-ml tubes positive*	*MPN/100 ml*
0	0	0	0
0	0	1	2
0	0	2	4
0	1	0	2
0	1	1	4
0	1	2	6
0	2	0	4
0	2	1	6
0	3	0	6
1	0	0	2
1	0	1	4
1	0	2	6
1	0	3	8
1	1	0	4
1	1	1	6
1	1	2	8
1	2	0	6
1	2	1	8
1	2	2	10
1	3	0	8
1	3	1	10
1	4	0	11
2	0	0	5
2	0	1	7
2	0	2	9
2	0	3	12
2	1	0	7
2	1	1	9
2	1	2	12
2	2	0	9
2	2	1	12
2	2	2	14
2	3	0	12
2	3	1	14
2	4	0	15
3	0	0	8
3	0	1	11
3	0	2	13
3	1	0	11
3	1	1	14
3	1	2	17
3	1	3	20
3	2	0	14
3	2	1	17
3	2	2	20
3	3	0	17
3	3	1	20
3	4	0	20
3	4	1	25
3	5	0	25
4	0	0	13
4	0	1	17

Table 10.3 (*continued*)

10-ml tubes positive	*1-ml tubes positive*	*0.1-ml tubes positive*	*MPN/100 ml*
4	0	2	20
4	0	3	25
4	1	0	17
4	1	1	20
4	1	2	25
4	2	0	20
4	2	1	25
4	2	2	30
4	3	0	25
4	3	1	35
4	3	2	40
4	4	0	35
4	4	1	40
4	4	2	45
4	5	0	40
4	5	1	50
4	5	2	55
5	0	0	25
5	0	1	30
5	0	2	45
5	0	3	60
5	0	4	75
5	1	0	35
5	1	1	45
5	1	2	65
5	1	3	85
5	1	4	115
5	2	0	50
5	2	1	70
5	2	2	95
5	2	3	120
5	2	4	150
5	2	5	175
5	3	0	80
5	3	1	110
5	3	2	140
5	3	3	175
5	3	4	200
5	3	5	250
5	4	0	130
5	4	1	170
5	4	2	225
5	4	3	275
5	4	4	350
5	4	5	425
5	5	0	250
5	5	1	350
5	5	2	550
5	5	3	900
5	5	4	1600
5	5	5	1800+

Table 10.4 MPN/100 ml, using three tubes each inoculated with 10, 1.0 and 0.1 ml of sample

Tubes positive			*MPN*	*Tubes positive*			*MPN*	*Tubes positive*			*MPN*
10 ml	*1.0 ml*	*0.1 ml*		*10 ml*	*1.0 ml*	*0.1 ml*		*10 ml*	*1.0 ml*	*0.1 ml*	
0	0	1	3	1	2	0	11	2	3	3	53
0	0	2	6	1	2	1	15	3	0	0	23
0	0	3	9	1	2	2	20	3	0	1	39
0	1	0	3	1	2	3	24	3	0	2	64
0	1	1	6	1	3	0	16	3	0	3	95
0	1	2	9	1	3	1	20	3	1	0	43
0	1	3	12	1	3	2	24	3	1	1	75
0	2	0	6	1	3	3	29	3	1	2	120
0	2	1	9	2	0	0	9	3	1	3	160
0	2	2	12	2	0	1	14	3	2	0	93
0	2	3	16	2	0	2	20	3	2	1	150
0	3	0	9	2	0	3	26	3	2	2	210
0	3	1	13	2	1	0	15	3	2	3	290
0	3	2	16	2	1	1	20	3	3	0	240
0	3	3	19	2	1	2	27	3	3	1	460
1	0	0	4	2	1	3	34	3	3	2	1100
1	0	1	7	2	2	0	21	3	3	3	1100+
1	0	2	11	2	2	1	28				
1	0	3	15	2	2	2	35				
1	1	0	7	2	2	3	42				
1	1	1	11	2	3	0	29				
1	1	2	15	2	3	1	36				
1	1	3	19	2	3	2	44				

From Jacobs and Gerstein's *Handbook of Microbiology*, D. Van Nostrand Company, Inc., Princeton, NJ (1960). Reproduced by permission of the authors and publisher

References

Bloomfield, S. F. (1991) Method for assessing antimicrobial activity. In *Mechanism of Action of Chemical Biocides* (eds S. P. Denyer and W. B. Hugo) Society for Applied Bacteriology Technical Series No. 20, Blackwells, Oxford, pp. 1–22.

BSI (1991) BS 5763: Part 1: *Enumeration of microorganisms: colony count technique at 30°C*, British Standards Institution, London

DHSS (1985) *The Bacteriological Examination of Drinking Water Supplies*, Reports on Public Health and Medical Subjects No. 71, HMSO, London

McCrady, M. H. (1918) Tables for rapid interpretation of fermentation test results. *Public Health Journal, Toronto*, **9**, 201–210

Miles, A. A. and Misra, S. S. (1938) The estimation of the bactericidal power of the blood. *Journal of Hygiene (Cambridge)*, **38**, 732–749

Swaroop, S. (1951) Range of variations of Most Probable Numbers of organisms estimated by dilution methods. *Indian Journal of Medical Research*, **39**, 107–131

Chapter

11

Clinical material

All clinical material should be regarded as potentially infectious. Blood specimens may contain hepatitis B and/or human immunodeficiency virus. For details of special precautions see NIH (1988); ACDP (1990); WHO (1991); Collins (1993).

The laboratory investigation of clinical material for microbial pathogens is subject to a variety of external hazards. Fundamental errors and omissions may occur in any laboratory. Vigorous performance control testing programmes and the routine use of controls will reduce the incidence of error but are unlikely to eliminate it entirely. See Chapter 2.

Most samples for microbiological examination are not collected by the laboratory staff, and it must be stressed that however carefully a specimen is processed in the laboratory, the result can only be as reliable as the sample will allow. The laboratory should provide specimen containers that are suitable, i.e. leak-proof, stable, easily opened, readily identifiable, aesthetically satisfactory, suitable for easy processing and preferably capable of being incinerated. It will be apparent that some of these qualities are mutually exclusive and the best compromise to suit particular circumstances must be made. Containers are discussed on p. 34. The specimen itself may be unsatisfactory because of faulty collection procedures, and the laboratory must be ready to state its requirements and give a reasoned explanation if cooperation is to be obtained from ward staff and doctors.

Specimens should be:

(1) collected without extraneous contamination;
(2) collected, if possible, before starting antibiotic therapy;
(3) representative; pus rather than a swab with a minute blob on the end; faeces rather than a rectal swab. If this is not possible, then a clear note to this effect should be made on the request form, which must accompany each sample.
(4) ideally collected and sent to the laboratory in a clear, sealed plastic bag to avoid leakage during transit. The request form accompanying each specimen should not be in the same bag (bags with separate compartments for specimens and request forms are available) and clearly displayed to minimize sorting errors;
(5) delivered to the laboratory without delay. If this is not possible, appropriate specimens must be delivered in a transport medium (see p.65). For other specimens, storage at 4°C will hinder bacterial overgrowth. If urine samples cannot be delivered promptly and refrigeration is not possible, boric acid preservative may be considered as a last resort. Dip slide culture is to be preferred. Sputum and urine specimens probably account for the greater part of the laboratory work load and the problems associated with their collection are particularly difficult to overcome because the trachea, mouth and urethra have a normal flora.

It must be stressed that in any microbiological examination, only that which is sought will be found and that it is essential to examine the specimen in the light of the available clinical information (which, regrettably, is often minimal). This is even more important with modern techniques which rely on the direct detection of bacterial antigens or nucleic acids in clinical specimens. The examination of pathological material for viruses and parasites is not within the scope of this book, but the possibility of infection by these agents should be considered and material referred to the appropriate laboratory.

The laboratory has an obligation to ensure that its rules are well publicized so that the periods of acceptance of specimens are known to all. Any tests that are processed in batches should be arranged in advance with the laboratory. This may be particularly important for certain tests, e.g. microbiological assays of antibiotics for which the laboratory may need to subculture indicator organisms in advance. In the UK it is now a requirement for clinical laboratory accreditation that the laboratory produces a user manual to enable the clinicians to use its services in an appropriate way.

Blood cultures

Blood must be collected with scrupulous care to avoid extravenous contamination. In patients with a recurrent fever, blood culture is an important diagnostic procedure and the highest success rate is associated with the collection of cultures just as the patient's temperature begins to increase rather than at the peak of the rigor. In patients with septicaemia, the timing is less important and a rapid answer and a susceptibility test result are essential. The yield from blood cultures may be increased by culturing a larger volume of blood, although it is rarely necessary to culture more than three sets from each patient.

Castenada blood culture bottles are biphasic, i.e. they have a slope of solid medium and a broth which may also contain Liquoid (sodium polyethanol sulphonate: Roche) which neutralizes antibacterial factors including complement. Large amounts of blood (up to 50% of the total volume of the medium) may be examined in Liquoid cultures; otherwise, it is necessary to dilute the patient's blood at least 1:20 with broth. The bottles are tipped so that the blood-broth mixture washes over the agar slope every 72 h. Colonies will appear first at the interphase and then all over the slope. The bottles should be retained for at least 4 and preferably 6 weeks. The Castenada system avoids repeated subculture and lessens the risk of laboratory contamination. A similar principle is used in a commercial system (Roche SeptiChek), in which a plastic paddle is attached to the top of the culture bottle, permitting subculture to more than one medium. It is important to avoid contamination because patients receiving immunosuppressive drugs are likely to be infected with organisms not usually regarded as pathogens. No organism can therefore be automatically excluded as a contaminant.

Vacuum blood collection systems also help to reduce contamination. Some of these ensure a 10% carbon dioxide atmosphere in the container. This is important when organisms such as *Brucella* spp. are sought. If other media are used they should be incubated in an atmosphere containing at least 10% carbon dioxide.

Evacuated containers or those containing additional gas may need 'venting' with a cotton wool plugged hypodermic needle (*caution*) or a commercial device. This is important, especially for the recovery of *Candida* spp. and

Pseudomonas spp. from neonatal samples where only a small volume of blood is available for culture.

A variety of enriched media is available for blood culture

Blood cultures should be checked for growth at least daily and subcultured to appropriate media when growth is evident, either by visual inspection or by one of the current automated detection systems (Chapter 8).

Rapid systems for the detection of growth in blood cultures are now available. The 'Bactec System' (Becton Dickinson) depends on the detection of radioactive CO_2 in the headspace, released during bacterial growth from labelled glucose in the medium. This instrument saves much time and labour in subculturing 'negative' bottles. The system has recently been extended to include a non-radiometric method for detecting bacterial growth. This uses infrared analysis of microbially-generated CO_2.

Other promising new commercial systems for the detection of growth in blood cultures rely on the visual recognition of gas production (Oxoid Signal) or electrical impedance. Other systems in which blood is cultured directly on solid media after concentration (e.g. the Dupont Isolator) may have advantages for the detection of specific organisms, e.g. *Candida*.

For more information about automated methods for blood culture see Chapter 8.

Likely pathogens

Staphylococcus aureus, Escherichia coli and other enterobacteria including salmonellas, streptococci and enterococci, *Neisseria meningitidis, Haemophilus influenzae, Brucella* spp, clostridia and pseudomonads. In ill or immunocompromised patients, any organisms, especially if recovered more than once, may be significant (e.g. coagulase-negative staphylococci, coryneforms). In some circumstances special media may have to be used to look for specific organisms (e.g. mycobacteria in AIDS patients). In patients who have had surgery, a mixed growth of other organisms may be found.

Commensals

None. Common contaminants (from skin) include coagulase-negative staphylococci, coryneforms. Transient bacteraemia with other organisms may occur occasionally.

All manipulations with open blood culture bottles should be done in a microbiological safety cabinet.

Cerebrospinal fluid (CSF)

It is important that these samples should be examined with the minimum of delay. As with other fluids, a description is important. It is essential to note the colour of the fluid; if it is turbid, then the colour of the supernatant fluid after centrifugation may be of diagnostic significance. A distinctive yellow tinge (xanthochromia) usually confirms an earlier bleed into the CSF (as in, for example, a subarachnoid haemorrhage as opposed to traumatic bleeding during the lumbar puncture) (it is necessary, of course, to compare the colour of the supernatant fluid with a water blank in a similar tube).

Before any manipulation, the specimen must be carefully examined for clots, either gross or in the form of a delicate spider's web that is usually associated with a significantly raised protein and traditionally is taken as an indication that one should look particularly for *Mycobacterium tuberculosis*. The presence of a clot will invalidate any attempt to make a reliable cell count.

If the cells and/or protein are raised (> three cells $\times 10^6$/l and/or > 40 mg/100 ml), a CSF sugar determination should be compared with the blood sugar level collected at the same time. This and the type of cell may indicate the type of infection.

Raised polymorphs with very low sugar – usually indicate bacterial infection or a cerebral abscess.
Raised lymphocytes with normal sugar – usually indicate viral infection.
Raised lymphocytes with lowered sugar – usually indicate tuberculosis.
Raised polymorphs or mixed cells with lowered sugar – usually indicate early tuberculosis. It is important, however, to note that there are many exceptions to these patterns.

Make three films from the centrifuged deposit, spreading as little as possible. Stain one by the Gram method, one with a cytological stain and retain the third for examination if indicated for acid-fast bacilli. A prolonged search may be necessary if tuberculosis is suspected.

Plate on blood agar and incubate aerobically and anaerobically and on a chocolate agar plate for incubation in a 5–10% carbon dioxide atmosphere. Examine after 18–24 h. If organisms are found, set up a direct sensitivity test including penicillin, ampicillin, chloramphenicol and sulphonamides and third-generation cephalosporins (e.g. cefotaxime). Culture for *M. tuberculosis* as indicated.

Commercial systems for the detection (e.g. by latex agglutination) of the common bacterial causes of meningitis (*Streptococcus pneumoniae, Neisseria meningitidis, Haemophilus influenzae*) are now available. They should be used in all cases of suspected bacterial meningitis but they are particularly useful when the Gram stain is unhelpful or when the patient has already received antibiotics.

Likely pathogens

Haemophilus influenzae, N. meningitidis, Streptococcus pneumoniae, Listeria monocytogenes, M. tuberculosis and, in very young babies, coliform bacilli, Group B streptococci and *Pseudomonas aeruginosa. Staphylococcus epidermidis* and some micrococci are often associated with infections in patients with devices (shunts) inserted to relieve excess intracranial pressure.

Opportunists

Any organism introduced during surgical manipulation involving the spinal canal may be involved in meningitis. Infections with other agents may occur in the immunocompromised patient (e.g *Cryptococcus neoformans* in AIDS).

Commensals

None.

Dental specimens

These may be whole teeth, when there may have been an apical abscess, or scrapings of plaque or carious material. Plate on selective media for staphylococci, streptococci and lactobacilli.

Likely organisms

S. aureus, streptococci, especially *S. mutans*, *S. milleri*, lactobacilli.

Ear discharges

Swabs should be small enough to pass easily through the external meatus. Many commercially available swabs are too plump and may cause pain to patients with inflamed auditory canals.

A direct film stained by the Gram method is helpful, as florid overgrowths of faecal organisms are not uncommon and it may be relatively difficult to recover more delicate pathogens. Plate on blood agar and incubate aerobically and anaerobically overnight at 37°C. Tellurite medium may be advisable if the patient is of school age. A medium that inhibits spreading organisms is frequently of value; CLED, or MacConkey, chloral hydrate or phenethyl alcohol agar are the most useful.

Likely pathogens

S. pyogenes, S. pneumoniae, S. aureus, Haemophilus, Corynebacterium diphtheriae, P. aeruginosa, coliform bacilli, anaerobic Gram-negative rods, fungi.

Commensals

Micrococci, diphtheroids, *S. epidermidis*, moulds and yeasts.

Eye discharges

It is preferable to plate material from the eye directly on culture media rather than to collect it on a swab. If this is not practicable then it is essential that the swab is placed in transport medium.

Examine a direct Gram-stained film before the patient is allowed to leave. Look particularly for intracellular Gram-negative diplococci and issue a tentative report if they are seen so that treatment can be started without delay; this is of particular importance in neonates. Infections occurring a few days after birth may be caused by *Chlamydia trachomatis*, which should be suspected if films have excess numbers of monocytes or the condition does not resolve rapidly. Detection of chlamydial infections may be attempted by microscopy (with Giemsa stain), cell culture or immunological methods, e.g. immunofluorescence (see p.120).

Plate on blood agar plates and incubate aerobically and anaerobically and on a chocolate agar or 10% horse blood Columbia agar plate for incubation in a 5% carbon dioxide atmosphere.

Likely pathogens

S. aureus, S. pneumoniae, viridans streptococci, *N. gonorrhoeae, Haemophilus, Chlamydia, Moraxella*, rarely coliform organisms, *C. diphtheriae, P. aeruginosa, Candida, Aspergillus, Fusarium*.

Commensals

S. epidermidis, micrococci, diphtheroids.

Faeces and rectal swabs

It is always better to examine faeces than rectal swabs. Swabs that are not even stained with faeces are useless. If swabs must be used, they should be moistened with sterile saline or water before use and care must be taken to avoid contamination with perianal flora. It may be useful to collect material from babies by dipping the swabs into a recently soiled napkin (diaper). Swabs should be sent to the laboratory in transport medium unless they can be delivered within 1 h.

Because of the random distribution of pathogens in faeces, it is customary to examine three sequential specimens.

Plate on DCA, XLD and MacConkey agar and inoculate selenite or tetrathionate broth. Incubate for 18–24 h and subculture selenite and tetrathionate broths on to DCA. It is essential that the MacConkey agar will support the growth of *S. aureus*, which may be causative in staphylococcal enterocolitis.

Plate stools from children, young adults and diarrhoea cases on campylobacter medium plus antibiotic supplements. Incubate for 48 h at 42°C and at 37°C in 5% oxygen and 10% carbon dioxide.

If the patient is under 3 years old, plate on CLED, MacConkey agar and blood agar and look for enteropathogenic *Escherichia coli*. In cases of haemorrhagic colitis or haemolytic uraemic syndrome culture for *E. coli* 0157 on sorbitol MacConkey agar (this organism ferments sorbitol). Culture also for *Aeromonas* and *Plesiomonas*.

Plate on TCBS and inoculate alkaline peptone water if cholera is suspected, if the patient has recently returned by air from an area where cholera is endemic or if food poisoning due to *Vibrio parahaemolyticus* is suspected.

Yersinia enterocolitica is now recognized as an enteric pathogen. Plate on Yersinia Selective Agar and incubate at 32°C for 18–24 h. See also p.346.

Plate on cycloserine cefoxitin egg yolk fructose agar (CCFA) for *Clostridium difficile* and test the stool for *Clostridium difficile* toxins in either cell culture or with a commercial kit.

In cases of food poisoning due to *S. aureus*, plate faeces on blood agar and on one of the selective staphylococcal media (p.64) and inoculate salt meat broth. Incubate all cultures overnight at 37°C and subculture the salt meat broth to blood agar and selective staphylococcal medium. It should be noted that here the symptoms relate to a toxin so the organisms may not be recovered.

In food poisoning where *Clostridium perfringens* may be involved make a 1:10 suspension of faeces in nutrient broth and add 1 ml to each of two tubes of Robertson's cooked meat medium. Heat one tube at 80°C for 10 min and then cool. Incubate both tubes at 37°C overnight and plate on blood agar with and without neomycin and on egg yolk agar. Place a metronidazole disc on the streaked-out inoculum. Incubate at 37°C anaerobically overnight.

Emulsify faeces in ethanol (industrial grade) to give a 50% suspension. Mix well and stand for 1 h. Inoculate media and incubate as above.

Likely pathogens

Salmonella, Shigella, Vibrio, enteropathogenic *E. coli, C. perfringens* (both heat resistant and non-heat resistant), *S. aureus, Bacillus cereus, Campylobacter, Yersinia, Aeromonas, C. botulinum*.

Commensals

Coliform bacilli, *Proteus, Clostridium, Bacteroides, Pseudomonas*.

Nasal swabs (see also Throat swabs)

The area sampled will influence the recovery of particular organisms, i.e. the anterior part of the nose must be sampled if the highest carriage rate of staphylococci is to be found. For *N. meningitidis*, it may be preferable to sample further in and for the recovery of *Bordetella pertussis* in cases of whooping cough a pernasal swab is essential.

Inoculate as soon as possible on charcoal agar containing 10% horse blood and 40 mg/l cephalexin and incubate at 35°C in a humid atmosphere.

Direct films are of no value. If *M. leprae* is sought, smears from a scraping of the nasal septum may be examined after staining with ZN stain or a fluorescence stain.

Plate on blood agar and tellurite media and incubate as for throat swabs.

Likely pathogens

S. aureus, S. pyogenes, N. meningitidis, B. pertussis, C. diphtheriae and other pathogens of uncertain significance.

Commensals

Diphtheroids, *S. epidermidis, S. aureus, Branhamella*, aerobic spore bearers and small numbers of Gram-negative rods (*Proteus* and coliform bacilli).

Pus

As a general rule, pus rather than swabs of pus should be sent to the laboratory. Swabs are variably lethal to bacteria within a few hours because of a combination of drying and toxic components of the cotton wool released by various sterilization methods. Consequently, speed in processing is important; a delay of several hours may allow robust organisms to be recovered while the more delicate pathogens may not survive. Swabs should be moistened with broth before use on dry areas.

Examine Gram-stained films and films either direct or of concentrated material (p.412), for acid-fast rods. If actinomycosis is suspected wash the pus with sterile water and look for 'sulphur granules'. Plate these on blood agar and incubate aerobically and anaerobically. Include a chloral hydrate plate in case *Proteus* is present. Prolonged anaerobic incubation of blood agar cultures, in a 90% hydrogen–10% carbon dioxide atmosphere possibly with neomycin or nalidixic acid may allow the recovery of strictly anaerobic *Bacteroides* spp. These organisms may be recovered frequently if care is taken and the number of 'sterile' collections of pus consequently reduced. Culture in thioglycollate broth may be helpful. The dilution effect may overcome specific and non-specific inhibitory agents.

Culture pus from post-injection abscesses for mycobacteria (p.412).

The recovery of anaerobes will be enhanced by the addition of menadione and vitamin K to the medium. Wilkins and Chalgren's medium is also of value.

If the equipment is available gas-liquid chromatography is useful for the detection of volatile fatty acids which may indicate anaerobes (but the bacteriologist's nose is cheaper, even if less sensitive).

Likely pathogens

S. aureus, S. pyogenes, anaerobic cocci, *Mycobacterium, Actinomyces, Pasteurella, Yersinia, Clostridium, Neisseria, Bacteroides, B. anthracis, Listeria, Proteus, Pseudomonas, Nocardia*, fungi, other organisms in pure culture.

Commensals

None.

Serous fluids

These fluids include pleural, synovial, pericardial, hydrocele, ascitic and bursa fluids.

Record the colour, volume and viscosity of the fluid. Centrifuge at 3000 rev/min for at least 15 min preferably in a sealed bucket and note the colour of the supernatant. Transfer the supernatant fluid to another bottle and record its appearance. Make Gram-stained films of the deposit and also stain films for cytological evaluation. Plate on blood agar and incubate aerobically and anaerobically at 37°C for 18–24 h. Culture on blood agar anaerobically plus 10% carbon dioxide atmosphere and incubate for 7–10 days for *Bacteroides*. Use additional material as outlined for pus. Prolonged incubation may be necessary for these strictly anaerobic organisms.

Gonococcal and meningococcal arthritis must not be overlooked; inoculate one of the GC media and incubate in a 10% carbon dioxide atmosphere for 48 h at 22 and 37°C. Culture, if appropriate, for mycobacteria.

Likely pathogens

S. pyogenes, S. pneumoniae, anaerobic cocci, *S. aureus, N. gonorrhoeae, M. tuberculosis, Bacteroides*.

Commensals

None.

Sputum

Sputum is a difficult specimen. Ideally, it should represent the discharge of the bronchial tree expectorated quickly with the minimum of contamination from the pharynx and the mouth, and delivered without delay to the laboratory.

Because of the irregular distribution of bacteria in sputum it may be advisable to wash portions of purulent material to free them from contaminating mouth organisms before examining films and making cultures.

Gram-stained films may be useful, suggesting a predominant organism.

Homogenization of the specimen with Sputolysin, *N*-acetylcysteine or pancreatin, for example, followed by dilution enables the significant flora to be assessed. In sputum cultures not so treated initially small numbers of contaminating organisms may overgrow pathogens.

The following method is recommended.

Add about 2 ml of sputum to an equal volume of Sputolysin (Calbiochem) or Sputasol (Oxoid) in a screw-capped bottle and allow to digest for 20–30 min with occasional shaking, e.g. on a Vortex Mixer. Inoculate a blood agar plate with 0.01 ml of the homogenate, using a standard loop (e.g. a 0.01 ml plastic loop) and make a Gram-stained film. Add 0.01 ml with a similar loop to 10 ml of peptone broth. Mix this thoroughly, preferably using a vortex mixer. Inoculate blood agar and MacConkey agar with 0.01 ml of this dilution. Incubate the plates overnight at 35°C; incubate the blood agar plates in a 10% carbon dioxide atmosphere. Chocolate agar may be used but should not be necessary for growing *H. influenzae* if a good blood agar base is used.

Five to 50 colonies of a particular organism on the dilution plate are equivalent to 10^6–10^7 of these organisms/ml of sputum and this level is significant except for viridans streptococci when one to five colonies are equivalent to 10^5–10^6 organisms/ml and this level is equivocal.

Set up direct sensitivity tests if the Gram film shows large numbers of any particular organisms.

Methods of examining sputum for mycobacteria are described in Chapter 45.

For the isolation of *Legionella* and similar organisms from lung biopsies or bronchial secretions use blood agar or one of the commercial legionella media plus supplement.

Likely pathogens

S. aureus, S. pneumoniae, H. influenzae, coliform bacilli, *Klebsiella pneumoniae, Pasteurella, Mycobacterium, Candida* spp., *Moraxella* (*Branhamella*) *catarrhalis, Mycoplasma, Pasteurella*/*Yersinia* spp., *Y. pestis, Legionella pneumophila, Aspergillus, Histoplasma, Cryptococcus, Blastomyces* and *Pseudomonas,* particularly from patients with cystic fibrosis (also *Chlamydia pneumoniae, C. psittaci* and *C. burnettii*).

Commensals

S. epidermidis, micrococci, coliforms, *Candida* in small numbers, viridans streptococci.

Throat swabs

Collect throat swabs carefully with the patient in a clear light. It is customary to attempt to avoid contamination with mouth organisms. Evidence has been advanced that the recovery of *Streptococcus pyogenes* from the saliva may be higher than from a 'throat' or pharyngeal swab. Make a direct film, stain with dilute carbol fuchsin and examine for Vincent's organisms, yeasts and mycelium.

Plate on blood agar and incubate aerobically and anaerobically for a minimum of 18 but preferably 48 h in 7–10% CO_2 at 35°C. Plate on tellurite medium and incubate for 48 h in 7–10% CO_2 at 35°C.
NB. For details of the logistics of mass swabbing, e.g. in a diphtheria outbreak, see the paper by Collins and Dulake (1983).

Likely pathogens

S. pyogenes, Corynebacterium diphtheriae, C. ulcerans, S. aureus, Candida albicans, Neisseria meningitidis, Borrelia vincenti (*H. influenzae* Pittmans type b in the epiglottis is certainly a pathogen).

Commensals

Branhamella, viridans streptococci, *S. epidermidis*, diphtheroids, *S. pneumoniae*, probably *Haemophilus influenzae*.

Tissues, biopsy specimens, post-mortem material

As a general rule, retain all material submitted from autopsy until the forensic pathologists agree to their disposal. If it is essential to grind up tissue, obtain approval.

Tissue specimens may be ground up after suitable selection procedures in a blender or, if the specimen is very small, in a Griffith's tube or with glass beads on a Vortex mixer. The Stomacher Lab-Blender is best for larger specimens as it saves time and eliminates the hazard of aerosol formation. Examine the homogenate as described under 'Pus'.

Urethral discharges

Swabs of discharge should be collected into a transport medium or plated directly at the bedside or in the clinic. Examine direct Gram-stained films. Look for intracellular Gram-negative diplococci. In males a Gram-stained film of the urethral discharge is a sensitive and specific means of diagnosing gonorrhoea. In the female, the recovery of organisms is more difficult and the interpretation of films may occasionally present problems because over-decolorized or dying Gram-positive cocci may resemble Gram-negative diplococci.

Plate on chocolate agar and a selective medium for gonococci (e.g. Modified Thayer-Martin, Modified New York City medium) and incubate in a humid atmosphere enriched with 5% CO_2.

Non-gonococcal urethritis is increasingly associated with chlamydial infections. These organisms (not considered in this book) may be isolated by tissue culture techniques or detected by commercial ELISA or immunofluorescence systems. Special culture methods may be used for mycoplasmas and ureaplasmas (see Chapter 49).

Likely pathogens

N. gonorrhoeae, Mycoplasma, and *Ureaplasma*. Occasionally other pathogens may be isolated, particularly coliforms when a urinary catheter is present.

Commensals

S. epidermidis, micrococci, diphtheroids, small numbers of coliform organisms.

Urine

The healthy urinary tract is free from organisms over the greater part of its length; there may, however, be a few transient organisms present, especially at the lower end of the female urethra.

Collection

Catheters may contribute to urinary infections and are no longer considered necessary for the collection of satisfactory samples from females.

A 'clean catch' mid-stream urine (MSU) is the most satisfactory specimen for most purposes if it is delivered promptly to the laboratory. Failing prompt delivery, the specimen can be kept in a refrigerator at 4°C for a few hours or, exceptionally, a few crystals of boric acid may be added.

Dip slides are now widely used for assessing urinary tract infections. They do, however, preclude examination for cells and other elements unless a fresh specimen is also submitted.

Examination

This includes counting leucocytes, erythrocytes and casts, estimating the number of bacteria/ml and identifying them. Normally, there will be none or less than 20×10^6/l of leucocytes or other cells and a colony count of less than 10^4 organisms/ml.

Mix the urine by rotation, never by inversion. Count the cells with a Fuchs Rosenthal or similar counting chamber or use an inverted microscope. Inoculate a blood agar plate and a MacConkey, CLED or EMB plate with 0.001 ml using a standard loop (disposable loops or micropipettes with disposable tips are satisfactory alternatives). Also 0.1 ml of a 1 in 100 dilution of urine may be used. Spread with a glass spreader, incubate overnight at 37°C and count the colonies.

< 10 colonies	= 10^4 organisms/ml	— not significant
10–100 colonies	= 10^4 – 10^5 organisms/ml	— doubtful significance
> 100 colonies	= > 10^5 organisms/ml	— significant bacteruria

If dip slides are submitted, incubate them overnight and follow the manufacturer's directions for counting.

Note that the accuracy and usefulness of bacterial counts on urines is limited by the age of the specimen. Only fresh specimens are worth examining. Other limitations are the volume of fluid recently drunk and the amount of urine in the bladder.

Various other techniques, e.g. a filter paper 'foot', multipoint inoculator, dip sticks for leucocyte enzymes, or simple visual inspection have been advocated for screening urines. These save media and may be appropriate in laboratories where the majority of samples tested are 'negative'.

For methods of examining urine for tubercle bacilli, see p.413.

Automated urine examination

The Rapid Automatic Microbiological Screening system (RAMUS, Orbec), is a particle volume analyser, capable of counting leucocytes and bacteria. Counts of bacteria $>10^5$/ml are regarded as significant.

Multipoint inoculation methods are very useful in the examination of large batches of urines. (See Faiers *et al.* (1991) and Chapter 8.) The technique can allow for tests for the presence of antibacterial substances, direct inoculation of identification media and media for those fastidious organisms that are now considered to be of importance in urinary tract infections.

Likely pathogens

E. coli, other enterobacteria, *Proteus*, staphylococci including *S. epidermidis*, *S. saprophyticus*, enterococci, *Salmonella* (rarely), leptospira, mycobacteria.

Commensals

Small numbers ($<10^4$/ml) of almost any organisms. It is unwise to be too dogmatic about small numbers of organisms isolated from a single specimen: it is better to repeat the sample.

Vaginal discharges

Bacterial vaginosis is a condition in which there is a disturbance in the vaginal flora. Lactobacilli decrease in numbers while those of a variety of others, e.g. *Gardnerella vaginalis, Mobiluncus, Prevotella* increase. This condition should be differentiated from vaginitis caused by *Candida albicans*. A Gram-stained film may reveal 'clue cells' – epithelial cells whose margins are obscured by large numbers of adherent Gram-variable rods may be seen. Culture for *G. vaginalis* is less helpful.

Puerperal infections

A high vaginal swab is usually taken. Swabs that cannot be processed very rapidly after collection should be put into transport medium. Examine direct Gram-stained films. If non-sporing square-ended Gram-positive rods, not in pairs, and which appear to be capsulated are seen they should be reported without delay, as they may be *C. perfringens*, which may be an extremely invasive organism.

Plate on blood agar and incubate aerobically and anaerobically overnight at 37°C and also plate on MacConkey or other suitable selective medium. Plate any swabs suspected of harbouring *C. perfringens* on a neomycin half-antitoxin Nagler plate for anaerobic incubation.

Likely pathogens

S. pyogenes, L. monocytogenes, C. perfringens, Bacteroides, G. vaginalis, anaerobic cocci, *Candida* spp., excess numbers of enteric organisms.

Commensals

Lactobacilli, diphtheroids, micrococci, *S. epidermidis*, small numbers of coliform bacilli, yeasts.

Non-puerperal infections

Examine a Gram-stained smear and also a wet preparation for the presence of *Trichomonas vaginalis*. If the examination is to be immediate, then a saline suspension of vaginal discharge will be satisfactory, otherwise a swab in transport medium is essential. A wet preparation may still be examined for actively motile flagellates but culture in Trichomonas medium usually yields more reliable results, especially in cases of light infection. An alternative to the wet preparation is the examination under the fluorescence microscope of an air-dried smear stained by acridine orange.

Examine a Gram-stained smear for *Mobiluncus*. If *N. gonorrhoeae* infection is suspected in the female, then swabs from the urethra, cervix and posterior fornix should be collected, placed into transport medium or plated immediately on selective medium. Many workers do not make films from these swabs because the difficulties of interpretation are compounded by the presence of charcoal from the buffered swabs; they rely upon cultivation using VCN (vancomycin, colistin, nystatin) or VCNT (VCN + trimethoprim) medium as a selective inhibitor of organisms other than *N. gonorrhoeae*.

Plate on blood agar and incubate, aerobically and anaerobically, Sabouraud agar (when looking for *Candida*) and chocolate agar or another suitable gonococcal medium for incubation in a 5% carbon dioxide atmosphere to encourage isolation of *N. gonorrhoeae*.

If *L. monocytogenes* infection is suspected plate on serum agar with 0.4 g/l of nalidixic acid added.

Likely pathogens

N. gonorrhoeae, C. albicans and related yeasts in moderate numbers, *S. pyogenes, Gardnerella, L. monocytogenes, Haemophilus, Bacteroides*, anaerobic cocci, *Mobiluncus*, possibly *Mycoplasma, T. vaginalis*.

Commensals

Lactobacilli, diphtheroids, enterococci, small numbers of coliform organisms.

Wounds (superficial) and ulcers

Examine Gram-stained films and culture on blood agar and proceed as for pus, etc. Examine Albert-stained smears for corynebacteria. Culture also on Lowenstein–Jensen medium and incubate at 30°C because some mycobacteria that cause superficial infections do not grow on primary isolation at 35–37°C.

Likely pathogens

S. aureus, streptococci, *C. diphtheriae, C. ulcerans* (both in ulcers) *M. marinum, M. ulcerans, M. chelonei.*

Contaminants

Staphylococci, pseudomonas and a wide variety of bacteria and fungi.

Wounds (deep) and burns

The collection of these specimens and their consequent rapid processing are two vital factors in the recovery of the significant organisms. Burn cases frequently become infected with staphylococci and streptococci as well as with various Gram-negative rods, especially *Pseudomonas* spp. It may be difficult to recover Gram-positive cocci for the overgrowing Gram-negative organisms.

Plate on blood agar and incubate aerobically and anaerobically. Plate also on MacConkey agar and CLED medium (to inhibit the spreading of *Proteus* spp.). Robertson's cooked meat medium may be useful.

Plate swabs, etc. from burns on media selective for *S. pyogenes* and *Pseudomonas*. Anaerobes are rarely a problem in burns.

Likely pathogens

Clostridium, S. pyogenes, S. aureus, anaerobic cocci, *Bacteroides*, Gram-negative rods.

Contaminants

Small numbers of a wide variety of organisms.

References

ACDP (1990) *HIV – the Causative Agent of AIDS and Related Conditions*, Advisory Committee on Dangerous Pathogens, HMSO, London

Collins, C. H. (1993) *Laboratory-acquired Infections*, 3rd edn. Butterworth–Heinemann, Oxford, pp. 205–213
Collins, C. H. and Dulake, C. (1983) Diphtheria: the logistics of mass swabbing. *Journal of Infection*, **6**, 277–280
Faiers, M., George, R., Jolly, J. and Wheat, P. (1991) *Multipoint Methods in the Clinical Laboratory*, Public Health Laboratory Service, London
NIH (1988) *Working Safely with HIV in the Research Laboratory*, National Institutes of Health, Bethesda, MD
WHO (1991) *Biosafety Guidelines for Diagnostic and Research Laboratories Working with HIV*, WHO AIDS Series 9, World Health Organization, Geneva

Chapter

12

Antimicrobial susceptibility tests

The primary purpose of antimicrobial susceptibility testing is to guide the clinician in the choice of appropriate agents for therapy. In practice, agents are commonly used empirically and the laboratory test serves to explain treatment failures and to provide a range of suitable alternative agents. Routine testing also provides up-to-date accumulated data from which information on the most suitable agents for empirical use can be derived. Apart from routine clinical work, antimicrobial susceptibility tests are used to evaluate the *in vitro* activity of new agents.

Antimicrobial susceptibility may be reported qualitatively, as sensitive, intermediate or resistant, or quantitatively in terms of the concentration of the agent which inhibits the growth of the organism, the minimum inhibitory concentration (MIC), or that which kills it, the minimum bactericidal concentration (MBC). The concentrations of agents defining the 'breakpoints' between the different categories of susceptibility are based on clinical, pharmacological and microbiological considerations (Phillips *et al.*, 1991).

Laboratory staff and clinicians should be aware of the meanings of the categories of susceptibility.

Sensitive – indicates that the standard dose of the agent should be appropriate for treating the patient infected with the strain tested.

Intermediate – can also be referred to as moderately resistant, moderately susceptible or indeterminant. This indicates that the strain may be inhibited by larger doses of the agent or may be inhibited in sites where the agent is concentrated (e.g. urine). It may also be regarded as a 'buffer-zone' which reduces the likelihood of major errors, i.e. reporting a resistant isolate as susceptible or vice versa. Caution is necessary when the basis of resistance is the production of inactivating enzymes as zone sizes or MICs may indicate intermediate susceptibility rather than resistance. The routine use of the intermediate category on patient reports should be avoided as clinicians rarely find it helpful. In contrast, laboratory staff should be aware of the meaning and significance of this category of susceptibility.

Resistant – indicates that an infection caused by the isolate tested is unlikely to respond to treatment with that antimicrobial agent.

Breakpoint concentrations of antimicrobial agents, which may be used to separate susceptible from resistant and/or intermediate isolates on the basis of growth or no growth at the selected concentrations, have been published in several countries (Faiers *et al.*, 1991). There is currently considerable debate about the standardization of breakpoints in Europe (Baquero, 1990). Such breakpoint concentrations may be used to interpret MIC results, as test concentrations in breakpoint methods, or as the basis for establishing zone size breakpoints in diffusion tests once the relationship of zone sizes to MICs by closely defined methods has been established for the agents tested.

Technical methods

The three principal methods used for susceptibility testing are outlined briefly below. There follows a discussion of technical factors relevant to all methods and subsequently detailed consideration of the individual methods.

Diffusion tests

Diffusion tests with agents in paper discs are still the most widely used methods in routine clinical laboratories. They are technically straightforward and relatively cheap.

Minimum inhibitory concentration

In minimum inhibitory concentration (MIC) methods the susceptibility of organisms to serial twofold dilutions of agents in agar or broth is determined. The MIC is defined as the lowest concentration of the agent that inhibits visible growth. The methods are not widely used for routine testing in clinical laboratories in most countries, although commercial systems with restricted range MICs in microtitre format are frequently used in some. The most widespread use of MIC methods is in testing new agents. In clinical laboratories MIC methods are used to test the susceptibility of organisms which give equivocal results in diffusion tests, for tests on organisms where disc tests may be unreliable, as with slow-growing organisms, or when a more accurate result is required for clinical management, e.g. infective endocarditis. Dilution MIC methods are accepted as the standard against which other methods are assessed. In some cases it is necessary to know the minimum concentration of an agent that kills, rather than merely inhibits growth of, a bacterium. The minimum bactericidal concentration (MBC) is determined by subculturing from tubes showing no turbidity onto antimicrobial agent-free media and observing for growth after further incubation.

Breakpoint methods

Breakpoint methods are essentially MIC methods in which only one or two concentrations are tested. The concentrations chosen for testing are those equivalent to the breakpoint concentrations separating different categories of susceptibility. In recent years such methods have been used routinely as an alternative to disc diffusion methods. Breakpoint tests are suitable for automated inoculation and reading of tests, and may be combined with identification tests.

Factors affecting susceptibility tests

Many of the effects of technical variation on susceptibility testing are widely known (Barry, 1976 and 1991; Brown and Blowers, 1978; Phillips *et al.* 1991).

Medium

The medium should support the growth of organisms normally tested and should not contain antagonists of antimicrobial activity. For example, thymidine antagonizes the activity of sulphonamides and trimethoprim, leading to false reports of resistance to these agents. If the medium is not low in thymidine, lysed horse blood or thymidine phosphorylase should be added to remove free thymidine. Monovalent cations such as Na^+ increase the activity of bacitracin, fusidic acid and novobiocin against staphylococci, and of penicillins against *Proteus* spp. Divalent cations such as Mg^{2+} and Ca^{2+} reduce the activity of aminoglycoside antimicrobials and polymyxins against *Pseudomonas* spp., and of tetracyclines against a range of organisms.

Media specially formulated for susceptibility testing are recommended. They should not be antagonistic to antimicrobial activity and should be consistent in their performance. Iso-Sensitest agar (Oxoid) is popular in the UK, and Mueller-Hinton agar (various suppliers) is widely used in several other countries. There have been problems with consistency of batches of Mueller-Hinton which has led to the publication of a performance standard (NCCLS, 1987 and 1992). Even with Iso-Sensitest agar, which is produced by a single manufacturer, there have been problems with variation among batches (Andrews *et al.*, 1990).

Supplements are necessary for growth of some organisms. Add 5% defibrinated blood to media for testing fastidious organisms such as streptococci. Use chocolate agar for *Haemophilus* and *Neisseria* spp., or add commercially available supplements such as haemoglobin and Isovitalex if they do not interfere with antimicrobial activity (Phillips *et al.*, 1991). NCCLS (1990a,b) recommends a supplemented Mueller-Hinton medium for testing *Haemophilus* spp. and a supplemented GC agar for *Neisseria* spp.

pH

The activity of aminoglycosides, macrolides, lincosamides and nitrofurantoin increases with increasing pH. Conversely, the activity of tetracycline, fusidic acid and novobiocin increases as the pH is lowered.

Depth of agar medium

Plates should have a level depth of 3–4 mm (18–25 ml in a 9 cm petri dish). With diffusion tests the size of the zone increases as the depth of the medium decreases but this effect is most marked with very thin plates, which should be avoided.

Inoculum size

Increasing inoculum size reduces the susceptibility to agents in both diffusion and dilution tests. Tests of susceptibility to sulphonamides, trimethoprim and β-lactam agents where resistance is based on β-lactamase production in staphylococci, *Haemophilus* spp. and *Neisseria* spp., and methicillin resistance in staphylococci, are particularly affected by variation in inoculum size. Among Gram-negative organisms the effect of inoculum on β-lactam MICs depends on the amount of any β-lactamase produced, and the activity of the enzyme against the particular penicillin or cephalosporin under test. Standardization of the inoculum size is therefore necessary.

In most diffusion methods an inoculum size that gives a semi-confluent growth is used. This has the advantage that incorrect inocula can be readily identified and the test rejected. There is no evidence, however, that semi-confluent growth gives more reliable results than just confluent growth, as used in the NCCLS diffusion method.

In agar dilution (MIC or breakpoint) methods an inoculum of 10^4 cfu/inoculum spot is the accepted standard. This should be increased to 10^6 cfu/inoculum spot for methicillin susceptibility tests on staphylococci. With organisms producing extracellular β-lactamases the standard inoculum may result in MICs only slightly higher than those obtained with susceptible strains. This can be avoided by increasing the inoculum to 10^6 cfu/inoculum spot on agar dilution plates. For broth dilution methods the standard inoculum should be 10^5 cfu/ml of test broth at the beginning of the incubation period.

Use four or five colonies of a pure culture to avoid selecting an atypical variant. Prepare the inoculum by emulsifying overnight colonies from an agar medium, by diluting an overnight broth culture or by diluting a late log phase broth culture. The broth should not be antagonistic to the agent tested (e.g. media containing thymidine should not be used to prepare inocula for tests on sulphonamides and trimethoprim). Organisms should be diluted in distilled water to a concentration of 10^7 cfu/ml. A 0.5 McFarland standard gives a density equivalent to approximately 10^8 cfu/ml. Alternatively, inocula can be adjusted photometrically or by standard dilution of broth cultures if the density of such cultures is reasonably constant. If there is any doubt about whether a particular inoculum preparation procedure yields the requisite density of organisms it should be checked by performing viable counts on a small number of representative isolates before embarking on a series of MIC tests. Inocula should be used promptly to avoid subsequent changes in their density. Plates should be inoculated within 30 min of standardizing the inoculum.

Preincubation and prediffusion

In diffusion tests preincubation and prediffusion decrease and increase the sizes of zones respectively. In practice these effects are avoidable by applying discs soon after plates are inoculated and incubating plates soon after discs are applied.

Antimicrobial agents/discs

Modern commercial discs are generally produced to a high standard and problems caused by incorrect content or incorrect agents in discs are rare. Humidity will cause deterioration of labile agents and many of the problems with discs are related to incorrect handling in the laboratory. Store discs in sealed containers in the dark (some agents, e.g. metronidazole, are sensitive to light): the containers should include an indicating desiccant. Discs may be kept at $< 8°C$ within the expiry date given by the maker. To minimize condensation of water on the discs allow containers to warm to room temperature before opening them. For convenience, leave closed containers at room temperature during the day and refrigerate them overnight only.

Problems with humidity also affect pure antimicrobial powders used in MIC testing and similar precautions to those described for discs should be taken. If powders absorb water the amount of active substance per weighed amount will be reduced even if the agent is not labile because of the weight of the water absorbed.

Incubation

Incubate plates at 35–37°C in air. Use alternative conditions only if essential for growth of the organisms. Incubation in an atmosphere containing additional carbon dioxide may reduce the pH, which may affect the activity of some agents. Incubation at 30°C is usually necessary for methicillin/oxacillin susceptibility tests on staphylococci but should not be used for other agents as zone sizes will be increased.

Do not stack plates more than four high in the incubator in order to avoid uneven heating during incubation.

Reading results

Unless tests are read by an automated system the reading is somewhat subjective unless endpoints are clear. Criteria for reading and interpreting susceptibility tests – whatever basic method is used – should be written down and available to staff who do these tests.

Definitions of endpoints for agar dilution tests differ, although up to four colonies or a thin haze in the inoculated spot are commonly disregarded. Endpoints trailing over several dilutions may indicate contamination or a minority resistant population.

For broth dilution tests the MIC is the lowest concentration of the agent which completely inhibits growth.

In reading diffusion tests take the point of abrupt diminution of growth as the zone edge. If there is doubt read the point of 80% inhibition of growth. Measure zones at points unaffected by interactions with other agents or by proximity to the edge of the plate. Ignore tiny colonies at the zone edge, swarming of *Proteus* spp. and traces of growth within sulphonamide and trimethoprim zones. Large colonies within zones of inhibition may indicate contamination or a minority resistant population.

Diffusion methods

The basic technique for disc diffusion tests has not changed in 40 years. The methods of control of variation and the basis of interpretation have fallen into two groups, the comparative methods and the standardized methods.

In the UK, variation is controlled by comparison of the zone size of the test strain with that of a control strain set up at the same time. Measured comparison of the test and control zones is also the basis of interpretation. It is assumed that variation in the test affects the test and control strains equally and thereby cancels them out. This is the comparative approach. The comparison may be made with a control strain on a separate plate or both control and test may be inoculated on the same plate (the Stokes method). The potential limitations of this approach have been discussed (Brown, 1990)

In many countries the approach to controlling variation in the test is to standardize all details of the tests as recommended in the report of an international collaborative study of susceptibility testing (Ericsson and Sherris, 1971). Interpretation of zone diameters is from a table of zone size breakpoints. The zone size breakpoints equivalent to defined MIC breakpoints have been established by regression analysis of the relationship of zone size to MIC, by error minimization, by study of the distribution of susceptibility of different species and by clinical experience relating the results of tests to the outcome of treatment. These different approaches have been combined in some methods. An important feature of the standardized methods is that the zone size breakpoints are valid only if all aspects of the defined method are followed precisely. Hence there are several standardized methods that differ in detail and in the breakpoints used. Undoubtedly, the most widely used method of this type is the United States NCCLS method, which is a development of the Kirby–Bauer approach.

Stokes method

This method remains the most widely used in the UK. As originally described, the method required one set of discs with lower concentrations of agents for isolates from systemic infections and another set with higher concentrations for isolates from urinary tract infections. The control organism for isolates from systemic infections was *S. aureus* NCTC 6571 and the control for isolates from urinary tract infections was *E. coli* NCTC 10418. The higher disc concentrations and generally more resistant control organism used for isolates from urinary tract infections were allowances for the concentration of agents in the urine. Isolates of *Pseudomonas* spp. were tested against *P. aeruginosa* NCTC 10662, because the susceptibility to gentamicin is affected by the divalent cation content of the medium, and *Pseudomonas* spp. are typically intermediate in susceptibility to gentamicin and carbenicillin.

There has been a tendency in recent years to use a single set of disc concentrations and to use a wider range of controls more closely related to the type of isolate than the source of the isolate. There are also problems related to interpretation of tests on newer agents to which control strains are very sensitive, leading to incorrect reports of resistance or intermediate susceptibility. Further updating of the method will be necessary to overcome some of these problems.

Preparation of plates

Use any medium designed specifically for susceptibility testing. Prepare and sterilize as directed by the manufacturers. Defibrinated blood may be necessary for tests on fastidious organisms. Cool the medium to 50°C and add defibrinated horse blood to a concentration of 5%. If the medium is not free from thymidine, add lysed horse-blood for tests on sulphonamides and trimethoprim. Pour the medium into petri dishes on a flat, horizontal surface to a depth of 3–4 mm. Store poured plates at 4°C and use within 1 week of preparation.

Before use, dry inoculation plates with the lids ajar until there are no droplets of moisture on the agar surface. The time taken to achieve this depends on the drying conditions.

Preparation of inoculum

In general, an acceptable inoculum can be prepared from a fully grown nutrient-broth culture or from a suspension of several colonies emulsified in broth to give a density similar to a broth culture.

The inoculum should give semiconfluent growth of colonies on the plates after overnight incubation. Reject tests with a confluent growth or clearly separated colonies and repeat.

Inoculation

With the band plating method the control culture is applied in two bands on either side of the plate leaving a central area uninoculated. Use pre-impregnated control swabs or apply a loopful of an inoculum prepared as described above to both sides of the plate and spread it evenly in the two bands with a dry sterile swab. Seed the test organism evenly in the band across the centre of the plate by transferring a loopful of the broth culture or suspension to the centre of the plate and spreading it with a dry sterile swab. This method usually allows up to four discs to be applied to the plate.

If a rotary plating method is used apply the control to the centre of the plate leaving an uninoculated 1.5 cm band around the edge of the plate. Apply the test organism to the 1.5 cm band. This method usually allows up to six discs to be applied to the plate.

The control strains to use with specific organisms are listed in Table 12.1. Use additional control strains in particular situations: i.e. a methicillin-resistant strain of *S. aureus* to control methicillin tests; *Streptococcus faecalis* ATCC 25922 to check that the thymidine content of the medium is not too high for trimethoprim and/or sulphonamide tests and a β-lactamase producing strain of *E. coli* (NCTC 11560) to control discs containing both a β-lactam agent and a β-lactamase inhibitor, e.g. a mixture of amoxycillin and clavulanic acid (co-amoxiclav).

Antimicrobial discs

After the inoculum has dried apply discs to the inoculated medium with forceps, a sharp needle or a dispenser and press down gently to ensure even contact. Handle and store discs as recommended in the section on antimicrobial agents/discs (p.181). Suitable concentrations of agents in discs are listed in Table 12.2.

Place discs on the line between the test and control organisms Four discs can be accommodated on a 9 cm circular plate. If a rotary plating method is used apply six discs on a 9 cm circular plate.

Incubation

Incubate tests and controls overnight at 35–37°C (30°C is recommended for methicillin/oxacillin susceptibility tests of staphylococci).

Table 12.1 Control strains for comparative methods including Stokes' modification

Control strain	*Test organisms*
Escherichia coli NCTC 10418	Coliform organisms
Pseudomonas aeruginosa NCTC 10662	Pseudomonads
Haemophilus influenzae NCTC 11931	*Haemophilus* spp.
Neisseria gonorrhoeae (sensitive strain)	Gonococci
Staphylococcus aureus NCTC 6571	Other organisms that will grow aerobically
Clostridium perfringens NCTC 11229	*Clostridium* spp
Bacteroides fragilis NCTC 9343	Other anaerobes

Table 12.2 Suitable disc contents for the comparative methods, including Stokes' modification

Antimicrobial agent	*Disc content*
Ampicillin	
Enterobacteriaceae and enterococci	10 µg
Haemophilus and *Branhamella*	2 µg
Augmentin	
Enterobacteriaceae	30 µg
Haemophilus, *Branhamella* and staphylococci	3 µg
Penicillin	
Staphylococci	2 IU
Meningococci	0.25 IU
For assessing penicillin susceptibility of pneumococci the following are recommended	
Oxacillin	1 µg
or methicillin	5 µg
Piperacillin	30 µg
Mezlocillin	30 µg
Azlocillin	30 µg
Cephalothin	30 µg
Cephalexin	30 µg
Cephadroxil	30 µg
Cephradine	30 µg
Cefuroxime	30 µg
Ceftazidime	10 µg
Cefotaxime	10 µg
Cefsulodin	30 µg
Methicillin	5 µg
Carbenicillin	100 µg
Ticarcillin	75 µg
Gentamicin	10 µg
Amikacin	30 µg
Tobramycin	10 µg
Neomycin	30 µg
Netilmicin	10 µg
Erythromycin	5 µg
Clindamycin	2 µg
Tetracycline	10 µg
Fusidic acid	10 µg
Chloramphenicol	
Enterobacteriaceae	30 µg
Haemophilus, pneumococci and meningococci	10 µg
Colistin	10 µg
Nalidixic acid	30 µg
Nitrofurantoin	50 µg
Sulphafurazole	
Enterobacteriaceae and enterococci	100 µg
Meningococci	25 µg
Trimethoprim	2.5 µg
Cotrimoxazole	25 µg
Spectinomycin	100 µg
Vancomycin	5 µg
Rifampicin	5 µg
Ciprofloxacin	1 µg
Mupirocin	5 µg

Based on Phillips *et al.* (1991

Reading of zoness of inhibition

Measures zones of inhibition from the edge of the disc to the edge of the zone. As the control and test organism are adjacent on the same plate the difference between the respective zone sizes can be easily seen. If the test zones are obviously larger than the control or give no zone at all it is not necessary to make any measurement. If there is any doubt, measure zones with calipers (preferably), dividers or a millimetre rule.

Interpretation
Each zone size is interpreted as follows
Sensitive: zone diameter equal to, wider than, or not more than 3 mm smaller than the control.
Intermediate: zone diameter greater than 2 mm but smaller than the control by more than 3 mm.
Resistant: zone diameter 2 mm or less.
Exceptions:

(1) Penicillinase-producing staphylococci show heaped-up, clearly defined zone edges, and should be reported as resistant irrespective of zone size.
(2) Polymyxins diffuse poorly in agar so that zones are small and the above criteria cannot be applied. Tests with polymyxins should therefore be interpreted thus:
Sensitive: zone diameter equal to, wider than, or not more than 3 mm smaller than the control.
Resistant: zone diameter more than 3 mm smaller than the control.
(3) Zones around ciprofloxacin discs are large with some of the sensitive control strains so tests should be interpreted thus:
(a) If *S. aureus* or *P. aeruginosa* are used as controls.
(b) If *H. influenzae* or *E. coli* are used as controls.
Sensitive: zone diameter equal to, wider than, or not more than (a) 7 mm or (b) 10 mm smaller than the control.
Intermediate: zone diameter greater than 2 mm but smaller than the control by more than (a) 7 mm or (b) 10 mm.
Resistant: zone diameter 2 mm or less.

Agar dilution methods

The agar dilution method for determining MICs has been accepted as the standard against which other methods are assessed. It has advantages over broth dilution methods in that contamination is more easily seen and re-isolation of the required organism is usually not a problem.

Medium
As for diffusion tests, media optimized for susceptibility testing should be used.

Antimicrobial agents
Use generic rather than proprietary names of agents. Obtain antimicrobial powders directly from the manufacturer or from commercial sources. The agent should be supplied with a certificate of analysis stating the potency (μg or IU per mg powder, or as a percentage potency), an expiry date, details of recommended storage conditions and data on solubility. Store powders in sealed containers in the dark at 4°C in a desiccator unless otherwise recommended by the manufacturer. Ideally, hygroscopic agents should be dispensed in small amounts and one used on each test occasion. Allow containers to warm to room temperature before opening them to avoid condensation of water on the powder.

Range of concentrations tested

The range of concentrations will depend on the particular organisms and antimicrobial agent being tested. Recommended ranges for major groups of organisms are given by Phillips *et al.* (1991). The ranges should certainly include those breakpoint concentrations used for defining susceptibility, intermediate susceptibility and resistance (see below).

It is now accepted convention to use a serial twofold dilution series based on 1 mg/l as shown in Table 12.3. The $\log_2 + 9$ notation is simply a means of avoiding working with negative numbers as when the $\log_2$ notation is used.

Table 12.3 Twofold dilution series

Concentration (mg/l)	*Log_2*	*Log_2 MIC + 9*
0.004	−8	1
0.008	−7	2
0.015	−6	3
0.03	−5	4
0.06	−4	5
0.125	−3	6
0.25	−2	7
0.5	−1	8
1	0	9
2	1	10
4	2	11
8	3	12
16	4	13
32	5	14
64	6	15
128	7	16
256	8	17
512	9	18

Preparation of stock solutions

To ensure accurate weighing of agents, use an analytical balance and, whenever possible, weigh at least 100 mg of powder. With a balance weighing to five decimal places the amount of powder weighed may be reduced to 10 mg. Use the following formula to make allowance for the potency of the powder:

$$\text{Weight of powder (mg)} = \frac{\text{volume (ml)} \times \text{concentration (mg/l)}}{\text{potency of powder } (\mu\text{g/mg})}$$

Alternatively, given a weighed amount of antimicrobial powder, calculate the volume of diluent needed from the formula:

$$\text{Volume of diluent (ml)} = \frac{\text{weight (mg)} \times \text{potency } (\mu\text{g/ml})}{\text{concentration (mg/l)}}$$

Concentrations of stock solutions should be made up to 1000 mg/l or greater although the insolubility of some agents will prevent this. The actual concentrations of stock solutions will depend on the method of preparing working solutions (see below). Agents should be dissolved and diluted in sterile water where possible. Some agents require alternative solvents and these are listed in Table 12.4 Sterilization of solutions is not usually necessary. If required, sterilize by membrane filtration and compare samples before and after sterilization to ensure that adsorption has not occurred. Unless otherwise

Table 12.4 Suggested solvents and diluents for antimicrobial agents requiring solvents other than water

Antimicrobial agent	*Solvent*	*Diluent*
Amoxicillin	Phosphate buffer 0.1 mol/l, pH 8.0	Phosphate buffer 0.1 mol/l, pH 7.0
Ampicillin	Phosphate buffer 0.1 mol/l, pH 8.0	Phosphate buffer 0.1 mol/l, pH 7.0
Aztreonam	Saturated sodium bicarbonate solution	Water
Ceftazidime	Saturated sodium bicarbonate solution	Water
Chloramphenicol	95% ethanol	Water
Enoxacin	½ vol water; add min vol NaOH 0.1 mol/l; make up vol with water	Water
Erythromycin	95% ethanol	Water
Fusidic acid	95% ethanol	Water
Nalidixic acid	½ vol water; add min vol NaOH 0.1 mol/l; make up vol with water	Water
Nitrofurantoin	Dimethylformamide	Phosphate buffer 0.1 mol/l, pH 8.0
Rifampicin	Methanol	Water
Sulphonamides	½ vol water; add min vol NaOH 0.1 mol/l; make up vol with water	Water
Trimethoprim	½ vol water; add min vol lactic acid or HCl 0.1 mol/l; make up vol with water	Water

instructed by the manufacturer, store stock solutions frozen in small amounts at −20°C or below. Most agents will keep at −60°C for at least 6 months. Use stock solutions promptly on defrosting and discard unused solutions.

Preparation of working solutions

Twenty ml volumes of agar are commonly used in 9 cm petri dishes for agar dilution MICs. Two alternative dilution schemes are given in Tables 12.5 and 12.6. In both schemes 19 ml volumes molten agar are added to 1 ml volumes antimicrobial solution. The more conventional method (Table 12.5) is based on diluting a 10^4 mg/l stock solution, always measuring 1 ml volumes antimicrobial solution. The other method (Table 12.5) is based on diluting a 10^4 mg/l stock solution and involves measuring various volumes of antimicrobial solution with high precision variable-volume micropipettes, which are now universally available. Automated diluter/dispenser systems may also be used.

Preparation of plates

Prepare agar as recommended by the manufacturer. Allow the sterilized agar to cool to 50°C in a water-bath. Prepare a dilution series of antimicrobial agents in 1 ml volumes, 20× final concentration, in 25 ml screw-capped bottles. Include a drug-free control. Add 19 ml molten agar to each bottle, mix thoroughly and pour the agar into prelabelled sterile petri dishes on a level surface. Alternatively, add the antimicrobial agent to 19 ml volumes of agar dispensed in universal containers and held in a water-bath at 50°C. To prepare

Table 12.5 A method of preparation of dilutions of agents for use in agar dilution susceptibility tests

Antimicrobial conc. (mg/l)	*Volume (ml) + solution*	*Distilled water (ml)*	=	*Antimicrobial conc. (mg/l)*	*Final conc. at 1:20 diln. in agar*
10 240 (stock)	1	0		10 240	512
10 240 (stock)	1	1		5 120	256
10 240 (stock)	1	3		2 560	128
2 560	1	1		1 280	64
2 560	1	3		640	32
2 560	1	7		320	16
320	1	1		160	8
320	1	3		80	4
320	1	7		40	2
40	1	1		20	1
40	1	3		10	0.5
40	1	7		5	0.25
5	1	1		2.5	0.125
5	1	3		1.25	0.06
5	1	7		0.625	0.03
0.625	1	1		0.3125	0.015
0.625	1	3		0.1562	0.008
0.625	1	7		0.0781	0.004

Table 12.6 An alternative method of preparation of dilutions of agents for use in agar dilution susceptibility tests

Antimicrobial conc. (mg/l)	*Volume (ml) + solution*	*Distilled water (ml)*	=	*Antimicrobial conc. (mg/l)*	*Final conc. at 1:20 diln. in agar*
10 000 (stock)	1	9		1 000	—
10 000 (stock)	0.1	9.9		100	—
100	1	9		10	—
100	0.1	9.9		1	—
10 000 (stock)	1	0		10 000	500
10 000 (stock)	0.512	0.488		5 120	256
10 000 (stock)	0.256	0.744		2 560	128
10 000 (stock)	0.128	0.872		1 280	64
1 000	0.64	0.36		640	32
1 000	0.32	0.68		320	16
1 000	0.16	0.84		160	8
100	0.8	0.2		80	4
100	0.4	0.6		40	2
100	0.2	0.8		20	1
10	1	0		10	0.5
10	0.5	0.5		5	0.25
10	0.25	0.75		2.5	0.125
10	0.125	0.875		1.25	0.06
1	0.625	0.375		0.625	0.03
1	0.313	0.687		0.313	0.015
1	0.156	0.844		0.156	0.008
1	0.078	0.922		0.078	0.004

multiple dilution series increase the volumes of antimicrobial solution and molten agar as required. Pour the plates immediately after mixing the antimicrobial solution with the molten agar. Allow the plates to set at room temperature and dry them so that no drops of moisture remain on the surface of the agar. Do not overdry the plates.

Preferably use plates immediately, particularly if they contain labile agents such as β-lactams. If the plates are not used immediately, store them in a refrigerator (4–8°C) in sealed plastic bags and use within 1 week.

Preparation of inoculum

Standardization of the density of inoculum is essential if variation in results is to be avoided. Follow the procedures described on p.180.

Inoculation of plates

Mark the plates so that the orientation is obvious. Transfer diluted bacterial suspensions to the inoculum wells of an inoculum replicating apparatus. Use the apparatus to transfer the inocula to the series of agar plates, including control plates without antimicrobial agent at the beginning and end of the series. Inoculum pins 2.5 mm in diameter will transfer close to 1 μl, i.e. an inoculum of 10^4 cfu/spot if the bacterial suspension contains 10^7 cfu/ml. Inoculum pins supplied by different manufacturers may deliver more or less than 1 μl. If in doubt as to the volume delivered check with the supplier and adjust the inoculum preparation procedure to yield a final inoculum of 10^4 cfu/spot. Allow the inoculum spots to dry at room temperature before inverting the plates for incubation.

Incubation of plates

Incubate plates at 35–37°C in air for 18 h. To avoid uneven heating do not stack them more than four high. If the incubation period is extended for slow-growing organisms, assess the stability of the agent over the incubation period. Avoid incubation in an atmosphere containing 5% CO_2 unless absolutely necessary for growth of the organisms (e.g. *Neisseria* spp.). Incubation at 30°C is recommended for methicillin/oxacillin susceptibility tests on staphylococci.

Reading the results

The MIC is the lowest concentration of the agent which completely inhibits growth, disregarding up to four colonies or a thin haze in the inoculated spot. A trailing endpoint with a small number of colonies growing on concentrations several dilutions above that which inhibits most organisms should be investigated by subculture and retesting as this may be caused by contamination, resistant variants, β-lactamase producing organisms or, if incubation is prolonged, regrowth of susceptible organisms following deterioration of the agent. With sulphonamides and trimethoprim endpoints may be seen as only a reduction in growth if the inoculum is too heavy or antagonists are present.

Broth dilution susceptibility tests

The traditional macrodilution method involves the use of test tubes (usually 100 × 13 mm) and culture volumes of 1–2 ml. Microdilution methods involve the use of plastic microdilution trays with culture volumes of 0.1–0.2 ml. The macrodilution method is most suitable for small numbers of tests as preparation of many sets of tubes is tedious. In the microdilution method multiple plates may be conveniently produced by the use of apparatus for automatically diluting and dispensing solutions. Within a single twofold dilution the macrodilution and microdilution procedures generally give similar results.

Medium

The requirements for broth media are similar to those for agar media. There are fewer broth media than agar media designed specifically for antimicrobial susceptibility testing. Mueller-Hinton, Oxoid Iso-Sensitest and PDM Antibiotic Sensitivity media are available in both agar and broth formulations.

Low cation contents of some batches of Mueller–Hinton broth may result in lower MICs of aminoglycosides for *P. aeruginosa* than on agar media. The NCCLS (1990b) provides control limits for MICs in Mueller–Hinton broth and gives instructions for supplementation of broth with calcium and magnesium ions if necessary.

Supplements other than blood may be added to broth for fastidious organisms as for agar media.

Preparation of dilutions

The total volume of each dilution prepared will depend on the number of tests to be done. The schemes given in Tables 12.4 and 12.5 may be used with broth substituted for molten agar. The volumes should be adjusted according to the number of tests to be done, but should not be less than 10 ml. Dispense the solutions in 1 ml volumes in tubes for the macrodilution method. For the microdilution method dispense the solutions into the wells of the microdilution plate in 0.1 ml volumes. Include control tubes or wells containing broth without antimicrobial agent to act as growth and sterility controls.

Macrodilution tubes are commonly used immediately after preparation (preferable for labile agents such as β-lactam agents) but may be stored at 4°C for up to 1 week. Microdilution trays may be kept in sealed plastic bags stored at −60°C or below for up to 1 month. Do not refreeze trays once they have been defrosted.

Preparation of inoculum

Prepare the inoculum as for the agar dilution method (p.180) but adjust the density so that, after inoculation, the final volume of test broth contains 10^5 – 5×10^5/ml. The density of the inoculum required depends on the volume of broth in the tubes or wells and the inoculation system used.

Inoculation of tubes/plates

Inoculate macrodilution tubes containing 1 ml broth with 0.05 ml inoculum containing 5×10^6 cfu/ml. Alternatively, add 1 ml inoculum (5×10^5 cfu/ml) in broth to 1 ml volumes of broth in the dilution series. The latter method requires the preparation of a double strength dilution series of the agent to compensate for the 1 in 2 dilution on inoculation.

Inoculate microdilution wells containing 0.1 ml broth with 0.005 ml inoculum containing 5×10^6 cfu/ml. Inoculum replicators which deliver 0.001–0.005 ml may be used with adjustment of inoculum density to compensate for the volume delivered to wells. Manual inoculation with a pipette is easier if 0.05 ml inoculum is added to 0.05ml volumes of broth in the dilution series. As with the macrodilution method, inoculum density and antimicrobial concentrations are adjusted to compensate for the 1 in 2 dilution of agents on inoculation.

Incubation of tubes/plates

Seal microdilution plates in a plastic bag, with plastic tape or some other method to prevent the plates from drying. To avoid uneven heating, do not stack microdilution plates more than four high. Incubate tubes or trays at 35–37°C as with agar dilution plates.

Reading results
The MIC is the lowest concentration of the agent which completely inhibits growth. The control containing no agent should be turbid. Apparatus which assists in viewing microdilution plates may be used. Automatic microdilution plate readers may be used if they give results equivalent to visual inspection.

Breakpoint methods

Agar dilution breakpoint susceptibility testing methods are increasingly popular in diagnostic laboratories. These methods are essentially highly-abbreviated agar dilution MIC tests in which only one or two critical antimicrobial concentrations are tested rather than a full range of dilutions. These critical or breakpoint concentrations of antimicrobial agents may be used to classify the susceptibility of organisms as shown below:

With a single breakpoint concentration (e.g. 0.5 mg/l erythromycin for staphylococci and streptococci)

growth = *resistant*
no growth = *susceptible*

Using a lower and a higher breakpoint concentration, (e.g. 4 and 16 mg/l cefuroxime for Enterobacteriaceae)

no growth at either concentration = *susceptible*
growth at the lower concentration only = *intermediate*
growth at both concentrations = *resistant.*

The general considerations detailed at length in the section on dilution methods (p. 185) apply equally to agar incorporation breakpoint susceptibility testing methods.

It is essential to inoculate growth control plates of the same basal medium as the antimicrobial containing plates at the beginning and end of the inoculation run and to incubate them under the same conditions as the test. Adequate growth of the test isolates on these control plates allows a confident interpretation of susceptibility based on the absence of growth on antimicrobial containing plates as the antimicrobial content is the only variable.

When considering the introduction of agar incorporation breakpoint susceptibility testing to a diagnostic laboratory it is imperative that all aspects of the proposed method(s) are considered carefully before the methods are actually applied on a routine basis. This is particularly important for inoculum preparation. Time spent evaluating the actual cfu/spot achieved by a particular method of inoculum preparation and careful standardization of the method selected is a sound investment and will minimize variability of the results achieved when that method is applied routinely. It is important to note at this point that different species and genera may well require different inoculum preparation procedures to achieve the same final result in cfu/spot deposited on the test medium.

Purity plates
Purity plates (which may be half, third or quarter plates) are mandatory and should be of a suitable non-selective growth medium, e.g. blood agar. Prepare these by subculturing a standard 1 μl loopful from the inoculum 'pots' used for the inoculation 'run' after all antimicrobial test and growth control plates have been inoculated. In this way mixed cultures should be detected and the use of a

standard loop allows an approximate check on the actual inoculum density used for each test organism. As for any other susceptibility testing method, results obtained with inocula that are outside the expected range should be rejected and the tests repeated.

Antimicrobial agents

If an agar incorporation breakpoint susceptibility testing scheme is selected for use in a diagnostic laboratory there are certain methodological considerations that may affect the choice of antimicrobials to test. These include the availability or otherwise of pre-prepared antimicrobials at specific concentrations for addition to molten agar (e.g. Mast Adatabs) and the stability of certain agents in agar media during storage of prepared plates pending use.

Laboratory workers who choose to prepare their own antimicrobial stock solutions from which to produce plates for routine use should consider in detail methods and laboratory procedures for the storage and accurate weighing of powders as well as for the storage and accurate dilution of stock solutions. Thus, for example if stock solutions of antimicrobial agents are made up and stored at −20°C or −70°C for use in the preparation of weekly batches of plates, stability during such storage may affect the choice of agents. For many agents data on stability during storage of stock solutions are not readily available though some may be particularly unstable at −20°C, e.g. amoxycillin, clavulanic acid and imipenem. Similarly the long-term storage of antimicrobial powders for use in the preparation of stock solutions should be considered and action taken to ensure that they are stored appropriately (e.g. protection from moisture and light as well as suitable refrigeration).

Whether antimicrobial test plates are prepared from commercial reagents or home-produced stock solutions, the stability of the agents, once incorporated into agar will also need to be considered. β-lactamase inhibitors, e.g. clavulanic acid, which may be combined with antimicrobial agents such as amoxycillin or ticarcillin may undergo some loss of activity during normal storage of breakpoint plates pending use (usually up to 1 week in most laboratories), as may the carbapenem antibiotic imipenem. If significant loss of activity does occur during normal storage there could well be a spurious increase in resistance over the time period in which a particular batch of plates was in use. For this reason it is essential to include quality control strains of known susceptibility to such agents on a daily basis. For most agents likely to be tested, storage for up to 1 week in sealed plastic bags is usually satisfactory, and thus the requirement for daily microbiological quality control is less pressing.

Awareness of, and preparedness for, such potential problems coupled with reliable microbiological quality control for breakpoint antimicrobial susceptibility tests cannot be over-emphasized. In addition to those normally considered, these factors should be borne in mind when deciding the range of agents to be tested.

Breakpoint concentrations to be tested

In considering the use of breakpoints in *in vitro* susceptibility testing the most critical question is 'what are the appropriate concentrations to test'? Table 12.7 lists the recommended breakpoint concentrations for a wide range of clinically relevant antimicrobial agents. Use of these breakpoints allows the categorisation of organisms as susceptible, resistant or intermediate on the basis of growth or no growth at the concentrations recommended. These recommendations are based in large measure on those made by the British Society for Antimicrobial Chemotherapy (BSAC) (Phillips *et al.*, 1991) and take into account clinical, pharmacological and microbiological considerations. For

Table 12.7 Breakpoint recommendations based on those produced by the British Society for Antimicrobial Chemotherapy (Phillips *et al.*, 1991)

Antimicrobial	S	R
Amikacin		
Organism not specified	≤4	>16
Ampicillin/amoxycillin (1)		
Enterobact	≤8	>8
Staph (2)	≤1	>1
Enterococci	≤8	>8
Other streps	≤1	>1
Bran	≤1	>1
Hi	≤1	>1
Amoxycillin/clavulanate (3)		
Enterobact	≤8	>8
Staph (2)	≤1	>1
Enterococci	≤8	>8
Bran	≤1	>1
Hi	≤1	>1
Azlocillin		
Enterobact	≤16	>16
Pseudo	≤64	>64
Aztreonam		
Enterobact/Pseudo	≤8	>8
Carbenicillin		
Enterobact	≤32	>32
Pseudo	≤128	>128
Cefixime		
Enterobact	≤1	>4
Staph (2)	≤1	>1
Hi/Strep/Bran	≤1	>1
Cefotaxime		
Enterobact/Pseudo	≤1	>1
Staph (2)	≤8	>8
Strep/Bran	≤1	>1
Hi	≤1	>1
Cefodoxime		
Enterobact	≤1	>4
Staph (2)	≤1	>1
Hi/Strep/Bran	≤1	>1
Cefsulodin		
Pseudo	≤8	>8
Ceftazidime		
Enterobact/Pseudo	≤2	>2
Staph (2)	≤8	>8
Strep/Bran	≤2	>2
Hi	≤2	>2
Ceftizoxime		
Enterobact/Pseudo	≤1	>1
Staph (2)	≤8	>8
Strep/Bran	≤1	>1
Hi	≤1	>1
Cefuroxime		
Enterobact	≤4	>16
Staph (2)	≤4	>4
Strep/Bran	≤4	>4
Hi	≤4	>4
Cefuroxime axetil		
Enterobact	≤1	>4
Staph (2)	≤1	>1
Strep/Hi/Bran	≤1	>1
Cephalexin/Cephradine		
Enterobact	≤2	>8
Staph (2)	≤8	>8
Hi/Strep/Bran	≤2	>2
Chloramphenicol		
Enterobact	≤8	>8
Staph/Strep	≤8	>8
Bran	≤2	>2
Hi	≤2	>2
Ciprofloxacin		
Enterobact	≤1	>1
Pseudo	≤4	>4
Staph	≤1	>1
Bran	≤1	>1
Hi	≤1	>1
Clindamycin		
Organism not specified	≤0.5	>0.5
Erythromycin		
Organism not specified	≤0.5	>0.5
Fusidic acid		
Organism not specified	≤1	>1
Gentamicin		
Organism not specified	≤1	>4
Imipenem		
Organism not specified	≤4	>4
Meropenem		
Organism not specified	≤4	>4
Methicillin (2)		
Staph	≤4	>4
Mezlocillin		
Enterobact	≤16	>16
Pseudo	≤64	>16
Strep/Hi/Bran	≤2	>2
Netilmicin		
Organism not specified	≤1	>4
Penicillin G (1)		
Staph/Strep	≤0.12	>0.12
Piperacillin		
Enterobact	≤16	>16
Pseudo	≤64	>64
Rifampicin		
Staph/Strep/Hi/Bran	≤1	>1
Sulphamethoxazole		
Enterobact	≤32	>32
Temocillin		
Enterobact	≤32	>32
Tetracycline		
Organism not specified	≤1	>1

Table 12.7—Continued

Antimicrobial	S	R	Antimicrobial	S	R
Ticarcillin			Tobramycin		
Enterobact	⩽16	>16	Organism not specified	⩽1	>4
Pseudo	⩽64	>64	Trimethoprim		
Ticarcillin/ clavulanate (3)			Organism not specified	⩽0.5	>2
Enterobact	⩽16	>16	Vancomycin		
Pseudo	⩽64	>64	Organism not specified	⩽4	>4

Notes
S = breakpoint concentration (mg/l) used to define susceptibility
R = breakpoint concentration (mg/l) used to define resistance
Abbreviations:
Enterobact = Enterobacteriaceae
Pseudo = *Pseudomonas* spp.
Staph = Staphylococci
Strep(s) = Streptococci
Bran = *Branhamella catarrhalis*
Hi = *Haemophilus influenzae*
General:
Antimicrobial agents are listed in alphabetical order. Because new agents regularly appear on the market, no such listing can be up to date. Manufacturers may be able to provide information on breakpoints for newer agents, or comparisons may be made with existing similar compounds. Such comparisons need to take into account both the pharmacokinetics and the inherent antimicrobial activity of the new agent.
(1) For *H. influenzae* and *B. catarrhalis* tests for β-lactamase production are recommended in addition to dilution testing.
(2) For staphylococci recommendations refer to test conditions favourable to the expression of methicillin/oxacillin resistance, i.e. dense inoculum and/or incubation at 30°C, and/or addition of supplementary sodium chloride and/or prolonged incubation. If other penicillin and cephalosporin antimicrobials are to be tested against staphylococci (in situations where methicillin resistance may occur) and the results reported, it may be prudent to consider similar conditions of test for these agents as well. To avoid anomalous results methicillin-resistant staphylococci should be reported as resistant to all other β-lactam antimicrobials.
(3) Concentrations of clavulanate not shown.

these reasons there is some variation in the breakpoints recommended for particular organism/antimicrobial agent combinations. Breakpoint testing should take account of the inherent susceptibility of the species under test. For example, it seems entirely reasonable to specify a different ampicillin breakpoint for Enterobacteriaceae as opposed to non-enterococcal streptococci. For those agents where breakpoints vary by genus or species this information is included in Table 12.7. This is analogous to the use of different disc strengths and/or different control strains in the Stokes method of comparative disc diffusion susceptibility testing. Recommendations on breakpoint levels made by other workers in the USA and continental Europe, in addition to those made by the BSAC, are listed by Faiers *et al.* (1991).

Breakpoint concentration recommendations are usually defined as being appropriate for a specified method of testing and their application to other methods (with possible variations in the composition of the test medium and inoculum density) is to be avoided. Those proposed by the BSAC refer to the use of Isosensitest or DST agars and inocula of 10^4 cfu per spot except for staphylococci (10^6 cfu per spot).

Depending on the source of the recommendation of particular breakpoints, the antimicrobial concentrations lying between those defining full susceptibility and resistance are variously classified as *intermediate, moderately susceptible, moderately resistant, indeterminant,* etc. The use of these various intermediate categories on patient reports is rarely understood by clinicians though laboratory workers should appreciate their significance. The practice of recording and reporting such 'intermediate' results as resistant should be discouraged. This may falsely inflate and partially obscure the significance of laboratory data compiled for the analysis of trends in antimicrobial resistance. A more prudent policy is to record and store the results as intermediate and to avoid reporting the result for that agent to the clinicians involved. If necessary, and always after discussion between clinician and microbiologist, such results may be used to guide patient treatment where no suitable therapeutic alternative exists.

Before introducing breakpoint susceptibility testing as the routine method it is wise to carry out agar dilution MICs of the agents to be tested on the most commonly examined species to establish the local 'normal range'. For agents where the 'normal' MIC is 2 or more dilutions below the suggested breakpoint it is reasonable to consider the use of a second lower breakpoint, in addition to that recommended, in order to detect incremental changes in degree of susceptibility. If such a procedure is adopted, results may still be reported in relation to the recommended breakpoint but the laboratory will be better informed as to possible changes in the antimicrobial susceptibility of the population of micro-organisms under study. For individual patients and specific infections the extra information provided may be extremely useful.

Reading and interpretation of breakpoint susceptibility tests

The 'interpretation' of breakpoint sensitivity testing has two elements: the reading of the laboratory tests and the assessment of their clinical relevance. In practice, the selected breakpoints are those considered to provide clinically relevant results and are usually based on one or two concentrations of a given agent. The results can then be interpreted as 'susceptible' or 'resistant' when one concentration is used or as 'susceptible', 'intermediate' and 'resistant' when two are used. The results may differ from those obtained by the Stokes disc diffusion method but each method has a set of interpretation criteria and 'different' results may both be 'correct' when assessed by the relevant criteria. Interpretation requires knowledge of the methods used.

The reading of breakpoint plates is the final stage of a susceptibility testing process which should adhere to standard methods for preparation and storage of media and antimicrobial agents, for preparation of the inoculum and for conditions of incubation.

Breakpoint sensitivity tests are easier to read than disc tests performed according to conventional UK methods. There are no zones to measure, and the results are usually unambiguous and free from subjective interpretation. The rules that govern their reading are readily taught and this enables junior staff to undertake the task. The same considerations mean that the method lends itself to automation.

The 'intermediate' category of the comparative disc method of Stokes may be replaced in the breakpoint method (if lower and higher breakpoint concentrations are used) by a well-defined category which includes a range of MICs. If appropriate breakpoints are chosen and the methods standardized rigorously, this category becomes a meaningful one highlighting subtle changes in antimicrobial susceptibility which are not readily detected by the comparative method without modifying the procedure, e.g. penicillin resistance in pneumococci and gonococci and gentamicin resistance in staphylococci, where small shifts in MIC values are not detected readily by the standard comparative disc method. The ability to produce such results may be of therapeutic significance in some circumstances. Many laboratories report disc intermediate results as resistant 'to be on the safe side'. Breakpoint intermediate results provide a more informative result than those generated by conventional disc diffusion methods, enabling more confident clinical assessment of the appropriateness or otherwise of a particular antimicrobial therapy. These results may also be withheld by the laboratory, but kept available for future reference.

Reading breakpoint susceptibility tests

(1) Check that the inoculum is pure and of approximately the correct density – the use of a standard loop to inoculate the purity plate provides for both of these requirements.
(2) Check that the inoculum is of the expected species. Randomization of species within the inoculum order aids the recognition of incorrect species owing to operator errors occurring when the isolate was originally subcultured or at the inoculation stage. This is considered by some to be a major reason for not dividing isolates and antimicrobial agents into Gram-positive and Gram-negative sets for testing.
(3) Check that the control has grown adequately. It is useful to have at least one antimicrobial agent in the test set to which isolates are likely to be resistant, e.g. aztreonam for Gram-positive and vancomycin for Gram-negative isolates. This acts as an additional control of adequate growth and can aid the recognition of mixtures of Gram-positive and Gram-negative organisms. Computer printouts of data or manual lists are useful for this purpose. These lists should also be checked for unexpected sensitivity patterns which could be due to mixed cultures, incorrect subculture, inversion of the plates or simply reading the wrong plate. They are also useful to alert the control of infection and medical teams to potential epidemiological or treatment problems.
(4) Laboratory staff should adhere to precise criteria for reading and recording results. Judgment of growth/no growth can be made in the same way as the end point determination for agar dilution MIC testing.

The growth of one or two colonies may be the result of a detected or undetected mixed culture or a genuine minority population of resistant organisms within a pure culture of more susceptible organisms. It is not unusual, during or after therapy with some antimicrobial agents, to isolate organisms with an increased resistance which may be heterogeneous so that subculture of different colonies may give rise to different results. These situations need clinical interpretation and judgment. They cannot be classified arbitrarily into *susceptible, resistant* or *intermediate*: the only answer is to describe the finding and examine the possible reasons for it.

Organization of breakpoint susceptibility testing schemes

The range of antimicrobial agents selected for inclusion within a breakpoint testing scheme must be related to the species of organisms to be tested and the clinical site of infection from which they are isolated. Some laboratories only use breakpoint methods for infections where the usual causative organisms can be predicted with some accuracy and the range of agents to be tested is limited. The most common example is urinary tract infection. Others adopt a broader approach and test all aerobic Gram-negative bacilli whatever their clinical source against the same set of antimicrobial agents. Similarly a set of antimicrobial agents may be selected for breakpoint testing against staphylococci. If the susceptibility testing medium used is supplemented with growth additives, e.g. 5% lysed horse blood and nicotinamide adenine dinucleotide

(NAD), the set of antimicrobial agents may be extended to cover susceptibility testing of more fastidious species such as pneumococci, other streptococci, *H. influenzae* and *Branhamella catarrhalis*. It should, however, be remembered that methods of inoculum preparation for these different groups of organisms may vary even though they may all be tested on the same set of plates.

Some laboratories combine the varying clinical and microbiological requirements of their breakpoint susceptibility testing schemes into a single 'universal' set of tests. One such system (Franklin, 1988) utilizes a susceptibility testing medium capable of supporting the growth of virtually all bacteria of medical importance. This approach was used to examine all bacterial isolates requiring testing against the same wide range of agents, usually at more than one breakpoint concentration. It must be remembered that many of the results generated by such a scheme will not be relevant to a particular species/genus or site of infection. Selectivity in reporting is necessary, with the exclusion of inappropriate or misleading results from laboratory reports. Such a 'universal' approach is most readily applicable in laboratories with large workloads and that utilize computerized reporting schemes whereby the editing of raw data into relevant reports can be accomplished by inclusion of a number of preset criteria within the reporting programme(s). For example, the definition of 'susceptibility' to ampicillin will vary according to the species under test (Table 12.7), and for some organisms, e.g. staphylococci, it may be inappropriate to report ampicillin susceptibility at all. Such an approach might appear confusing to the laboratory worker although it is very similar in principle to the use of different sets of discs and disc concentrations depending on the organism and site of isolation in the widely used disc diffusion methods. Centralization of susceptibility testing and interpretation (though not necessarily of reporting), to a particular area of the laboratory with appropriately trained staff, should overcome some of these potential problems, particularly in laboratories in which breakpoint methods have only recently been introduced.

From the foregoing, it is clear that the 'universal' approach to breakpoint susceptibility testing requires the use of a medium capable of supporting growth of all likely species and the inclusion of sufficient plates to cover a broad range of antimicrobial concentrations. As with any susceptibility testing method, a species or genus identification should be available at the time of issue of the report. Similar considerations regarding the appropriateness of the antimicrobial agents and concentration(s) tested to the site of infection and the species being examined apply to more selective approaches, e.g. schemes for urine isolates or aerobic Gram-negative bacilli only. As stated above, the manageability of this interpretation and editing process will depend on the range of agents, their concentrations and species tested within the scheme. If a computer is utilized to assist with the reporting process it can also be programmed to highlight truly aberrant or unusual results for special attention.

Variations and extensions of dilution methods

The E test

This novel system for testing MICs represents a different approach from quantitative testing of antimicrobial activity and has some advantages over conventional methods in its simplicity, ease of use and robust nature. The commercial E test strips are 50-mm by 5-mm plastic carriers with an exponential antimicrobial gradient dried on one side and a graduated MIC scale on the other. To set up an E test, inoculate agar medium by flooding or with a swab in the same way that plates are inoculated for disc diffusion susceptibility tests. Up to six E test strips can be placed in a radial pattern on a 15-cm plate or a single strip may be placed on a 9-cm plate. On incubation, elliptical zones of inhibition are produced and the MIC is read directly from the graduated E test strip at the point of intersection of the zone of inhibition with the strip. Include a control strain with each E test by inoculating one half of a plate with the test organism and the other half of the plate with the control organism and apply the E test strip to the line between the two organisms.

Media
As for all susceptibility testing methods, media optimized for susceptibility testing should be used.

Inoculum preparation and inoculation
Make faintly turbid suspensions of the test and control organisms in a suitable broth. Mark a line across the centre of an agar plate. Dip a sterile swab into the suspension of the control organism and evenly inoculate the agar on one half of the plate. Similarly inoculate the test organism on to the other half of the plate, leaving a gap of 2 mm between the test and control organisms. Do not leave a larger gap as it is essential that there is no gap between the organisms and the E test strip.

E-test strips
Store E test strips at −20°C with a desiccant. Allow strips to warm up to room temperature for 30 min before opening containers. Use flamed forceps to transfer an E test strip, with the printing uppermost, to the plate along the line between the test and control organisms. Avoid moving the strips once they have been applied to the agar as diffusion of the antimicrobial agents is very rapid.

Incubation
Incubate the plate overnight at 37°C.

Reading the results
Read the MIC from the point of intersection of the growth with the E-test strip. MICs of control organisms should be within one twofold dilution step of those shown in Table 12.8.

Minimal bactericidal concentration

The minimal bactericidal concentration (MBC) is the lowest concentration of an agent which kills a defined proportion (usually 99.9%) of the population after incubation for a set time. The method is usually an extension of the broth dilution MIC. After reading the MIC, organisms are quantitatively subcultured from tubes or wells showing no growth to antimicrobial-free agar medium. After incubating the plates the proportion of non-viable organisms, compared with the original inoculum, is assessed.

The MBC is used to guide therapy in situations such as endocarditis, where the bactericidal activity of an agent is considered essential.

Table 12.8 MICs for control organisms

Antimicrobial agent	*E. coli* *NCTC 10418*	*P. aeruginosa* *NCTC 10662*	*S. aureus* *NCTC 6571*
Amikacin	1	2	1
Ampicillin	4	—	0.03
Amoxicillin	4	—	0.03
Amoxicillin/clavulanic acid	4/2	—	0.03/0.015
Azlocillin	8	4	0.25
Aztreonam	0.03	4	—
Benzyl penicillin	—	—	0.03
Cefalexin	4	—	2
Cefotaxime	0.015	1	1
Cefoxitin	4	—	2
Ceftizoxime	0.08	—	2
Cefradine	4	—	2
Cefsulodin	—	2	—
Ceftazidime	0.125	1	4
Cefuroxime	2	—	1
Chloramphenicol	2	—	2
Ciprofloxacin	0.015	0.125	0.125
Clindamycin	—	—	0.06
Erythromycin	—	—	0.125
Fusidic acid	—	—	0.06
Gentamicin	0.25	1.4	0.125
Imipenem	0.06	0.5	0.03
Methicillin	—	—	1
Mezlocillin	2	8	0.5
Netilmicin	0.5	1	0.25
Nalidixic acid	1	—	—
Nitrofurantoin	4	—	—
Norfloxacin	—	—	—
Oxacillin	—	—	0.5
Piperacillin	0.5	4	0.5
Rifampicin	4	—	0.03
Sulphamethoxazole	16	—	—
Tetracycline	1	—	0.125
Ticarcillin	1	16	0.5
Tobramycin	0.25	0.5	0.125
Trimethoprim	0.06	—	0.25
Vancomycin	—	—	0.5

NCTC strains were tested as described in this chapter. The medium used was Iso-Sensitest agar.
NCTC is the National Collection of Type Cultures, Central Public Health Laboratory, 61 Colindale Avenue, London NW9 5HT, UK.

Time-kill curves

These are an extension of the MBC and give information on the rates at which organisms are killed. The rate is measured by counting the number of viable cells at different time intervals after exposure to various concentrations of an agent.

Tolerance

This is the ability of strains to survive but not grow in the presence of an agent and relates particularly to Gram-positive organisms tested against agents acting on the terminal stages of peptidoglycan synthesis, i.e. β-lactam and glycopeptide agents. Strains which have an MIC/MBC ratio of 32 or greater after 24-h incubation are usually termed tolerant. However, MICs and MBCs are determined after fixed periods of incubation and tolerance may be a reflection of a lower rate of killing. Technical variation has a marked effect on the detection of tolerance (Thrupp, 1986; Sherris, 1986).

Methicillin/oxacillin susceptibility tests on staphylococci

The expression of resistance is markedly influenced by test conditions, particularly with coagulase-negative staphylococci, and there are considerable differences among strains

Diffusion tests

(a) Use 5 μg methicillin discs or 1–2 μg oxacillin discs.
(b) Incubate tests at 30°C (methicillin/oxacillin only).
(c) Tests may be more reliable if incubation is continued for 48 h.
(d) Tests may be reliable when incubated at 37°C if the medium contains 5% NaCl. This, however, is medium-dependent and 5% NaCl may be inhibitory to some strains.
(e) Resistance may appear as a markedly reduced zone diameter, gradual reduction in colony size up to the disc, or a film of growth or isolated colonies of various sizes within a zone of inhibition.
(f) Some strains which produce large amounts of β-lactamase may give reduced methicillin zone diameters when compared with fully sensitive control strains. Care is necessary to avoid reporting such strains as methicillin-resistant.

No single set of conditions has been shown to be 100% reliable, but resistance in most strains will be detected with Mueller-Hinton agar with 5% NaCl, a semi-confluent inoculum, methicillin discs, and incubation at 30°C for 24 h. Continue incubation for a further 24 h before reporting coagulase-negative staphylococci as susceptible to methicillin/oxacillin.

Breakpoint tests

As for diffusion tests, methicillin susceptibility testing of staphylococci by breakpoint methods requires some variation of routine susceptibility testing protocols and no single set of test conditions can be regarded as 100% reliable. The foregoing discussion of variations in media and incubation conditions is applicable equally to methicillin testing by breakpoint methods.

The recommended breakpoint concentration of methicillin is 4 mg/l and failure to grow at this concentration is regarded as indicating methicillin susceptibility. However, most fully methicillin-susceptible staphylococci are inhibited by ≤ 2 mg/l methicillin. For this reason it is recommended that tests should be performed at breakpoint concentrations of both 2 and 4 mg/l, a medium that the laboratory finds appropriate and incubation at 30°C. It is often necessary to continue incubation for up to 48 h especially for coagulase-negative staphylococci. Resistance may appear as a film of growth or isolated colonies of various sizes. Any growth on methicillin breakpoint test plates should be regarded with suspicion. Known methicillin-susceptible and methicillin-resistant control strains should be included with every set of tests.

Interpret results as follows:

	2 mg/l	*4 mg/l*	*Report*
After overnight incubation	growth	growth	methicillin-resistant
	no growth	no growth	methicillin-susceptible
	growth	no growth	delay reporting*

* Investigate whether methicillin resistance is likely on clinical or epidemiological grounds and *reincubate test plates.*

After further incubation (up to 48 h)	growth	growth	methicillin-resistant
	growth	no growth	methicillin-susceptible

β-Lactamase testing

Rapid biochemical tests may be used to detect β-lactamase production by *Haemophilus* spp., *Neisseria* spp., *Branhamella* spp. (not acidimetric tests) and staphylococci (β-lactamase production may require induction in staphylococci). Disc tests are still required to detect intrinsic resistance mechanisms.

Methicillin (5 μg) or oxacillin (1 μg) discs may be used to detect intrinsic resistance to penicillin of pneumococci.

Direct (primary) susceptibility tests

Tests in which the specimen is the inoculum can be useful as the results are available a day earlier than tests on pure cultures; isolation from mixed cultures may be easier if there are differences in susceptibility and small numbers of resistant variants may be seen within zones of inhibition in diffusion tests. There are some potential disadvantages (Waterworth and Del Piano, 1976) in that the inoculum cannot be controlled, care must be taken to avoid reporting results on commensal organisms, and the specimens on which tests are likely to be useful are limited to those from sites which are normally sterile and, in the case of urine, to specimens shown microscopically to contain organisms. Direct tests should not be undertaken with specimens from patients on antimicrobial therapy.

For direct tests by the Stokes method streak pus swabs may be streaked in place of the test organisms or well mixed urines used in place of broth cultures or suspensions of pure cultures of organisms. Use *E. coli* NCTC 10418 as a control for urines and *S. aureus* NCTC 6571 for other specimens. With breakpoint methods only urine specimens can be tested. With the important and common exception of direct susceptibility testing on urines, repeat all tests on pure cultures and only report direct tests as 'provisional'.

Tests on anaerobic organisms

The reliability of diffusion susceptibility tests on many anaerobic organisms is questionable and tests should be limited to rapidly-growing species, i.e. those which grow on overnight incubation on supplemented medium. Suitable anaerobic control organisms should be included.

Quality control

Departure from the usual range of routine susceptibility may indicate gross errors in the method. Routine quality control is necessary, however, to maintain high standards. The external quality control scheme provides independent assessment of the performance of a test, and allows performance to be compared with that of other laboratories, but internal quality control is necessary to detect day-to-day variation.

Standard strains are used to control the performance of the method. In the comparative methods, control organisms must be used if the method is to be interpreted correctly. Details of appropriate control strains are given in Tables 12.1 and 12.8.

Store working cultures of control strains on agar slopes and subculture fortnightly. Replace working cultures from freeze-dried or frozen cultures (–20°C or below) every 2 months, or earlier if contamination is suspected.

The degree of control exercised will depend on the interest and resources of the laboratory. Test colonies of control cultures in the same manner as test cultures. Include control cultures in each batch of tests and use them to evaluate new lots of agar or broth before they are put into routine use.

Diffusion tests
A rapid examination of control zone sizes will indicate major problems. Recording of control zone sizes on a chart allows problems to be more easily detected. Greater control may be exercised by establishing zone-size limits. Maximum and minimum zone sizes that should be observed may be set and the mean of a series of control zone sizes should be close to the midpoint of the acceptable range. Limits should be based on at least 30 sequential observations. An acceptable range is the 95% confidence limit (i.e. only 1 in 20 observations would be expected to fall outside the range). This may be calculated as the mean zone size plus or minus 2 standard deviations.

If control tests indicate problems the source of error should be investigated. Common problems are:

(a) Inactivation of labile agents in discs owing to inadequate storage or handling in the laboratory may be indicated by a gradual decrease in zone sizes. New batches should be tested before being put into routine use.
(b) Too heavy or too light an inoculum may be indicated by a general decrease or increase in zone sizes.
(c) Errors in transcription or measuring zone sizes. In particular, different observers reading zone edges differently will result in fluctuating zone sizes.
(d) Problems with medium, for example high pH, will produce larger zones with aminoglycoside antimicrobials and erythromycin, and smaller zones with tetracycline, methicillin and fusidic acid. The reverse may occur if the pH is too low. New batches of medium should be tested before being put into routine use. Variation in the depth of medium may be indicated by fluctuating zone sizes.
(e) Contamination or mutational changes in the control strain.

Minimum inhibitory concentrations
In general, the MICs of control organisms should be within plus or minus one dilution step of the target values, or within the ranges given in Table 12.7.

In addition:

(a) For broth dilution tests, incubate one set of tubes or one microdilution plate uninoculated to check for sterility (media contaminants on agar dilution plates are usually obvious).

(b) The control without antimicrobial agents should show adequate growth of test and control strains.
(c) Plate a sample of inoculum prepared for each strain on a suitable agar medium to ensure that the inoculum is pure.
(d) Occasionally check that the inoculum density is correct by counting organisms in inocula (taken from growth controls with broth dilution methods).
(e) Check that endpoints are read consistently by all staff independently reading a selection of tests.

Breakpoint methods
In addition to the foregoing comments on quality control the following points should be borne in mind by users of breakpoint methods. Monitor the concentrations of antimicrobial agents in the plates by utilizing control strains with known MICs for the agents. To achieve suitable control these strains should have MICs only one or two dilutions above (for the resistant control) or below (for the sensitive control) the concentration in the breakpoint plates. Control strains used in Stokes method of disc diffusion susceptibility testing, e.g. *E. coli* NCTC 10418, are not generally suitable for the control of breakpoint susceptibility tests. MICs of relevant agents for these strains are usually very much lower than the concentrations being tested. Thus, there could be very significant loss of antimicrobial activity in the test plates before such sensitive controls yielded a resistant result and alerted the operator to a problem.

When breakpoint control strains of known MIC are used the method of breakpoint testing used must be identical to that used for the initial selection of the control strains in order to ensure reliability of performance. However, a batch of 'breakpoint' plates with, say, six different antimicrobial agents, with two breakpoints each, would require up to 24 control organisms with appropriate MICs to check the levels of agents adequately in all plates. If performed every day, this extensive control procedure would leave little room for testing the clinical isolates! As plates can be stored without deterioration for up to 1 week, a sample of each batch of media may be subjected to this degree of control. In addition, new batches of media may be checked by initially running them in parallel with the previously controlled plates. For certain agents where there is some evidence that degradation of antimicrobial activity may occur during storage of plates (e.g. amoxycillin with clavulanic acid (co-amoxiclav) and imipenem) daily microbiological quality control may be advisable. Similarly it is advisable to include a 'difficult' control strain of methicillin-resistant *Staphylococcus aureus* (MRSA) on a daily basis to ensure reliable detection of methicillin heteroresistant staphylococci. (Control strains for some breakpoints are available from the Antibiotic Reference Laboratory, Public Health Laboratory Service, Colindale, London.)

Franklin (1990) described a method to overcome the practical difficulties of using numerous control strains. By cutting plugs of agar from a batch of the prepared plates and placing them on the surface of a seeded, non-inhibitory, culture plate, the antimicrobial content can be monitored by the size of the inhibition zone around the agar plug. Since deviation from the expected zone size indicates error, an acceptable range has first to be determined. Establishing and validating this method of quality control may be at least as complex as searching for a small number of strains with appropriate MICs.

Inoculum size is critical; therefore, the careful cleaning, maintenance and calibration of inoculating pins is necessary.

Transcription errors are always possible and their likelihood can be reduced by appropriate staff training. Automated reading of the plates reduces this risk; in its absence, the following are useful points.

(i) Plates containing antimicrobial agents with similar sounding names (e.g. the various cephalosporins) should be separated in the order of plate reading to avoid confusion.
(ii) A legibly labelled template showing pin positions should be available to the plate reader.
(iii) Worksheets should be simple and easy to fill in.
(iv) All isolates with unusual susceptibility patterns should be checked, preferably by another worker and, if necessary, investigated further.

References

Andrews, J. M., Ashby, J. P. and Wise, R. (1990) Problems with Iso-Sensitest agar. *Journal of Antimicrobial Chemotherapy*, **26**, 596–597

Baquero, F. (1990) European standards for antibiotic susceptibility testing: towards a theoretical concensus. *European Journal of Clinical Microbiology and Infectious Diseases*, **9**, 492–495

Barry, A. L. (1976) *The Antimicrobial Susceptibility Test: Principles and Practice*, Lea and Febiger, Philadelphia, pp.163–179

Barry, A. L. (1991) Procedures and theoretical considerations for testing antimicrobial agents in agar medium. In *Antibiotics in Laboratory Medicine*, 3rd Edn (ed. V. Lorian), Williams and Wilkins, Baltimore, pp.1–16

Brown, D. F. J. (1990) Comparative methods of antimicrobial susceptibility testing – time for a change? *Journal of Antimicrobial Chemotherapy*, **25**, 307–310

Brown, D. F. J. and Blowers, R. (1978) Disc methods of sensitivity testing and other semiquantitative methods. In *Laboratory Methods in Antimicrobial Chemotherapy* (eds. D. S. Reeves, I. Phillips, J. D. Williams and R. Wise), Churchill Livingstone, Edinburgh, pp.8–30

Ericsson, H. M. and Sherris, J. C. (1971) Antibiotic sensitivity testing. Report of an international collaborative study. *Acta Pathologica et Microbiologica Scandinavica*, Section B, Supplement 217

Faiers, M., George, R., Jolly, J. and Wheat, P. (1991) *Multipoint Methods in the Clinical Laboratory, a Handbook*, Public Health Laboratory Service, London

Franklin, J. C. (1988) Assessment of antibiotic-sensitive, intermediate or resistant status of bacteria by agar dilution. *Medical Laboratory Sciences*, **45**, 225–234

Franklin, J. C. (1990) Quality control in agar dilution sensitivity testing by direct assay of the antibiotic in the solid medium. *Journal of Clinical Pathology*, **33**, 93–95

NCCLS (1987) *Evaluating Production Lots of Dehydrated Mueller-Hinton agar; Proposed Standard*, Proposed Standard M6-P, NCCLS, National Committee for Clinical Laboratory Standards, Villanova, PA, USA

NCCLS (1990a) *Methods for Dilution Antimicrobial Susceptibility Tests for Bacteria That Grow Aerobically*, 2nd Edn, Approved Standard M7-A2, National Committee for Clinical Laboratory Standards, Villanova, PA, USA

NCCLS (1990b) *Performance Standards for Antimicrobial Disk Susceptibility tests*, 4th edn, Approved Standard M2-A4. National Committee for Clinical Laboratory Standards, Villanova, PA, USA

NCCLS (1992) *Performance standards for antimicrobial susceptibility tests*, Update, National Committee for Clinical Laboratory Standards, Villanova, PA, USA

Phillips, I., Andrews, J., Bint, A. J., Bridson, E., Brown, D. F.J., Cooke, E. M., Greenwood, D., Holt, H. A., King, A., Spencer, R. C., Williams, R. J. and Wise, R. (1991) A guide to sensitivity testing. Report of the Working Party on Antibiotic Sensitivity Testing of the British Society for Antimicrobial Chemotherapy. *Journal of Antimicrobial Chemotherapy*, **27**, Supplement D, 1–50

Sherris, J. C. (1986) Problems of *in vitro* determination of antibiotic tolerance in clinical isolates. *Antimicrobial Agents and Chemotherapy*, **30**, 633–637

Thrupp, L. D. (1986) Susceptibility testing of antibiotics in liquid media. In *Antibiotics in Laboratory Medicine* (ed. V. Lorian), Williams and Wilkins, Baltimore, pp.93–150

Waterworth, P. M. and Del Piano, M. (1976) Dependability of sensitivity tests in primary culture. *Journal of Clinical Pathology*, **29**, 179–184

Chapter

13

Food poisoning and food-borne disease

The incidence of cases and outbreaks of food-borne disease and food poisoning is still very high in spite of much new legislation and a greater public awareness of the problems.

The distinction between the two terms is blurred. Although food-borne disease includes food poisoning, a practical distinction is usually drawn:

Food poisoning may be regarded as an illness which results from the consumption of food in which certain microorganisms have multiplied during preparation and storage in kitchens and catering establishments before consumption. The organisms may have been present in the original food in small numbers or introduced during subsequent handling. Usually a number of people are affected at about the same time after eating a meal prepared at one particular place.

Food-borne disease is caused by the consumption of food containing microorganisms or their products and is independent of kitchen and catering activities.

Food poisoning is considered first in this chapter as it usually requires the most urgent laboratory investigation.

Food poisoning

This usually presents as an explosive outbreak, affecting a number of people all of whom develop similar symptoms at about the same time (the incubation period) after eating the same food. Not everyone who eats that food, however, will become ill. The agent responsible is rarely distributed evenly throughout the food and people vary in their susceptibility.

The symptoms may include diarrhoea, vomiting, abdominal pain, nausea, fever and headache, depending on which organism is responsible, the severity of the disease and whether it is caused by the organisms themselves or by a toxin that they have released into the food. To cause food poisoning the organisms must be present in large enough numbers (these vary according to that agent); the food must be physically and chemically suitable for their growth; the temperature must be optimal for rapid reproduction; and enough time must elapse between contamination and consumption of the food to allow the organisms to multiply to an infective dose or to produce their toxin.

Investigation of food poisoning

Ideally three individuals or groups will be involved: the environmental health officer to whom the incident is first reported; and epidemiologist; and the bacteriologist.

The following information should be obtained at the beginning of an investigation into food poisoning.

(1) number of people at risk, i.e. how many ate the suspected food;
(2) number of persons actually ill;
(3) nature of illness: (a) vomiting, (b) diarrhoea, (c) nausea, (d) headache, (e) disturbance of the central nervous system;
(4) exact time food was eaten and exact time of onset of symptoms;
(5) description of *all* food eaten in 24 h before the onset of symptoms and where it was eaten, and as much information as possible about the food consumed during the previous 4 days;
(6) details of foreign travel.

This information will indicate whether the incident merits laboratory investigation. A very small number of people alleged to have been ill when a very large number is at risk is probably not an outbreak of food poisoning. Similarly, stories of single cases of poisoning due to foreign bodies, canned fruit, fresh vegetables, unusual articles of diet or food which has 'gone off' can usually be discredited. In nurseries or institutions, a wide variation in the times of onset usually suggests viral gastroenteritis rather than bacterial food poisoning. When symptoms appear 30 min or so after eating, *B. cereus* or chemical food poisoning is indicated. It may not necessarily be the last meal which causes bacterial illness.

The information will suggest one of the types of food poisoning described below and thus indicate the bacteriological examinations required (Table 13.1).

Types of food poisoning

Salmonella food poisoning

This is due to *infection* with a salmonella organism and usually the food must contain several million organisms to initiate infection. Diarrhoea and vomiting occur from 12–36 h after eating infected food. Illness may last from 2–7 days. Not all the persons who ate the food will develop symptoms; some may become symptomless excreters. Foods to suspect include poultry, meat products, eggs and egg products. *S. enteritidis* (phage type 4) is currently causing much concern. These foods become infected from animal or human intestinal sources, directly or indirectly. The salmonellas responsible for enteric fever (*S. typhi* and *S. paratyphi* A, B and C) do not usually infect animals. They are transmitted from person to person by the faeces-food-oral route.

Campylobacter food poisoning

These organisms have been growing steadily in importance since 1977 and are now the most common cause of gastroenteritis. They are not a new cause of infection but efficient methods for their isolation have demonstrated their

Table 13.1 Food poisoning: agents, symptoms and likely foods

Incubation period	*Organisms*	*Vomiting*	*Diarrhoea*	*Cramps*	*Fever*	*Duration*	*Examples of foods*
30 min–4 h	*B. cereus* (emetic)	+ + +	(+)[a]	(+)[a]	—	short	Rice
4–6 h	*S. aureus*	+ +	+	+ +	—	6–12 h	Cold meats Dairy produce, salads, artificial cream fillings, eggs, custards, fermented sausages
8–22 h	*C. perfringens*	—	+ +	+ +	—	12–24 h	Reheated meat-based products
8–48 h	*V. parahaemolyticus*	—	+ + +	—	+	2–3 days	Seafoods, meat
8–12 h	*B. cereus* (diarrhoeal)	—	+ +	+ +	—	1–2 days	Soups, milk products
12–48 h	*Salmonella*	(+)	+ +	—	+	48 h–7 days	Poultry, meat, duck eggs
24–72 h	*C. botulinum*	—	—	—	—[b]	3 days	(Home) canned/bottled vegetables, meat, fish
2–11 days	*Campylobacter*	—	+ + +	+ +	+ +	3 days–3 weeks	Milk, water, poultry, raw and undercooked meat and fish, unpasteurized milk, mushrooms

[a](+) may or may not be present [b] Disturbed CNS

ubiquity. Campylobacters are part of the normal flora of many mammals and birds. Strains responsible for food poisoning do not grow below 30°C in the laboratory and are known as 'thermophilic campylobacters'. Poultry appear to be responsible for some sporadic outbreaks but outbreaks associated with inadequately pasteurized milk have been reported. Water supplies are also a source of infection. The organisms are heat sensitive, are destroyed by pasteurization, do not survive in acid foods and, in common with salmonellas, control measures such as adequate cooking and prevention of recontamination are indicated. They do not normally multiply in food, except possibly in vacuum packs.

The incubation period has been estimated to be between 2 and 11 days. Symptoms commonly include abdominal pain, diarrhoea, fever, headache, malaise, musculoskeletal pain, rigors and delirium. Vomiting occurs in less than 30% of cases. Sometimes 'flu-like' symptoms are present.

Clostridium perfringens food poisoning

Our knowledge of the role of *C. perfringens* in food poisoning has altered in recent years. Originally, only *C. perfringens* forming spores resistant to boiling for 1 h was thought to cause this type of food poisoning. These strains were invariably non-haemolytic. Recently, both haemolytic and non-haemolytic heat-sensitive strains have been incriminated in food poisoning outbreaks. The incubation period varies from 8–22 h. Young and long-stay patients and hospital staff tend to become colonized with *C. perfringens*. Counts of 10^5/g in the faeces of sufferers are not uncommon.

Because the heat resistance of *C. perfringens* spores varies, the organisms survive some cooking processes. Much depends upon the amount of heat that reaches the cells and the period of exposure. If the spores survive and are given suitable conditions, they will germinate and multiply. It is a common practice to cook a large joint of meat, allow it to cool, slice it when cold and re-heat it before serving. This is hazardous unless cooling is rapid and re-heating is thorough and above 80°C. Growth of heat-sensitive strains in cooked food may be the result of too low a cooking temperature or may be due to recontamination after cooking. The usual vehicles for *C. perfringens* food poisoning are meat and poultry. Stews, made-up meat dishes and large deep-frozen birds are suspects. *C. perfringens* is commonly found in soil; small numbers are normally present in human faeces and in some foods. The symptoms are abdominal cramps and diarrhoea, lasting for 12–24 h, and are caused by the production of an enterotoxin.

Staphylococcal food poisoning

Staphylococcus aureus produces several enterotoxins which will withstand heating at 100°C for 30 min. An enterotoxin, present in food in which the organisms have grown, causes an *intoxication* type of disease with onset in 4–6 h and symptoms of diarrhoea and vomiting lasting 6–8 h. Foods likely to be infected are those which contain a high proportion of salt, meat such as ham, synthetic creams, sauces (particularly Hollandaise) which are never heated above 110°F (40°C), custards, trifles and occasionally fresh fish or canned vegetables (due to leaker spoilage). Infection occurs usually in the kitchen from infected cuts and abrasions, boils and other lesions or from nasal carriers.

Bacillus cereus food poisoning

Cases of food poisoning associated with eating fried rice have been reported from Scandinavia, the USA and Europe. Large numbers of *B. cereus* were isolated from the rice.

Rice for fried rice dishes is often boiled in bulk and left to cool at room temperature. Refrigeration causes the grains to stick together. Before serving, egg is added to the cooked rice, which may have been standing in a warm kitchen for many hours. The mixture is then lightly cooked in a frying pan and served. During the initial cooking, vegetative organisms are destroyed; during standing, spores germinate in ideal conditions for growth. Subsequent heating does not destroy any toxins that may have formed.

B. cereus food poisoning can be confused with staphylococcal or perfringens food poisoning. There are two types, diarrhoeal and emetic.

	Diarrhoeal	*Emetic*
Incubation	8–12 h	30 min–4 h
Symptoms	diarrhoea, nausea	vomiting (v. common) diarrhoea (common)
Duration of illness	12–24 h	6–24 h
Isolation from faeces	sometimes (low numbers)	sometimes
Isolation from vomit	—	large numbers
Nature of toxin	protein	small peptide

The emetic type is more common. Occasionally other *Bacillus* spp. may be responsible for similar symptoms. See Gilbert *et al.* (1981).

Botulism

This form of intoxication, caused by the pre-formed toxin of *Clostridium botulinum*, is rare in the UK, but more prevalent in Europe and the USA where home canning of meat and vegetables is common practice. Between 12 and 36 h after eating food containing the toxin, patients develop symptoms including thirst, vomiting, double vision and pharyngeal paralysis, lasting 2–6 days and usually fatal. Vehicles are usually home-preserved meat or vegetables, contaminated with soil or animal faeces and insufficiently heated. Correct commercial canning processes destroy the organisms. Conditions must be optimum for toxin formation: complete anaerobiosis, neutral pH, absence of competing organisms.

Recently there have been several reports (from outside the UK) of botulism caused by the Type E organism that is usually associated with fish and fish products. This type is frequently present in estuarine mud.

Yersinia enterocolitica food poisoning

This is an uncommon cause of food poisoning but its ability to grow at 4°C makes it unusual and a potential hazard. Foods implicated include milk, ice cream and seafoods. The symptoms are diarrhoea, fever, abdominal pain and vomiting.

Vibrio parahaemolyticus food poisoning

This is rare in the UK but it is one of the commonest causes of food poisoning in Japan and has been incriminated in outbreaks in the USA and Australia.

It is closely associated with eating raw and processed fish products. The organism has been frequently isolated from sea foods, particularly those harvested from warm coastal waters. Symptoms of infection may appear from 2 to 96 h after ingestion, depending on the size of the infecting dose, type of food and acidity of the stomach. They vary from acute gastroenteritis with severe abdominal pain to mild diarrhoea.

Escherichia coli food poisoning

Large numbers of *E. coli* are present in the human and animal intestines. They are often found in raw foods where they may indicate poor quality. At least four different types (p. 308) may be involved in food poisoning. *E. coli* may cause travellers' diarrhoea when the source may be contaminated food or water.

Shigella food poisoning

This is rare in developed countries. It is the result of contamination of food because of unhygienic practices by persons who are suffering from shigellosis or who are symptomless excreters.

Food-borne disease

Agents of food-borne disease include *Brucella* spp., *Mycobacterium bovis* (both milk-borne), *Bacillus anthracis*, *Aeromonas* and enterobacteria other than those mentioned above. Some viruses, e.g. that of hepatitis A, and parasites, not considered in this book, are also known to be associated with food-borne disease, as are toxins produced by several fungi. The most important agents at present are *Listeria* spp. and mycotoxins.

Listeriosis

This food-borne disease, caused by *Listeria monocytogenes*, affects mainly young children, pregnant women and the elderly. The organisms are present in small numbers in many foods, but are unlikely to cause disease unless they have the opportunity to multiply. This may happen at quite low temperatures, e.g. in improperly maintained cold storage cabinets and refrigerators.

Methods for the examination of foods are given in Chapter 14.

Mycotoxins and aflatoxins

Certain fungi, mainly *Aspergillus* spp., form toxins in nuts and grains during storage. These toxins can cause serious disease in humans and animals. The toxins are detected chemically and commercial kits are available. For information about these toxins see Moss *et al.* (1989).

Examination of pathological material and food

Do a total viable count on all foods suspected of causing food poisoning. In most cases this will be high. Relate the findings to conditions of storage after serving. If possible count the numbers of causative organisms.

Table 13.1 (p.208) suggests the causative organism. Isolation and identification methods are given in the relevant chapters:

Salmonella	Ch. 28	*Campylobacter*	Ch. 33
Staphylococcus	36	*Y. enterocolitica*	34
B. cereus	42	*V. parahaemolyticus*	24
C. botulinum	44	*E. coli*	26
C. perfringens	44	*Listeria*	40

For further information on food poisoning see Corry *et al.* (1982), Hobbs and Roberts (1987), Cliver (1990) and Harrigan and Park (1991).

References

Cliver, D. O. (ed.) (1990) *Foodborne Disease*, Academic Press, New York

Corry, J. E. L., Roberts, D. and Skinner, F. A. (eds) (1982) *Isolation and Identification Methods for Food Poisoning Organisms*, Society of Applied Bacteriology Technical Series No. 17, Academic Press, London

Gilbert, R. J., Turnbull, P. C. B., Parry, J. M. and Kramer, J. M. (1981) *Bacillus cereus* and other bacillus species: their part in food poisoning and other clinical infections. In *The Aerobic Endosporing Bacteria: Classification and Identification* (eds R. C. M. Berkeley and M. Goodfellow), Academic Press, London, pp.297–314

Harrigan, W. F. and Park, R. W. A. (1991) *Making Food Safe*, Academic Press, London

Hobbs, B. C. and Roberts, D. (1987) *Food Poisoning and Food Hygiene*, 5th ed, Edward Arnold, London

Moss, M. O., Jarvis, B. and Skinner, F. A. (eds) (1989) Filamentous Fungi in Foods and Feeds. *Journal of Applied Bacteriology* Symposium Supplement No.18, **67**, 1S–144S

Chapter

14

Food microbiology: general principles

The microbiology of food is under consideration by several EU committees and by the Codex Alimentarius Commission of the World Health Organization (WHO) and the Food and Agriculture Organization (FAO). In the UK there are statutory methods for testing milk and recognized methods for the examination of water. The British Standards Institute (BSI, 1986, 1987, 1991 a–e) has published a number of standard methods for the examination of foods and animal feeding stuffs. These include general methods for the preparation of dilutions, total viable counts, for coliforms by the pour-plate method and the MPN, *E.coli* (presence/absence of Enterobacteriaceae), a variety of pathogens, yeasts and moulds. Most of these methods are also published by the International Standards Organization (ISO, 1991a–e). Some laboratories use these methods but others use simplified versions or follow their own techniques.

In the USA standard methods are published by the American Public Health Association (APHA) (Speck, 1992), and the Food and Drug Administration produces a *Bacteriological Analytical Manual* (FDA, 1992). There are also the publications of the International Commission on the Microbiological Specification of Foods (ICMSF, 1980, 1986, 1988) and the Meat Industry Research Institute of New Zealand (MRINZ, 1991).

With the development of quality assurance (QA) schemes such as those of the BSI (1987) in industrial laboratories and the National Measurement Accreditation Service (NAMAS) scheme for consultant and contract laboratories more attention has been given to standard methods and/or validation of alternative methods. Standard methods are useful for resolving conflicting results but are often slow. 'In house' methods are sometimes more satisfactory because results depend on the skill of the worker and familiarity with the method. There are no standard methods, however, for some organisms (e.g. *L. monocytogenes*).

In addition to QA schemes there are proficiency testing and quality control schemes organized by the Public Health Laboratory Service (PHLS; see Snell *et al.*, 1991) and industries. Participating laboratories receive shelf-stable, simulated food samples together with information about the type of food and the illness associated with it. Presence or absence of pathogens, indicator and spoilage organisms are then determined.

Scope of microbiological investigations

Four kinds of investigations are usually carried out:

(1) estimation of the viable count (colony-forming units, cfu/g or ml);
(2) estimation of the numbers of *E. coli*, coliform bacilli (Enterobacteriaceae) and/or *Enterococcus* spp. ('faecal streptococci'). These may indicate the standard of hygiene during the preparation of the foods, especially where processing (e.g. heating) would be expected to kill these organisms (see Indicator organisms, below);
(3) detection of specific organisms known to be associated with spoilage. This may be more important than the viable count in determining the potential shelf-life;
(4) detection of food-poisoning organisms.

No single method is completely satisfactory for the examination of all foodstuffs. The methods selected will depend on local conditions, e.g. availability of staff, space and materials, numbers of samples to be examined and the time allowed to obtain a result. To these ends, rapid, automated methods have much to offer (see Chapter 8).

It may be desirable to examine a large number of samples by a slightly suboptimal method than a smaller number by an exhaustive method.

Many of the microorganisms in foods can be sublethally damaged, e.g. by heat, cold or adverse conditions such as low pH, low water content, high salt or sugar content. On the other hand these conditions may favour the growth of yeasts and moulds, when mycotoxins may be formed.

Indicator organisms

These are organisms that may indicate evidence of contamination or pollution, especially of a faecal nature, e.g. *E. coli* and coliform bacilli. The family term Enterobacteriaceae is sometimes used because 'coliforms' are poorly defined taxonomically. Enterobacteriaceae includes other organisms, like important pathogens such as salmonellas and various non-lactose fermenters that may be present in human and animal faeces. It should be noted, however, that the presence of these organisms is not *prima facie* evidence of faecal contamination: many of them are normally present in soil and on plant material.

Microbiological criteria (standards)

Limits for numbers of organisms (or their toxins) are given various names. The terms used by the ICMSF, 1986) are:
Standard – criteria set out in a law or regulation controlling foods produced, processed or stored in the area of jurisdiction, or imported into that area.
Guidelines – criteria used by manufacturers or regulatory agencies in monitoring a food ingredient, process or system (can be applied at points during manufacture).
Microbiological purchasing specifications – criteria that determine the acceptance of a specific food or food ingredient by a food manufacturer or other private or public purchasing agency.

Sampling plans

Limits are normally set in terms of numbers of organisms (cfu/g or ml) or presence/absence of a pathogen in a given amount of the food. The exact number of organisms can never be determined with precision and it is therefore inappropriate to set criteria with exact limits beyond which a food is unacceptable. For this reason it is customary to use three-class sampling plans for the numbers of organisms. These classes are: 'Acceptable' – under a specified number (**m**); 'Marginally acceptable' under a higher specified number (**M**); 'Unacceptable' when numbers exceed **M**. **M** is usually at least 10 times higher than **m**. For example, the total count limit for liquid pasteurized whole egg products suggested by ICMSF is

$$n = 5, c = 2, \mathbf{m} = 5 \times 10^4, \mathbf{M} = 10^6$$

where n is the total number of samples tested, c is the number that may exceed **m**. None should exceed **M**.

Two-class plans are usual for salmonella testing. For example, 5, 10, 20 or >25 g samples might be tested for the presence/absence of salmonellas and the food would be unacceptable if any sample was positive (n = 5, 10, 20 or more, c = 0, **m** = 0. The number of samples examined would depend on the perceived risk if the food was contaminated: foods with the highest risk would be those eaten without prior heating and by a very susceptible class of consumer, e.g. dried milk for babies.

Limits in microbiological standards that are applied by enforcement authorities normally include those for pathogens and indictor organisms, not viable counts. The philosophy of end-point testing *vs* application of good manufacturing practices, and the statistics of sampling are outside the scope of this book but are detailed by ICMSF (1986). It is enough here to state that end-product testing alone is not sufficient to check, with a reasonable degree of certainty, that a given batch of food is satisfactory unless an impracticable number of samples is examined. End-product testing is useful only as a check on the monitoring of the whole process – although it may detect an unsatisfactory batch. If a sample of a perishable food is taken at a point after manufacture and after an unknown period on the shelf the viable count result is often difficult to interpret and is of little use in assessing the quality at the point of manufacture. It is more appropropiate to test for pathogens.

Sampling

The usefulness of laboratory tests depends largely on correct sampling procedures. The sampling plan (see above) must be applied to each situation so that the samples submitted to the laboratory are fully representative of the batch. In factory sampling it is more useful to take small numbers of samples at different times during the day than larger numbers at any one time. Where resources are limited sampling may concentrate on the finished product, but if there is an increase in counts it is necessary to go back through the process and take 'in line' samples.

Containers and samplers

Use screw-capped aluminium or polypropylene jars or plastic bags; never use glass containers where any breakages could contaminate the product or cause injury to personnel. Instruments for sampling from bulk material vary: for frozen samples use a wrapped sterilized brace and bit or a chopper that has been washed in methylated spirit and flamed. Take care to avoid cuts from ice splinters. Use individually wrapped spoons, spatulas or wooden tongue depressors for soft materials. With prepacked food take several packages as offered for sale.

When sampling carcase meat and fish, where contamination occurs only on the surface, use a surface swab over a defined area or excise a surface layer 2–3 mm thick (see p.222).

Transport and storage of samples

Carry samples (other than canned and bottled foods) in insulated containers, cooled with ice packs ('Coolkeepers'). The ice should not have melted by the time the samples arrive at the laboratory. The samples should then be transferred to a refrigerator and examined as soon as possible. There is a code of practice for sampling by authorized officers under the Food Safety Act 1990. See also Harrigan and Park (1991).

Pre-testing considerations

Before examining any particular food check the type of spoilage and/or pathogens that might be expected. In general, products with low pH and/or low water activity (a_W) tend to be spoiled by yeasts and moulds. At intermediate pH levels and low a_W lactic acid bacteria and other Gram-positive organisms such as micrococci and enterococci may predominate. At neutral pH and/or high a_W spoilage is mostly due to Gram-negative bacteria such as *Pseudomonas* spp. In some heat-processed foods, such as meat pies, spoilage is due to outgrowth of surviving spores of *Bacillus* and/or *Clostridium* spp., especially in the absence of competing flora.

General methods

If the organisms are expected to be distributed throughout the material homogenize 10 g of the food (weighed aseptically) in 90 ml of maximum recovery diluent in a Stomacher. Make serial dilutions, e.g. 10^{-1}–10^{-4}, do viable counts and inoculate culture media (see below).

If the organisms are likely to be on the surface only weigh 50 or 100 g into a sterile jar and add 100 ml of diluent. Shake for 10 s, stand for 30 min and shake again. The organisms are assumed to have been washed into the diluent, that is, 100 ml now contains the organisms from the original weight of sample. Make dilutions and proceed as above.

Total and viable counts

Methods are given in Chapter 10. For total counts Breed smears and DEFT counts are useful and suggest appropriate dilutions for viable counts. For counting mould hyphae and yeasts the Neubauer chamber, used in haematology, is better than the Helber chamber because it is deeper.

The methods and incubation temperatures for viable counts depend on the nature and storage conditions of the food. For most foods use pour plates and incubate at 30°C for 24–48 h.

For chilled and frozen meat and frozen fish use surface plating as the psychrophiles and psychrotrophs that predominate in such foods may be damaged if exposed to the temperature of melted agar used in pour plating. Incubate at 20–25°C. To count psychrophiles incubate at 1°C for 14 days.

Use the 'black tube' method (p. 407) for counting anaerobes.

Report counts as cfu per g or ml. The tables on p. 152 show the amount of food in grammes contained in the various dilutions.

Presence/absence (P/A) tests

It may not be necessary to set up full viable counts. If a standard is set on the basis of experience the technique can be modified to give a present or absent ('pass' or 'fail') response. If, for example, an upper limit of 100 000 cfu/g is set, then a plate count method that uses 1 ml of a 1:1000 dilution will suffice. More than 100 colonies, easily observed without laborious counting, will fail the sample. Similarly, more than 100 colonies in roll tubes containing 0.1 ml of a 1:100 dilution, or more than 40 per drop in Miles and Misra counts from a 1:50 dilution will suggest rejection.

Enterobacteriaceae, coliforms and E. coli

There is a wide choice of media for these tests. Most 'standard' tests specify EE broth (which is brilliant green glucose broth), violet red bile glucose agar (VRBG) and violet red bile lactose agar (VRBL or VRB). VRBG and VRB should not be autoclaved: see manufacturers' instructions. If these are not available MacConkey broth and agar may be used. Lauryl sulphate tryptose broth is also specified for some tests. Some workers prefer minerals modified glutamate medium (MMGM) instead of EE broth.

If small numbers are anticipated use the MPN or P/A methods, but for heavily contaminated material plate or surface counts are better.

Enterobacteriaceae

Presence/absence
If resuscitation of sublethally damaged organisms is indicated add dilutions of the original suspension (see above) to maximum recovery diluent, incubate at room temperature for 6 h.

Add 10 ml of suspension to 10 ml of double strength EE broth and 1 ml of suspension to 10 ml of single strength EE broth. Incubate at 30°C for 18–24 h and then plate on VRBG. Incubate plates at 30°C and count colonies.

MPN method
See p. 157.

Pour plates
If resuscitation is necessary incubate the dilutions in maximum recovery diluent for 90 min. Use freshly prepared VRBG for the counts.

Surface counts
Spread 0.1 volumes of the dilutions on non-selective medium (e.g. plate count

agar). Incubate for 6 h at room temperature to resuscitate damaged cells, overlay with 5 ml of VRBG and incubate at 30°C overnight.

Coliforms

Use the methods outlined above but with lauryl sulphate tryptose broth. Subculture positive tubes to brilliant green lactose broth to check gas production.

Incubation temperatures
Incubate coliform tests on dairy products, etc. at 30°C. For other products use 37°C.

Membrane method
Place the membranes on MMGA or non-selective media. Spread 1 ml volumes of the diluted suspension on the membranes. Incubate for 4 h at 30°C and then for 24 h at 37°C. Transfer membranes to tryptose bile agar and incubate at 44°C for 18–24 h. Count colonies and subculture if necessary.

Escherichia coli ('faecal coli')
Subculture suspect colonies or broth tubes to brilliant green broth and peptone water and incubate at 44 ± 0.2°C overnight. Only *E. coli* produces gas and indole at 44°C. For other confirmation tests see Chapter 26.

Counting enterococci

Do surface counts as described on p. 153 with kanamycin azide aesculin agar, azide blood agar, MacConkey or Slanetz and Bartley medium. Consult manufacturers' manuals for colonial appearance on azide media.

Counting clostridia

As counting these organisms usually incurs identifying them as well, the techniques are described under *Clostridium*, in Chapter 44.

Spoilage organisms

See Table 14.1. Methods for culturing them are given under the headings of the foods concerned. See also ICMSF (1980).

Food poisoning bacteria

'Routine' examination of all foods for salmonellas and staphylococci, or other food poisoning organisms, is hardly worthwhile. Only those foods and ingredients known to be vehicles need be tested. They are given in Chapter 13 and under the appropriate headings in the following chapters.

Table 14.1 Types of organisms associated with various non-sterile foods

pH	a_w	*Food*	*Comments*	*Microbial flora*
7.4–5.4	1.00–9.97	Animal protein, milk, whole liquid egg, raw red meat, fish, poultry	Chilled aerobic	*Pseudomonas*, *Acinetobacter*, *Aeromonas*, *Serratia*, *Hafnia*, *Enterobacter*, *Yersinia*, *Lactobacillus*, *Carnobacterium*, *Enterococcus* (in milk and egg), *Bacillus* (in milk and egg), *Brochothrix thermosphactum* (in meat), *Shewanella* (in meat and poultry) (pH 6.0) and fish
			Vacuum or gas packed (high CO_2)	*Lactobacillus*, *Carnobacterium*, *B. thermosphactum*, Pseudomonads, Enterobacteriaceae
		Cooked non-cured meats (including slices)		As raw meats
		Raw salads, bean sprouts	High counts normal	*Pseudomonas*, *Erwinia*, other Enterobacteriaceae, *E. coli*
		Raw bacon and ham	Aerobic	*Micrococcus*, *Acinetobacter*, *Vibrio* (at pH > 5.9), yeasts, moulds
		Raw bacon and ham with NaCl and nitrite	CO_2 or vacuum-packed	Lactic acid bacteria, *Micrococcus*, *Enterococcus*, Enterobacteriaceae (low salt, high temperature and high pH favour Gram-negative bacteria)
		Semi-preserved, cooked cured meats, e.g. ham	Pasteurized at 65–70 °C and packed in cans or flexible film	*Enterococcus* other lactic acid bacteria that survive pasteurization, often in large numbers, occasionally causing spoilage by gelatin liquefaction, gas production or souring. Low salt and high spoilage due to *Clostridium* and *Bacillus*
< 4.5	0.95–0.80	Cheeses, including cottage cheese, yoghurt	Fermented products contain lactic acid bacteria (starters)	Yeasts and moulds (spoiling starters), yoghurts containing fruit are very prone to yeast spoilage, causing lids to 'dome'. Surface-ripened cheese (e.g. Brie), or mould-ripened with localized high pH are prone to contamination with *L. monocytogenes*. May also contain low numbers of Enterobacteriaceae, including *E. coli*
		Fermented meats, e.g. salami	Contain starters which may include enterococci and/or micrococci, staphylococci and lactic acid bacteria	Starters, yeasts and moulds on surface. Low numbers of surviving flora from raw materials, e.g. meat
4.5–3.0		Pasteurized, chilled fruit juices		Yeasts, lactic acid bacteria
		Tomatoes and other fruit		Yeasts and moulds
		Mayonnaise		Yeasts, lactobacilli
	0.70–0.85	Dried fruits, jams	Higher a_W products contain preservatives – SO_2, sorbate	Moulds and yeasts
		Fermented foods, e.g. sauerkraut, pickles		Yeasts, moulds, lactic acid bacteria, *Bacillus* spp. if pH > 4.0
	0.8–0.6	Grains and flour		Moulds

Yeasts and moulds

Sampling
Circumstances vary. Obtain as much 'background' information as possible before deciding on the method of examination. The number of samples should be as large as is practicable.

Sample whole packages where possible and examine the casing for damage and water staining.

Visual inspection
Open carefully and look for evidence of moulds. Microscopy is rarely helpful.

Direct examination
Pre-incubation may be useful (Jarvis *et al.*, 1983). Place a filter paper soaked with glycerol in a large petri dish and sterilize it. Suspend a sample above the filter paper, e.g. on glass rods, replace the lid and incubate for up to 10 days. Examine daily for moulds.

To estimate shelf-life relative to mould growth under adverse conditions incubate unopened packages in a controlled humidity cabinet.

Culture
Prepare a 1:10 suspension in peptone water diluent containing 0.1% Tween 80. Use a Stomacher if possible. Make serial dilutions in the same diluent and surface plate (in duplicate) on Rose Bengal chloramphenicol agar (p. 65). Incubate at 22°C for 5 days. For low a_w foods use DG18 agar.

References

BSI (1986) BS 5763: *Methods for the Microbiological Examination of Foods and Feeding Stuffs, Part 10, Enumeration of Enterobacteriaceae*, British Standards Institution, Milton Keynes

BSI (1987) BS 5750: *Quality Assurance: Management System*, British Standards Institution, Milton Keynes

BSI (1991a) BS 5763: *Methods for the Microbiological Examination of Foods and Feeding Stuffs, Part 1, Enumeration of microorganisms – colony count technique at 30°C*, British Standard Institution, Milton Keynes

BSI (1991b) BS 5763: *Methods for the Microbiological Examination of Foods and Feeding Stuffs, Part 2, Enumeration of coliforms – colony count techniques*, British Standard Institution, Milton Keynes

BSI (1991c) BS 5763: *Methods for the Microbiological Examination of Foods and Feeding Stuffs, Part 3, General guidance for the enumeration of coliforms – most probable number techniques at 30°C*, British Standards Institution, Milton Keynes

BSI (1991d) BS 5763: *Methods for the Microbiological Examination of Foods and Feeding Stuffs, Part 13, Enumeration of Escherichia coli: colony count technique using membranes*, British Standards Institution, Milton Keynes

BSI (1991e) BS 5763: *Methods for the Microbiological Examination of Foods and Feeding Stuffs. Part 15. Detection of Enterobacteriaceae with pre-enrichment*, British Standards Institution, Milton Keynes

FDA (1992) *Bacteriological Analytical Manual*, 7th edn, Food and Drug Administration, AOAC International, Arlington, VA

Harrigan, W. F. and Park, R. W. A. (1991) *Making Food Safe*, Academic Press, London

ICMSF (1980) Cereal and Cereal Products. International Commission of Microbiological Standards for Foods. In *Microbial Ecology of Foods*, Vol. 2, Academic Press, New York

ICMSF (1986) *Microorganisms in Foods. 2. Sampling for Microbiological Analysis. Principles and Specific Applications*, Blackwells, Oxford

ICMSF (1988) *Microorganisms in Foods. 4. HACCP in Microbiological Safety and Quality*, International Commission on the Microbiological Specifications for Foods, Blackwells, London

ISO (1991a) 4831: see BSI (1991) Part 3

ISO (1991b) 4831: see BSI (1991) Part 13

ISO (1991c) 7250: see BSI (1986) Part 10

ISO (1991d) 4832: see BSI (1991) Part 2

ISO (1991e) 8523: see BSI (1991) Part 15

Jarvis, B., Seiler, D. A. L., Ould, S. J. L and Williams, A. P. (1983) Observations on the enumeration of moulds in food and feedingstuffs. *Journal of Applied Bacteriology*, **55**, 325–326

MRINZ (1991) *Microbiological Methods for the Meat Industry*, 2nd ed, Meat Industry Research Institute of New Zealand, Hamilton

Snell, J. J. S., Farrell, I. D. and Roberts, C. (eds) (1991) *Quality Control. Principles and Practice in the Microbiology Laboratory*, Public Health Laboratory Service, London

Speck, M. (ed.) (1992) *Compendium of Methods for the Microbiological Examination of Foods*, American Public Health Association, Washington, DC

Chapter

15

Meat and poultry

Fresh and frozen carcase meat

Table 14.1 (p.219) summarizes the types of bacteria that cause spoilage. A wide variety of bacteria is present immediately after slaughtering. Gram-negative organisms multiply and predominate during storage. The deep muscle tissue is usually sterile but may be contaminated by instruments used in butchery. Meat from stressed animals may be of poor keeping quality because the pH is higher than normal.

Carcase sampling

Microbial contamination is usually confined to the skin or surface of a carcase. Gram-negative, motile bacteria show a greater adherence to it than do Gram-positive species. Although deep muscle is usually sterile the meat may have a higher pH if the animal has suffered stress and it may be contaminated.

The most reliable methods are destructive, involving scraping off the top 3 mm of a measured area (e.g. 25 or 50 cm^2). The most practical non-destructive method is swabbing.

To sample the whole carcase by a non-destructive technique use the method devised by Kitchell *et al.* (1973). Sterilize large cotton wool pads, wrapped in cotton gauze in bulk and transfer them to individual plastic bags. Moisten a pad with a 0.1% peptone water and wipe the carcase with it, using the bag as a glove. Take a dry pad and wipe the carcase again. Place both pads in the same bag, then seal and label it. Add 250 ml of diluent to the pads and knead, e.g. in a Stomacher, to extract the organisms. Prepare suitable dilutions from the extract and do total counts at 20°C. Inoculate other media as desired.

The other areas on the carcase that are most likely to be contaminated after butchering are the rump, brisket and forelegs. Murray (1969) recommended swabbing a 16-cm^2 area on each part. With good hygiene, counts of less than 150 000 on the brisket, 50 000 on the rump and 25 000 on the forelegs per 16 cm^2 at 20°C may be achieved immediately after dressing and after 3 to 4 days in chill.

'Hot boning' is now becoming more common and this may give rise to different problems.

Routine examination of boned-out meat in the laboratory

It is important to take equal quantities of both fat and lean tissues. Deep tissue near the bone should be examined if bone taint is suspected. To avoid contamination during examination for bone taint, sear or paint the surface with an antiseptic dye. If an overall picture is required, take core samples from boned-out joints, macerate, dilute and plate for aerobic and anaerobic culture. Express counts as the number of cfu/g of tissue. Most counts should be less than 100 000 cfu/g.

Most spoilage organisms will be growing on the surface of the meat under aerobic conditions. Scrape a convenient area, e.g. 10 cm^2 with a sterile scalpel, shake the scrapings in 90 ml of warm diluent. Prepare dilutions, plate, incubate and count.

In a meat factory, where the hygiene is good, more than 70% of samples may be expected to have count of less than 1000 cfu/cm^2.

Microbial content

In moist, chill conditions the predominant spoilage organisms are *Pseudomonas*, *Acinetobacter*, *Enterobacter* and *Brochothrix thermosphactum*. *Shewanella putrefasciens* grows in meat with a pH > 6.0. Anaerobes do not appear to be important except where meat is held at temperatures above 25°C, when *Clostridium* spp. (sometimes *C. perfringens*) predominate. About 10% of pork carcases contain *C. botulinum*, which fortunately is a poor competitor. About 60% of them carry *C. perfringens*.

Salmonellas are not often found in raw red meats, except raw minced products. They are more common in meat sold as pet food, frozen boneless beef and horseflesh.

In dry (*ca* 80% relative humidity) chill conditions mould spoilage may occur on chilled (−1°C) beef and frozen mutton (−5°C or less) but moulds do not grow below −10°C. 'Green spots' are usually due to *Penicillium* spp., 'white spots' to *Sporotrichum* spp., 'black spots' to *Cladosporium* spp. and 'whiskers' to *Mucor* and *Thamnidium* spp.

Comminuted meat

This is fresh meat, minced or chopped and with no added preservative. Surface organisms are therefore distributed unevenly throughout the mass and further contamination may occur during the process.

Take several random samples.

Viable counts
Do plate counts on dilutions 10^{-4} to 10^{-8}. Counts are usually very high, but standards must be set by experience. It may be desirable to estimate the number of coliforms present and to determine what proportion of these are *E. coli*. Surface counts, MPN and membrane filter methods may all be used, but the results may not be comparable.

Microbial content
This is similar to that in fresh meat. Salmonellas and staphylococci may be present.

British fresh sausages

A fresh sausage contains comminuted meat, cereals, spices and sulphur dioxide up to a limit of 450 ppm. This checks the growth of Gram-negative species.

This product should not be confused with European or American sausages, which are smoked.

Viable counts
Remove the casing, if present, and do total viable counts using dilutions from 10^{-4} to 10^{-8}. Incubate at 30°C for 48 h and at 22°C for 3 days. Counts of several million cfu/g may be expected. Skinless sausages may have lower counts than those with casings as the process involves blanching with hot water to remove the casings in which they are moulded.

Sausage casings, whether natural (stripped small intestine of pig or sheep) or artificial, usually do not present any bacteriological problems.

Microbial content
The predominant spoilage organisms are yeasts, and B. *thermosphactum* which is associated with souring. Gardner's (STAA) medium is useful for the isolation and enumeration of this organism. Salmonellas may be present.

We have not found the use of surfactants particularly advantageous. Sometimes they are inhibitory.

Prepacked fresh meat

In gas or vacuum packs with non-permeable wrappers pseudomonads are suppressed and lactobacilli, enterobacteria and *B. thermosphactum* predominate.

Meat pies

Meat pies may be 'hot eating', e.g. steak and chicken pies, or 'cold eating', e.g. pork, veal and ham pies and sausage rolls.

'Hot eating' pies

The meat filling is pre-cooked, added to the pastry casing and the whole is cooked again.

Viable counts
Examine the meat content only. Open the pie with a sterile knife to remove the meat. Do counts on dilutions of 10^{-1} and 10^{-2} and incubate at 30°C for 48 h. A reasonable level is not more than 100 cfu/g.

Clostridia
Add 1-ml amounts of the 10^{-1} dilution to 15-ml bottles of reinforced clostridial medium melted and at 50°C. Cool, seal with petroleum jelly and incubate at 37°C for 48 h. Examine tubes for blackening. Subculture any black tubes in purple milk. Plate out on blood agar, incubated anaerobically to identify any other clostridia.

'Cold eating' pies

These contain cured meat, cereal and spices that are placed in the uncooked pastry cases. The pies are then baked and jelly made from gelatin, spices, flavouring and water is added. Provided that the jelly is heated to and maintained at a sufficiently high temperature until it reaches the pie, there should be no problem.

Bad handling, insufficient heating and re-contamination of cool jelly can cause gross contamination by both spoilage organisms and pathogens.

The pies are cooled in a pie tunnel through which cold air is blown. At this stage mould contamination is likely if the air filters are not properly cleaned. Air in pie tunnels can be monitored by exposing plates of malt agar or similar medium. Bad storage may result in outgrowth of spores of *Bacillus* and *Clostridium* spp.

Viable counts

Examine only the filling. Do counts on homogenized material using dilutions of 10^{-1} and 10^{-2}. Incubate at 30°C for 48 h. A reasonable level is less than 1000 cfu/g.

Coliform counts

Examine meat and jelly separately. Use dilutions 10^{-1} and 10^{-2}.

Cured meat

Cured raw meat

This is pork or beef that has been treated with salt and nitrite. The haemoglobin is altered so that the characteristic pink colour is produced when the meat is cooked. This is sold as bacon, gammon, cured shoulder, salt beef or brisket.

Viable counts

Do counts on dilutions of 10^{-2} and 10^{-3}. Incubate at 30°C for 48 h. Also do counts using diluent and media containing 4% of sodium chloride to detect halophiles. Incubate these for 3 days.

Do direct counts for staphylococci on Baird-Parker medium.

Microbial content

Besides halophilic denitrifying bacteria, lactobacilli, micrococci, staphylococci, *B. thermosphactum* and moulds may be present under chill conditions. *Pseudomonas* spp. are inhibited by salt. Above 25°C, *C. putrefasciens* may be found. This organism gives a characteristic sweet/sour odour. 'Mild cure', 'sweet cure', 'tender cure', etc., are trade terms which may indicate some degree of heat or the addition of sweetening substances. The balance of the flora may be altered by such processes.

Bone taint

This may occur in the meat close to the bone in carcase leg joints or rib areas. The organisms most often responsible are vibrios, micrococci, proteus and clostridia.

Vacuum packaging

Vacuum packaging is used for a variety of products in an effort to present them to the customer in as fresh a condition as possible.

Vacuum-packed bacon

The packs used for bacon are not usually permeable to oxygen because oxidation causes fading of the colour. The initial bacterial load, the salt content and storage temperatures have a marked influence on the shelf-life.

Low salt content (5–7%) spoilage of bacon gives a characteristic scented sour odour due to the action of lactobacilli, pediococci, streptococci and leuconostoc. High salt content (12%) spoilage gives a cheesy odour which is associated with micrococci or *B. thermosphactum*. Spoilage by enterobacteria or vibrios gives a sulphurous smell. Table 14.1 (p. 219) indicates the types of spoilage organisms.

Viable counts
Swab the outside of the pack with alcohol and open with sterile scissors. Do counts as for cured raw meat.

Microbial content
The main groups of organisms found immediately after packaging are micrococci and coagulase-negative staphylococci. During storage at 20°C, lactobacilli, enterococci and pediococci become dominant. Yeasts may also be found. If large numbers of Gram-negative rods are present it is usually an indication that the initial salt content was low.

Vacuum-packed cooked meats

These meats are invariably cured products and, as with bacon, the absence of oxygen is important. These products, however, always carry a risk of infection with C. *botulinum*. Spores of this organism may survive cooking and during storage germinate and grow to produce toxin in an otherwise sterile and oxygen-free environment. Type E and some strains of Group B are known to grow at low temperatures. Fortunately, the salt and nitrite contents combined with low-temperature storage help to reduce the danger. In addition, many producers pasteurize their vacuum-packed cooked meat products in the package. This allows for spore germination after the initial cooking and killing of the vegetative forms on pasteurization. Nevertheless, C. *botulinum* has been reported in temperature-abused vacuum-packed frankfurters and cooked ham.

Laboratory examination
Do surface counts on blood agar and on MacConkey agar for enterococci. These are useful organisms for assessing the efficiency of pasteurization. Total counts should be less than 1000 cfu/g.

Cured, cooked meats

These meats include ham, luncheon meats, brawns, tongues and continental-type sausages. Except in delicatessen stores, these are usually sold in vacuum packs.

Viable counts
Do plate counts on dilutions of 10^{-1} to 10^{-3}, and use medium to which 5% of horse blood has been added immediately before pouring. This enables the streptococci that cause 'greening' to be counted in addition to the other organisms. Counts of 1000 or less cfu/g are reasonable.

Test for *S. aureus* (p.353) and *L. monocytogenes* (p.384).

Microbial content

Contaminants of products cooked in packs are usually heat resistant and include spore-bearers and enterococci.

Coliforms, micrococci, staphylococci and lactobacilli may be introduced after heating or during slicing. Canned hams may contain coryneforms, lactobacilli and yeasts. Staphylococci may proliferate just beneath the casing.

Brines

Injection brine

This is prepared freshly for each batch of meat.

Sampling

Collect from storage tank, store at < 7°C before counting. Do total count using nutrient agar containing 4% NaCl, incubate at 22°C for 72 h. (Gardner, 1983)

Advisory standard ($\times 10^3$ cfu/ml)

Good	< 0.5
Fair	0.5–1.0
Poor	1.1–5.0
V. poor	> 5.0

Cover brines

These are used repeatedly.

Sampling

Sample when fresh and between batches of meat.

Direct microscopical count (DMC)

Use a Helber haemocytometer (p. 150) and examine by dark field or phase contrast microscopy, or use the DEFT method (p. 156).

Total viable count

Add 4% NaCl to the medium.

E. coli count

Dilute the sample 1:10 with 4% NaCl in 0.1% peptone and use the membrane filter technique (Gardner, 1983).

Advisory standard

	DMC ($\times 10^6$ cfu/ml)	*Total count* ($\times 10^3$ cfu/ml)	*E. coli* ($\times 10^3$ cfu/ml)
Good	< 50	< 50	< 1
Fair	50–100	50–100	1–10
Poor	101–150	101–500	11–100
V. poor	> 150	> 500	> 100

(Gardner, 1983)

Pseudomonas and vibrio counts may also be a useful monitor of cover brine quality.

Poultry

Intensive rearing produces birds with very tender flesh but the close confinement necessary in this method of production often results in cross-infection. Inadequate thawing of frozen birds followed by relatively little cooking has, on several occasions, resulted in outbreaks of food poisoning.

Poultry meat is thought to be the source of a large proportion of cases of salmonellosis, and most cases of campylobacter diarrhoea in the UK, either as a result of undercooking or of raw → cooked food contamination. Campylobacter infections probably occur by the hand-to-mouth route after handling the raw product. Table 14.1 (p. 219) indicates the type of spoilage organisms.

Sampling

The neck skin is the best sampling site because (1) it is the lowest point during processing and all fluids pass over it, (2) it is easy to sample and may be removed without affecting the value of the remainder of the carcase.

Viable counts

Discard the subcutaneous fat, homogenize a known weight of the material, prepare dilutions as described in Chapter 10 and use plate count agar. Incubate at 30°C or 37° C for 24–48 h.

Express the counts as cfu/g of neck skin. These counts may be about 10 times higher than those obtained from the skin on other parts of the carcase. Normal ranges ($\log_{10}$ cfu/g neck skin) are: viable count, 4.5–5.3; coliforms, 2.7–3.8; pseudomonads, 2.9–3.9 (Mead *et al.*, 1993).

Pathogens

Examine the whole carcase or 25 g of neck skin. Culture for salmonellas, staphylococci and campylobacters as described in Chapters 28, 33, 36. High *S. aureus* counts are sometimes found; these result from colonization of plucking machines. They are rarely toxin-producing food-poisoning strains.

References

Gardner, G. A. (1983) Microbiological examination of curing brines. In *Sampling – Microbiological Monitoring of the Environment* (eds R. G. Board and D. W. Lovelock), Society for Applied Bacteriology Technical Series No. 7, Academic Press, London, pp.21–27

Kitchell, A. G., Ingram, G. C. and Hudson, W. R. (1973) Microbiological sampling in abattoirs. In *Sampling – Microbiological Monitoring of the Environment* (eds R. G. Board and D. W. Lovelock), Society for Applied Bacteriology Technical Series No. 7, Academic Press, London, pp.43–54

Mead, G. C., Hudson, W. R. and Hinton, M. H. (1993) Microbiological survey of five poultry processing plants in the UK. *British Poultry Science*, **17**, 71–82

Murray, J. G. (1969) An approach to microbiological standards. *Journal of Applied Bacteriology*, **32**, 123–135

Chapter

16

Fresh, preserved and extended shelf-life foods

Fruit and vegetables

Fresh fruit and vegetables

Spoilage of fresh fruit is mostly caused by moulds. Mould and yeast counts may be indicated in fruit intended for jam making or preserving. Orange serum agar is useful for culturing citrus fruit.

Washed, peeled and chopped vegetables are common in supermarkets. Viable counts are often high—up to 10^7–10^8; products such as bean shoots, cress, watercress and spring onions have the highest counts (Roberts *et al.*, 1981; Lund, 1988, 1992, 1993). Some vegetables are 'blanched' (exposed to boiling water for 1–2 min) to destroy enzymes. This also helps to reduce the bacterial load. Spoilage is caused by pseudomonads, lactobacilli, enterococci and leuconostocs. Salad crops, e.g. watercress, which may be grown in polluted streams, should be tested for *E. coli* and salmonellas if indicated.

Dehydrated fruit and vegetables

Counts

Examine washings (p.216). An *E. coli* count of more than 5 cfu/g is suspicious. If repeat tests give the same result the batch is unsatisfactory. Counts of lactic acid bacteria may be useful: use Rogosa agar medium and incubate at 30–32° C for 3 days.

Culture

Inoculate glucose tryptone agar and tomato juice, or MRS medium and incubate at 30°C. Incubate glucose tryptone cultures also at 55–60°C aerobically and anaerobically for thermophiles.

Microbial content

Counts should be low. Lactic acid bacteria, flat sour thermophiles (*B. stearothermophilus*) and hydrogen sulphide-producing anaerobes may be present.

Blakey and Priest (1980) examined red and brown lentils, yellow and green peas, black-eyed, kidney, mung and soya beans, scotch broth mix, rice, pearled barley and chapatti flour. They found *B. cereus* at levels ranging from 1×10^2 to 6×10^4/g.

Frozen vegetables and fruit

Frozen peas offer the greatest problem as they deteriorate rapidly on thawing. Slime usually contains large numbers of leuconostocs, which imparts a yellow appearance (acid pH) but sometimes a heavy growth of coryneforms produces ammonia, which neutralizes the acid formed by the leuconostocs. If sucrose is used in the medium instead of glucose, a levan is produced by leuconostocs.

Counts on other frozen vegetables are usually low (about 100 000 cfu/g). Coliforms and enterococci are commonly found on vegetables, *E. coli* type 1 may have public health significance.

Fruits may have low yeast and lactobacilli counts. In general, they keep well at −15°C (0°F) for 2 to 3 years. Vegetables remain in good condition at this temperature for 6–12 months.

Pickles, ketchups and sauces

Pickles

Vegetables are first pickled in brine. The salt is then leached out with water and they are immersed in vinegar (for sour pickles) or vinegar and sugar (for sweet pickles). Some products are pasteurized. Spoilage is due to low salt content of brine, poor quality vinegar, underprocessing and poor closures.

Counts
Use glucose tryptone agar at pH 6.8 and 4.5 and do total counts and also counts of acid-producing colonies (these have a yellow halo due to a colour change of indicator). Count lactic acid bacteria on Rogosa or other suitable medium. Estimate yeasts either by the counting chamber method (stain with 1 : 5000 erythrosin) or on Rose Bengal chloramphenicol agar.

Culture
Use glucose tryptone and Rogosa or MRS agar for lactic acid bacteria. Grow suspected film-yeasts in a liquid mycological medium, containing 5 and 10% of sodium chloride for 3 days at 30°C. For obligate halophiles, use a broth medium containing 15% of sodium chloride.

Microbial content
Pickles are high-acid foods. Counts are usually low, for example 1000 cfu/g. Yeasts are a frequent source of spoilage and may be either gas-producing or film-producing. In the former case, enough gas may be generated to burst the container. Bacterial spoilage may be due to acid-producing or acid-tolerant bacteria such as acetic acid bacteria, lactic acid bacteria and aerobic spore-bearers. Infected pickles are often soft and slimy.

Fermented pickles of the sauerkraut type contain large numbers of lactic acid bacteria (*Lactobacillus* and *Leuconostoc*) which are responsible for their texture and flavour.

Acid-forming bacteria are active at salt concentrations below 15%. Above this concentration, obligate halophiles are found.

Ketchups and sauces

The most common cause of spoilage is *Zygosaccharomyces* (*Saccharomyces*) *baillii*. This organism grows at pH 2, at < 5°C up to 37°C, in 50–60% glucose and is heat resistant to 65°C.

Sugar and confectionery

Sugars, molasses and syrups

The importance of microorganisms is related to the intended use of the products, for example flat-sour bacteria are more important in bakery than in canning. Do total counts, examine for spore-forming thermophiles (*B. stearothermophilus, C. nigrificans, C. thermosaccharolyticum*) and for yeasts and moulds. Osmophilic yeasts, aspergillus, penicillium, etc., may cause inversion. Spoilage is often caused by osmotolerant fungi, usually yeasts; culture for these on a low medium (DG18).

Chocolate

This has been shown to be the vehicle of salmonella infection in a number of cases. (See the review by D'Aoust, 1977.)

Shave the chocolate into nutrient broth and selenite, incubate overnight and plate out on DCA medium. Do not place large lumps of chocolate into liquid medium. Automated methods, especially those employing impedimetric principles, are useful for salmonella screening in industry. Examine spoiled soft-centred chocolates for yeasts and moulds by direct microscopy and culture on DG18 medium.

Cake mixes and instant desserts

Spores are likely to be present, e.g. of *C. perfringens* and *B. cereus*. Salmonellas and staphylococci may survive processing or be post-processing contaminants. Any of these organisms may multiply if the product is reconstituted and then kept under unsuitable conditions. The powdered product may also cross-contaminate other products which could provide suitable conditions for growth.

Do total viable counts; test for clostridia, *B. cereus*, coliforms and *E. coli*, and staphylococci.

Salmonellas and *E. coli* are potentially much more hazardous in instant desserts than in cake mixes which will be cooked before consumption.

Cereals and protein additives

It is important that these materials should not contribute unduly to the bacterial load of the product. Do total counts and, if the material is to be used in canned foods, do spore count. The total count at 30°C should not exceed 20 000 cfu/g. The spore counts at 30 and 55°C should not be greater than 100/g.

Flour

Contaminated flour may be responsible for the spread of infection by spoilage organisms in kitchens as well as spoilage of the bread, pastry, etc., for which it is used. Grain is naturally infected by soil, dust and rodent and bird faeces during ripening, harvesting and storing. During transport and handling, this contamination is distributed throughout the bulk. Before milling, the grain is washed, sometimes with polluted water.

Counts

Weigh 10 g of flour into a sterile jar containing coarse sand or small glass beads. Add 100 ml of 0.1% peptone water diluent and shake mechanically for 10 s, stand for 30 min, and shake again. If moisture absorption is great this dilution may need to be adjusted.

Do viable counts by surface plating on glucose tryptone agar for bacteria and on Rose Bengal agar for yeasts and moulds. If appropriate culture for *S. aureus* and *B. cereus*. If the flour is to be used in canning test for anaerobic thermophiles. Incubate at 32°C for 3 days.

Count coliform bacilli by the MPN method.

To count 'rope spores', heat the homogenate for 20 min at 90°C to kill vegetative bacteria, make dilutions and do MPN counts in glucose tryptone broth. Alternatively, inoculate two tubes of glucose tryptone broth each with 1 ml of the serial dilutions of heated homogenate. Incubate at 30°C for 3 days and record as positive tubes that show a pellicle. Examine the pellicle to identify '*B. mesentericus*' or *B. subtilis*. Record as rope spores/g or as present in so much of 1 g.

Culture

Use glucose tryptone agar for aerobic and glucose tryptone agar and iron sulphite medium for anaerobic culture. Inoculate malt extract or Rose Bengal agar for yeasts and moulds.

Microbial content

Counts of 5000–500 000 are usual and *E. coli* is often present in 1 g or less depending on processing. Flat-sour bacteria (*B. stearothermophilus*) and hydrogen sulphide-producing clostridia may be found in varying numbers. 'Rope organisms' ('*B. mesentericus*', but see p. 394) are important and cause 'ropy bread'. Mould counts may be 2000 or more per g.

Pastry

Uncooked pastry, prepared for factory use or for sale to the public, may suffer spoilage due to lactobacilli. Do counts on Rogosa or MRS agar and incubate at 30°C for 4 days. Counts of up to 10 000/g are not unreasonable.

Pasta products

These are man-made, from wheat flour, semolina, farina and water. Egg (powdered or frozen), spinach, vitamins and minerals may be added. The egg in particular may contain salmonellas and while these may be destroyed in the subsequent cooking there is the possibility of cross-contamination from uncooked to cooked products.

High levels of *S. aureus* and performed toxin have been reported in lasagne and high levels of those organisms in dried pasta (ICMSF, 1980).

Examination
As contamination is likely to occur during manufacture it is necessary to liberate the organisms from within the pasta.

Add 25 g of pasta to 225 ml of peptone water diluent and allow it to soften at room temperature for about 1 h. Macerate, e.g. in a stomacher, and do total and coliform counts.

Inoculate salt meat broth. Incubate at 37°C for 24 h and plate on Baird-Parker medium for staphylococci. If staphylococcal enterotoxin is suspected, in the absence of viable *S. aureus* (e.g. if microscopy reveals large numbers of non-viable Gram-positive cocci) use one of the commercial kits (p.109).

Inoculate selenite medium. Incubate at 37°C for 24 h and plate on DCA or other medium for salmonellas.

Inoculate *B. cereus* selective agar and incubate at 37°C for 24 h.

If the presence of moulds is suspected it is best to obtain help from a reference expert because of the possibility of aflatoxins.

Extrusion-cooked products

These include breakfast cereals and crispbreads and may be textured with vegetable protein and bread crumbs. The liquid and solid ingredients are blended, shaped and cooked within 1–2 min.

Examine as for pasta.

Gelatin (dry product)

Gelatin is often used to top-up pastry cases of cold-eating pies, in canned ham production and in ice-cream manufacture. It should be free from spores and coliforms.

If the process involves low temperature reconstitution examine for *S. aureus*, salmonellas and clostridia.

Laboratory examination
Weigh 5 g of gelatin into a bottle containing 100 ml of sterile water and allow to stand at 0–4°C for 2 h. Place the bottle in a water-bath at 50°C for 15 min and then shake well. Mix 20 ml of this solution with 80 ml of sterile water. This gives a 1:100 dilution. Use 1.0 and 0.1 ml for total counts by the pour-plate method. Incubate at 35°C for 48 h.

Gelatin for ice-cream manufacture
Do a semi-quantitative coliform estimation using MacConkey or similar broth. Add 10 ml of 1:100 gelatin to 10 ml of double-strength broth; add 1 ml of 1:100 gelatin to 5 ml of single-strength broth and 0.1 ml of 1:100 gelatin to 5 ml of single-strength broth. Incubate at 35°C for up to 48 h and do confirmatory tests where indicated (see p.306). Thus, the presence or absence of coliforms and *E. coli* in 0.1, 0.01 or 0.001 g of the original material can be determined. It is desirable that coliforms should be absent from 0.01 g and *E. coli* absent from 0.1 g. The total count in gelatin to be used for ice-cream manufacture should not exceed 10 000 cfu/g.

Gelatin for canned ham production
This should have a low spore count. After doing the total counts, heat the remaining 1:100 solution of gelatin at 80°C for 10 min. Plate 4 × 1 ml of this solution on standard plate count medium. Incubate two plates at 35°C and two plates at 55°C for 48 h. There should be not more than one colony per plate, i.e. 100 g of the original gelatin. The total count should not exceed 10 000 cfu/g.

Spices and onion powder

A plastic bag inverted over the hand is a satisfactory way of sampling spices. Some are toxic to bacteria and the initial dilution should be 1/100; for cloves use 1/1000, in broth. Do total viable counts and if indicated by intended usage consider inoculating *B. cereus* selective medium and one of the media for staphylococci and salmonellas.

The total counts on these materials vary widely. Total viable counts of 10^8/g are generally acceptable but low spore counts are important if these materials are to be used for canned foods: an acceptable level at 30 and 55°C is less than 100/g.

Coconut (desiccated)

In the 1950s and 1960s this product was often contaminated with salmonellas. In spite of improved processing it is still a potential hazard in the confectionery trade.

Examine samples for salmonellas by the method described on p.318.

Oily material

Mix 1.5 g of tragacanth with 3 ml of ethanol and add 10 g glucose, 1 ml of 10% sodium tauroglycocholate and 96 ml of distilled water. Autoclave at 115°C for 10 min. Add 25 g of the oily material under test and shake well. Make dilutions for examination in warm diluent.

Salad creams
These contain edible oils and spoilage may be due to the lipolytic bacteria. Test for these with tributyrin agar. Examine for coliform bacilli and thermophiles. All these should be absent. There should not be more than five yeasts or moulds/g.

Mayonnaise-based salads
Examine for lactobacilli. Yeasts will also grow so colonies should be examined by Gram staining. Alternatively do parallel counts on both MRS medium (lactobacilli plus yeasts) and on MRS containing 100 mg/ml chloramphenicol (yeasts only) (Rose, 1985).

Canned, prepacked and frozen foods

The practice of prolonging the shelf-life of foods by sealing them in metal or glass containers and heat processing them has long been established but modern packaging materials have given the consumer a wider choice.

Eating habits have changed. Single portions which can be taken from the freezer and heated in a microwave oven are popular. With these, packaging is important. It should

(1) prevent contamination by microorganisms,
(2) preserve quality and nutritional value,
(3) be inert and offer no hazard in use,
(4) be economic to manufacture and distribute,
(5) be easily labelled.

Routine control of the product in the factory is the responsibility of the quality assurance and laboratory departments of the manufacturer but general laboratories may be asked to help when defects or spoilage have developed after the products have left the factory or where the product is suspected of having caused food poisoning or enteric fever. Occasionally, arbitration is needed between suppliers of ingredients and manufactuers of finished goods.

Sealed containers intended to be stored at ambient temperature for long periods pose the highest risk if errors occur during or after processing. In particular, 'low acid' (pH > 4.5), in which *C. botulinum* can grow and produce toxin are potentially the most hazardous. This type of product is usually given at least a '12D' heat treatment, i.e. 12 times the time at a particular temperature to kill 90% of the population sufficient to kill 10^{12} spores of *C. botulinum*.

The following sections provide information about the defects that can occur during or after canning. Faults are uncommon and cannot be detected by post-production quality control sampling—the canning process needs to be run on a *quality assurance* system such as the Hazard Analysis Critical Control Point system (HACCP) (ICMLS, 1988).

Type of containers

Cans

Tin plate and aluminium are widely used. Improved lacquering gives resistance to corrosion. Welded three-piece cans have largely superseded the soldered type. Base metal thickness has been reduced, giving lighter and cheaper cans but these have to be 'beaded' to withstand processing. Two-piece cans with 'easy-open ends' are popular for carbonated beverages. Self-heating cans are convenient for camping and picnics.

Jars and bottles

Glass closed with metal cap and resilient seal, sometimes under vacuum. They should be tamper-proof.

Trays

Aluminium or aluminium/polypropylene laminated, with lids.

Semi-rigid flexible plastic and laminate containers

These are easy to stack, light and allow a shelf-life of up to 2 years. Plastic-sided metal-ended containers are used for fluids.

For a review of containers see Dennis (1987).

Physical defects of cans

These defects may be due to improper processing; during exhausting and autoclaving incorrect stresses may be imposed and cause 'peaking' or 'panelling', which are distortions of the ends (distinct from 'swelling') and of the body, respectively. The ends of the cans have concentric rings impressed in them to absorb the normal strain and to permit swelling in normal stresses.

Rusting of cans, causing pinholes and consequent spoilage, is revealed by inspection. Hydrogen swell is caused by hydrogen formed when acidic foods attack the metal in places where the lacquer is defective.

Faulty can manufacture and improper closure of seams, either side or lid, can lead to spoilage.

Inadequate drying may result in contamination of the can contents by bacteria which enter in water droplets through minute pinholes in the seams. These holes are usually self-sealing when the can is dry. Wet cans also tend to rust.

Investigation of canning faults, including seam examination, and microbiology of cans, should be done by specialists.

Spoilage due to microorganisms

Leaker spoilage

When this occurs, usually only a small number of cans in each batch is affected. Minute faults may be present in some cans (see above) particularly where the end seams cross the body seams. After autoclaving and during cooling there is a negative pressure in the can and cooling water containing bacteria may be drawn in, and canned foods may thus be contaminated with pathogens. Only very few organisms, pathogens or spoilers need be drawn into a can, as they will multiply rapidly. Gas producers will manifest themselves by 'blowing' the can but organisms that produce gas in normal culture may not do so in cans. Other spoilage will be obvious when the can is opened but food contaminated with pathogens, e.g. typhoid bacilli, may appear sound and wholesome. Care must also be taken that there is not a build-up of organic matter in the production line which can overcome the disinfectant and lead to bacterial multiplication in the water remaining on the cans. Personnel with septic conditions must be excluded from handling cans. Dirty or infected hands can contaminate the surface water on the cans before it is drawn into the cans through these defective seams or pinholes. Water used for cooling is usually chlorinated but some organisms are relatively resistant to the process. When the seams are dry, the chances of contamination are slight.

Underprocessing

Gross underprocessing will usually have been found by the manufacturer's tests. If this is not the case, it is common to find only one type of organism in this situation.

Some cured meats are deliberately underprocessed because they are rendered less palatable by autoclaving. Catering-size cans of ham and mixtures of ham and other meat are usually salted and spiced and given the minimum of heat treatment. The manufacturers do not claim sterility and the label on the can invariably recommends cold or cool storage. The conditions of pH, salt content and storage temperature should be such that the bacteria in the can are prevented from multiplying. There is evidence that some species of enterococci produce antibacterial substances in canned hams, which act antagonistically on some species of clostridia, lactobacilli and members of the genus *Bacillus*. This is most likely a factor in the successsful preservation of commercial products of this nature.

Poor plant hygiene, faulty design or careless operation of equipment, e.g. bad stacking of retorts, may be a contributory cause of underprocessing. Vegetative organisms are killed, except in very rare cases of gross underprocessing, but spores can be unaffected; they subsequently germinate and cause spoilage. The spores of *C. botulinum* are very heat resistant and they may germinate and produce toxin in low acid foods (pH > 4.5). Spores of thermophiles such as *B. stearothermophilus* are more resistant than those of *C. botulinum* (a poor competitor). *B. stearothermophilus* is one of a variety of sporebearers that might survive. Production of safe canned foods, however, cannot depend on the detection of faulty cans by fortuitous means (see p.239).

Gas-producing organisms cause the can to swell. The first stage is the 'flipper' when the end of the can flips outward if the can is struck sharply. A 'springer' is caused by more gas formation. Pressing the end of the can causes the other end to spring out in a bulge. The next stage is the 'swell' or 'blower' when both ends bulge. A 'soft swell' can be pressed back but bulges again when the pressure is released. 'Hard swells' cannot be compressed.

Spoilage may not result in gas formation and may be apparent only when the can is opened. This kind of spoilage includes the 'flat-sour' defect.

Inadequate cooling

If cooling is too slow or inadequate or the cans are stored at very high temperatures, e.g. in tropical countries, there may be sufficient time for the highly resistant spores of *B. stearothermophilus* to outgrow and multiply. The optimum growth temperature for this organism is between 59 and 65°C. It will not grow at 28°C or at a pH of less than 5. Together with *B. coagulans* it is the chief cause of 'flat-sour' spoilage in canned, foods. Most strains of *B. coagulans* will grow at 50-55°C. They will also grow in more acid conditions.

Preprocessing spoilage

This can occur when the material to be canned is mishandled before processing, e.g. if precooked meats with a large number of surviving spores are kept for too long at a high ambient temperature this may result in the outgrowth of spores and the production of gas. When canned, the organisms will be killed leaving the gas to give the appearance of a blown can.

Type of food, and organisms causing spoilage

High-acid foods

Spoilage is rare in processed foods with a pH of 3.7 or less, for example pickles and citrus fruits. Yeasts may occur when there has been serious underprocessing. These foods are usually not pressurized during heating.

Acid foods

When the pH is 3.7–4.5, as in most canned fruits, aerobic and anaerobic spore-bearers may cause spoilage but this is not common. Lactobacilli and *Leuconostoc* have been reported. Osmophilic yeasts and the mould *Byssochlamys* are sometimes found. These foods are usually not pressurized. They are too acidic for the growth of most bacteria and leaker spoilage is uncommon.

Low-acid foods

If the pH is 4.5 or above, as in canned soups, meat, vegetables and fish (usually about pH 5.0), one of the following thermophiles is usually found in spoilage due to underprocessing.

(1) *B. stearothermophilus*, causing 'flat-sour' spoilage,
(2) *C. thermosaccharolyticum*, causing 'hard swell',
(3) *C. nigrificans*, causing 'sulphur stinkers',
(4) mesophilic spore-bearers, obligate or facultative anaerobes, causing putrefaction.

Leaker spoilage may be due to a variety of organisms; aerobic and anaerobic spore-bearers, Gram-negative non-sporing rods and various cocci, including *Leuconostoc* and *Micrococcus* may be found. S. *aureus* (food poisoning type) has been isolated.

Methods of examination

Sampling
If packs are blown or swollen, examine six and take six normal ones from another batch as controls. In suspected underprocessing examine 6–12 packs from each batch. Leaker-spoilage is likely to occur in only a very small number of packs in a batch; therefore, examine as many as possible.

Physical examination
Inspect the seams and can surfaces. A jeweller's saw is useful for cutting across seams. Note the batch or code numbers printed on the labels or stamped on the lid.

Jars
Examine cap for perforations and note code.

Trays and pouches
Inspect the seals. These are formed from a continuous weld and are less likely to leak than the double seams of cans. Furthermore there is no headspace or vacuum which could cause organisms to be sucked in if there was a small hole in the seal. Note code.

Pre-incubation
Incubate apparently sound packs at 35–37°C for 7 days. This encourages the multiplication of small numbers of organisms which might otherwise be missed in sampling the contents.

Sampling contents: cans normal in appearance
This should be done in laminar flow clean air cabinets. Swab the top with cotton wool and methylated spirit. Pour 1 ml of spirit on the swabbed area and flame it. Allow the spirit to burn out.

If the contents of the can are liquid, puncture the flamed surface with a 100-mm wire nail (sterilized in tins containing 10–12 nails in a hot air oven) by a sharp blow with a hammer. Remove a sample of the contents with a pasteur pipette into culture media and into a screw-capped bottle for viable counts if required.

If the contents are solid, use a punch made from brass rod 9–10 mm in diameter with one end drawn to a point. Sterilize these individually. Drive the punch well in to make a large hole. Remove a core sample with a length of glass tubing of 7–8 mm outside diameter by pushing it right to the bottom of the can. Push the core sample from the glass tube into a screw-capped bottle with a piece of glass rod of suitable thickness. These core and rod samplers can be sterilized together in copper pipette drums. In addition to taking core samples, it is desirable to sample jelly adjacent to the seam of the can. To do this, remove the end of the can, previously punctured, with a sterile domestic can opener and tip the contents on to a sterile tray. Take care not to disturb the material under the seam. Note the appearance of this material and sample with a cotton wool swab.

Jars normal in appearance
Sterilize as for cans and pierce cap with a sterile nail. It is very important to release any vacuum. Carefully remove the cap so that the sealing gasket is undisturbed. This should be examined thoroughly for evidence of improper seating or twisting. Liners are sometimes misplaced causing inadequate closure of the jar. Look for damage to the rim of the jar. Examine contents as for cans.

Flexible pouches and trays normal in appearance
Support in suitable racks. Sterilize with 50:50 alcohol/ether and allow to dry. Open with sterile scissors and sample with a sterile spoon. Examine contents as for cans.

Sampling blown, swollen or leaking packs
These contain gas under pressure and the contents may be offensive. Chill before opening. Place the container on a metal tray with the seam facing away from the operator. Swab with 4% iodine in 70% alcohol, allow to stand for a few minutes then dry with a sterile towel. Do not flame.

Invert a previously sterilized metal funnel or new plastic bag over pack. The diameter of the funnel should be slightly larger than that of the can. Pass a sterile brass rod with a point at one end down the funnel spout until it rests on the can; hold both firmly and puncture the can by tapping the rod with a hammer, then withdraw the rod slightly. The contents of the can may be ejected with some force but the funnel and tray will prevent broadcast. Before removing the funnel and brass rod, push the latter in and out of the hole in the can several times. Sometimes a piece of food is forced against the hole by internal gas pressure and when a sampler is inserted more gas and food are ejected.

Take samples with a pasteur pipette or core sampler as described above. After sampling, open the can with a domestic can opener and inspect the contents.

Direct film examination
Make Gram films of sample. The presence of Gram-positive rods may suggest underprocessing while cocci, yeasts, etc., indicate leaker spoilage.

Note: The organisms seen may be dead (killed during processing), so too much reliance must not be placed on this examination.

Autosterilization may also account for this phenomenon. In this instance the organisms die out during storage. When this has occurred the organisms appear degenerate and poorly stained.

Culture
For general examination inoculate glucose tryptone agar (with bromocresol purple indicator) and incubate aerobically and anaerobically at 22–25, 35–37 and 55–60°C for 24–36 h.

For high-acid foods, i.e. pH 4.5 or less, inoculate four tubes of Rogosa medium. Incubate two tubes at 55°C for 48 h and two tubes at 30°C for 96 h.

Inoculate two tubes of malt extract broth and incubate these at 30°C for 96 h. Subculture and make Gram-stained smears as necessary.

Also inoculate the following media if indicated: iron sulphite medium (sulphur stinkers), blood agar and MacConkey (for putrefactive organisms, micrococci, leuconostocs, etc.), Crossley Milk Medium (putrefactive aerobic or anaerobic spore-bearers), reinforced Clostridial agar, malt extract or other mycological medium (for yeasts and fungi).

Make Gram films of colonies.

Microbial content
Identify as follows:

(1) Gram-positive rods
  (a) Thermophiles:
    (i) Aerobic: *B. stearothermophilus* (flat-sour). See p. 394.
    (ii) Anaerobic: *C. thermosaccharolyticum* (hard swell). See p. 407.
    (iii) Anaerobic: black colonies in iron sulphite medium: *C. nigrificans* (sulphur stinkers). See p. 407.
  (b) Mesophiles:
    (i) Aerobic: *Bacillus*. See Chapter 42.
    (ii) Anaerobic: *Clostridium*. See Chapter 44.
(2) Gram-negative rods
  Pseudomonas alcaligenes or enterobacteria. See Chapters 22 and 25.
(3) Gram-positive cocci
  Micrococci, leuconostoc. See Chapters 36 and 37.
(4) Yeasts and moulds
  See Chapters 50 and 51.

Pathogens in canned food

Outbreaks of enteric (typhoid) fever and staphylococcal disease have caused food bacteriologists, public health authorities and canners to revise their opinions on the safety of canned foods, although these outbreaks are very few in proportion to the enormous amount of canned food consumed. Random or 'routine' sampling of canned foods for pathogens is an unrewarding procedure. Only low-acid foods, such as meat and dairy products and certain canned vegetables can support the growth of enteric organisms, staphylococci and *C. botulinum*..

Certain moulds and other organisms can raise the pH of some acid fruits, e.g. tomato juice and pears to a level at which *C. botulinum* will grow.

Examination for pathogens
When this is indicated, open the cans with sterile can openers and if the food is solid take samples under the seams, particularly where the end seams cross the side seam. Culture in selenite medium for salmonellas (continue as on p. 319), in salt meat for staphylococci (continue as on p. 353) For botulism, see p. 400

Frozen foods

Some frozen convenience foods are mentioned here, others – frozen meat, fish and ice-cream – are included under their appropriate headings in Chapters 15, 17 and 18. For frozen vegetables see p. 230.

Storage

Unless the cabinets are maintained at −18 to −20°C there will be difficulties due to the different melting points of the stored products, the presence of water films and the variety of microclimates. Two kinds of spoilage occur:

(1) low-temperature spoilage due to enzymes and, less often, to psychrophiles if the temperature is at or about the freezing point of water;
(2) unfreeze spoilage, when bacteria can grow because of gross temperature fluctuations. Off-odours and off-flavours and spoilage losses are likely to be doubled for each 2–3°C rise in temperature.

Counts

Count total bacteria growing on at 5–7°C in 5–7 days and 20–30°C in 2–3 days. Use enriched media because organisms will be cold damaged. Count coliforms, lactic acid bacteria (except in fish), moulds and yeasts.

Culture

For total counts, use enriched media as the organism will be cold shocked and media for lactobacilli, fungi, staphylococci, salmonellas and enterococci. There is evidence that enterococci survive longer than coliforms in frozen foods.

Frozen pies and complete meals

Frozen pies often have low counts (5000–30 000 cfu/g) but may contain enterobacteria and staphylococci. 'Complete meals' vary enormously in their counts and flora, usually reflecting factory conditions.

The practice of keeping complete meals in cold storage for several months appears to reduce the counts but this may be a false effect and reflect the failure to resuscitate cold damaged organisms.

Storage life

Foods should be labelled with a 'use by . . .' date. Chicken, pies and complete meals keep for 2–6 months. Storage life depends on the maintenance of a constant low temperature.

Arbitrary standards

Standards that have no legal status are used by the trade and public health authorities as a guide in the frozen food industry:

Plate count at 35°C in 48 h – not more than 100 000 cfu/g.
Coliform bacilli NOT present in 0.1 g.
S. *aureus* NOT present in 0.01 g.

For further information on the microbiology of frozen foods, see Roberts *et al.* (1981).

Precooked chilled ('cook-chill') products

The safety and shelf-life of these foods depends on sublethal treatment and aseptic packaging. Major advances in transportation and handling of chilled and frozen products have increased the availability of these foods. Chilled foods account for over 40 per cent of expenditure on foods in the UK. Temperatures should be controlled between −1 and −4°C throughout handling and storage. As many of these foods are capable of supporting the growth of food poisoning organisms it is important that they are not mishandled.

The packaging and the food are sterilized separately and united under 'commercially sterile' conditions. Flexible pouches and semi-rigid pots and cartons are used for low acid foods, e.g. ice-cream, custards, soups and sauces.

Cook-chill catering
This system is used by hospitals, works canteens, prisons, schools, banqueting and travel organizations. It demands a high level of technical control to ensure safety and quality. If handled and stored according to the guidelines (DoH, 1989) they should be safe.

Bacteriological examination
These products may be examined by the same method as other foods, and the same standards should apply.

These foods should conform to the following standards:
Viable count after 48 h at 37°C – < 100 000/g
E. coli – < 10/g
S. aureus – < 100/g
Salmonella spp. not present in 25 g
Listeria monocytogenes not present in 25g
Clostridium perfringens not present in 100 g.

Baby foods

Direct breast feeding allows little chance of infection of the infant with enteric pathogens. Testing of human milk is described on p.261. With other feeds contamination may occur during the time lag between production and consumption.

There are many milk preparations and weaning formulas, as well as dried, bottled and canned foods designed for small children. Dehydrated milk may contain organisms that have survived processing and these may multiply if the product is not stored correctly after it is reconstituted. Dirty equipment may contribute bacteria to an otherwise sterile material. Unsuitable water may be used for reconstitution. Since an outbreak of salmonellosis in the UK, from a comtaminated dried milk-based formula, attention has been paid to sampling plans for salmonellas in dried milks. The ICMSF (1986) recommends examining 60 samples of 25 g (which may be pooled). None should be positive.

Central milk kitchens in maternity units should have very high standards of hygiene (see p.263 for sampling surfaces, etc.). In-bottle terminal heating, with teat in place, is good practice.

Viable counts should be low and pathogens should be absent. See Robertson (1974) and Collins-Thompson *et al.* (1980).

Nasogastric feeds should conform the same standards.

Heat-treated jars and cans should be examined as described for extended shelf-life foods on p.234.

Soft drinks

Total counts should be low and coliform bacilli absent, as in drinking water. Membrane filters can be used to test water intended for soft drink production. Millipore publish a very useful booklet on the microbiological examination of soft drinks. Yeast and mould spoilages are not uncommon and they may raise the pH and allow other less acid-tolerant organisms to grow.

In non-carbonated fruit drinks, yeasts are not inhibited by the amounts of preservatives that are permitted by law. Spoilage is generally controlled by acidity (except *Z. baillii*) (see Ketchups, p.231). In both these cases and carbonated drinks the microbial count diminishes with time.

Fruit juices

Lactic acid and acetic acid bacteria may grow at pH 4.0 or less in some fruit juices.

Heat resistant fungi can cause problems in concentrated juices. Screen by heating at 77°C for 30 min. Cool and pour plates with 2% agar. Incubate for up to 30 days. Dip slides, including those with medium for yeasts and fungi, are useful.

Automated counting methods (e.g. ATP assay) are useful. Fruit juices are incubated at 25°C for 24–48 h (72 h for tomato juice), the reagent is added and incubated for 45 min when the instrument gives the result.

Bottled waters

See p. 276.

Milk-based drinks

See p. 256.

Vending machines

Water and flavoured drinks in vending machines may be of poor quality, with high counts (> 1000 cfu/ml) and coliforms. This may be due to inadequate cleaning (see Hunter and Burge, 1986).

Useful references

Food microbiology is a very large subject and in addition to the references cited in this chapter and those preceding it the following are recommended: Hersom and Hulland (1980), ICMSF (1980), Jowitt (1980), Harrigan and Park (1981), Roberts and Skinner (1983), Hayes (1992), Jay (1992), Speck (1992).

References

Blakey, L. J. and Priest, F. G. (1980) The occurrence of *Bacillus cereus* in some dried foods, including pulses and cereals. *Journal of Applied Bacteriology*, **48**, 297–302

Collins-Thompson, D. L., Weiss, K. F., Riedel, G. W. and Charbonneau, S. (1980) Microbiological guidelines and sampling plans for dried infant cereals and dried infant formulae. *Journal of Food Potection*, **43**, 613–616

D'Aoust, J. Y. (1977) Salmonella and the chocolate industry: a review. *Journal of Food Protection*, **40**, 718–726

Dennis, C. (ed.) (1987) *Symposium on the Microbiological and Environmental Health Problems in Relation to the Food and Catering Industries*, Campden Food Preservation Association, Chipping Campden

DoH (1989) *Chilled and Frozen Food. Guidelines on Cook-Chill and Cook-Freeze Catering Systems*, Department of Health, HMSO, London

Harrigan W. F. and Park, R. W. A. (1991) *Making Food Safe*, Academic Press, London

Hayes, R. (1992) *Food Microbiology and Hygiene*, 2nd edn, Elsevier, London

Hersom, A. C. and Hulland, E. D. (1980) *Canned Foods. Thermal Processing and Microbiology*, Churchill-Livingstone, London

Hunter, P. R. and Burge, S. H. (1986) Bacteriological quality of drinks from vending machines. *Journal of Hygiene* (*Cambridge*), **97**, 497–504

ICMSF (1980) Cereal and Cereal Products. International Commission of Microbiological Standards for Foods. In *Microbial Ecology of Foods*, Vol.2, Academic Press, New York

ICMSF (1986) *Microorganisms in Foods, 2, Sampling for Microbiological Analysis. Principles and Specific Applications*, International Commission on the Microbological Specification for Food, Blackwells, Oxford

ICMSF (1988) *Microorganisms in Foods, 4, HACCP in Microbiological Safety and Quality*, International Commission on the Microbiological Specifications for Foods, Blackwells, London

Jay, J. M. (1992) *Modern Food Microbiology*, 4th edn, Van Nostrand Reinhold, New York

Jowitt, R. (ed.) (1980) *Hygienic Design and Operation of Food Plants*, Ellis Horwood, Chichester

Lund, B. M. (1988) Bacterial contamination of food crops. *Aspects of Applied Biology*, **17**, 71–82

Lund, B. M. (1992) Ecosystems in vegetable foods. In *Ecosystems: Microbes: Food* (eds D. Jones, R. G. Kroll and G. L., Pettipher), *Journal of Applied Bacteriology*, **73** (Suppl.), pp.115S–126S

Lund, B. M. (1993) The microbiological safety of prepared salad vegetables. *Food Technology International Europe*, 196–200

Roberts, T. A. and Skinner, F. A. (eds) (1983) *Food Microbiology*, Society for Applied Bacteriology Symposium Series No. 11, Academic Press, London

Roberts, T. A., Hobbs, G., Christian, J. H. B. and Skovgaard, N. (eds) (1981) *Psychrotrophic Microorganisms in Spoilage and Pathogenicity*, Academic Press, London

Robertson, M. H. (1974) The provision of bacteriologically safe infant feeds in hospitals. *Journal of Hygiene* (*Cambridge*), **73**, 297–303

Rose, S. A. (1985) A note on yeast growth in media used for the culture of lactobacilli. *Journal of Applied Bacteriology*, **59**, 53–56

Speck, M. (ed.) (1992) *Compendium of Methods for the Microbiological Examination of Foods*, American Public Health Association, Washington

Chapter

17

Fresh fish, shellfish and crustaceans

Fresh fish

The quality of raw fish is best assessed by appearance and odour. Counts are of limited value and the flora is mainly halophilic and psychrophilic.

Counts

These are of limited value, but if necessary sample the whole surface of the fish or fillet either by swabbing a defined area (see p.263) or by washing the sample with sterile 0.1% peptone + 1% NaCl in a plastic bag. Use the washings to prepare dilutions in salt peptone and do counts on Marine agar, or use an impedance method (p.129). Incubate at 20°C for 5 days. Coliform counts are sometimes useful. Examination for salmonellas (p.318) and *V. parahaemolyticus* (p.294) may be desirable.

Surface slime is usually heavily infected. The count is increased by careless handling and contact with dirty ice and decreased by salting in barrels, hypochlorite and ice mixtures.

Culture

Inoculate glucose tryptone media, MacConkey media, and media containing 5–10% of sodium chloride for halophilic bacteria.

Inoculate PPPA, MSA or other media for staphylococci and also plate out the salt broths on these media. *S. aureus* is not uncommon in fish.

Microbial content

Large numbers of pseudomonas, flavobacteria, coryneforms, acinetobacter, aeromonas and cytophaga, often associated with slime, and micrococci may be found. Photobacteria, luminescent in the dark, are often present. The flesh count increases rapidly after filleting but in good plants may be as low as 100 000 cfu/g.

Smoked fish

Since 1976 there has been a marked increase in the incidence of scombroid fish poisoning (non-bacterial). The symptoms resemble those of food poisoning, i.e. diarrhoea and vomiting, headache, giddiness, rash on head and neck, 10 min to a few hours after ingestion of scombroid fish, e.g. mackerel, tuna, etc. Histamine compounds and some spoilage of the fish is always involved. The condition can result from eating canned fish.

Shellfish and raw crustaceans

Shellfish include bivalves, e.g. oysters, mussels, cockles and clams, as well as gastropods, e.g. whelks and winkles. Crustaceans include crabs, shrimps, prawns and lobsters. These animals are often found in inshore waters or river estuaries that are prone to sewage pollution. The bivalve molluscs, in particular, are filter feeders and hence tend to concentrate bacteria and viruses from their environment. There are regulations about the quality of the water from which they may be harvested as well as for the numbers of faecal indicator organisms in the fish.

Viable counts

Roll-tube method

This is applicable to both shellfish and raw crustaceans.

Place oysters or mussels in the freezing compartment of a refrigerator overnight. This makes them easy to open and no fluid will be spilled. Scrub and clean the outside of the shell and rinse in boiled water. Hold with the concave shell down and open with a sterile oyster knife. Cut the frozen flesh into small pieces (to release intestinal contents) with a sterile scalpel and place it in a sterile 200-ml Pyrex measuring cylinder. Push the flesh down with a sterile glass rod and continue until there is about 100 ml of mush. Similarly, obtain about 100 ml of the flesh of winkles, cockles or whelks. Add an equal volume of diluent, stopper and shake well, For prawns and shrimps, place 100 g in a blender with 100 ml of diluent and homogenize.

Allow the gross material to settle for 15–20 min. Melt three roll-tubes each containing 2 ml of MacConkey roll-tube agar and cool to 45–50°C. To each tube add 1 ml of the supernatant liquor and roll the rubes. When cool incubate overnight in a water-bath at 44°C inverted and completely submerged (inversion prevents water of syneresis from washing off colonies developing near the bottom of the tube).

Count the large red coliform colonies. Less than five colonies on each of the tubes inoculated with 1 ml (i.e. less than 5/g of shellfish) is regarded as satisfactory, from 5–15 colonies suspicious and more than 15 colonies as unsatisfactory.

To detect salmonellas in oysters, add 100 ml of mush to 100 ml of double-strength selenite broth (see p.318).

To detect sulphite-reducing clostridia and group D streptococci in oysters the pour plate method is recommended (Easterbrook and West, 1987).

To detect *S. aureus* which may be present in imported frozen prawns and shrimps (infected during handling), add 10-ml amounts of the mush to 50-ml tubes of broth containing 10% of sodium chloride. Incubate overnight at 37°C and plate on blood agar and PPPA or MSA medium (see Chapter 36). For *V. parahaemolyticus* add 10 g emulsified flesh to 100 ml of alkaline peptone water containing 3% salt and proceed as on p.294.

The 'percentage clean' method for oysters and mussels

Take 10 shellfish, scrub and clean the outside; and wash in sterile water. Hold with the concave shell down and open with a sterile oyster knife. Take care not to spill any liquor. Cut the flesh into small pieces with a sterile scalpel and mix with the liquor already present. Add 0.2 ml of liquor from each shellfish to tubes of single-strength MacConkey broth. Add 0.2 ml from three shellfish to a tube of glucose broth and 1 ml from three others to a tube of litmus milk.

Incubate at 44°C for 24 h and examine the MacConkey tubes for acid and gas, the glucose broth microscopically for streptococci (enterococci) and the litmus milk for *C. perfringens*.

If coliform bacilli (44°C ± 0.2) are absent from all 10 shellfish, they are '100% clean'. If they are absent from eight out of 10, they are '80% clean', and so on. It is generally regarded that '80–100% clean' is satisfactory, '70% clean' is suspicious and '60% clean' or less is unsatisfactory.

Alternative methods of isolating enterococci are given in on p.366. For more information about the bacteriology of shellfish see Codex Alimentarius (1978a, b) and West and Coleman (1986).

EC Standards

There is an EC Directive (EEC, 1991): bivalve molluscs must not contain more than 300 faecal coliforms or more than 230 *E.coli* per 100 g of flesh. This limit is sometimes achieved by harvesting the fish and then keeping them in clean sea-water for a few days (depuration), when they will eliminate many bacteria.

The possibility of contamination by viruses should be considered. Molluscs are filter feeders and depuration, which removes bacteria, may not remove viruses, nor are they always killed by heat treatment. Bivalves are therefore often associated with outbreaks of viral diseases such as hepatitis A and winter vomiting disease. Molluscs are sometimes vehicles for algal toxins produced by dinoflagellates. Consumption may result in paralytic or diarrhoetic shellfish poisoning. Tests for viruses and toxins are the province of specialist laboratories.

Frozen sea-food

Most contamination is introduced in the factory in cutting, battering, packing, etc. Enterobacteria and staphylococci also may be introduced at this stage. If the food is pre-cooked, the bacterial count is reduced but this treatment is usually not enough to kill all enterobacteria, staphylococci and anaerobes. Counts may be very high. *S. aureus*, coliforms, *V. parahaemolyticus* and salmonellas may be present. The flora is usually mixed and reflects processing rather than the raw materials.

Ready cooked deep-frozen prawns and shrimps

These are imported from the Far East and are an increasingly common article of diet in the West.

The following method, devised by Mitchell (1970) allows a simple quantitative examination to be made on this product. Chisel 20 g of fish from a frozen block into a sterile screw-capped jar. Place in a water-bath at 44°C for 10–30 min to hasten release of the juice. Prepare 1 : 50 and 1 : 500 dilutions and plate on well dried blood agar and MacConkey plates using the Miles and Misra method. Incubate overnight at 35°C. This gives the total viable and coliform counts/ml of extruded juice. Homogenize the remaining tissue preferably in a Stomacher and inoculate salt meat broth for *S. aureus*, alkaline peptone water for *V. parahaemolyticus* and selenite broth for salmonellas. Incubate overnight at 37°C and plate on suitable solid media.

The bacterial population of these crustaceans will consist of spore-forming bacteria that have survived boiling and organisms introduced after cooking.

Counts on isolated samples may not give significant results. Ideally 5–10 samples per batch should be obtained at the port of entry. Suggested standards, based on a weighing and macerating technique and using an incubation temperature of 35°C are: counts up to 100 000 cfu/g, release unconditionally; 100 000 to 1 000 000 cfu/g, release with a warning to use immediately on thawing, and over 1 000 000 cfu/g detain.

Vinegar-pickled fish

Test pH. If this is 4.5 or less no further action is necessary.

References

Codex Alimentarius Commission (1978a) *Recommended International Code of Hygiene Practice for Shrimps and Prawns*, Food and Agricultural Organization, Rome

Codex Alimentarius Commission (1978b) *Recommended International Code of Hygiene Practice for Molluscan Shellfish*, Food and Agricultural Organization, Rome

EEC (1991) Council Directive laying down the health standards for the production and placing on the market of bivalve molluscs. 91/492/EEC. *Official Journal of the European Communities*, 24/9/91, No L268/1

Easterbrook, J. and West, P. A. (1987) Comparison of most probable number and pour plate procedures for the isolation and enumeration of sulphite-reducing *Clostridium* spores and Group D faecal streptococci from oysters. *Journal of Applied Bacteriology*, **62**, 413–419

Mitchell, N. J. (1970) A simplified method for the quantitative examination of deep-frozen seafood. *Journal of Applied Bacteriology*, **33**, 523–527

West, P. A. and Coleman, M. R. (1986) A tentative national preference procedure for the isolation and enumeration of *Escherichia coli* from bivalve molluscan shellfish by the most probable number method. *Journal of Applied Bacteriology*, **61**, 505–516

Chapter

18

Milk, dairy produce, eggs and ice-cream

Milk

In Europe, microbiological tests and standards for milk are laid down by the European Commission (EC, 1992). Similar tests and criteria are used in non-European Union states and in the USA.

Raw milk

Raw milk, which includes that from cows, goats, ewes and buffaloes, is milk that is:

(a) intended for the production of heat-treated drinking milk, fermented milk, junket, jellies and flavoured milk drinks;
(b) intended for the manufacture of milk-based products other than those specified in (a);
(c) cows' milk, after packing for drinking in that state.

The bacteriological tests required for these milks are:

(a) the plate count (total viable count, TVC) at 30°C;
(b) for *Staphylococcus aureus* and salmonellas.

Pasteurized milk

This is raw milk that meets the above requirements and has been heated at 62.9–65.6°C for 30 min or 71.7 °C for 15 s, and then immediately cooled to below 10°C. The tests are:

(a) the plate count at 21 °C
(b) the coliform test
(c) tests for pathogens (*Listeria monocytogenes* and salmonellas)
(d) the phosphatase test

Ultra-heat treated (UHT) milk

This is raw milk, as above, that has been heated at 135°C for 1 s. The bacteriological test is the plate count.

Sterilized milk

This is raw milk, as above, that has been heated at 100°C for such a time that it will pass the turbidity test. The tests are:

(a) the plate count at 30°C after preincubation at 30°C for 15 days,
(b) the turbidity test.

Collection and transport of samples

Raw milk for (a) and (b) above should be sampled at least once every 2 months at the point of production. Cows' milk packed for drinking should be sampled at the point of sale. Samples should be transported to the laboratory at 0–4°C.

Statutory testing

For legal purposes the techniques specified in the EC Directive and national standards must be used, but for normal practice the methods described below are adequate.

Plate (total viable) count, TVC

The method is given on p. 152. Test raw milks on arrival, use standard milk plate count agar and for raw milks incubate at 30 °C for 72 h. Store pasteurized milks at 6°C for 5 days before testing and then incubate plate counts at 21 °C. The targets levels, TVC/ml, are:

Raw milk not intended for drinking in that state	100 000
Raw milk packed for drinking	50 000

Colony count, UHT and sterilized milk

Store samples at 30°C for 15 days after sampling at point of production. Do plate counts in duplicate with 0.1 ml volumes of milk and plate count agar. Incubate at 30 °C for 72 h. The target level is:

UHT and sterilized milk	< 10 per 0.1 ml

Coliform count

Make serial tenfold dilutions and do plate counts with violet red bile lactose agar. Incubate at 30°C for 24 h. Count only red colonies that are 0.5 mm in diameter or larger. Confirm if necessary by subculture in lactose bile brilliant green broth at 37°C. It is useful to test confirmed coliforms by the 44°C (Eijkman) (p.111) and indole tests (p.113) to determine if they are *Escherichia coli*. The target level is:

Pasteurized milk.	5 cfu/ml

Phosphatase test

The Aschaffenberg–Mullen test is used.

Test at once or refrigerate the milk overnight. Make up the reagent as follows. Dissolve 3.5 g of anhydrous sodium carbonate (AnalaR) and 1.5 g of sodium bicarbonate (AnalaR) in 1 litre of water. Store in a refrigerator. To 100 ml of this solution add 0.15 g of disodium *p*-nitrophenyl phosphate. Keep in the dark and in a refrigerator and use within 7 days. There must be no yellow colour.

To 5 ml of the reagent in a 152 × 16 mm test-tube add 1 ml of milk. Stopper with a rubber stopper and mix by inversion. Incubate for 2 h at 37°C in a water-bath. Include controls of boiled milk and also boiled milk containing about 2% of raw milk. Determine the actual amount by titration, i.e. by adding varying proportions of raw milk to pasteurized milk. Do a phosphatase test on these and make the control in the proportion which gives a reading of slightly less than 42 μg. Preserve this mixture with 0.5 mg/100 ml of saturated mercuric chloride. Store in a refrigerator. Label the bottle 'Poison'.

Compare the test sample with the boiled sample with the Lovibond Comparator and disc designed for this purpose. This is an all-purposes model supported on a stand on which the test-tubes are placed in a sloping position and viewed by reflected light. Unheated milks and insufficiently treated milks give a yellow colour caused by *p*-nitrophenol released from the substrate by the action of phosphatase. In properly pasteurized milk, phosphatase has been destroyed and the reading on the disc will be 10 μg or less of *p*-nitrophenol/ml of milk.

Clean the glassware used in this test in chromic acid solution.

Sterilized milk: the turbidity test

Weigh 4.0 g of ammonium sulphate (AnalaR) into a small flask or bottle. Add 20 ml of milk and shake for 1 min. Stand for 5 min and filter through a 12.5-cm Whatman No. 12 folded filter-paper into a test-tube. When 5 ml filtrate have collected, place the tube in a boiling water-bath for 5 min and then cool. A properly sterilized milk gives no turbidity.

Sterilization alters the protein constituents and all on-heat coagulable protein is precipitated by the ammonium sulphate. If heating is insufficient, some protein remains unaltered and is not precipitated by the ammonium sulphate. It coagulates, giving turbidity when the filtrate is boiled.

US standard methods for milk examination (Speck, 1992)

Three routine tests are prescribed:

(1) plate count at 32°C;
(2) direct microscopical count, for raw milks if a high count is expected; and
(3) the coliform test.

Plate count
Use Standard Plate Count agar to prepare pour-plates. Incubate at 32 ± 1°C for 48 ± 3 h.

Direct microscopical count
This is recommended only for raw milks with a fairly high count. The technique is similar to the Breed count (see p.150).

Coliform test
This test is used after pasteurization in order to detect re-contamination. There are two methods.

(1) *Solid media method*. Use a pour-plate technique with either violet red bile or deoxycholate lactose agar. Mix 1.0 and 0.1 ml of milk with melted and cooled agar, allow it to set and overlay it with a further 3–4 ml of sterile medium. Cover the surface of the inoculated agar completely to

prevent surface growth. Incubate at 32°C for 24 ± 2 h and count all dark red colonies that are 5 mm or more in diameter as coliforms.

(2) *Liquid media method.* Add 10 ml to each of five tubes of double-strength brilliant green bile broth and 1 ml to each of five tubes of single-strength brilliant green bile broth. Incubate at 32°C for 48 ± 3 h. Streak a loopful from each tube showing fermentation on to eosin methylene blue or Endo agar. Inoculate typical colonies grown on this medium to nutrient agar for Gram stain and a lactose broth to demonstrate gas production. Use the most probable number (MPN) tables on p.158–162 to report.

Microbial content of milk

Bacteria enter milk during milking and handling. Even with the most hygienic production, some bacteria gain access. The cooling that is normal practice after milking retards the multiplication of these bacteria. Pasteurization, intended to kill pathogenic bacteria, does not necessarily reduce the count of other organisms. It may increase the numbers of thermophiles.

The 'normal microflora' depends on temperature: at 15–30°C *Streptococcus lactis* predominates and many streptococci and coryneform bacteria are present, but at 30–40°C they are replaced with lactobacilli and coliform bacilli. All of these organisms ferment lactose and increase the lactic acid content, which causes souring. The increased acid content prevents the multiplication of putrefactive organisms. Spoilage during cold storage is due to psychrophilic pseudomonads and *Alcaligenes*, psychrotrophic coliforms, e.g. *Klebsiella aerogenes* and *Enterobacter liquefaciens* which are anaerogenic at 37°C and in pasteurized milk, thermoduric coryneform organisms (*Microbacterium lactis*) may be significant. These coryneforms probably come from the animal skin or intestine and from utensils.

At temperatures above 45°C, thermophilic lactobacilli (*Lactobacillus thermophilus*) rapidly increase in numbers.

Gram-negative bacilli are rarely found in quarter samples collected aseptically. These enter from the animal skin and dairy equipment during milking and handling. *Alcaligenes* species are very common in milk. *P. fluorescens* gels UHT milk.

Undesirable microorganisms responsible for 'off flavours' and spoilage include psychrophilic *Pseudomonas*, *Achromobacter*, *Alcaligenes* and *Flavobacterium* spp., which degrade fats and proteins and give peculiar flavours. Coliform bacilli produce gas from lactose and cause 'gassy milk'. *S. cremoris*, *Alcaligenes viscosus* and certain *Aerobacter* species, all capsulated, cause 'ropy milk'. *Oospora lactis* and yeasts are present in stale milk. *P. aeruginosa* is responsible for 'blue milk' and *Serratia marcescens* for 'red milk' (differentiate from bloody milk). *B. cereus* can cause rapid decolorization of methylene blue.

Pathogenic organisms which may be present include *S. aureus*, *Campylobacter*, salmonellas, *Y. enterocolitica*, listerias, *S. pyogenes* and other streptococci from infected udders in mastitis. Tubercle bacilli may be found by culture of centrifuged milk deposits and gravity cream. *B. abortus* is excreted in milk and can be isolated by culture, or antibodies can be demonstrated by the ring test or whey agglutination test. *Rickettsia burnetti*, the agent of Q fever, may be found by animal inoculation of milk from suspected animals.

Mastitis

Looking for evidence of mastitis in bulk milk is obviously unrewarding in any but in farm or other small samples. Centrifuge 50 ml of milk and make films of the deposit. Dry in air, treat with xylene to remove fat, dry, fix and stain with methylene blue. Examine for pus cells. For bacteriological examination and identification of causative organisms, see p.361).

Examination of dairy equipment

See Chapter 19 (Environmental microbiology) for methods.

Microbial content of dairy equipment

The types of organisms found will be similar to those in milk. Soil organisms, e.g. *Bacillus* spp., may be present. *B. cereus* reduces methylene blue very rapidly and may also cause food poisoning. Organisms originating from faeces, e.g. various coliforms and *Pseudomonas* spp., are found. Coliform bacilli also rapidly reduce methylene blue. Pseudomonads may produce oily droplets in the milk.

Goat and sheep milk

At present there are no specific regulations in the UK for the examination of these milks. The Ministry of Agriculture, Fisheries and Food has issued a code of practice (MAFF, 1988) that gives target total viable count (TVC) values for fresh goat milk of < 100/ml after collection, < 1000/ml after packaging and with coliforms < 10/ml.

It would seem wise to use the standards currently used for cows' milk as target levels for raw and pasteurized goat and sheep milk respectively:

	Raw	*Pasteurised*
TVC, 30°C, 72 h cfu/ml	< 2000	< 3000
Coliforms cfu/ml	< 100	< 1

Additional tests and target levels

S. aureus	< 20/ml
Listeria	None in 25 ml
E. coli	< 10/ml

Pathogens

Methods for these are given elsewhere: salmonellas, p.317; *S. aureus*, p.353; *Listeria*, p.384 and campylobacters (p.340).

Natural cream

Most cream on sale in the UK is heat treated and there are statutory tests (SI, 1983a). Pasteurized cream must satisfy the three-tube coliform test and the phosphatase test. UHT and sterilized cream the colony count test after preincubation at 37°C for 24 h.

There are no statutory tests for untreated cream but we suggest that it is tested for total count, coliforms, *E. coli, S. aureus* and possibly for campylobacters and salmonellas.

Coliform test for pasteurized cream

Place 90 ml of 2% sodium citrate in a bottle and warm it to 37°C in a water-bath. Warm the cream to the same temperature. Weigh the bottle, add 10 ml of the warm cream and mix well.

Add 10 ml of this 1:10 dilution to each of three bottles of 10 ml double strength brilliant green lactose bile broth (BGLBB) each containing a Durham's tube. Incubate at 30°C for 48 h.

The test is satisfactory if two of the three bottles show no gas production.

This is an unsatisfactory test as the opacity of the cream suspension makes it difficult to see gas in the Durham's tube.

If there is gas in at least two tubes subculture to brilliant green broth or similar medium to verify the presence of coliforms. Other organisms, e.g. *Bacillus* spp., may also give gas.

Special colony count for cream

Use the test for UHT milk.

Viable counts

Do plate counts or Miles and Misra counts on serial dilutions ranging from 10^{-2} to 10^{-6} in a 0.1% peptone water solution, using milk plate count agar.

If the cream is solid or difficult to pipette, make the initial 1:10 dilution by weight as described on p.152. The 1:10 dilution may be used, with appropriate media, for spiral plating for total viable, coliform and S. *aureus* counts.

Culture for pathogens

Plate serial dilutions on Baird-Parker medium for S. *aureus*.

For salmonellas pre-enrich a 1:10 mixture of cream (preferably 25 g) in buffered peptone water and proceed as on p.317.

For campylobacters use a 10 g sample and the method on p.340.

Microbial content

Fresh cream may contain large numbers of organisms, including coliforms, B. *cereus*, *E. coli*, S. *aureus*, other staphylococci, micrococci and streptococci. Most fresh cream has been pasteurized and therefore contamination by many of these organisms is the result of faulty dairy hygiene.

Cream should not be used in catering in circumstances where these organisms can multiply rapidly

Phosphatase test

Place 15 ml of the buffer reagent (p.250) into each of two tubes. Stopper and bring to 37°C in a water-bath. Weigh the tubes and add 2 g of test cream to one tube and 2 g of boiled cream to the other (control). Mix well and incubate (water-bath) for 2 h, remove and add 0.5 ml of 30% (w/v) zinc sulphate. Mix well and stand for 3 min. Add 0.5 ml of 15% (w/v) potassium ferrocyanide. Mix well and filter through a Whatman No. 40 paper. Read test, if yellow coloured, against blank using a Lovibond Comparator and APTW disc. Record results in µg of *p*-nitrophenol/ml of cream.

Less than 10 µg is satisfactory. If the reading is greater than 10 µg do the verification test for reactivation of phosphatase.

Place 10 g of cream into each of two tubes. To one add 40% (w/v) magnesium chloride according to fat content: double cream, 0.25 ml; whipping cream, 0.35 ml; single cream, 0.5 ml. To the other add nothing. Stopper and mix. Incubate both tubes at 37°C for 60 min with occasional shaking.

Remove 2 g of cream from each tube and repeat the phosphatase test. If the test (magnesium chloride treated) sample gives a higher reading than the control dilute it 1:4. If the colour is now equal or still more intense than the control reactivation has occurred and the test is reported as satisfactory. If the colour is less than that of the control the positive phosphatase test has been verified.

Imitation cream

Do plate counts as for milk, at 30°C, coliform counts, and also examine for *S. aureus* (p.353). Examine for staphylococci only, as for natural cream.

Processed milks

Dried milk

This often has a high count and, if stored under damp conditions, is liable to mould spoilage. Spray-dried milk may contain *S. aureus*.

Examination
Reconstitute in distilled water before testing. Examine direct films, do total counts as for fresh milk but prepare extra plates for incubation at 55°C for 48 h for thermophiles. Do coliform counts to detect contamination after processing. Examine for salmonellas, using at least 100 g samples. Examine for yeasts and moulds as described for butter examination (p.357).

The number of spore bearers is important in milk and milk powder intended for cheese making. Inoculate 10 ml volumes of bromocresol purple milk in triplicate with 10, 1.0 and 0.1 g amounts of milk. Heat at 80°C for 10 min, overlay with 3 ml of 2% agar and incubate for 7 days. Read by noting gas formation and use Jacobs and Gerstein's MPN tables to estimate the counts (p.162).

Condensed milk

This contains about 40% of sugar and is sometimes attacked by yeasts and moulds that form 'buttons' in the product. Examine as for butter (p.257).

Evaporated milk

This is liable to underprocessing and leaker spoilage. *Clostridium* spp. may cause hard swell and coagulation. Yeasts may cause swell. Open aseptically and examine as appropriate (see p.238).

Fermented milk products

Yoghurt, leben, kefir, koumiss, etc., are made by fermentation (controlled in factory-made products) of milk with various lactic acid bacteria, and streptococci and/or yeasts. The pH is usually about 3.0–3.5 and only the intended bacteria are usually present, although other lactobacilli, moulds and yeasts may cause spoilage. The presence of large numbers of yeasts and moulds can be indicative of poor hygiene. Plate on potato glucose agar (pH adjusted to 3.5) and incubate at 23 ± 2°C for 5 days.

Yoghurt is made from milk with *Lactobacillus bulgaricus* and *Streptococcus thermophilus*. In the finished product there should be more than 10^8/g of each and they should be present in equal numbers. Lower counts and unequal numbers predispose off-flavours and spoilage.

Make doubling dilutions of yoghurt in 0.1% peptone water and mix 5 ml of each with 5 ml of melted LS Differential Medium. Pour into plates, allow to set and incubate at 43°C for 48 h. Both organisms produce red colonies, because the medium contains triphenyl tetrazolium chloride (TTC). Those of *L. bulgaricus* are irregular or rhizoid, surrounded by a white opaque zone; those of *S. thermophilus* small, round and surrounded by a clear zone. Count the colonies and calculate the relative numbers of each organism. Post-pasteurization contamination with coliforms may occur. Check pH, which is usually very acid. If this is pH > 5.6, examine for coliforms, *E. coli*, salmonellas and other enteric pathogens.

Milk-based drinks

There are statutory tests in the UK (SI, 1983b).

Pasteurized drinks	Coliform and or phosphatase tests
Sterilized drinks	Colony count after preincubation at 37 °C for 24 h
UHT drinks	Colony count
	Coliform test

Coliform test
Dilute 1:10 in quarter strength Ringer solution. Add 1 ml to each of three tubes containing 5 ml of single strength brilliant green lactose bile broth (BGLBB) containing Durham's tubes. Incubate at 37°C for 48 h and examine for gas production. The test is satisfactory if two of three tubes show no gas.

Subculture from tubes showing gas to brilliant green broth or similar medium. Test typical colonies on this medium for gas production in BGLBB as the original gas may have been produced from sugar in the product.

Phosphatase test
As for milk (p.250).

Colony count
As for UHT milk (p.250).

Buttermilk

Test for coliforms only. Use 10 ml of a 1:10 dilution in a pour-plate as for creams.

Butter

Soured cream is pasteurized and inoculated with the starter, usually *S. cremoris* or a *Leuconostoc* spp. Salted butter contains up to 2% of sodium chloride by weight.

Examination and culture
Emulsify 10 g of butter in 90 ml of warm (40–45°C) diluent and do plate counts. Culture on a sugar-free nutrient medium and tributyrin agar.

Microbial content
Correct flavours are due to the starters, which produce volatile acids from the acids in the soured raw material.

Rancidity may be due to *Pseudomonas* and *Alcaligenes* spp., which degrade butyric acids. Anaerobic butyric organisms (*Clostridium*) cause gas pockets. Some lactic acid bacteria (*Leuconostoc*) cause slimes and others give cheese-like flavours. Coliforms, lipolytic psychrophiles and casein-digesting proteolytic organisms, which give a bitter flavour, may be found. To detect these, plate on Standard Plate Count agar containing 10% sterile milk and incubate at 23 ± 2°C for 48 h. Flood the plates with 1% hydrochloric acid or 10% acetic acid. Clear zones appear round the proteolytic colonies.

Yeasts and moulds may be used as an index of cleanliness in butter. Adjust the pH of potato glucose agar to 3.5 with 10% tartaric acid and inoculate with the sample. Incubate at 23 ± 2°C for 5 days. Report the yeast and mould count/ml or /g. Various common moulds grow on the surface but often these grow only in water droplets or in pockets in the wrapper.

Cheese

Many different cheeses are now available including varieties that contain herbs which may contribute to the bacterial load. Some are traditionally made from raw milk and many contain pathogens such as *S. aureus*, brucellas, campylobacters and/or listerias.

Milk is inoculated with a culture of a starter, for example *S. lactis, S. cremoris, S. thermophilus* and rennet added at pH 6.2–6.4. To make hard cheeses, the curd is cut and squeezed free from whey, incubated for a short period, salted and pressed. The streptococci are replaced by lactobacilli (e.g. *L. casei*) naturally at this stage and ripening begins. Some cheeses are inoculated with *Penicillium* spp. (e.g. *P. roquefortii*), which give blue veining and characteristic flavours due to the formation under semi-anaerobic conditions of caproic and other alcohols.

Soft cheeses are not compressed, have a higher moisture content and are inoculated with fungi (usually *Penicillium* spp., which are proteolytic and flavour the cheese).

Types and flavours of cheeses are due to the use of different starters and ripeners and varying storage conditions.

Examination and culture
Take representative core samples aseptically, grate with a sterile food grater, thoroughly mix and sub-sample. Fill the holes, left as a result of boring, with Hansen's paraffin cheese wax to prevent aeration, contamination and texture deterioration.

Make films, de-fat with xylol and stain. Bacteria may be present as colonies:

examine thin slices under a lower-power microscope. Weigh 10 g and homogenize it in a blender with 90 ml of warm diluent. Do plate counts on yeast extract milk agar and coliform counts. Culture on whey and tributyrin agar and on media for staphylococci.

Microbial content

Apart from streptococci, lactobacilli and fungi that are deliberately inoculated or encouraged the following organisms may be found: contaminant moulds, *Penicillium, Scopulariopsis, Oospora, Mucor* and *Geotrichum* give colours and off-flavours. Putrefying anaerobes (*Clostridium* spp.) give undesirable flavours. *Rhodotorula* gives pink slime and *Torulopsis* yellow slime. Gassiness (unless deliberately encouraged by propionibacteria in Swiss cheeses) is usually due to *Enterobacter* spp., but these are not found if the milk is properly pasteurized.

Gram-negative rods, some of which may hydrolyse tributyrin, may also be present.

Psychrophilic spoilage is common and is due to *Alcaligenes* and *Flavobacterium* spp. Counts may be very high.

Bacteriophages which attack the starters and ripeners can lead to spoilage.

Butyric blowing

Clostridium butyricum and *C. tyrobutyricum* may cause gas pocket spoilage during the ripening of hard cheeses.

Homogenize cheese samples in peptone water diluent and add dilutions to melted SPS agar in tubes. Layer paraffin wax on the agar and incubate at 30°C. Butyric clostridia will produce gas bubbles in the medium. The organisms may be enumerated by the three or five tube MPN method with inocula of 10, 1 and 0.1 ml of homogenate.

Propionibacteria

Although deliberately introduced into some cheeses these can cause spoilage in others. Accurate counts are difficult to do because the organisms are microaerophilic.

Prepare doubling dilutions of cheese homogenate in 0.1% peptone water and add 1 ml of each to tubes of yeast extract agar containing 2% sodium lactate or acetate. Mix and seal the surface with 2% agar. Incubate anaerobically or under carbon dioxide at 30°C for 7–10 days. Propionibacteria produce fissures and bubbles of gas in the medium. Choose a tube with a countable number of gas bubbles and calculate the numbers of presumptive propionibacteria (some other organisms may produce gas bubbles).

Staphylococcus aureus

Outbreaks of food poisoning caused by this product are occasionally reported. Cheddar cheeses may have a high staphylococcus count. In the USA cottage cheeses have been incriminated and in some states there is a statutory limit of 50 *S. aureus*/g. Enterotoxin may be produced during the long setting period, although the organisms themselves tend to die out on subsequent storage.

Listeria monocytogenes

This is known to survive some cheese making processes and increase during storage and to cause human disease. For method of isolation see p.384. Quantitative methods may be needed.

Eggs

Eggs contaminated with *Salmonella enteritidis* have been responsible for many human infections and it is now thought necessary to test eggs, both externally (shell) and internally (yolk and albumen), for salmonellas.

Satisfactory results may be obtained with samples of, for example, 60 eggs, and by testing pools of six eggs separated into shell and contents.

Disinfect the shells by immersion in, or swabbing with, industrial alcohol. Break the eggs in batches of six, keeping the shells and contents in separate, sterilized containers, e.g. screw-capped jars. Add to each an equal volume of buffered peptone water (BPW) (pre-enrichment) and incubate at 37°C for 18–24 h. Then add 0.1 ml volumes to 10 ml amounts of Rappaport-Vasiliadis broth. Incubate these at 42°C for 24 h and subculture to at least two of the following selective media: xylose lysine deoxycholate (XLD) agar, brilliant green agar and bismuth sulphite agar. Incubate at 37°C for 14 h and examine for salmonellas as described on p.317.

Pasteurized liquid egg

Test these for salmonellas as above and as for α-amylase (p.317), the absence of which indicates effective pasteurization.

Microbial content

Shell eggs

Externally, various salmonellas, especially *S. enteritidis*, coliforms, *Proteus, Pseudomonas, B. cereus*, yeasts, moulds and putrefactive anaerobes may be present. Internally, *S. gallinarum* and *S. pullorum*, coliforms, mycoplasmas, avian strains of mycobacteria and *Pasteurella anatipestifer* may be found. The degree of contamination is related to the rate of cooling and age of the egg, the porosity of the shell and humidity of the environment. *Proteus* causes coloured rots, e.g. 'black rot', in eggs.

Frozen egg

Total counts are high. Cultures may contain *Pseudomonas, Aeromonas, Achromobacter, Alcaligenes*, enterobacteria, *Serratia*, micrococci and putrefactive anaerobes. Counts less than 50 000 cfu/g are rare; usually the count is several million and three million has been suggested as a reasonable maximum.

Frozen egg is invariably pasteurized nowadays, although this presents heat penetration problems and the α-amylase test is done by analytical chemists to confirm adequate heat treatment. Egg is incubated with a standard starch solution and the blue colour produced when iodine is added is measured. Adequate pasteurization, which reduces the bacterial content and destroys salmonellas, destroys most of the amylase. This test is less subject than bacteriological examination to sampling errors.

Dried egg

The total count is usually low (5000–10 000 cfu/g) in dried whole egg. Coliforms are rarely present in 1 g.

Dried albumen gives variable results. High-quality spray-dried material gives counts usually not exceeding 10 000 cfu/g but in some albumens counts may be as high as 10 million. Cross contamination may occur in tray drying of flake albumen.

Frozen and dried egg are recognized vehicles of salmonellas. Culture replicate samples of up to 20 g in not less than 100 ml of medium (because of lysozyme). See standard method for salmonellas, p.317.

Ice-cream and ice-lollies (water-ices)

These products vary in their composition and may be ice-cream, as defined in the various food regulations, mixtures of ice-cream and fruit or fruit juices, or contain no ice-cream. In general, plate counts, coliform tests and tests for pathogens should be applied to ice-cream and commodities that are said to contain it whether as a core or a covering. Otherwise the pH should be ascertained. If this is 4.5 or less there is no point in proceeding further.

Sampling

Send samples to the laboratory in a frozen condition, preferably packed in solid carbon dioxide. Remove the wrappers of small samples aseptically and transfer to a screw-capped jar. Sample loose ice-cream with the vendor's instruments (which can be examined separately if required, see Chapter 19). Keep larger samples in their original cartons or containers.

Hold ice-lollies by their sticks, remove the wrapper and place the lolly in a sterile screw-capped 500-ml jar. Break off the stick against the rim of the jar.

Reject samples received in a melting condition in their original wrappers.

Allow the samples to melt in the laboratory but do not allow their temperature to rise above 20°C.

Viable counts

The difficulty with bacterial counts is the measurement of volumes of a product that contains a variable amount of air. For the best results, weigh about 10 ml of ice-cream into a screw-capped jar, multiply the weight to the nearest 0.1 g by nine and add this volume of 0.1% peptone water. Shake well. From this initial 1:10 dilution make two further tenfold serial dilutions and do plate counts on 1-ml samples with Yeastrel agar. Incubate at 37°C for 48 h.

Alternatively, and for routine work, make the initial 1:10 dilutions by adding 10 ml of melted ice-cream to 90 ml of Ringer solution. Smaller amounts are less reliable.

Miles and Misra counts (p.154) are very convenient. Dilute the ice-cream 1:5 and drop three 0.02-ml amounts with a 50-dropper on blood agar plates. Allow the drops to dry, incubate at 37°C overnight and count the colonies with a hand lens.

Multiply the average number of colonies per drop by 250 (i.e. dilution factor of 5 and volume factor of 50) and report as count/ml. If there are no colonies on any drops, the count can be reported as less than 100/ml. It is difficult to count more than 40 colonies per drop; therefore, report such counts as greater than 10 000 cfu/ml.

The spiral plating method (p.95) may be used for total counts.

In counts, note the presence of *B. cereus*. This spore-bearer may be present in the raw materials. Its spores resist pasteurization.

Counts at 0–5°C for psychrophiles may be useful in soft ice-cream.

Coliform test
To test for coliforms, dilute the sample 1 : 10 and use the pour-plate technique described for milks (p.250).

In view of the non-dairy content of some ice-creams a range of tests and target levels appropriate for chilled foods should be used:

TVC, 30°C	< 10 000/g
Coliforms	< 10/g
E. coli	< 10/g
S. aureus	< 100/g
Salmonellas	none in 25 g
Listeria monocytogenes	none in 25 g

Ingredients and other products

Do plate counts at 5, 20 and 37°C. Although ice-cream is pasteurized and then stored at a low temperature, only high-quality ingredients with low counts will give a satisfactory product.

In cases of poor grading or high counts on the finished ice-cream, test the ingredients and also the mix at various stages of production. Swab-rinses of the plant are also useful (see p.265).

Human milk

The milk is usually tested before and after pasteurization. Cooling water in the pasteurization unit is potentially hazardous. There should be 10 ppm residual free chlorine to eliminate pseudomonads, etc. Mains water should be used.

Raw milk

Use dip slides (nutrient agar, blood agar, MacConkey and CLED). Incubate at 37°C overnight and count colonies on all media. Note presence and numbers of coliforms, *S. aureus*, streptococci, pseudomonads

$$\frac{\text{Colony count}}{\text{Medium surface area (cm)}} \times 100 = \text{cfu/ml}$$

Reincubate dip slides at room temperature for 24 h to detect psychrotrophs. Confirm identity of potential pathogens.

The levels for acceptance and pasteurization are:

Total count:	not exceeding 10 cfu/g
Normal skin flora:	not exceeding 10 cfu/g

Absence of *S. aureus* and other potential pathogens.

Milk not satisfying these criteria should be discarded.

Pasteurized milk

Do phosphatase test as for cows' milk (p.250) using boiled human milk controls.

For total count, spread-plate 0.1 ml (standard loopful) on two plates of blood agar. Incubate both at 37°C overnight, one aerobically, the other anaerobically. Count colonies and multiply by 100 to give count/ml of original milk. Satisfactory samples contain less than 100 cfu/g.

For more information see Hewitt (1981–2) and DHSS (1982).

References

DHSS (1982) *The Collection and Storage of Human Milk*, Reports on Health and Social Subjects No. 29, Department of Health and Social Security, HMSO, London

EC (1992) Council Directive laying down the health rules for the production and placing on the market of raw milk, heat-treated milk and milk-based products, 92/46/EEC, Brussels

Hewitt, J. H. (1981–2) Possible bacteriological hazards associated with the use of raw (untreated) human milk for special care infants. In *Human Milk Banking. Report Series: Refrigeration and Science Technology*, International Institute of Refrigeration, Paris

MAFF (1988) *The hygienic production of goats milk*, MAFF Publication 1988 BL 5677, D/M 8971006, HMSO, London

SI (1983a) *The Milk and Dairies (Heat Treatment of Cream) Regulations 1983*, Statutory Instrument No.1509 Food, HMSO, London

SI (1983b) Milk-based Drinks (Hygiene and Heat Treatments) Regulations 1983, Statutory Instrument 1508, HMSO, London

Speck, M. (ed.) (1992) *Compendium of Methods for the Microbiological Examination of Foods*, American Public Health Association, Washington DC

Chapter

19

Environmental microbiology

An estimate of the numbers of viable organisms on surfaces and in the air of buildings, in containers and on process equipment may be required:

(1) to test the standard of hygiene and efficiency of cleaning procedures;
(2) to trace contamination from dirty to clean areas;
(3) to test for the escape of 'process' organisms in industrial microbiology and biotechnology plants;
(4) to assess the level of bacterial contamination in 'sterile' environments;
(5) to investigate premises where humidifier fever, sick building syndrome, etc. are suspected.

Surface sampling

Agar contact plates

'Rodac' (Replica Organism Direct Contact) plates are available commercially. These are similar to petri dishes but are filled to the brim with culture medium and the lid fits well above the surface of the agar. Non-selective or selective media may be used. If the surface to be sampled has been cleaned with a phenolic or quaternary ammonium agent add 0.5% Tween 80 and 0.7% soya lecithin to the medium.

To sample a surface remove the lid and press the agar firmly on to the surface. Replace the lid and incubate the plate.

Only about 10–20% of the organisms on a surface are likely to be recovered on these plates. They may be used to show staff the extent of local surface contamination. To avoid the risk of infection spray the plates on which colonies have developed with a solution of vinyl acetate in methylated spirit (or use a domestic hair spray) before they are shown to staff. Return plates to the laboratory for autoclaving immediately after display.

Surface swab counts

It is generally accepted that the swabbing technique gives a count approximately 10 times higher than that obtained by agar surface contact when sampling smooth surfaces.

Cut card or cellophane squares with 10-cm sides and cut squares in them with 5-cm sides. Sterilize these templates in envelopes. Place a template on the surface to be examined and swab the area within the 5 × 5 cm square with a cotton wool or alginate swab. Soak and squeeze cotton-wool swabs in nutrient broth to recover the organisms or place alginate swabs in nutrient broth containing 1% sodium hexametaphosphate in which the alginate will dissolve. If the surface has been treated with a phenolic or quaternary ammonium (QAC) disinfectant add 0.5% Tween 80 and 0.7% soya lecithin to the medium. The count/25 cm^2 is given by the number of cfu/ml of rinse or solvent multiplied by 10. With Miles and Misra counts, it is given by the number of cfu/5 drops multiplied by 100.

Air sampling

Contamination of the air in hospitals and other workplaces may responsible for the spread of infection. In addition, in offices and factories, especially where there is 'air conditioning', it may be related to allergic illnesses such as humidifier fever and the 'sick building syndrome'. It may also lead to contamination of products such as foods and pharmaceuticals.

Air-borne contaminants may be released by people, industrial processes and poorly maintained air-conditioning plant. Bacteria and fungal spores may be suspended in the air singly, in large clumps, or on the surfaces of skin scales and dust from the materials being processed. The smaller of these particles, e.g. ≤ 5 μm) may remain suspended in the air for long periods and be moved about by air currents. Filtration, as in some air-conditioning plants, may not remove them. Larger particles settle rapidly and contaminate surfaces.

Settle plates

Settle plates supplement surface sampling for assessing potential surface contamination. Several plates containing appropriate media are exposed for a given time and incubated. The colonies are counted. Settle plates are favoured by those who need to monitor the air over long periods, e.g. in hospital cross-infection work. Expose blood agar to test for the presence of 'presumptive' *S. aureus* and selective media, e.g. for enterococci. This method is less satisfactory than using slit samplers for testing for very small suspended particles.

Mechanical air samplers

In principle, known volumes of air are drawn through the equipment and microorganisms and other particles are deposited on the surface of agar media, on membrane filters or into broth media where they may be enumerated. Some samplers, e.g. the Andersen, May and cascade samplers not only permit colony counting but provide information about the sizes of the particles from which the colonies develop. Particles deposited on membranes may be analysed chemically or immunologically. At least one ATP detection kit is available commercially.

Table 19.1, adapted from Hambleton *et al.* (1992), lists the characteristics of several commonly-used devices. Choice depends on the volume of air to be sampled and convenience of handling, e.g. portability and power supply. The SAS and RCS models are hand-held and may be battery operated. Personal samplers are attached to the individual's belt or clothing and are battery operated.

None of these samplers gives the total number of viable organisms per unit volume. Not all the bacteria or spores are trapped on to the medium; some

Table 19.1 Characteristics of commonly used air-sampling devices

Sampler type	*Flow rate (litres/min)*	*Advantages*	*Disadvantages*
Impactors		No sample manipulation	Semi-quantitative, not suitable for high concentrations (> 10 organisms/litre)
Casella slit	30–700	Efficient, easy to use	Static, bulky
Andersen	27	Particle size information	Difficult to assemble
SAS Cherwell	180	Portable, easy to use, directable	Inefficient at small particle sizes (< 5 μm)
Biotest RCS	40	Portable, easy to use, directable	Inefficient at small particle sizes (< 5 μm)
Cascade	175	Portable, particle size information	Low volume sample, processing may be necessary
Impingers		Suitable for high concentrations (> 10 organisms/ litre)	Sample processing necessary. Not suitable for low concentrations (< *10* organisms/ litre)
Porton	11	Easy to use, small	Fragile, sample evaporation, violent sampling
May three-stage	55	Particle size information, gentle sampling	Fragile, standardized manufacture difficult
Others			
Cyclone (impinger)	750	Suitable for wide range of concentrations	No particle size information
Personal filter samplers	1–4	Suitable for high concentrations, estimates worker exposure	Low volume, sample recovery and processing difficulties
Settle plates (impactor)	N/A	Easy to use	Inefficient, qualitative

Compiled from May (1945), Leaver *et al.* (1987), Bennett *et al.* (1991) and Deans (1991)
Reproduced from Hambleton *et al.* (1992) by permission of the publishers, TIBTECH, Elsevier Science Publishers UK

adhere to other parts of the equipment. In addition, viability may be impaired by physical conditions or inactivated by impact. Nevertheless they are extremely useful in the assessment of the microbiological quality of air.

For more detailed information on air sampling see Bennet *et al.* (1991), Ashton and Gill (1992) and Hambleton *et al.* (1992).

Containers and catering equipment

The swab rinse method for crockery, cartons and containers

Dip a swab in sterile 0.1% peptone water and rub it over the surface to be tested, for example the whole of the inside of the cup or glass or the whole surface of a plate. Use one swab for five such articles. Use one swab for a predetermined area of a cutting table, chopping board, etc. Swab both sides of knives, ladles, etc. Return this swab to the tube and swab the same surfaces again with another, dry, swab.

To the tube containing both swabs add 10 ml of 0.1% peptone water. Shake and stand for 20–30 min. Do plate counts with 1.0 and 0.1 ml amounts using yeast extract agar. Divide the count/ml by 5 to obtain the count per article.

Inoculate three tubes of single-strength MacConkey broth with 0.1-ml amounts.

There appear to be no current standards for the bacteriology of crockery and utensils used by the public in catering establishments but during the 1940s the USPHS required counts to be less than 100 per article, with no coliform bacilli.

Alginate swabs and drop counts

Proceed as for surface drop counts (p.153). Do Miles and Misra counts (p.154) with 6 drops from each pair of swabs on one blood agar and one MacConkey plate. Two colonies or less per 6 drops on the blood agar plate is within the USPHS limits. This method also allows the organisms to be identified. Apart from coliform bacilli, the presence of respiratory organisms such as viridans and salivary streptococci, staphylococci and neisserias is evidence of inefficient sanitation.

Rinse method

For churns, bins and large utensils

Add 500 ml of 0.1% peptone water to the vessel. Rotate the vessel to wash the whole of the inner surface and then tip the rinse into a screw-capped jar.

Do counts with 1.0- and 0.1-ml amounts of the rinse in duplicate. Incubate one pair at 37°C and the other at 22°C for 48 h.

Take the mean of the 37°C and 22°C counts/ml and multiply by 500 to give the count per container.

For milk churns, counts of not more than 50 000 cfu per container are regarded as satisfactory, between 50 000 and 250 000 as fairly satisfactory and over 250 000 as unsatisfactory.

For milk bottles, soft drink bottles and jars

If bottles are sampled after they have been through a washing plant in which a row of bottles travels abreast through the machine, test all those in one such row. This shows if one set of jets or carriers is out of alignment. In any event, examine not less than six bottles. Cap or stopper them immediately. To each bottle add 20 ml of 0.1% peptone water. Close with sterile rubber bungs and roll the bottles on their sides so that all of the internal surface is rinsed. Leave them on their sides and roll at intervals for half an hour.

Pipette 5 ml from each bottle into each of two petri dishes. Add 15–20 ml of molten yeast extract agar (at 45–50°C), mix and incubate one plate from each bottle at 37°C and one at 22°C for 48 h. Pipette 5 ml from each bottle into 10 ml of double-strength MacConkey broth and incubate at 37°C for 48 h.

Take the mean of the 37°C and 22°C plate count and multiply by 4 to give the count per bottle. Find the average of the counts, omitting any figure that is 25 times greater than the others (indicating a possible fault in that particular line).

The figure obtained is the average colony count (ACC) per container. Milk bottles giving average colony counts of 200 cfu or less are satisfactory; counts from 200 to 600 cfu are regarded as fairly satisfactory, but over 600 as unsatisfactory.

Coliform bacilli should not be present in 5 ml of the rinse.

Membrane filter method

Examine clean bottles and jars by passing the rinse through a membrane filter. Place the membrane on a pad saturated with an appropriate medium, e.g.

double-strength tryptone soya broth for total count. Incubate at 35°C for 18–20 h. Stain the membrane with methylene blue to assist in counting the colonies under a low-power lens. Examine another membrane for coliforms using MacConkey membrane broth or membrane enriched lauryl sulphate broth. Incubate at 35°C for 18–24 h. Staining to reveal colonies is unnecessary. Subculture suspect colonies into lactose peptone water and incubate at 37°C for 48 h to confirm gas production.

Roll-tube method for bottles

Use nutrient or similar agar for most bacteria, deMan, Rogosa and Sharp (MRS) medium for lactobacilli, reinforced clostridial medium (RCM) for clostridia and buffered yeast agar for yeasts. Increase the agar concentration by 0.5%. Use the roll-tube MacConkey agar for coliforms.

Melt the medium and cool it to 55°C. To 1-quart or 1-litre bottles add 100 ml of medium, and to smaller bottles proportionally less. Stopper the bottles with sterile rubber bungs and roll them under a cold water tap to form a film of agar over all the inner surface (see Roll-tube counts, p.153). Incubate vertically and count the colonies. For lactobacilli and clostridia, replace the bung with a cotton-wool plug and incubate anaerobically.

Vats, hoppers and pipework

Large pieces of equipment are usually cleaned in place (CIP). Test flat areas by agar contact or swabbing. Take swab samples of dead ends of pipework and crevices. The dairy industry has devised a simple and effective method for testing filling equipment. Sample the first, 100th and 200th container. The first will contain any residual bacteria not killed by CIP treatment. If this has not been effective, sample 1 will give a higher count than samples 100 and 200. If all three samples are satisfactory then the cleaning was efficient.

Examination of sink waters, cloths, towels, etc.

To demonstrate unhygienic conditions and the necessity for frequent changes of washing water and cloths, examine by agar contact or sample and examine as follows.

Take 100 ml of washing or rinse water by immersing a water sample bottle (containing sodium thiosulphate in case hypochlorites are used in washing-up) into the sink and allow it to fill. Stopper and cool under a tap. Do plate and coliform counts. Test at the beginning and at intervals during the washing-up.

Spread a wash-cloth or drying cloth over the top of a screw-capped jar of known diameter. Pipette 10 ml of 0.1% peptone water on the cloth so that the area over the jar is rinsed into it. Do plate counts and compare with freshly laundered cloths. If a destructive technique is possible, cut portions of cloths, sponges or brushes with sterile scissors and add to diluent.

Mincers, grinders, etc.

After cleaning, rinse with 500 ml of 0.1% peptone water. Treat removable parts separately by rinsing them in a plastic bag in diluent. Do colony count on rinsings as for milk bottles.

Chopping blocks

Organisms are sometimes deeply embedded in these blocks. Sample by taking scrapings from representative areas, e.g. 100 cm^2, with a sterile scalpel. Disperse the scrapings in warm diluent and do counts.

Biotechnology plant

To detect the release of aerosols sample the air by one of the methods described on p.264. To detect leakage around pipe joins and valves use the swab rinse method (p.265). Test surface contamination with Rodac plates (p.263). For more details on sampling and monitoring in process biotechnology see Bennet (1991) and Tuijnenburg-Muijs (1992).

Pharmaceutical, etc. 'clean rooms'

Use one of the mechanical air samplers (p.264), preferably placed outside the room and sampling through a probe. There are several different levels of clean rooms and the standards, i.e. the number of particles per unit volume, are set out in a British Standard (BSI, 1976). Settle plates are also useful but do not give comparable results.

Humidifier fever, etc.

Use one of the mechanical samplers (p.264) to test the air in premises where there are suspected cases of humidifier fever, sick building syndrome, extrinsic allergic alveolitis, etc. Do counts for both fungi and bacteria.

Sample the water and the biofilm in the humidifier reservoir for bacteria and fungi (pp.275, 276). See also Collins and Grange (1990), Flannigan *et al.* (1991) and Ashton and Gill (1992).

References

Ashton, I. and Gill, F. S. (1992) *Monitoring Health Hazards at Work*, 2nd ed, Blackwells, Oxford

Bennet, A. M., Hill, S. E., Benbough, J. E. and Hambleton, P. (1991) Monitoring safety in process biotechnology. In *Genetic Manipulation: Techniques and Applications* (eds. J. M. Grange, A. Fox and N. L. Morgan), Society for Applied Bacteriology Technical Series No. 28, pp.361–376

BSI (1976) BS 5295: '*Environmental cleanliness in enclosed spaces. Part 1. Specification for controlled environment clean rooms, work stations and clean air devices*, British Standards Institution, London

Collins, C. H. and Grange, J. M. (1990) *The Microbiological Hazards of Occupations*, Occupational Hygiene Monograph No. 17, Science Reviews, Leeds

Deans, J. S. (1991) *WSL Report*, Warren Springs Laboratory, Stevenage

Flannigan, B., McCabe, M. E. and McGarry, F. (1991) Allergenic and toxigenic microorganisms in houses. In *Pathogens in the Environment* (ed. B. Austin), Society for Applied Microbiology Symposium Series No. 20, Vol. 20, pp.S61–S74

Hambleton, P., Bennett, A. M., Leaver, J. and Benbough, J. E. (1992) Biosafety monitoring devices for biotechnology processes. *Trends in Biotechnology*, **10**, 192–199

Leaver, T., Salusbury, T. T. and Stewart, I. W. (1987) *State of the Art Report No. 1*, Industrial Biosafety Project Warren Springs Laboratory, Stevenage

May, K. R. (1945) The Cascade Impactor. An instrument for sampling coarse aerosols. *Journal of the Science Institute*, **22**, 187–195

Tuijnenburg-Muijs, G. (1992) Monitoring and validation in biotechnological processes. In *Safety in Industrial Microbiology and Biotechnology* (eds. C. H. Collins and A. J. Beale), Butterworth-Heinemann, Oxford, pp.214–238

Chapter

20

Water microbiology

The quality of water, for both drinking and recreation purposes, is now a matter of national and international concern. The European Commission has issued a council directive relating to the quality of water supplies (EC, 1980) and the UK has embodied this in its Water Act (1989) and the Water Supply (Water Quality) Regulations (1989). The UK Department of the Environment has also published guidance on water quality (DoE, 1989). In the UK bottled water is subject to the Natural Mineral Water Regulations (1985). For a review of these and the associated issues see Lee (1991).

Officially approved methods for the bacteriological examination of water are given by the UK Department of Health (DHSS, 1985) and in the USA by the American Public Health Association (APHA, 1986).

The UK Public Health Laboratory Service has a Water Microbiology External Quality Assessment Unit.

In relation to public health the principal tests applied to water are the viable plate count and those for coliform bacilli, faecal coli, faecal streptococci and sulphite-reducing clostridia. The terms used for the microorganisms may be defined as follows:

Coliform bacilli are members of the Enterobacteriaceae and include the genera *Citrobacter, Enterobacter, Escherichia, Hafnia, Klebsiella* and *Serratia*. These grow at 37°C and possess a β-galactosidase enzyme.

Faecal coli, also known as thermotolerant coli refers to *Escherichia coli*, which grows and produces indole at 44°C.

Faecal streptococci are members of the genus *Enterococcus*, and include *E. faecalis, E. faecium* and *E. durans*. They grow at 10°C and 45°C, in the presence of 40% bile, 6% NaCl, and on standard azide media, and hydrolyse aesculin.

Sulphite-reducing clostridia refer to *Clostridium perfringens*.

Sampling

Water samples are usually collected by environmental health officers and water engineers, who use sterile 300 ml or 500 ml bottles supplied by the laboratory. Plastic is replacing glass bottles because of concern about glass in food preparation and recreational areas.

For samples of chlorinated water the bottles must contain sodium thiosulphate (0.1 ml of a 1.8% solution per 100 ml capacity) to neutralize chlorine.

Viable plate counts

These are now required under UK regulations and EC Directives and are standard procedures in the USA. The technique is described in Chapter 10. Yeast extract agar is used and tests are done in duplicate with undiluted and serially diluted samples. One set is incubated at 20–22°C for 3 days and the other at 37°C for 24 hours.

The target values are < 100 cfu/ml at 22°C and < 10 cfu/ml at 37°C.

Coliform test: MPN method with Minerals Modified Glutamate broth

Select the range according to the expected purity of the water:

Mains chlorinated water	*A* and *B*
Piped water, not chlorinated	*A*, *B* and *C*
Deep well or borehole	*A*, *B* and *C*
Shallow well	*B*, *C* and *D*
No information	*A*, *B*, *C* and *D*

A: 50 ml of water to 50 ml of double-strength broth.
B: 10 ml of water to each of five tubes of 10 ml of double-strength broth.
C: 1 ml of water to each of five tubes of 5 ml of single-strength broth.
D: 0.1 ml of water to each of five tubes of 5 ml of single-strength broth.

Incubate at 35–37°C and note the numbers of tubes showing acid and gas at 48 h. Tap any tubes showing no gas. A bubble may then form in the Durham's tube. Consult the MPN tables (Tables 10.1–10.3) and read the most probable number of *presumptive* coliform bacilli/100 ml of water. Small amounts of gas occurring after 48 h in *presumptive* tubes are disregarded unless the presence of coliform bacilli is confirmed by plating.

From each tube showing acid and gas, inoculate a tube of brilliant green bile salt broth and a tube of peptone water. Incubate these at 44°C for 24 h in a reliable water-bath (Eijkman test) along with controls of known strains of *E. coli* (which grows at 44°C) and *K. aerogenes* (which does not). Plate also from positive tubes on MacConkey agar and nutrient agar.

Observe gas formation at 44°C and test the peptone–water culture for indole. Do oxidase test on growth from nutrient agar. *Aeromonas* spp., which may give acid and gas in lactose broth, are positive. Only *E. coli* produces gas *and* indole at 44°C.

Read the most probable numbers of *E. coli* ('faecal coli') from Tables 10.1–10.3.

Translation from the MPN tables of the 44°C positive tubes sometimes causes difficulty. Remember that the organisms cultured from any positive 37°C tube and grown at 44°C represent coliforms cultured from the volume of water placed in the 37°C tube. For example:

	50 ml	*10 ml*	*1 ml*	*MPN/100 ml*
Tubes positive at 37 °C	1	2	2	10 'presumptive coli'
Tubes positive at 44 °C	1	1	0	3 *E. coli*

For further investigation subculture colonies from the MacConkey plate for biochemical tests, e.g. with an API kit (see pp.108 and 306). The IMViC (indole, methyl red, Voges-Proskauer) tests are no longer used in water microbiology.

Acid and gas in MMG medium, as in MacConkey broth, may occasionally be due to spore bearers, e.g. *C. perfringens* at both 37 and 44°C. These organisms do not grow in brilliant green broth or on the MacConkey plate.

Most raw waters in the UK showing acid and gas do in fact contain coliform bacilli but in about 5% of chlorinated waters acid and gas are caused by *C. perfringens*.

The target levels for coliforms and *E. coli* are absence from 100 ml.

Coliform test: membrane filter method

Advantages of using membrane filter techniques for waters

(1) Speed of obtaining results.
(2) Saving of labour, media, glass and cost of materials if the filter is washed and re-used.
(3) Sample can be filtered on site, if the filter is placed on transport medium and posted to the laboratory, thus avoiding delay in transporting the sample.
(4) Organisms can very easily be exposed to pre-enrichment media for a short time at an advantageous temperature.

Disadvantages of using membrane filter techniques for waters

(1) There is no indication of gas production (some waters contain large numbers of non-gas producing lactose fermenters capable of growth in the medium).
(2) Membrane filtration is unsuitable for waters with high turbidity and low counts because the filter will become blocked before sufficient water can pass through it.
(3) Large numbers of non-coliform organisms capable of growing on the medium may interfere with coliform growth.

If large numbers of water samples are to be examined and much field work is involved the membrane method is undoubtedly the most convenient. The booklets published by Millipore and Gelman give valuable advice and information about all aspects of the application of this technique.

Pass two separate 100-ml volumes of the water sample through 47-mm diameter membrane filters. If the supply is known or is expected to contain more than 100 coliform bacilli/100 ml, use 10 ml of water diluted with 90 ml of quarter-strength Ringer's solution.

Place sterile Whatman No. 17 absorbent pads in sterile petri dishes and pipette 2.5–3 ml of enriched lauryl sulphate broth over the surface. Place a membrane face up on each pad. For both chlorinated and unchlorinated samples incubate (in dual temperature water-baths) one membrane at 30°C for 4 h followed by 37°C for 14 h and one at 30°C for 4 h followed by 44°C ± 0.2°C for 14 h.

Counting

Count the yellow colonies only and report as *presumptive* coliform and *E. coli* count/100 ml of water. Membrane counts may be higher than MPN counts because they include all organisms producing acid, not only those producing acid and gas. *C. perfringens* does not grow.

Confirmatory tests for *Escherichia coli*
Subculture colonies to brilliant green bile broth and tryptone water for gas production and indole test. Incubate broth at 44 ± 0.2°C overnight. 4-Methyl umbelliferyl-β-D-gluconate (MUG) may be added to the tryptone water to give an additional test for β-glucuronidase activity which is positive only for *E. coli* (*ca.* 90% of strains) and some shigellas. MUG is hydrolysed to give a fluorescent compound, detected by exposure to UV light. The indole reagent may then be added. The API RAPIDEC coli uses glucopyranosiduronic acid as substrate. This is hydrolysed to a yellow compound by β-glucuronidase. Tests for indole and β-galactosidase are also used in this kit and improve the quality of the results.

Membrane filter: US method

The methods are very similar to those above, but the filter is rinsed with three volumes of 20–30 ml of sterile buffered water before the sample is passed through it. Membrane agar medium (MMA) or membrane fluid (MMB) media are used. Pre-enrichment, where necessary, is effected on pads saturated with lactose broth at 35°C for 2 h.

Full details are given in the Standard Methods (APHA, 1986).

Coliform test: US method (APHA, 1986)

In the US Standard Method, 15 fermentation tubes, each containing 20 ml of 0.5% lactose broth are used. They are inoculated with 5 × 10 ml, 5 × 1 ml and 5 × 0.1 ml of water sample. These are incubated at 35°C and examined for gas production after 24 h and 48 h. Gas within 48 h is presumptive evidence of coliform bacilli.

Confirmatory test
Tubes showing gas are subcultured on eosin methylene blue (EMB) agar, incubated at 35°C for 24 h and examined for typical colonies of *E. coli*. If atypical colonies are seen, the 'completed test' is carried out.

Completed test
Several colonies from the EMB plate are subcultured into lactose broth fermentation tubes and on a nutrient agar slope. Both are incubated at 35°C for 24 h. Gas in the broth and a Gram-negative non-sporing rod on the slope is evidence of coliform bacilli.

Faecal coli test
Tubes showing gas are subcultured into EC or similar medium and incubated at 44 ± 0.2°C for 48 h to test for gas production, which indicates faecal coli. Colonies from solid medium are subcultured to lactose broth and if gas is produced in these they are further subcultured into EC broth as above. For full technical methods and directions, see Standard Methods (APHA, 1986).

Faecal streptococci in water

These organisms are useful indicators when doubtful results are obtained in the coliform test. They are more resistant than *E. coli* to chlorine and are therefore useful when testing repaired mains. Group D organisms only are significant.

MPN method

Use one of the azide broths, e.g. azide glucose broth or Enterococcus Presumptive Broth.

Add 50 ml of water to 50 ml of double-strength medium.

Add 10 ml of water to each of five tubes of 10 ml of double-strength broth.

Add 1 ml of water to each of five tubes of 5 ml of single-strength broth.

Incubate at 37°C for 48 h. Subculture any tubes showing acid production to tubes of single-strength medium and incubate at 44–45°C for 18 h. Record tubes showing acid and consult the MPN tables (Tables 10.1–10.3). Confirm by microscopie examination for short-chain streptococci. Subculture each presumptive positive tube to ethyl violet azide broth and incubate at 37°C for 24–48 h. Turbidity and a purple-stained button of growth at the bottom of the tube indicate enterococci.

Membrane method

Always use a new membrane when testing for faecal streptococci because these organisms are not always removed when membranes are cleaned. Pass 100 ml of water through a membrane filter and place the filter on a pad of one of the enterococcus membrane broths or a plate of membrane enterococcus agar or Slanetz and Bartley agar. Incubate at 37°C for 4 h and then at 44–45°C for 44 h. All red or maroon colonies are presumptive positives. Carefully remove the filter and place colony-side down a plate of aesculin bile agar to imprint the colonies.
Incubate at 37°C for 6 h. A black zone appears under colonies of faecal streptococci.

The target level for faecal streptococci is absence from 100 ml.

US methods for faecal streptococci

These methods (APHA, 1986) are very similar to the above. In the membrane method, suspected colonies are subcultured for catalase tests and into broth incubated at 45°C for confirmation.

Sulphite-reducing clostridia

MPN method

Heat the sample to 75°C in a water-bath and hold it at this temperature for 10 min. Culture as follows in reinforced clostridial medium containing 70 μg/ml polymyxin to inhibit facultative anaerobes:

Add 50 ml of the sample to a 50 ml bottle of double strength medium, 10 ml to each of five 10 ml tubes of double strength medium, and 1 ml to each of five 10 ml tubes of single strength medium.

Fill the bottles almost to the neck with single strength medium to exclude as much air as possible. Cap and incubate at 37°C for 48 h. Tubes showing blackening are presumptive positives but other clostridia may give this reaction.

Confirm by subculture in purple milk medium. Incubate at 37°C overnight and record tubes showing stormy fermentation. Consult the MPN tables (Tables 10.1–10.3) and identify as on p.402.

Membrane method

Pass 100 ml of the heated sample (see above) through a 47 mm 0.45 μm filter. Place the filter face upwards on the surface of membrane clostridial agar. Pour 20 ml of the same medium, cooled to 50°C, over the surface and when this has solidified incubate the plates at 44°C anaerobically. Count the black colonies with haloes. These are probably *Clostridium perfringens*. If too many are present all the medium will be blackened.

There is a British Standard (BSI, 1993) for determination of sulphite-reducing anaerobes. The target levels appear to be < 20 per ml.

Pathogens in water

Methods for the detection of certain bacterial pathogens are given below, but other pathogens, outside the scope of this book, such as viruses (e.g. Norwalk-like agents, rotaviruses, hepatitis A virus) and parasites (e.g. cryptosporidium, giardia, amoebae) should not be overlooked.

It is pointless examining water samples for pathogens unless there is evidence of pollution. Pass several 100–200 ml volumes of water through 47 mm 0.2 μm membrane filters.

Salmonellas

Cut the membranes into pieces, place them in 10 ml of a pre-enrichment medium, e.g. buffered peptone water containing 5 mg/ml cefsulodin. Vortex mix to dislodge the organisms and incubate at 37°C overnight. Add portions of this to selenite broth for incubation at 30°C and Rappaport medium for incubation at 42°C, both overnight. Plate these on DCA, XLD and brilliant green agar. Proceed as on p.317.

Cholera and other vibrios

Cut the membranes into pieces, place them in 10 ml of alkaline peptone water and vortex mix to dislodge the vibrios. Incubate at 37°C and subculture to TYCBS medium at 4, 8 and 12 h. Incubate at 37°C overnight and proceed as on p.295.

Legionellas

Legionellas may be present in the water in cooling towers, humifier tanks, etc. and also in the biofilm which is usually attached to the walls of the containers. Examination of water alone may give a false impression.

Membrane filtration is the usual method of testing the water for legionellas.

Examine several samples. If the water is cloudy pre-filter it through glass fibre pads. Pass the sample through 47 mm 0.45 μm membrane filters. Cut the filters into small pieces and place them in 10 ml of peptone saline diluent in a screw-capped bottle. Vortex mix for 5 min to detach the microorganisms. Decant the diluent and centrifuge it for 30 min at 3500 rev/min.

Inoculate 118 BCYE or other legionella selective media with 100 μl amounts of deposit. Incubate under high humidity at 37°C. Include stock control cultures.

If the samples are highly contaminated use one of these decontamination methods:

(1) Heat the deposit at 50°C for 30 min and then culture 100 μl amounts.
(2) Suspend 0.1 ml of the deposit in 0.9 ml of acid buffer:

KCl (0.2 mol/l, 14.91 g/l)	25 ml
HCl (0.2 mol/l, 17.2 ml conc. acid/l)	3.9 ml
Sterilize by autoclaving	

Plate out 100 μl amounts every min for 5 min.

Identify legionellas as described on p.350. Centrifuge biofilm suspensions and treat the deposit as described for contaminated samples (above).

Campylobacters

Campylobacters may not be retained by membrane filters, especially those of pore size >0.45 μm. Place the membrane filters face upwards on plates of campylobacter selective media and incubate under microaerophilic conditions at 37°C overnight. Remove the membrane to another plate and continue to incubate both plates for another 48 hours. Proceed as on p.341.

Pseudomonas aeruginosa

Place the membranes on pads soaked with modified King's A broth in a petri dish and incubate at 37°C for 48 h. Examine under UV light. Confirm fluorescent colonies by subculture on cetrimide agar incubated at 42°C. Only *P. aeruginosa* grows on this medium at that temperature.

Microfungi and streptomycetes

These organisms cause odours and taints and often grow in scarcely used water pipes, particularly in warm situations, e.g. basements of large buildings where the drinking water pipes run near to the heating pipes.

Sample the water when it has stood in the pipes for several days. Regular running or occasional long running may clear the growth.

Centrifuge 50 ml of the sample and plate on Rose Bengal agar and malt agar medium. Add 100 μg/ml of kanamycin to the malt medium to suppress most eubacteria.

Alternatively, pass 100 ml or more of the sample through a membrane filter and apply this to a pad soaked in liquid media of similar composition.

Bottled natural waters

These may be 'still' or carbonated. The tests used, with guidance levels at the point of sale and within 12 h of bottling*, are:

* The total viable count levels are permitted to increase after bottling to allow for normal multiplication in untreated waters.

Total viable count, 20–22°C, 72 h	< 100 cfu/ml
Total viable count, 37°C, 24 h	< 20 cfu/ml
P. aeruginosa	absent from 250 ml
Sulphite-reducing clostridia	absent from 50 ml

Environmental mycobacteria have been found in moderate numbers in some bottled waters.

Swimming pools

Use a commercial kit to determine the chlorine content. This may be more useful than bacteriological examination but, if the latter is required, take 300 ml samples in thiosulphate bottles (p.270) from both the deep and shallow ends. Do plate counts and test for coliforms, faecal coli and pseudomonads. In satisfactory pools target levels are as follows:

Total viable count, 37°C, 24 h	< 100 cfu/ml
Coliforms and *E. coli*	absent from 100 ml
P. aeruginosa	absent from 100 ml

Staphylococci are more resistant than coliforms to chlorination. They tend to accumulate on the surface of the water in the 'grease film'. To find staphylococci take 'skin samples'. Open the sample bottle so that the surface water flows into it. Add 20-ml volumes to each of five tubes containing 20 ml of Robertson's meat medium plus 10% of sodium chloride. Incubate overnight and plate on one of the staphylococcal media.

Spa, jacuzzi and hydrotherapy pools

These have been associated with human infections with pseudomonads and legionellas. As they are maintained at 37°C the chlorine (or bromine) levels should be closely monitored.

Do plate counts and test for coliforms, *E. coli* and pseudomonads. Target levels are:

Total viable count, 37°C, 24 h	< 10 cfu/ml
Coliforms and *E. coli*	absent from 100 ml
P. aeruginosa	absent from 100 ml

Pools with viable counts > 100 cfu/ml need further investigation.

Bathing beaches and other recreational waters

Test undiluted and serially diluted samples for coliforms, faecal coliforms and faecal streptococci.

The EU guide levels are as follows (per 100 ml of water):

	Excellent	*Satisfactory*	*Poor*
Total coliforms	500 or less	500–10 000	> 10 000
Faecal coli	100 or less	100–2 000	> 2000
Faecal streptococci	100 or less	100–2 000	> 2000

Other organisms of public health importance

These are not considered in this book. Some references are cited.
Viruses (especially rotaviruses) have been implicated in outbreaks associated with recreational waters (see Sellwood, 1991).
Cyanobacteria Blue-green algae produce a variety of dangerous toxins. Serious illnesses have been associated with their presence in recreational and potable waters and fish (see NRA, 1990).
Cryptosporidium Oocysts of this organism may be present in recreational and potable waters and offer a serious hazard to health (see Badenoch, 1990).

Water testing in remote areas

Problems arise in the bacteriological examination of water in remote parts of tropical and subtropical countries. Some of the culture media described above are expensive or may not be available. It may be necessary to use simpler media such as MacConkey and litmus lactose broths and agars. These will affect results and standards. In addition, the 44°C and 45°C tests may be unreliable as some indigenous, non-faecal organisms may grow at those temperatures.

Water samples collected in remote areas and transported over long distances to laboratories are unlikely to give reliable results. A portable kit, the Del Agua, devised for the WHO project for drinking water in rural areas, will give reliable faecal coli counts, turbidity and free chlorine levels. It is available from Del Agua Ltd (Guildford, UK). See Lloyd and Helmer (1991).

Sewage

Effluents and sludge are usually monitored microbiologically to ensure that treatment has reduced the loads of salmonellas, listerias and campylobacters, all of which can survive in water, and if untreated sludge is spread on land.

References

APHA (1986) *Standard Methods for the Examination of Water and Waste-water*, 16th edn, American Public Health Association, Washington

Badenoch, J. (1990) *Cryptosporidium in Water Supples.* Report of a group of experts, HMSO, London

BSI (1993) BS EN 2641: Parts 1 and 2. *Water Quality — Detection and Enumeration of the Spores of Sulfite-Reducing Anaerobes* [*Clostridia*], British Standards Institution, London

DHSS etc, (1985) *The Bacteriological Examination of Water Supplies*, Report No 71, HMSO, London

DoE (1989) Guidance on Safeguarding the Quality of Public Water Supplies, Department of the Environment, London

EC (1980) European Commission. Council Directive relating to the quality of drinking water, 80/78/EEC, Brussels

Lee, R. J. (1991) The microbiology of drinking water. *Medical Laboratory Sciences*, **48**, 303–313

Lloyd, B. and Helmer, R. (1991) *Surveillance of Drinking Water in Rural Areas*, Longman, London

Natural Mineral Water Regulations (1985) SI 71, HMSO, London
NRA (1990) *Toxic Blue-Green Algae*, National Rivers Authority, London
Sellwood, J. (1991) Human viruses in water, *Current Topics in Clinical Virology*, Public Health Laboratory Service, London.
Water Act (1989) HMSO, London
Water Supply (Water Quality) Regulations (1989) SI 1147, HMSO, London

Chapter

21

Key to common aerobic non-sporing Gram-negative bacilli

Aerobic Gram-negative non-sporing bacilli that grow on nutrient and usually on MacConkey agars are frequently isolated from human and animal material, from food and from environmental samples. Some of these bacilli are confirmed pathogens; others, formerly regarded as non-pathogenic, are now known to be capable of causing human disease under certain circumstances, e.g. in 'hospital infections', after chemotherapy or treatment with immunosuppressive drugs; others are commensals; many are of economic importance.

The organisms described in this and Chapters 22 and 23 have received much attention from taxonomists in recent years. Names of species and genera have been changed and new taxa created. The nomenclature used here is that in common use at the time of going to press, mostly as in the current edition of *Cowan and Steel's Manual* (1993) and *Stedman's Bergey's Bacterial Words* (1993).

Table 21.1 lists the general properties of aerobic Gram-negative rods that grow on nutrient agar.

Table 21.1 Genera of Gram-negative rods that grow on nutrient agar

	O/F	*Oxidase*	*Arginine hydrolysis*	*Gelatin liquefaction*	*Motility*
Pseudomonas	O/A	+	v	v	+
Acinetobacter	O/N	−	−	−	−
Alcaligenes	N	+	−	−	+
Flavobacteria	O	+	−	+	−
Chromobacterium	O	+	−	+	+
Janthinobacterium	F	+	+	+	+
Aeromonas	F	+	+	+	+
Listonella	F	+	+	v	+
Plesiomonas	F	+	+	−	+
Vibrio	F	+	+	−	+
Shewanella	N	+	−	+	+
Enterobacteria	F	−	v	v	v
Pasteurella	F	+	−	−	−
Yersinia	F	−	−	−	v[a]
Moraxella	N	+	−	−	v
Bordetella	N	v	−	−	v

O, oxidative; A, alkaline; N, no reaction; F, fermentative; v, variable
[a] Some species are motile at 22 °C

References

Cowan and Steel's Manual for the Identification of Medical Bacteria (1993) 3rd edn (eds G. I. Barrow and R. K. A. Feltham), Cambridge University Press, Cambridge

Stedman's Bergey's Bacterial Words (1993) (eds J. G. Holt, M. A. Bruns, B. J. Caldwell and C. D. Pease), Williams and Wilkins, Baltimore

Chapter

22

Pseudomonas

There are approximately 27 species in this genus but only the common ones can be described here. They are aerobic Gram-negative non-sporing rods about 3 μm × 0.5 μm, which are motile by polar flagella, are oxidase positive, utilize glucose oxidatively (or produce no acid) and do not produce gas. Some produce a fluorescent pigment.

They commonly occur in soil and water. Some species are recognized human and animal pathogens but some others, formerly regarded as saprophytes and commensals, have been incriminated as opportunist pathogens in hospital-acquired infections and have colonized distilled water supplies, soaps, disinfectants, intravenous infusions and other pharmaceuticals. Species colonizing swimming pools have been associated with ear infections.

Isolation and identification

Inoculate nutrient agar, MacConkey agar, nutrient agar containing 0.1% cetrimide, Pseudomonas agar with supplements and King's Medium A. Incubate at 20–25°C (food, environment, etc.) or 35–37°C (animal material). Temperature and length of incubation are crucial for accurate identification.

Colonies on nutrient agar are usually large (2–4 mm), flat, spreading and pigmented. A greenish yellow or bluish yellow fluorescent pigment may diffuse into the medium. Occasionally, melanogenic strains are encountered. A brown pigment is formed around the colonies. Mucoid (encapsulated) strains are sometimes found in clinical material, particularly in sputum from patients with cystic fibrosis, and may be confused with klebsiellas.

Do oxidase test, reaction in Hugh and Leifson medium, inoculate Thornley's arginine broth (Moeller's method is too anaerobic), MacConkey agar and test for gelatin liquefaction (5 days' incubation). Inoculate nutrient broths and incubate at 4 and 42°C. Use a subculture from a smooth suspension for this test. Examine growth in King's A medium with a Wood's lamp for fluorescence. If this is seen test for reduction of nitrate, urease and acid production from glucose and mannitol in ammonium salt sugars and gelatin liquefaction (Table 22.1).

The API 20NE kit, designed for non-fermenting Gram-negative rods, is useful for identifying pseudomonads. Other kits include Minitek NF and Oxiferm.

Pseudomonads are motile (occasionally non-motile strains may be met), oxidative, non-reactive or produce alkali in Hugh and Leifson medium. The oxidase test is usually positive except for *P.* (now *Xanthomonas*) *maltophilia, P. cepacia, P. stutzer* and *P. alcaligenes*. Those pseudomonads which give a positive arginine test are common and give a positive arginine test more rapidly

Table 22.1 Fluorescent pseudomonads[a]

	Growth at		NO$_3$	Urease	Acid from		Gelatin liquefaction
	4°C	42°C			Mannitol	Maltose	
P. aeruginosa	–	+	+	+	+	–	+
P. fluorescens	+	–[a]	+	–	+	+	+
P. putida	+[a]	–	–	v	–	–	–

[a] Most strains; v, variable

than other Gram-negative rods. This test is very useful for recognizing non-pigmented strains. *Alcaligenes* and *Vibrio* spp. do not hydrolyse arginine. Pseudomonads vary in their ability to liquefy gelatin and grow at 4 and 42°C (see Table 22.2).

Table 22.2 Species of *Pseudomonas*

	Growth on MacConkey	Arginine hydrolysis	O/F	Gelatin liquefaction	Growth at	
					4°C	42°C
Oxidizers						
P. aeruginosa	+	+	O	+	–	+
P. fluorescens	+	+	O	v	+	–
P. putida	+	+	O	–	v	–
P. cepacia	+	–	O	v	–	–
P. maltophilia[a]	+	–	v	+	–	–
P. stutzeri	v	v	N	v	v	+
P. paucimobilis	–	–	v/N	–	–	–
Alkali producers						
P. alkaligenes	+	–	A	–	–	+
P. pseudo-alkaligenes	+	+	A	–	–	+
Group 3 pathogens						
P. mallei	–	+	O	v	–	–
P. pseudomallei	+	+	O	–	–	+

O, oxidative; N, no reaction; A, alkaline; v, variable
[a] Now *Xanthomonas maltophilia*

Some fluorescent strains produce an opalescence (egg yolk reaction) on Willis and Hobbs' medium and also a lipase (pearly layer). To demonstrate the latter, flood the plate with saturated copper sulphate solution; the fatty acids released by lipolysis give a greenish blue precipitate of copper soaps.

Some varieties associated with food spoilage are psychrophiles and may be lipolytic. Brine-tolerant strains and phosphorescent strains are not uncommon.

Species of Pseudomonas

The pseudonads may be divided conveniently into four groups: fluorescent species (Table 22.1); oxidisers (Tables 22.1, 22.2); alkali producers (Table 22.2); and dangerous pathogens (Tables 22.2, 22.3).

For detailed identification methods for medically-important pseudomonads see Costas *et al.* (1992) and *Cowan and Steel's Manual* (1993).

Table 22.3 *Pseudomonas* spp. in Group 3 (Dangerous Pathogens)

	Growth on MacConkey	*Oxidase*	*O/F*	*Motility*	*Citrate utilization*
Ps. mallei	–	–	N	–	–
Ps. pseudomallei	+	+	O	+	+

N, no change; O, oxidative

P. aeruginosa

Typical colonies on agar medium after incubation for 18–24 h at 25–30°C, are large, flat, spreading and irregular, greyish green in colour. A variety of other colonial forms, including mucoid, dwarf and non-pigmented have been observed. The greenish pigment diffuses into the medium. Broth cultures are blue-green in colour. Two pigments are formed, pyocyanin and fluorescein; both are soluble in water but only pyocyanin is soluble in chloroform. Arginine is hydrolysed rapidly. Gelatin is liquefied. Growth takes place at 42°C but not at 4°C. The egg yolk reaction is negative. This organism is very resistant to antibiotics except aminoglycosides, some recent β-lactams (e.g. ceftazidime), other β-lactams (e.g. carbenicillin, piperacillin) and fluoroquinolones.

It is a saprophyte, found in soil and water, causes spoilage of foods, including 'blue milk', is pathogenic for humans and animals ('blue pus'), often as a secondary infection, and is often found in clinical material, e.g. in cases of cystic fibrosis.

It is an important agent in cross-infection in hospitals and strains may be typed by pyocin production and antisera. This is usually done at reference laboratories.

P. fluorescens

Colonies on agar and its biochemical properties are similar to those of *P. aeruginosa* but only one pigment (fluorescein) is produced. Growth occurs at 4°C. Some strains do not grow at 37°C; none grow at 42°C. The egg yolk reaction is positive. This organism is a saprophyte found in soil, water and sewage. It is a food spoilage organism. It may gel UHT milk if this is stored above 5°C. It is important in patients with burns and cystic fibrosis. It is particularly resistant to antibiotics, including imipenem.

P. putida

Colonies resemble those of *P. aeruginosa* but only fluorescein is formed. It is a psychrotroph that may grow at 22°C and may grow at 37°C. There are two biotypes; one grows at 4°C, the other does not; both hydrolyse arginine. The egg yolk reaction is negative, gelatin is not liquefied and old cultures have a marked odour of trimethylamine (bad fish). It is an important fish pathogen and fish spoilage organism and has been isolated from human material.

P. (Xanthomonas) maltophilia

Colonies resemble those of *P. aeruginosa* but a yellow or brown diffusible pigment may be produced. It is usually oxidase negative when tested in clinical laboratories. It fails to hydrolyse arginine and does not grow on cetrimide agar. It is the only pseudomonad that gives a positive lysine decarboxylase reaction. It has been isolated from a variety of sources and is the third most common non-lactose fermenter found in clinical material.

P. cepacia
Colonies may produce a yellow, water-soluble, non-fluorescent pigment, occasionally may be purple, or may be non-pigmented. Growth at 42°C has been reported. Widely distributed in nature, it occurs in hospital infections and in cystic fibrosis, is found in pharmaceuticals, cosmetics and is a particular hazard in water used for pharmaceutical purposes.

P. stutzeri
Colonies are rough, dry and wrinkled and can be removed entire from the medium. Older colonies turn brown. It does not hydrolyse arginine and may not grow at 4°C but does grow at 42°C. It is salt tolerant but not halophilic. A ubiquitous organism, it has been found in clinical material but is not regarded as a pathogen.

P. paucimobilis
Colonies may be mucoid or butyrous, producing a non-diffusible yellow pigment. Motility is poor and best seen in cultures incubated at room temperature. It has been found in clinical material and recovered from hospital equipment.

P. alcaligenes and P. pseudoalcaligenes
Both of these species grow at 42°C but neither grow at 4°C. They are ubiquitous and are opportunist pathogens. They do not liquefy gelatin and may or may not grow on cetrimide agar. *P. alcaligenes* does not hydrolyse arginine.

P. mallei
In exudates this organism may appear granular or beaded and may show bipolar staining. It gives 1-mm, shining, smooth, convex, greenish yellow buttery or slimy colonies which may be tenacious on nutrient agar, no haemolysis on blood agar and does not grow on MacConkey agar in primary culture. Growth may be poor on primary isolation. It grows at room temperature, does not change Hugh and Leifson medium, produces no acid in peptone carbohydrates, is nitrate positive and gives a variable urease reaction. The oxidase test gives variable results.

This is the causative organism of glanders

P. pseudomallei
Cultures on blood agar and nutrient agar at 37°C give mucoid or corrugated, wrinkled, dry colonies in 1–2 days, and an orange pigment may develop on prolonged incubation. There is no growth on cetrimide agar. This organism is not easy to identify. It must be distinguished from non-pigmented strains of *P. aeruginosa, P. stutzeri* and *mallei* (Table 22.3).

This is an important pathogen of humans (melioidosis) and farm animals in tropical and subtropical areas, where it is endemic in rodents and is found in moist soil, on vegetables and on fruit. Cultures should be sent to a reference laboratory. See Dance (1990) and Ashdown (1992).

Caution — Pseudomonas mallei and P. pseudomallei are in Hazard (Risk) Group 3 and should be handled only under full Biosafety Containment Level 3 conditions. Although odour is said to be a useful characteristic cultures should never be sniffed. Laboratory-acquired infections are not uncommon.

For futher information about the pseudomonads see Gilardi (1985) King and Phillips (1985) and *Cowan and Steel's Manual* (1993).

References

Ashdown, L. R. (1992) Melioidosis and safety in the clinical laboratory. *Journal of Hospital Infection*, **21**, 301–306

Costas, M., Holmes, B., On, S. L. W. and Stead, D. E. (1992). Identification of medically important *Pseudomonas* species using computerized methods. In *Identification Methods in Applied and Environmental Microbiology* (eds G. G. Board, D. Jones and F. A. Skinner), Society for Applied Bacteriology Technical Series No. 29, Blackwells, Oxford, pp.1–20

Cowan and Steel's Manual for the Identification of Medical Bacteria (1993) 3rd edn (eds G. I. Barrow and R. K. A. Feltham), Cambridge University Press, Cambridge

Dance, D. A. B. (1990) Melioidosis. *Review of Medical Microbiology*, **1**, 143–150.

Gilardi, G. L. (1985) Pseudomonas. In *Manual of Clinical Microbiology*, 4th edn (eds E. H. Lennet, A. Balows, W. J. Hauser and H. J. Shadomy), Association of American Microbiologists, Washington, pp.429–445

King, A. and Phillips, I. (1985) Pseudomonads and related species. In *Microorganisms of Medical and Veterinary Importance* (eds C. H. Collins and J. M. Grange), Society for Applied Bacteriology Technical Series No. 21, Academic Press, London, pp.1–12

Chapter

23

Acinetobacter, Alcaligenes, Flavobacterium, Chromobacterium, Janthinobacterium and acetic acid bacteria

Acinetobacter

Organisms have been moved into and out of this genus for several years. At present it seems to contain two species, *Acinetobacter calcoaceticus* (formerly *A. anitratum*) and *A. lwoffii*, but some workers use the name *A. beaumannii* for both or for various 'subspecies'. It is difficult to distinguish between the two species.

Isolation and identification

Plate material on nutrient, blood and MacConkey agar (and Bordetella or similar media if indicated). Incubate at 22°C and at 37°C for 24–48 h. Do oxidase and catalase tests. Inoculate nutrient broth for motility, Hugh and Leifson medium, glucose peptone water, nitrate broth and urea medium.

Acinetobacter are non-motile, obligate aerobes usually oxidase and nitratase negative, catalase positive and are oxidative or give no reaction in Hugh and Leifson medium (see below and Table 23.1).

Species of Acinetobacter

A. calcoaceticus ss. anitratus
This gives large non-lactose fermenting colonies on MacConkey and DC agars and grows at room temperature. On Hugh and Leifson medium it is oxidative, producing acid but not gas from glucose, lactose and xylose. It is nitratase negative and gives a variable urease reaction.

Infections in a wide variety of sites in humans have been reported. Many strains are resistant to several antibiotics. Hospital cross-infections with resistant strains cause problems. Sometimes in direct films this organism may resemble *Neisseria*.

A. calcoaceticus ss. lwoffii
Colonies on blood agar are small, haemolytic and may be sticky, but this organism also grows on nutrient and MacConkey agars. There is no change in Hugh and Leifson medium; it fails to attack carbohydrates, is urease negative. It can cause conjunctivitis. This and the other two biotypes are distinguished by the API-20NE.

For more information about infections with *Acinetobacter* see Vivian *et al.* (1981).

Table 23.1 ***Acinetobacter***

	Growth on MacConkey	*Oxidase*	*HL test*	*Motility*	*Acid from glucose*	*Nitrate reduction*	*Urease*
A. calcoaceticus							
ss. anitratus	+	–	Ox	–	+	–	v
ss. lwoffii	+	–	none	–	–	–	–

Ox, oxidative, v, variable

Alcaligenes

These organisms are widely distributed in soil, fresh and salt water and are economically important in food spoilage (especially of fish and meat). Many strains are psychrophilic. Some species have been isolated from human material and suspected of causing disease.

Isolation and identification

Plate on blood and MacConkey agars. Incubate at 20–22°C (foodstuffs) or at 35–37°C (pathological material) for 24–48 h. Subculture white colonies of Gram-negative rods on agar slopes for oxidase test, in peptone water for motility, in Hugh and Leifson medium, glucose peptone water, bromocresol purple milk, arginine broth and gelatin.

Species of Alcaligenes

The genus *Alcaligenes* is restricted here to motile, oxidase positive, catalase positive Gram-negative rods which give an alkaline or no reaction in Hugh and Leifson medium, an alkaline reaction in bromocresol purple milk and grow on MacConkey agar. Arginine is not hydrolysed, gelatin is not usually liquefied (Table 23.1). Most strains are resistant to penicillin (10 IU disc) There is one valid species, *A. faecalis*, which now includes *A. odorans* and several strains of doubtful taxonomic status.

A. faecalis
Culturally resembles *B. bronchiseptica* but is urease negative. It is a widely distributed saprophyte and commensal. Cultures of some strains, especially those associated with food spoilage, have a fruity odour, e.g. of apples or rotting melons. At least one strain of *Alcaligenes* is associated with ropy milk and ropy bread.

Achromobacter

This is no longer officially recognized, but food bacteriologists find it a convenient 'dustbin' for non-pigmented Gram-negative rods which are non-motile, oxidative in Hugh and Leifson medium, oxidase positive, produce acid but no gas from glucose, acid or no change in purple milk, do not hydrolyse arginine and usually liquefy gelatin.

Flavobacterium

The genus *Flavobacterium* contains many species that are difficult to identify. Gram-negative bacilli forming yellow colonies are frequently isolated from food and water samples and although some undoubtedly belong to other genera it suits the convenience of many food bacteriologists to call them 'flavobacteria'. Some of these organisms are proteolytic or pectinolytic and are associated with spoilage of fish, fruit and vegetables.

There is one species of medical importance.

Flavobacterium meningosepticum

This has been isolated from CSF in meningitis in neonates, from other human material and from hospital intravenous and irrigation fluids.

It grows on ordinary media; colonies at 18 h are 1–2 mm, smooth, entire, grey or yellowish and butyrous. There is no haemolysis on blood agar, no growth on DCA and none on primary MacConkey cultures, but growth on this medium may occur after several subcultures on other media. A yellowish green pigment may diffuse into nutrient agar media.

F. meningosepticum is oxidative in Hugh and Leifson medium, non-motile, oxidase and catalase positive, liquefies gelatin in 5 days, does not reduce nitrates and does not grow on Simmons' citrate medium. Rapid hydrolysis of aesculin is a useful character. The API ZYM system is useful. Acid is produced from 10% glucose, mannitol and lactose in ammonium salt medium but not from sucrose or salicin. Indole production seems to be positive by Ehrlich's but negative by Kovac's methods (Snell, 1973). See Table 23.2 for differentiation from pseudomonads and *Acinetobacter*. For more information see Holmes (1987).

Table 23.2 *Flavobacterium meningosepticum, Pseudomonas* and *Acinetobacter*

Species	*Oxidase*	*Nitrate reduction*	*Citrate*	*Acid from 10% solution in ammonium salt medium of*		
				Glucose	*Mannitol*	*Lactose*
F. meningosepticum	+	−	−	+	+	+
Pseudomonas	+	+	+	+	+	+
Acinetobacter	−	−	+/−	v	v	v

v, varies with subspecies, see text

Chromobacterium and Janthinobacterium

This group includes two species, both formerly in the genus *Chromobacterium*.

Table 23.3 *Chromobacterium violaceum* and *Janthinobacterium lividum*

Species	*Growth at*		*HL test*	*Hydrolysis of*		
	5°C	*37°C*		*Arginine*	*Casein*	*Aesculin*
C. violaceum	−	+	F	+	+	−
J. lividum	+	−	Ox	−	−	+

F = fermentative; Ox = oxidative

An important characteristic of both is the production of a violet pigment (unpigmented strains are not unknown), but this may not be apparent until the colonies are several days old. They grow well on ordinary media, giving cream or yellowish colonies that turn purple at the edges. The best medium to demonstrate the pigment is a potato slice. The pigment is soluble in ethanol but not in chloroform or water and can be enhanced by adding mannitol or meat extract to the medium. Citrate but not malonate is utilized in basal synthetic medium, the catalase test is positive and the urease test negative. Ammonia is formed. Gelatin is liquefied. The oxidase test is positive but difficult to do if the pigment is well formed. Both species are motile (see Table 23.3).

C. violaceum

This species is mesophilic, growing at 37°C but not at 5°C. It is fermentative in Hugh and Leifson medium, hydrolyses arginine and casein but not aesculin. It is a facultative anaerobe. The violet pigment is best seen in cultures on mannitol yeast extract agar incubated at 25°C.

Although usually a saprophyte, cases of human and animal infection have been described in Europe, the USA and the Far East.

Janthinobacterium lividum

This organism is psychrophilic, growing at 5°C but not at 37°C. It is oxidative in Hugh and Leifson medium, does not hydrolyse arginine or casein and hydrolyses aesculin. It is an obligate aerobe.

The violet pigment is best seen in cultures on mannitol yeast extract agar incubated at 20°C.

More information about the identification of these organisms is given by Logan and Moss (1992).

Acetic acid bacteria

These bacteria are widely distributed in vegetation. They are of economic importance in the fermentation and pickling industries, e.g. in cider manufacture (Carr and Passmore, 1979), as a cause of ropy beer and sour wine, and of off-odours and spoilage of materials preserved in vinegar. Ethanol is oxidized to acetic acid by *Gluconobacter* spp. but *Acetobacter* spp. continue the oxidation to produce carbon dioxide and water. They also differ in that *Acetobacter* has peritricate and *Gluconobacter* polar flagella. Seven species of *Acetobacter* and three of *Gluconobacter* are recognized but identification is difficult.

Isolation and identification to genus

Identification to species is rarely necessary. It is usually sufficient to recognize acetic acid bacteria and to determine whether the organisms produce acetic acid and/or destroy it.

Inoculate wort agar or Universal Beer Agar or yeast extract agar containing 10% glucose filter-sterilized and added at 50°C and 3% calcium carbonate at pH 4.5. Incubate at 25°C for 24–48 h. Colonies are large and slimy.

Subculture into yeast broth containing 2% ethanol and congo red indicator at pH 4.5. Acid is produced.

Test the ability to produce carbon dioxide from acetic acid in 2% acetic acid broth. Use either a Durham tube or the method described on p.374 for heterofermentative lactic acid bacilli. Inoculate lactose and starch peptone water 'sugars' at pH 4.5 with congo red indicator.

Plant bacterial masses (heavy inoculum, do not spread) on yeast extract agar at pH 4.5 containing 5% glucose and at least 3% finely divided calcium carbonate in suspension. Clear zones occur around masses in 3 weeks but if less chalk is used some pseudomonads can do this.

Culture on yeast extract agar containing 2% calcium lactate. *Acetobacter* grow well and precipitate calcium carbonate. *Gluconobacter* grow poorly and give no precipitate.

Carr and Passmore (1979) describe a medium to differentiate the two genera. It is yeast extract agar containing 2% ethanol and bromocresol green (1 ml of 2.2% solution/litre) made up in slopes. Both *Gluconobacter* and *Acetobacter* produce acid from ethanol and the indicator changes from blue-green to yellow. *Acetobacter* then utilize the acid and the colour changes back to green (See Table 23.4).

Table 23.4 Acetic acid bacteria

Methods	*Acid from ethanol*	*CO_2 from acetic acid*	*Carbonate from lactate*
Gluconobacter	+	−	−
Acetobacter	+	+	+

For more information, see Carr and Passmore (1979) and Swings *et al.* (1992).

References

Carr, J. G. and Passmore, S. M. (1979) Methods for identifying acetic acid bacteria. In *Identification Methods for Microbiologists*, 2nd edn (eds F. A. Skinner and D. W. Lovelock), Society for Applied Bacteriology Technical Series No.14, Academic Press, London, pp.33–45

Holmes, B. (1987) Identification and distribution *Flavobacterium meningosepticum* in clinical material. *Journal of Applied Bacteriology* **43**, 29–42

Logan, N. A. and Moss, M. O. (1992) Identification of *Chromobacterium, Janthinobacterium* and *Iodobacter* species. In *Identification Methods in Applied and Environmental Microbiology* (eds R. G. Board, D. Jones and F. A. Skinner), Society for Applied Bacteriology Technical Series No. 29, Blackwells, London

Snell, J. J. S. (1973) *The Distribution and Identification of Non-fermenting Bacteria.* Public Health Laboratory Service Monograph No. 4, HMSO, London

Swings, J., Gillis, M. and Kersters, K. (1992) Phenotypic identification of acetic acid bacteria. In *Identification Methods in Applied and Environmental Microbiology* (eds R. G. Board, D. Jones and F. A. Skinner), Society for Applied Bacteriology Technical Series No. 29, Blackwells, London

Vivian, A., Hinchcliffe, E. J. and Fewson, C. A. (1981) *Acinetobacter calcoaceticus:* some approaches to a problem. *Journal of Hospital Infection,* **2**, 199–204

Chapter

24

Vibrio, Plesiomonas, Shewanella and Aeromonas

The differential properties of the organisms in this group and those of the enterobacteria are summarized in Table 24.1.

Table 24.1 General properties of *Vibrio, Plesiomonas, Shewanella, Aeromonas, Pseudomonas* and enterobacteria

	0/129 150 μg	*Oxidase*	*O/F test*	*Salt enhancement*
Vibrio	S	+	F	+
Plesiomonas	R	+	F	−
Shewanella	R	+	Alk/ −	+
Aeromonas	R	+	F	−
Pseudomonas	R	+	O	−
Enterobacteria	R	−	F	−

S, sensitive; R, resistant; v, variable; F, fermentative; O, oxidative

The genus *Vibrio* includes at least 20 species. Two have been transferred to the new genus *Listonella.*

Vibrios and Listonellas

Vibrios are Gram-negative, non-sporing rods, motile by polar flagella enclosed within a sheath. Some have lateral flagella and may swarm on solid media. They are catalase positive, utilize carbohydrates fermentatively and rarely produce gas. All but one species (*Vibrio metschnikovii*) are oxidase and nitratase positive. They are all sensitive to 150 μg discs of the vibriostatic agent 2,4-diamino-6,7-di-isopropyl pteridine (0/129). Most strains liquefy gelatin and hydrolyse deoxyribonucleic acid. Vibrios occur naturally in fresh and salt water. Species include some human and fish pathogens. See Table 24.1 for differentiation between *Vibrios, Plesiomonas, Shewanella, Aeromonas, Pseudomonas* and enterobacteria.

Isolation

Vibrios grow readily on most ordinary media but enrichment and selective media are necessary for faeces and other material containing mixed flora.

Faeces

Add 2 g of faeces to 20-ml amounts of alkaline peptone water (APW). Inoculate thiosulphate citrate bile salt sucrose agar (TCBS) medium.

Incubate APW at 37°C for 5–8 h or at 20–25°C for 18 h and subculture to TCBS.

Incubate TCBS cultures at 37°C overnight.

Other pathological specimens

Vibrios may occur in wounds and other material particularly if there is a history of sea bathing. Examine primary blood agar plates (see below for colony appearance).

Water and foods

Add 20 ml to water to 100 ml of APW. Add 10 ml of food to 100 ml of APW and emulsify in a Stomacher. Incubate at 20–30°C for 18 h and subculture on TCBS as for faeces.

Enumeration of vibrios

Use liquid samples neat and diluted 1:10. Make 1:10 homogenates of solid samples. Spread 50 or 100 μl on TCBS medium, incubate 20–24 h and count vibrio-like colonies. Report as cfu/g.

For the three-tube Most Probable Number method prepare dilutions of 10^{-1}, 10^{-2} and 10^{-3}. Add 1 ml of each dilution to each of three tubes of APW, GSTB or SCB and incubate at 30°C overnight. Subculture to TCSB agar and incubate at 30°C for 20–24 h. Examine for vibrio-like colonies, identify them and use the three-tube table (p.162).

Colonial morphology

Table 24.2 shows the colony appearances of vibrios and related organisms on TCBS medium.

On non-selective media colonial morphology is variable. The colonies may

Table 24.2 Growth of some vibrios and other bacteria on thiosulphate citrate bile salt agar (TCBS), 37 °C for 18 h

Species	*Growth*	*Sucrose fermented*	*Colour*	*Colony diameter (mm)*
V. cholerae	+	+	Y	2–3
V. metschnikovii	+	+	Y	2–4
V. fluvialis	+	+	Y	2–3
V. furnissii	+	+	Y	2–3
V. mimicus	+	−	G	2–3
V. parahaemolyticus	+	−	G	2–5
L. anguillarum	+	+	Y	0–3
L. hollisae	−	−		—
L. damsela	+	−	G	2–3
V. vulnificus	+	−	G	2–3
Aeromonas hydrophila	v	+	Y	0–3
Plesiomonas shigelloides	v	−	G	1
V. nigropulchritude	−	−		—

+, growth or sucrose fermented G, green
−, no growth or sucrose not fermented Y, yellow
v, variation within strains

be opaque or translucent, flat or domed, haemolytic or non-haemolytic, smooth or rough. One variant is rugose and adheres closely to the medium.

Identification

The following tests will enable the common species to be identified. For more detailed tests use the API 20E and 2ONE systems (Austin and Lee, 1992) but include 0/129 sensitivity and growth on CLED medium.

Ox/Ferm (O/F) test
Use Hugh and Leifson medium.

Oxidase test
Use Kovac's method (p.113) and test colonies from non-selective medium. Do not use colonies from TCBS or other media containing fermentable carbohydrate because changes in pH may interfere with the reaction.

Arginine dihydrolase
Use arginine broth or Moeller's medium (p.111) with 1% NaCl.

Decarboxylase tests
Use Moeller's medium containing 1% additional NaCl, but do not read the results too early as there is an initial acid reaction before the medium becomes alkaline. The blank should give an acid reaction; failure to do so may suggest poor growth and the electrolyte supplement should be added.

Acid from mannitol
Add 1% NaCl to the peptone water sugar medium.

Aesculin hydrolysis
Use the method on p.109.

Citrate utilization
Use Koser medium.

Hydrogen sulphide production
Use TSI medium. *Shewanella putrefaciens* is the only organism in this group that produces H_2S.

Indole
Use peptone water containing 2% NaCl and Kovac's method.

VP test
Use a semisolid medium under controlled conditions. If incubation is prolonged and a sensitive method is used almost any vibrio may give positive results.

Gelatin liquefaction
Use gelatin agar.

ONPG
Use ONPG sodium potassium magnesium broth or discs.

Growth at 43°C
Inoculate nutrient broth and incubate in a water-bath.

Sensitivity to 0/129 (2, 4-diamino-6, 7-di-isopropyl pteridine)
This was originally used as a disc method to distinguish between *Vibrio* (sensitive) and *Aeromonas* (resistant) (NI Re *et al.*, 1978), but the use of two discs (150 μg and 10 μg) enables two groups of vibrios to be recognized. Make a lawn of the organisms on nutrient agar, and place one disc of each concentration of 0/129 on it and incubate overnight. Do not use any special antibiotic sensitivity testing medium because the growth of vibrios and the diffusion characteristics of 0/129 differ from those on nutrient agar.

Growth on CLED medium
Some vibrios require the addition of sodium chloride to the medium while others do not. Culture on an electrolyte deficient medium, e.g. CLED. This permits two groups, halophilic and non-halophilic vibrios to be distinguished. Inoculate CLED lightly with the culture and incubate at 30°C overnight.

Salt tolerance
This varies with species. Inoculate peptone water containing 0, 3, 6, 8 and 10% NaCl. This technique must be standardized to obtain consistent results.

Toxin detection
There is a kit (Oxoid) for detecting cholera toxin in culture filtrates.

Properties of vibrios, etc.

See Tables 24.2 and 24.3.

Species of vibrios

V. cholerae
Is sensitive to 0/129, oxidase positive, decarboxylates lysine and ornithine but does not hydrolyse arginine. It produces acid but no gas from glucose (fermentative in Hugh and Leifson's medium) and sucrose, but no acid from arabinose or lactose. It is non-halophilic in that it grows on CLED medium.

All strains possess the same heat-labile H antigen but may be separated into serovars by their O antigens.

Serovar O:I
Is the causative organism of epidemic or Asiatic cholera. It is agglutinated by specific O:I cholera antiserum. It is possible, by using carefully absorbed sera, to distinguish two subtypes of *V. cholerae* O:I. These are known as Ogawa and Inaba but as they are not completely stable and variation may occur *in vitro* and *in vivo* subtyping is of little epidemiological value, unlike phage typing (Lee and Furniss, 1981) which is of epidemiological value. Non-toxigenic strains of *V. cholerae* O:I have been isolated and have shown distinctive phage patterns.

Although there are two biotypes of *V. cholerae* O:I, the 'classic' (non-haemolytic) and the 'eltor' (haemolytic) the former are now virtually non-existent. Biotyping is not, therefore, a useful epidemiological tool (see Table 24.4).

Serovars other than O:I
Have identical biochemical characteristics as *V. cholerae* O:I and the same H antigen but possess different O antigens and are not agglutinated by the O:I serum. They have been called 'non-agglutinating' (NAG) vibrios. This is an obvious misnomer; they are agglutinated both by cholera H antiserum and by

Table 24.3 Properties of *Vibrio* species and allied genera likely to be encountered in clinical laboratories

	Arginine dihydrolase	*Lysine decarboxylase*	*Acid from mannitol*	*Aesculin hydrolysis*	*Citrate utilization*	H_2S	*Indole*	*VP*	*Gelatin liquefaction*	*ONPG*	*Growth at 43°C*	*0/129*		*Growth on CLED*	*NaCl*	
												10 µg	*150 µg*		*0%*	*6%*
V. cholerae	−	+	+	−	v	−	+	+	+	+	+	S	S	+	+	−
V. fluvialis	+	−	+	+	−	−	+	v	−	+	v	R	S	−	−	+
L. hollisae	−	−	−	−	−	−	+	−	−	−	·	S	S	v	−	v
V. metschnikovi	+	+	+	·	v	−	v	v	+	+	+	S	S	+	v	+
V. mimicus	−	+	+	·	v	−	+	−	+	−	+	S	S	+	+	−
V. parahaem-olyticus	−	+	+	−	v	−	+	−	v	+	+	R	S	−	−	+
V. vulnificus	−	v	v	+	+	−	+	+	+	+	−	S	S	−	−	+
L. anguillarum	+	−	−	·	+	−	+	+	+	+	−	S	S	v	v	v
L. damsela	+	−	+	−	−	−	−	+	−	−	·	S	S	v	−	v
Plesiomonas	+	+	−	−	−	−	+	−	−	+	+	v	R	+	+	−
Shewanella	−	·	−	v	+	+	−	−	+	−	·	R	R	+	·	·
Aeromonas	+	v	+	+/−	+/−	−	+/−	v	+	+/−	+	R	R	+	+	−

S, Sensitive; R, resistant; v, variable; +/− depends on species; ·, no information.
Compiled from various sources, including Lee and Donovan (1985), Austin and Lee (1992) and Cowan and Steel (1993).

antisera prepared against the particular O antigen they possess. Another term which has been used is 'Non-cholera vibrio' (NCV) and as both terms have been used in different ways this has resulted in considerable confusion. It is best if all these vibrios are referred to as non-O:I *V. cholerae*, or, if a strain has been serotyped, by that designation.

Some of these strains are undoubtedly potential pathogens and produce a toxin similar to, if not identical with, that of the cholera vibrio. Some outbreaks have occurred but most isolates have been from sporadic cases. These vibrios are widespread in fresh and brackish water in many parts of the world, including the UK. They do not, however, cause true epidemic cholera.

Table 24.4 Recognition of classic and eltor strains of *Vibro* cholerae

	Classic	*Eltor*
Haemolysis	−	+
VP	−	+
Chick cell haemagglutination	−	+
Polymyxin (50 IU)	S	R
Classic phage IV	S	R
Eltor phage 5	R	S

S, sensitive; R, resistant

V. mimicus

Resembles non-O:I *V. cholerae* but is VP negative and does not ferment sucrose. Colonies on TCBS media are therefore green. Antigenically it appears to be *V. cholerae* but it was given specific rank because of its low degree of DNA homology with *V. cholerae*. On the other hand it could be retained as a subspecies or biovar of *V. cholerae*. It has been isolated from the environment and is associated with sea-foods. It has also been isolated from human faeces. Some strains produce a cholera-like toxin.

V. parahaemolyticus

Is a halophilic vibrio and will not grow on CLED. It does not ferment sucrose and therefore gives a (large) green colony on TCBS agar. O and K antigens may be used to serotype strains. It is recognized as the commonest cause of food poisoning in Japan. It is usually present in coastal waters, although only in the warmer months in the UK. The Kanagawa haemolysis test is said to correlate with pathogenicity. Controlled conditions for testing the haemolysis of the human red blood cells are essential. Only laboratories with sufficient experience should do this test. Most environmental strains are negative.

V. vulnificus

Resembles *V. parahaemolyticus* but hydrolyses aesculin and is sensitive to 10 μg 0/129. It has been isolated from blood cultures from patients (mostly in the USA) who are immunologically compromised or suffering from liver disease. Also isolated from wound infections associated with exposure to sea-water, particularly in patients with underlying conditions.

V. fluvialis

First reported as Group F vibrios, this is common in rivers, particularly in the brackish water of estuaries and may cause gastroenteritis in humans. *V. fluvialis* needs to be distinguished from *L. anguillarum* (Table 24.3).

V. furnissi
Was previously classified as *V. fluvialis* biovar II. It has been associated with human gastroenteritis in Japan and other parts of the East and has been isolated from the environment and sea-food.

V. metschnikovii
Is both oxidase and nitrate reduction negative but its other characters are typical of vibrios. It includes the now illegitimate species *V. proteus*. Frequently isolated from water and shellfish, but there is little evidence of pathogenicity for humans.

Listonella

These are marine organisms, formerly in the genus *Vibrio*. See Table 24.3

L. anguillarum
Is phenotypically far from being a uniform species. It is uncommon in rivers. Some strains may be pathogenic for fish, although not especially for the eel, as its name might suggest. Many strains will not grow at 37°C and would be missed in laboratories which confine their incubation temperatures to about 37°C. There is no evidence that *L. anguillarum* has ever been pathogenic for humans.

L. hollisae
A halophilic vibrio isolated from human cases of gastroenteritis. It does not grow on TCBS. It is distinguished from other halophilic vibrios from human sources by its negative lysine, arginine and ornithine tests and the limited range of carbohydrates fermented.

It has not been recovered from the environment and a clear relation with human disease has not been established.

L. damsela
A marine bacterium isolated from human wounds and from water and skin ulcers of damsel fish. It is arginine positive, produces gas from glucose, requires NaCl for growth, is not bioluminescent and ferments glucose, mannose and maltose. It is distinguished from other named vibrios by DNA-DNA hybridization.

Plesiomonas shigelloides

There is no specific enrichment method or selective medium for *Plesiomonas*. It grows poorly on TCBS but well on deoxycholate and MacConkey agar. Colonies appear to be *Shigella*-like—hence the specific name—and some strains are agglutinated strongly by *Shigella sonnei* antiserum.

This organism has been variously classified and included within the genus *Vibrio*, because of its sensitivity to 0/129. It is now considered to be distinct enough to be placed in a separate genus containing one species.

It is now thought to be a cause of gastroenteritis, because most clinical isolates are from patients with diarrhoea; it is rarely isolated otherwise. Many of the isolates in the UK are from people returning from abroad.

Shewanella

There are several species in this new genus but only *Shewanella putrefaciens*, formerly *P. putrefaciens*, is considered here. A spoilage organism, it has been found in clinical material. It grows on the same media as the other organisms in this chapter and is distinguishable by its salmon-pink colonies and production of H_2S (Table 24.3).

Aeromonas

These are small Gram-negative rods which are motile by polar flagella. They are oxidase and catalase positive and resistant to 0/129. They attack carbohydrates fermentatively (gas may be produced from glucose) liquefy gelatin and reduce nitrates. The arginine dihydrolase test using arginine broth is positive. Indole and VP reactions vary with species. Some strains are psychrophilic and most grow at 10°C. See Tables 24.2 and 24.3 for differentiation from other groups.

The genus is divided into two groups. The Salmonicida group contains psychrophilic, non-motile aeromonads. They may be further divided into three subspecies which are associated with diseases of fish (see Frerichs and Hendrie, 1985). Human infections have been reported—see below. The Hydrophila group contains mostly motile mesophiles which are found in food.

Isolation

Aeromonas species will grow on non-selective media. Plate on tryptone agar or Aeromonas medium and incubate at 22°C for Salmonicida group and 37°C for Hydrophila group.

Identification

Test for growth at 20 and 37°C. Do oxidase and nitrate reduction tests, OF test, incubate glucose, arabinose, salicin and aesculin media. Do lysine decarboxylase, arginine dihydrolase and VP tests. Look for brown pigment on agar medium containing 1% tyrosine and incubated at 20°C (see Tables 24.5 and 24.6) if the Salmonicida group is suspected. Plate on nutrient or blood agar and test susceptibility to cephalothin (30 μg disc). The API 20 and 20NE series are useful (Austin and Lee, 1992).

Table 24.5 ***Aeromonas groups***

	Motility	*Growth at 37°C*	*Brown pigment*
Salmonicida group	–	–	+
Hydrophila group	+	+	v

Table 24.6 Hydrophila group

	Aesculin hydrolysis	*Acid from*		*Gas from glucose*	*Lysine decarboxylase*	*VP reaction*	*Cephalothin 30 µg disc*
		Arabinose	*Salicin*				
A. hydrophila[a]	+	+	+	+	+	+	R
A. caviae[b]	+ (most)	+	+	−	−	−	R
A. sobria[a]	− (most)	−	−	+	+	v	S

[a] Enterotoxigenic
[b] Not enterotoxigenic

Species of Aeromonas

There is still uncertainty about the status of the species but we favour the recognition of the four mentioned below.

Salmonicida group
Produces a brown pigment on media containing tyrosine but does not grow at 37°C and is non-motile. It causes furunculosis of fish, a disease of economic importance in fish farming.

Hydrophila group
A *hydrophila, A. caviae* and *A. sobria* are motile and usually grow at 37°C. They are more commonly isolated from the stools of patients with diarrhoea than from normal faeces in the UK but in countries where aeromonads are common in drinking water they are as frequently isolated from normal as abnormal stools. It appears that enteropathogenicity, if it does occur, is confined mainly to strains of *A. hydrophila* and *A. sobria*, as judged by virulence factors, e.g. cytotoxigenicity, haemolysis and enterotoxin production (Turnbull *et al.*, 1984).

It is now generally accepted that members of this group can cause food-borne gastroenteritis. It has been found in the gut of leeches which are used in plastic surgery to reduce haematomas; there have been reports of wound infections from this source. They also cause red leg disease in frogs. They grow readily, even at refrigeration temperatures, and are associated with spoilage.

For more information about the identification of organisms mentioned in this chapter see Lee and Donovan (1985) and Austin and Lee (1992).

References

Austin, B. and Lee, J. V. (1992) Aeromonadaceae and Vibrionaceae. In *Identification Methods in Applied and Environmental Microbiology* (eds R. G. Board, D. Jones and F. A. Skinner), Society for Applied Bacteriology Technical Series No. 29, Blackwells, Oxford, pp. 163–183

Frerichs, G. N. and Hendrie, M. S. (1985) Bacteria associated with disease of fish. In *Isolation and Identification of Micro-organisms of Medical and Veterinary Importance* (eds C. H. Collins and J. M. Grange), Society for Applied Bacteriology Technical Series No. 21, Academic Press, London, pp. 355–371

Furniss, A. L., Lee, J. V. and Donovan, T. J (1978) *The Vibrios*, Public Health Laboratory Service Monograph No. 11, HMSO, London

Lee, J. V. and Donovan, T. J. (1985) Vibrio, Aeromonas and Plesiomonas. In *Isolation and Identification of Micro-organisms of Medical and Veterinary Importance* (eds C. H. Collins and J. M. Grange), Society for Applied Bacteriology Technical Series No. 21, Academic Press, London, pp. 13–34

Lee, J. V. and Furniss, A. L. (1981) Discussion 1. The phage-typing of *Vibrio cholerae* serovar O1. In *Acute Enteric Infections in Children. New Prospects for Treatment and Prevention* (eds T. Holme *et al.*), Elsevier, Amsterdam, pp. 191–222

Turnbull, P. C. B. *et al.* (1984) Enterotoxin production in relation to taxonomic grouping and source of infection of *Aeromonas* species. *Journal of Clinical Microbiology*, **19**, 175–180

Chapter

25

Key to the enterobacteria

There are at least 17 genera in this group of Gram-negative non-sporing rods. They are aerobic, facultatively anaerobic, oxidase negative and fermentative in Hugh and Leifson medium.

The genera considered here and in the following chapters may be conveniently divided into four groups:

(1) *Escherichia, Citrobacter, Klebsiella* and *Enterobacter*, which usually produce acid and gas from lactose. For historic reasons they are known collectively as the coliform bacilli (Chapter 26).

(2) *Edwardsiella, Erwinia, Hafnia* and *Serratia*, which usually do not ferment lactose, although some workers include them in the coliforms (Chapter 27).

(3) *Salmonella* and *Shigella*, which do not ferment lactose and are important intestinal pathogens (Chapter 28).

(4) *Proteus, Providencia* and *Morganella*, which do not ferment lactose and differ from the other groups in their ability to deaminate phenylalanine (Chapter 29).

Table 25.1 shows the general biochemical properties of these genera.

The genus *Yersinia* is included among the enterobacteria by some workers but for convenience is considered in this book in Chapter 34.

The newer and less important genera *Buttiauxella, Kluyvera* and *Tatumella* are not included and information about them may be found in *Cowan and Steel's Manual* (1993).

Reference

Cowan and Steel's Manual for the Identification of Medical Bacteria (1993) 3rd edn (eds G. I. Barrow and R. K. A. Feltham), Cambridge University Press, Cambridge

Table 25.1 General properties of enterobacteria

	Indole	Motility	Acid from			Citrate	Urea	H_2S	Gelatin liquef.	PPA	ONPG	
			Lactose	Mannitol	Inositol							
Escherichia	+	+	+	+	–	–	–	–	–	–	+	Ch. 26
Citrobacter	v	+	v	+	–	+	–	+	–	–	+	
Klebsiella	v	+	v	+	+	+	(+)	–	–	–	+	
Enterobacter	–	+	v	+	–	+	–	–	–	–	+	
Erwinia	–	+	v	+	–	v*	–	–	v	–	+	Ch. 27
Edwardsiella	+	+	–	–	–	v*	–	+	–	–	–	
Hafnia	–	+	–	+	–	+	–	–	–	–	+	
Serratia	–	+	–	+	–	+	–	–	+	–	+	
Salmonella	–	+	–	+	+	+	–	+	–	–	–	Ch. 28
Shigella	v	–	–	v	–	–	–	–	–	–	+	
Proteus	v	+	–	v	v	v	+	v	v	+	–	Ch. 29
Providencia	+	+	–	v	+	+	–	–	–	+	–	
Morganella	+	+	–	–	–	v	+	–	–	+	–	

v, variable, (+) slow
v*, varies with method

Chapter

26

Escherichia, Citrobacter, Klebsiella and Enterobacter

These genera, known as 'coliform bacilli' are members of the Enterobacteria (see also Chapters 27–29).

They are Gram-negative, oxidase-negative, fermentative non-sporing rods, are nutritionaly non-exacting and grow well on ordinary culture media. They also grow in the presence of bile, which inhibits many cocci and Gram-positive bacilli.

Table 25.1 shows their general properties and those that distinguish them from the other enterobacteria in Chapters 27–29.

Isolation

Pathological material

Plate stools from children under 3 years old, intestinal contents of animals, urine deposits, pus, etc., on MacConkey, eosin methylene blue (EMB) or Endo agar and on blood agar. Cystine lactose electrolyte deficient medium (CLED) is useful in urinary bacteriology. Proteus does not spread on this medium. Incubate at 37°C overnight.

Foodstuffs

The organisms may be damaged and may not grow from direct plating. Make 10% suspensions of the food in tryptone soya broth. Incubate at 25°C for 2 h and then subculture into brilliant green bile broth and lauryl tryptose broth. MacConkey broth should not be used because it supports the growth of *Clostridium perfringens*, which produces acid and gas at 37°C. Minerals modified glutamate broth may be useful for the recovery of 'damaged' coliforms. Incubate at 37°C overnight and plate on MacConkey, Endo, EMB, violet red bile agar or CLED media.

Identification

Coliform bacilli show pink or red colonies, 2–3 mm in diameter, on MacConkey agar. Klebsiella colonies may be large and mucoid. On Levine EMB agar, colonies of *E. coli* are blue-black by transmitted light and have a metallic sheen by incident light. Colonies of klebsiellas are larger, brownish, convex and mucoid and tend to coalesce. On Endo medium, the colonies are deep red and colour the surrounding medium. They may have a golden yellow sheen.

Table 26.1 *Escherichia*, *Citrobacter*, *Klebsiella*, *Enterobacter* and *Erwinia*

	Indole		*Lysine decarboxylase*	*Ornithine decarboxylase*	*Arginine dihydrolase*	*Citrate*	H_2S	*Acid from*		*Urease*	*ONPG*	β-*GUR*
	37°C	44°C						*Adonitol*	*Inositol*			
E. coli	+	+	+	v	v	−	−	−	−	−	v	+
C. freundii	v	−	−	v	+	+	+	−	−	v	+	−
C. koseri	+	−	−	v	+	+	−	+	−	v	+	−
K. pneumoniae	−	−	+	−	−	+	−	+	+	+	+	−
K. oxytoca	+	−	+	−	−	+	−	+	+	+	+	−
K. ozaenae	−	−	v	−	v	+	−	+	v	v	+	−
K. rhinoscleromatis	−	−	−	−	−	−	−	+	+	−	−	−
E. aerogenes	−	−	+	+	−	+	−	+	+	−	+	−
E. cloacae	−	−	−	+	+	+	−	v	−	v	−	−
E. herbicola	−	−	−	−	−	+	−	−	−	−	+	−

v, variable

Strains from clinical material
Examine a Gram-stained film: some cocci and Gram-positive bacilli grow on selective media such as MacConkey and CLED.

Inoculate peptone water to test for indole production and motility and test for β-glucuronidase (β-GUR) activity on medium containing *p*-nitrophenyl-β-D glucopyranosideroic acid (Mast ID37).

Or use 4-methylumbelliferyl-β-D-gluconide (MUG) which is hydrolysed to a compound which fluoresces under UV light.

Typical *E. coli* strains are indole positive, motile and give a positive β-GUR and MUG reactions. Only *E. coli* and some shigella strains give a positive β-GUR.

Non-clinical strains
Examine a Gram-stained film. Inoculate lactose broth containing an indicator and a Durham's tube, and tryptone peptone water for indole test: incubate in a water bath at 44 + 0.2°C. Include these controls:

(1) Stock *E. coli*, which produces gas and is indole positive.
(2) Stock *K. aerogenes* which produces no gas and is indole negative.

The β-GUR test is also useful.

Confirm lactose-fermenting colonies on membrane filters and in primary broth cultures by subculturing to tryptone peptone water and lactose broth, testing for indole and gas production at 37 and 44°C. Plate on MacConkey agar for colony appearance.

For identification by conventional methods (other than for *E. coli*) do indole, lysine and ornithine decarboxylase, arginine dihydrolase, citrate, H_2S, urease and ONPG tests. Fermentation of adonitol and inositol may be useful. (See Table 26.1.)

Rapid and kit methods
There are two API systems: RAPIDEC coli and RAPIDEC GUR, both of which include a β-GUR test.

Multipoint inoculation tests are cost-effective for testing large number of strains. A 'short set' from the MAST range may be used, e.g. β-GUR, inositol, aesculin, hydrogen sulphide or the full MAST 15 system. There are two that are aimed at *E. coli*: the Coli strip (LabM, UK) which uses a fluorogenic substrate and an indole test; and the Bactident *E. coli* (Merck, Germany) which uses β-GUR and indole.

Other tests
A modification of Donovan's (1966) system uses the above 'short set' in tubed media (Table 26.2).

Thermotolerant coliforms
This term is used to describe coliform bacilli that produce gas at 44°C but do not produce indole or have not been tested for it.

Antigens of enterobacteria
There are three kinds of antigens. The O or somatic antigens of the cell body are polysaccharides and are heat stable, resisting 100°C. The H or flagellar antigens are protein and destroyed at 60°C. The K and Vi are envelope, sheath or capsular antigens and heat labile. There are three kinds of K antigens: L, which is destroyed at 100°C and is an envelope, occasionally capsular; A, which is destroyed at 121°C and is capsular; and B, which is destroyed at

Table 26.2 'Short set' for coliforms

	β-GUR	*Motility*	*Inositol*	*Aesculin*	*H_2S*	*Indole*
E. coli	+	+	−	−	−	+
Klebsiella	−	−	+	+	−	v
Enterobacter	−	+	−	v	−	−
C. freundii	−	+	−	−	+	−
C. koseri	−	+	−	−	−	+

100°C and is an envelope. These Vi and K antigens mask the O, and agglutination with O sera will not occur unless the bacterial suspensions are heated to inactivate them.

Serological testing of E. coli
Screen first by slide agglutination with the (commercially available) antisera. Heat saline suspensions of positives in a boiling water bath for 1 h to destroy H and K antigens. Cool, and do tube O agglutination tests. A single tube test at 1 : 50 (with a control) is adequate. Read at 3 h and then leave on the bench overnight and read again.

Species of coliform bacilli

Escherichia coli
This species is motile, produces acid and gas from lactose at 44°C and at lower temperatures, is indole positive at 44 and 37°C, fails to grow in citrate, is H_2S negative and usually decarboxylates lysine.

These are the so-called 'faecal coli' that occur normally in the human and animal intestine and it is natural to assume that their presence in food indicates recent contamination with faeces. *E. coli* is, however, widespread in nature and although most strains probably had their origin in faeces, its presence, particularly in small numbers, does not necessarily mean that the food contains faecal matter. It does suggest a low standard of hygiene. It seems advisable to avoid calling them 'faecal coli' and to report the organisms as *E. coli.*

It is associated with human and animal infections and is the commonest cause of urinary tract infections in humans and is also found in suppurative lesions, neonatal septicaemias and meningitis. In animals it causes mastitis, pyometria in bitches, coli granulomata in fowls and white scours in calves.

As least four types of *E. coli* cause gastrointestinal disease in humans. According to Gorbach (1986, 1987) they may be described as: enteropathogenic (EPEC), enterotoxigenic (ETEC), enteroinvasive (EIEC) and verotoxigenic (VTEC). See Table 26.3.

Table 26.3 Common serotypes of *E. coli* causing gastrointestinal infections

		Common O serotypes
EPEC	Enteropathogenic (infantile)	18, 26, 44, 55, 86, 111, 114, 119, 124, 125, 126
ETEC	Enterotoxigenic	6, 8, 15, 25, 27, 63, 78, 115, 148, 153, 154
EIEC	Enteroinvasive	28ac, 112ac, 124, 136, 143, 144, 152, 164
VTEC	Verotoxigenic	157

(After Gorbach, 1986)

The EPEC strains have been associated with outbreaks of infantile diarrhoea, also travellers' diarrhoea, and are identified serologically (see above) but this is necessary only in outbreaks. It is not recommended that single cases of EPEC are investigated in this way.

ETEC strains are thought to cause gastroenteritis in both adults (travellers' diarrhoea) and children (especially in developing countries). The vehicles are contaminated water and food. They produce enterotoxins, one of which is heat labile (LT) and other heat stable (ST).

EIEC strains cause diarrhoea similar to that in shigellosis. The strains associated with invasive dysenteric infections are less reactive than typical *E. coli*. They may be lysine negative, lactose negative, anaerogenic and resemble shigellas. The correlation between serotype and pathogenicity is imperfect. Reference laboratories use tissue culture and other specialized tests.

VTEC strains derive their name from their cytotoxic effect on Vero cells in tissue culture. Their alternative name is enterohaemorrhagic *E. coli* (EHEC). They have been associated with haemolytic uraemic syndrome and haemorrhagic colitis and are known to be responsible for food-borne disease (see Chapter 13). See also Levine (1987) and Doyle and Cliver (1990). Serotype O:157 H7 is the most frequently reported. They do not ferment sorbitol and can be recognized on Sorbitol MacConkey agar (Oxoid) and tested with commercial O:157 antisera (Oxoid latex). They are also β-GUR negative.

Several commercial kits are available for detecting the toxins of these organisms, e.g. ETCLT (Phadebact, UK), VET-RPLA (Oxoid), E. COLI ST EIA (Oxoid).

Citrobacter freundii

Is motile, indole variable, and citrate positive. Hydrogen sulphide is produced but the lysine decarboxylase test is negative. Non- or late-lactose fermenting strains occur (see below).

This organism occurs naturally in soil. It can cause urinary tract and other infections in humans and animals.

Citrobacter koseri

Is indole positive, citrate positive but does not produce hydrogen sulphide.

It is found in urinary tract infections and occasionally in meningitis.

Klebsiella pneumoniae

Also known as Friedlander's pneumobacillus, this is indole negative, gives a positive lysine decarboxylase reaction but does not produce H_2S. It is urease and ONPG positive. It may produce large, mucoid colonies on media that contain carbohydrates.

It is associated with severe respiratory infections and can cause urinary tract infections (Table 26.1).

Klebsiella aerogenes

There has always been some confusion between this organism and *K. pneumoniae* because the latter name was used in the USA. It is now regarded as a subspecies of *K. pneumoniae*. *K. aerogenes* is gluconate positive but *K. pneumoniae* is gluconate negative (Table 26.1).

K. oxytoca
Differs from K. *aerogenes* in being indole positive (Table 26.1).

K. ozoenae
Is also difficult to identify, but is not uncommonly found in the respiratory tract associated with chronic destruction of the bronchi (Table 26.1)

K. rhinoscleromatis
Also difficult to identify, is rare in clinical material and may be found in granulomatous lesions in the upper respiratory tract (Table 26.1).

Klebsiellas may be classified serologically by their O and K antigens. For more information on this genus and associated organisms, see Shinebaum and Cooke (1985).

Enterobacter cloacae
Is motile, indole negative, MR negative, VP positive, grows in citrate and liquefies gelatin (but this property may be latent, delayed and lost). It does not produce hydrogen sulphide and the lysine decarboxylase test is negative. Growth in malonate is variable and the gluconate test is positive. It is found in sewage and in polluted water (Table 26.1).

E. aerogenes
Resembles *E. cloacae* but the lysine decarboxylase test is positive. Gelatin liquefaction is late. It is often confused with *K. aerogenes. E. aerogenes* is motile and urease negative; *K. aerogenes* is non-motile and urease positive (Table 26.1).

Erwinia herbicola
This grows in citrate and is ONPG positive but is negative for lysine decarboxylase, H_2S and urease (Table 26.1).

The non-lactose fermenting enterobacteria

For detailed information about the characterization and taxonomy of these organisms see Cowan and Steel (1993). See also Farmer *et al.* (1985) and Shinebaum and Cook (1985).

References

Cowan and Steel's Manual for the Identification of Medical Bacteria (1993) 3rd edn (eds G. I. Barrow and R. K. A. Feltham), Cambridge University Press, Cambridge

Donovan, T. J. (1966) A Klebsiella screening medium. *Journal of Medical Laboratory Technology*, **23**, 194–196

Doyle, M. P. and Cliver, D. O. (1990) *Escherichia coli*. In *Foodborne Diseases* (ed. D. O. Cliver), Academic Press, New York, pp.210–216

Farmer, J. J. Davis, B. R., Hickman-Brenne, F. W. *et al.* (1985) Biochemical identification of new species and biogroups of Enterobacteriaceae. *Journal of Clinical Microbiology*, **21**, 46–76

Gorbach, S. L. (1986) *Infectious Diarrhoea*, Blackwells, London

Gorbach, S. L. (1987) Bacterial diarrhoea and its treatment. *Lancet*, **ii**, 1378–1382

Levine, M. M. (1987) *Escherichia coli* that cause diarrhoea: enterotoxigenic, enteropathogenic, enterohaemorrhagic and enteroadherent. *Journal of Infectious Diseases*, **155**, 241–244

Shinebaum, R. and Cooke, M. E. (1985) Klebsiellas. In *Isolation and Identification of Micro-organisms of Medical and Veterinary Importance* (eds C. H. Collins and J. M. Grange), Society for Applied Bacteriology Technical Series No. 21, Academic Press, London, pp.35–42

Chapter

27

Edwardsiella, Erwinia, Hafnia and Serratia

Although most species in these genera do not usually ferment lactose some workers include then in the coliforms.

They are Gram-negative, motile, oxidase negative, fermentative (O/F test), non-sporing rods that are catalase and citrate positive, urease negative (with exceptions), and arginine dihydrolase and PPA negative. See Table 25.1.

Identification

Test for indole, lysine decarboxylase, fermentation of mannitol and inositol, H_2S (with TSI), gelatin liquefaction and ONPG. See Table 27.1.

Table 27.1 ***Edwardsiella, Erwinia, Haffnia*** **and** ***Serratia*** **species**

			Acid from				
	Indole	*Lysine* decarboxylase	*mannitol*	*inositol*	H_2S^a	*Gelatin liquefaction*	*ONPG*
Edwardsiella tarda	+	+	−	−	+	−	−
Erwinia herbicola	−	−	+	v	−	+	+
Haffnia alvei	−	+	+	−	−	−	+
Serratia marcescens	−	v	+	+	−	+	+

[a], TSI; v, variable

Species

Edwardsiella tarda

This is the only species of medical interest in the genus *Edwardsiella*. It is the only member of the group that produces indole and H_2S (in TSI) and that does not ferment mannitol. It is lysine decarboxylase positive, ONPG negative and does not liquefy gelatin.

It is a fish pathogen and a rare pathogen of humans, when it may cause wound infection. Although it is occasionally isolated from patients with diarrhoea its causative role is uncertain.

Erwinia herbicola
Also known as *E. agglomerans*, this may ferment lactose late. It produces acid from mannitol, liquefies gelatin, is ONPG positive but is indole negative, does not decarboxylate lysine nor form H_2S. It is a plant pathogen and is sometimes found in clinical material, mostly from the human respiratory tract.

Hafnia alvei
This gives reliable biochemical reactions only at 20–22°C. It ferments mannitol, decarboxylates lysine, is ONPG positive, but is indole and H_2S negative and does not liquefy gelatin. It is a commensal in the intestines of animals, present in water, sewage and soil but is rarely, if ever, pathogenic.

Serratia marcescens
Once known as *Chromobacterium prodigiosus* this is known for the red pigment it sometimes produces on agar media. It ferments mannitol and inositol, liquefies gelatin and is ONPG positive, but does not produce indole or H_2S.

Originally thought to be a saprophyte, it was used in aerosol and filter testing experiments, it is now known to be associated with hospital-acquired infections such as meningitis, endocarditis and urinary tract infections.

For more information about the biochemical reactions, etc. of these organisms see Farmer *et al.* (1985) and *Cowan and Steel's Manual* (1993).

References

Cowan and Steel's Manual for the Identification of Medical Bacteria (1993) 3rd edn (eds G.I. Barrow and R.K.A. Feltham), Cambridge University Press., Cambridge

Farmer, J.J., Davis, B.R., Hickman-Brenner, F.W. *et al.* (1985) Biochemical identification of new species and biogroups of Enterobacteriaceae. *Journal of Clinical Microbiology*, **21**, 46–76

Chapter

28

Salmonella and Shigella

These are important intestinal pathogens. The very large numbers of different salmonellas have been given species names but there have been attempts to reduce these to a few species with many serotypes. This can cause confusion in clinical and public health bacteriology so the original 'species' names are used here. Fortunately, there are far fewer species of shigellas.

Identification to species is ultimately achieved by serological methods. An outline of the antigenic structures of these organisms is therefore presented first.

Salmonella antigens

The system used was initiated by Kauffmann and White and the tables which are used bear their names.

There are more than 60 somatic or O antigens and these occur characteristically in groups. Antigens 1–50 are distributed among Groups A-Z. Subsequent groups are labelled 51–61. This enables more than 1700 *Salmonella* 'species' or serotypes to be divided into about 40 groups with the commoner organisms in the first six groups.

For example, in Table 28.1 each of the organisms in Group B possesses the antigens 4 and 12 but, although 12 occurs elsewhere, 4 does not. Similarly, in Group C all the organisms possess antigen 6; some possess in addition antigen 7 and other antigens 8. In Group D, 9 is the common antigen. Unwanted antigens may be absorbed from the sera produced against these organisms and single-factor O sera are available that enable almost any salmonella to be placed by slide agglutination into one of the groups. Various polyvalent sera are also available commercially.

To identify the individual organisms in each group, however, it is necessary to determine the H or flagellar antigen. Most salmonellas have two kinds of H antigen and an individual cell may possess one or the other. A culture may therefore be composed of organisms all of which have the same antigens or may be a mixture of both. The alternative sets of antigens are called *phases* and a culture may therefore be in phase 1 or in phase 2 or in both phases simultaneously. The antigens in phase 1 are identified by lower-case letters; thus, the H antigen of *S. paratyphi A* was called *a*, of *S. paratyphi B*, *b*, of *S. paratyphi C*, *c*, and *S. typhi*, *d*, and so on. Unfortunately, after *z* was reached more antigens were found and so subsequent antigens were named z_1, z_2, z_3, etc. It is important to note that z_1 and z_2 are as different as are *a* and *b*; they are not merely subtypes of *z*.

Phase 2 antigens are given the arabic numerals *1–7*.

Table 28.1 Antigenic structure of some of the common salmonellas (Kauffmann-White classification)

Absorbed O antisera available for identification	*Group*	*Name*	*Somatic (O) antigen*	*Flagellar (H) antigen*	
				Phase 1	*Phase 2*
Factor 2	A	*S. paratyphi A*	1, 2, 12	a	—
Factor 4	B	*S. paratyphi B*	1, 4, 5, 12	b	1, 2
		S. stanley	4, 5, 12	d	1, 2
		S. schwarzengrund	4, 12, 27	d	1, 7
		S. saintpaul	1, 4, 5, 12	e, h	1, 2
		S. reading	4, 5, 12	e, h	1, 5
		S. chester	4, 5, 12	e, h	e, n, x
		S. abortus equi	4, 12	—	e, n, x
		S. abortus bovis	1, 4, 12, 27	b	e, n, x
		S. agona	1, 4, 12	gs	—
		S. typhimurium	1, 4, 5, 12	i	1, 2
		S. bredeney	1, 4, 12, 27	l, v	1, 7
		S. heidelberg	1, 4, 5, 12	r	1, 2
		S. brancaster	1, 4, 12, 27	z_{29}	—
Factor 7	C1	*S. paratyphi C*	6, 7, Vi	c	1, 5
		S. cholerae-suis[a]	6, 7	c	1, 5
		S. typhi-suis[a]	6, 7	c	1, 5
		S. braenderup	6, 7	e, h	e, n, z_{15}
		S. montevideo	6, 7	g, m, s	—
		S. oranienburg	6, 7	m, t	—
		S. thompson	6, 7	k	1, 5
		S. infantis	6, 7, 14	r	1, 5
		S. virchow	6, 7	r	1, 2
		S. bareilly	6, 7	y	1, 5
Factor 8	C2	*S. tennessee*	6, 8	z	—
		S. muenchen	6, 8	d	1, 2
		S. newport	6, 8	e, h	1, 2
		S. bovis morbificans	6, 8	r	1, 5
		S. hadar	6, 8	z_{10}	e, n, z
Factor 9	D	*S. typhi*	9, 12, Vi	d	—
		S. enteritidis	1, 9, 12	g, m	—
		S. dublin	1, 9, 12	g, p	—
		S. panama	1, 9, 12	lv	1, 5
Factors 3, 10	E1	*S. anatum*	3, 10	e, h	1, 6
		S. meleagridis	3, 10	e, h	I, w
		S. london	3, 10	l, v	1, 6
		S. give	3, 10	l, v	1, 7
Factor 19	E4	*S. senftenburg*	1, 3, 19	g, s, t	—
Factor 11	F	*S. aberdeen*	11	i	1, 2
Factors 13, 22	G	*S. poona*	13, 22	z	1, 6

Reproduced by permission of Dr F. Kauffman, formerly Director of the International Salmonella Centre. The complete tables, revised regularly, are too large for inclusion here. They are obtainable from Salmonella Centres.
[a] Identical serologically but differ biochemically.

The Kauffmann-White scheme (Table 28.1) is brought up to date every few years by the International Salmonella Centre.

Some of the lettered antigens in phase I also occur in phase 2 as alternatives to other lettered antigens. Whereas *S. paratyphi B* has antigens *b*, and *1* and *2*, *S. worthington* has z, and *l* and *w*, and *S. meleagridis* has *e* and *h*, and *l* and *w*. The lettered antigens in phase 2 do, however, occur mostly in groups, e.g. as *e, h; e, n,* z_{15}; *e, n, x; l, y*, etc., and lettered antigens in phase 2, apart from these, are uncommon.

It is therefore necessary to identify both the phase 1 and the phase 2 antigens as well as the O antigens. It is usual to find the O group first. Various polyvalent and single-factor O antisera are available commercially for this (see note below). The H antigens are then found with a commercial Rapid Salmonella Diagnostic (RSD) sera. These are mixtures of antisera that enable the phase 1 antigens to be determined. The phase 2 antigens are found with a polyvalent serum containing factors *1–7* and then individual sera.

Another antigen is the Vi, found in *S. typhi* and a few other species. This is a surface antigen which masks the O antigen. If it is present the organisms may not agglutinate with O sera unless the suspension is boiled.

Shigella antigens

The genus *Shigella* is divided into four antigenic and biochemical subgroups (Table 28.2).

Table 28.2 *Shigella* classification (see also Table 28.3)

Subgroup	*Species*	*Serotypes*
A: Mannitol not fermented	*S. dysenteriae*	1–10 all distinct
B: Mannitol fermented	*S. flexneri*	1–6 all related; 1–4 divided into subserotypes
C: Mannitol fermented	*S. boydii*	1–15 all distinct
D: Mannitol fermented, lactose fermented late	*S. sonnei*	1 serotype

Subgroup A

S. dysenteriae
Contains 10 serotypes that have distinct antigens and do not ferment mannitol.

Subgroup B

S. flexneri
Contains six serotypes (1–6) that can be divided into subserotypes according to their possession of some group factors designated 3,4; 4; 6; 7; and 7,8 (Table 28.3).

The X variant is an organism that has lost its type antigen and is left with the group factors 7,8.

The Y variant is an organism that has lost its type antigen and is left with the group factors 3,4.

Table 28.3 Flexner subserotypes

Subserotype	*Agglutination with serotype*	*Abbreviated antigenic formula*
1	1a	I:4
1 and 3	1b	I:6
2	2a	II:3, 4
2 and X (7, 8)	2b	II:7, 8
3 and X (7, 8)	3a	III:6, 7, 8
3 and Y (3, 4)	3b	III:3, 4, 6
3	3c	III:6
4	4a	IV:3, 4
4 and 3	4b	IV:6
5	5	V:7, 8
6	6	VI:—

Subgroup C

S. boydii
Contains 15 serotypes with distinct antigens; all ferment mannitol.

Subgroup D

S. sonnei
Contains only one, distinct serotype; this ferments mannitol. It may occur in phase 1 or in phase 2, sometimes referred to as 'smooth' and 'rough'. The change from phase 1 to phase 2 is a loss of variation and phase 2 organisms are often reluctant to agglutinate or give a very fine, slow agglutination. Phase 2 organisms are rarely encountered in clinical work but the sera supplied commercially agglutinate both phases.

Isolation of salmonellas and shigellas

Choice of media

Liquid media
These are generally inhibitory to coliforms but less so to salmonellas. Selenite F media are commonly used. If they are not overheated selenite F media will permit the growth, but not the enrichment, of *S. sonnei* and *S. flexneri* 6. Mannitol selenite and cystine selenite broths are preferred by some workers for the isolation of salmonellas from foods. Tetrathionate media enrich only salmonellas: shigellas do not grow.

Cultures from these media are plated on solid media.

Solid media
Many media exist for the isolation of salmonellas and shigellas but in our experience the Hynes modification of deoxycholate citrate agar (DCA) and Xylose Lysine Deoxycholate agar (XLD) give the best results for faeces. For the isolation of salmonellas from foods, XLD and Brilliant Green agar (BGA) are superior to DCA.

Colony appearance

Deoxycholate citrate agar (DCA)
On the ideal DCA medium, colonies of salmonellas and shigellas are not sticky. Salmonella colonies are creamy brown, 2–3 mm in diameter at 24 h and usually have black or brown centres. Unfortunately, colonies of proteus are similar. Shigella colonies are smaller, usually slightly pink and do not have black centres. *S. cholerae-suis* grows poorly on this medium. White, opaque colonies are not significant. Some *Klebsiella* strains grow.

Xylose lysine decarboxylase agar (XLD)
Salmonella colonies are red with black centres, 3–5 mm in diameter. Shigella colonies are red, 2–4 mm in diameter.

Brilliant green agar (BGA)
Salmonella colonies are pale pink with a pink halo. Other organisms have yellow-green colonies with a yellow halo.

Faeces, rectal swabs and urines
Plate faeces and rectal swabs on DCA and XLD agars. Inoculate selenite F broth with a portion of stool about the size of a pea or with 0.5–1.0 ml of liquid faeces. Break off rectal swabs in the broth. Add about 10 ml of urine to an equal volume of double-strength selenite F broth.

Incubate plates and broths at 37°C for 18–24 h. Plate broth cultures on DCA, XLD or BGA and incubate at 37°C for 18 h.

Drains and sewers
Fold three or four pieces of cotton gauze (approx. 15 × 20 cm) into pads of 10 × 4 cm. Tie with string and enclose in small-mesh chicken wire to prevent sabotage by rats. Autoclave in bulk and transfer to individual plastic bags or glass jars. Suspend in the drain or sewer by a length of wire and leave for several days. Squeeze out the fluid into a jar and add an equal volume of double-strength selenite F broth. Place the pad in another jar containing 200 ml of single-strength selenite F broth. Incubate both jars at 43°C for 24 h and proceed as for faeces described by Moore (1948) (see also Vassiliades *et al.*, 1978).

Foods, feedingstuffs and fertilizers
The method of choice depends on the nature of the sample and on the total bacterial count. In general, raw foods such as comminuted meats have very high total counts and salmonellas are likely to be present in large numbers. On the other hand, heat-treated and deep-frozen foods should have lower counts and if any salmonellas are present they may suffer from heat shock or cold shock and require resuscitation and pre-enrichment.

Pre-enrichment
Make a 1:10 suspension or solution of the sample in buffered peptone water. Use 25 g of food if possible. For dried milk make the suspension in 0.002% brilliant green in sterile water. Use a stomacher if necessary. Incubate at 37°C for 16–20 h.

Enrichment
Add 10 ml of the pre-enrichment to 100 ml of Muller-Kauffman tetrathionate broth (MK-TB), mixing well so that the calcium carbonate stabilizes the pH.

Incubate at 43°C for up to 48 h.

Selective plating
At 18–24 h and at 42–48 h subculture from the surface of the MK-TB to brilliant green phenol red agar (BGA). Incubate at 37°C for 22–24 h.

Salmonella colonies are pink, smooth and low convex. Subculture at least two colonies for identification.

Additional procedures
Add 0.1 ml of the pre-enrichment to 10 ml of Rappaport-Vasiliades (RV) enrichment medium. Subculture at intervals as with MK-TB. This method may increase isolation rates, as will additional cultures on DCA and XLD.

Many rapid and automated methods have been developed for determining the presence of salmonellas in foods. There is a variety of commercially available equipment and instrumentation. An up-to-date and comprehensive review, with an extensive bibliography, is provided by de Blackburn and Stannard (1989). See also Chapter 13. There are also immunological (ELISA) techniques (Clayden *et al.*, 1987).

Identification

Kit methods are useful in screening for salmonellas but they do not necessarily identify the serotype. They include the Tecra (Tecra Diagnostics, USA) which is an immunocapture method; Spectate (Rhône Poulenc, France), coloured latex particles with specific antibodies; Salmonella Rapid (Oxoid, UK) which uses a purpose-designed culture vessel; Rapid Salmonella Latex (Oxoid), latex agglutination; Microscreen (Mercia, UK), another latex method (Hadfield *et al.*, 1987) that uses coloured latex; and MUCAP, which uses a fluorogenic substrate for testing colonies.

For full identification of salmonellas and shigellas to species/serotype there are two approaches: agglutination tests supported by screening or biochemical tests; and biochemical tests followed by agglutination tests. The first of these requires considerable expertise in the recognition of colonies and in agglutination procedures but gives an answer much earlier, often within 24 h from primary plates but certainly by 48 h after enrichment.

Agglutination tests for salmonellas

Polyvalent, specific and 'rapid salmonella diagnostic' (RSD) sera are available commercially. The latter consist of three or four mixtures of specific and non-specific sera; agglutination by combinations indicates the presence of commoner antigens or groups of antigens.

Test the suspected organism by slide agglutination with Polyvalent O sera. If positive, do further slide agglutinations with the relevant single-factor O sera; only one should be positive. If in doubt, test against 1:500 acriflavine; if this is positive, the organisms are unlikely to be salmonellas.

Subculture into glucose broth and incubate in a waterbath at 37°C for 3–4 h, when there will be sufficient growth for tube H agglutinations. Tube H agglutinations are more reliable than slide H agglutinations because cultures on solid media may not be motile. The best results are obtained when there are about $6–8 \times 10^8$ organisms/ml. Dilute, if necessary, with formol saline. Add 0.5 ml of formalin to 5 ml of broth culture and allow to stand for at least 10 min to kill the organisms.

Place one drop of Polyvalent (phase 1 and 2) Salmonella H serum in a 75 ×

9 mm tube. Add 0.5 ml of the formolized suspension and place in a water-bath at 52°C for 30 min. If agglutination occurs, the organism is probably a salmonella.

Identify the H antigens as follows. Do tube agglutinations against the Rapid Salmonella Diagnosis (RSD) sera, *S. typhimurium* (*i*) if it is not included in the RSD set, phase 2 complex 1–7 EN (e,n) complex and L (l) complex. The manufacturer's guide will indicate which phase 1 antigens are present according to which sera or groups of sera give agglutination. Check the result, if possible by doing tube agglutinations with the indicated monospecific serum. If the test indicates EN (e,n), G or L (l) complexes, send the cultures to a Public Health or Communicable Diseases Laboratory.

If agglutination occurs in the phase 2 *1–7* serum, test each of the single factors, *2, 5, 6* and *7*. Note that with some commercial sera, factor *1* is not absorbed and it may be necessary to dilute this out by using one drop of 1:10 or weaker serum.

Changing the H phase

If agglutination is obtained with only one phase, the organism must be induced to change to the other phase.

Cut a 50 × 20 mm ditch in a well-dried nutrient agar plate. Soak a strip of previously sterilized filter-paper (36 × 7 mm) in the H serum by which the organism is agglutinated and place this strip across the ditch at right-angles. At one end, place one drop of 0.5% thioglycollic acid to neutralize any preservative in the serum. At the other end, place a filter-paper disc, about 7 mm in diameter, so that half of it is on the serum strip and the other half on the agar. Inoculate the opposite end of the strip with a young broth culture of the organism and incubate overnight. Remove the disc with sterile forceps, place it in glucose broth and incubate it in a water-bath for about 4 h, when there should be enough growth to repeat agglutination tests to find the alternative phase. Organisms in the original phase demonstrated will have been agglutinated on the strip. Organisms in the alternate phase will not be agglutinated and will have travelled across the strip.

The phase may also be changed with a Craigie tube. Place 0.1 ml of serum and 0.1 ml of 0.5% thioglycollic acid in the inner tube of a Craigie tube and inoculate the inner tube with the culture. Incubate overnight and subculture from the outer tube into glucose broth.

Some organisms, e.g. *S. typhi* and *S. montevideo*, have only one phase. These should be sent to a reference laboratory.

Antigenic formulae

List the O, phase 1 and phase 2 antigens in that order and consult the Kauffmann-White tables (Table 28.1) for the identity of the organism. In the table, the antigenic formula is given in full, but as we use single factor sera for identification it is usually written thus:

S. typhimurium	4, *i*, 2
S. typhi	9, *d*, —
S. newport	6, 8, *e, h*, 2

(The – sign indicates the organism is monophasic)

The Vi antigen

If this antigen is present and prevents the organisms agglutinating with O group sera, make a thick suspension in saline from an agar slope and boil it in a water-bath for 1 h. This destroys the *Vi* and permits the O agglutination.

Salmonella subgenera

The genus *Salmonella* has been divided into four subgenera, numbered I, II, III and IV. Subgenus I is the most important because it contains the majority of the human and animal pathogens. Serotypes ('species') are named. Subgenus II contains serotypes commonly found in reptiles but rarely found in humans. Some are named; others known only by their antigenic formula. Subgenus III contains the organisms previously described as the Arizona group. Subgenus IV contains rare serotypes, only some of which are named.

Identification of subgenus is by biochemical tests but as this is of limited practical value it is not pursued here (see Ewing 1986).

Biochemical screeening tests

Pick representative colonies on a urea slope as early as possible in the morning. Incubate in a water-bath at 35–37°C. In the afternoon, reject any urea positive cultures and from the others inoculate the following: Triple Sugar Iron (TSI) medium by spreading on slope and stabbing the butt; broth for indole test; lysine decarboxylase medium. Incubate overnight at 37°C (see Table 28.4) and proceed to serological identification if indicated.

Table 28.4 Biochemical reactions of *Salmonella*, *Shigella*, etc.

	TSI[a]		H_2S	*Indole*	*ONPG*	*Lysine decarboxylase*
	Butt	*Slope*				
S. typhi	A	–	+[b]	–	–	+
S. paratyphi A	AG	–	–	–	–	–
Other salmonellas	AG	–	+/–	–	–	+
Shigella	A	–	–	+/–	+/–	–
Citrobacter	AG	–	+	–	+	–

A = acid; AG = acid and gas.
[a] In TSI medium acid in the butt indicates the fermentation of glucose and in the slope that of lactose and/or sucrose.
[b] H_2S production by *S. typhi* may be minimal.

Full biochemical characters

For salmonella serotypes and subgenera it is best to use kits such as the API 20E, Mast ID-15 (Mast, UK), Cobas ID and Enterotube (Roche, UK). See Holmes and Costas (1992) for more details of these.

Salmonella species and serotypes

Caution
Salmonella typhi, and in some countries S. paratyphi A [see national lists] are in Hazard Group 3. If tests suggest that a suspect organism is one of these all further investigations should be conducted under Containment Biosafety Level 3 conditions.

Salmonella typhi
Produces acid but no gas from glucose and mannitol. It may produce acid from dulcitol but fails to grow in citrate media and does not liquefy gelatin. It

decarboxylates lysine but not ornithine. It is ONPG negative. Freshly isolated strains may not be motile at first. This organism causes enteric fever.

Other serotypes

Produce acid and usually gas from glucose, mannitol and dulcitol (anaerogenic strains occur) and rare strains ferment lactose. They grow in citrate and do not liquefy gelatin. They decarboxylate lysine and ornithine (except *S. paratyphi A*) and are ONPG negative. *S. paratyphi A, B* and *C* may also cause enteric fever. Strains with the antigenic formula 4; *b; 1, 2* may be *S. paratyphi B*, usually associated with enteric fever, or *S. java* which is usually associated with food poisoning. Many other serotypes cause food poisoning.

Arizona organisms

These are now included with salmonellas. They have the antigenic formula 23; z_4, z_{23}, z_{26}. There are a number of subserotypes. They fail to ferment dulcitol, may ferment lactose late, slowly liquefy gelatin, decarboxylate lysine and ornithine and are malonate and ONPG positive.

Reference laboratories

It is not possible to identify all of the 1700 or more salmonella species or serotypes with the commercial sera available and in the smaller laboratory. The following organisms should be sent to a Reference or Communicable Diseases Laboratory.

Salmonellas placed by RSD sera into Groups E, G, or L.

Salmonellas agglutinated by Polyvalent H or monospecific H sera but not by phase II sera.

Salmonellas agglutinated by Polyvalent H but not by phase I or phase II sera.

Salmonellas giving the antigenic formula 4; *b; 1, 2*, which may be either *S. paratyphi B* or *S. java*.

S. typhi, S. paratyphi B, S. tyhpimurium, S. thompson, S. virchow, S. hadar and *S. enteritidis* should be sent for phage typing.

Note that *Brevibacterium* spp., which are sometimes found in foods, possess Salmonella antigens. These organisms are Gram positive.

Agglutination tests for shigellas

Do slide agglutinations with *S. sonnei* (phase 1 and 2), polyvalent Flexner and polyvalent Boyd antisera. If these are negative test with *S. dysenteriae* antiserum. If there is agglutination with the polyvalent antisera test with the individual antisera: Flexner 1–6 and Boyd 1–5.

Sometimes there is no agglutination although biochemical tests indicate shigellas. This may be due to masking by a surface (K) antigen. Boil a suspension of the organisms and repeat the tests.

Strains that cannot be identified locally, i.e. give the correct biochemical reactions but are not agglutinated by available sera, should be sent to a reference laboratory.

Note that the motile organism *Plesiomonas shigelloides* (p.299) may be agglutinated by *S. sonnei* antiserum.

Table 28.5 Shigellas

Species	*Acid from*					*Indole*	*Lysine decarboxylase*	*Ornithine decarboxylase*
	Glucose	*Lactose*	*Sucrose*	*Dulcitol*	*Mannitol*			
S. dysenteriae	+	−	−	−	−	−/+	−	−
S. flexneri 1–5	+	−	−	−	+	v	−	−
S. flexneri 6 (Boyd 88)	+ (G)	−	−	−	+/−	−	−	−
S. sonnei	+	(+)	(+)	−	+	−	−	+
S. boydii 1–15	+	−	−	−	+	+/−	−	−

+ (G), a small bubble of gas may be formed; −/+, most strains negative; +/−, most strains positive.
None of these organisms grow in citrate or ferment salicin; none produce H_2S, liquefy gelatin or are motile.

Biochemical tests

Test for acid production from glucose, lactose, sucrose, dulcitol and mannitol, indole production, lysine and ornithine decarboxylases. See Table 28.5.

Alternatively use one of the kits for enteric pathogens.

Species of shigella

S. dysenteriae (S. shiga)
Forms acid but no gas from glucose but does not ferment mannitol. Types 1, 3, 4, 5, 6, 9 and 10 are indole negative. Types 2, 7 and 8 are indole positive and are also known as *S. schmitzii* (*S. ambigua*). *S. dysenteriae* is responsible for the classic bacillary dysentery of the Far East.

S. flexneri
Serotypes 1–5 form acid but no gas from glucose and mannitol and are indole positive. Some type 6 strains (Newcastle strains) may not ferment mannitol and may be indole negative. A bubble of gas may be formed in the glucose tube. This serotype is not inhibited by selenite media. Serotypes 1–6 are widely distributed, especially in the Mediterranean area, and often cause outbreaks in mental and geriatric hospitals, nurseries and schools in temperate climes.

S. flexneri subserotypes
Wellcome Reagents do not absorb group 6 factor from their type 3 serum and they also provide X and Y variant sera containing group factors 7, 8 and 3, 4. These sera may therefore be used to determine the Flexner subserotypes for epidemiological purposes (Table 28.5).

S. sonnei
Forms acid but no gas from glucose and mannitol, is indole negative and decarboxylates ornithine. It causes the commonest and mildest form of dysentery, which mostly affects babies and young children and spreads rapidly through schools and nurseries. It may be typed for epidemiological purposes by colicin production (reference laboratory). It survives in some selenite media.

S. boydii
Forms acid but no gas from glucose and mannitol; indole production is variable. The 15 serotypes of this species are widely distributed but not common and cause mild dysentery.

References

Blackburn, C. de W. and Stannard, C. J. (1989) Immunological detection methods for salmonellas in foods. In *Rapid Microbiological Methods for Foods, Beverages and Pharmaceuticals* (eds C. J. Stannard, S. B. Pettit and F. A. Skinner), Society for Applied Bacteriology Technical Series No. 25, Blackwells, London, pp. 249–264

Clayden, J. A., Alcock, S. J. and Stringer, M. F. (1987) Enzyme linked immunosorbent assays for the detection of salmonellas in food. In *Immunological Techniques in Microbiology* (eds J. M. Grange, A. Fox and N. L. Morgan), Society for Applied Bacteriology Technical Series No. 24, Blackwells, London, pp. 217–229

Ewing, W. H. (ed.) (1986) *Edwards and Ewing's Identification of the Enterobacteriaceae*, Elsevier, New York

Hadfield, S. G., Jouy, N. F. and McIllmurray, M. B. (1987) The application of a novel coloured latex test to the detection of *Salmonella*. In *Immunological Methods in Microbiology* (eds J. M. Grange, A. Fox and N. L. Morgan), Society for Applied Bacteriology Technical Series No. 24, Blackwells, London, pp. 145–151

Holmes, B. and Costas, M. (1992) Methods and typing of Enterobacteriacea by computerized methods. In *Identification Methods in Applied and Environmental Microbiology* (eds R. G. Board, D. Jones and F. A. Skinner), Society for Applied Bacteriology Technical Series No. 29, Blackwells, Oxford, pp. 127–150

Moore, B. (1948) The detection of paratyphoid B carriers by means of sewage examination. *Monthly Bulletin Ministry of Health and PHLS*, **6**, 241–251

Vassiliades, P. Trichopoulos, D., Kalandidi, A. and Xirouchaki, E. (1978) Isolation of salmonellas from sewage with a new enrichment method. *Journal of Applied Bacteriology*, **44**, 233–239

Chapter

29

Proteus, Providencia and Morganella

These organisms are widely distributed in nature. Some *Proteus* strains spread over the surface of ordinary agar media ('swarming'), obscuring colonies of other organisms. This may be prevented by using chloral hydrate agar. Swarming does not take place on CLED and media containing bile salts.

The important characteristic that distinguishes these genera from other enterobacteria is their ability to deaminate phenylalanine (positive PPA test).

Table 29.1 ***Proteus, Providencia*** **and** ***Morganella*** **species**

	Motility	*Indole*	*Acid from*		*Citrate utiliza-tion*	H_2S[a]	*Gelatin lique-faction*	*Urease*
			Mannitol	*Inositol*				
Proteus								
vulgaris	v	+	–	–	v	+	v	+
mirabilis	+	–	–	–	+	+	+	+
Providencia								
alcalifaciens	v	+	–	–	+	v	–	–
rettgeri	+	+	+	+	+	+	–	+
stuartii	v	+	–	+	+	v	–	–
Morganella								
morgani	+	+	–	–	v	+	–	+

[a] Lead acetate paper; v, variable

Isolation

Plate material on nutrient or blood agar to observe swarming and on CLED and chloral hydrate agar to prevent swarming and reveal any other organisms. Incubate at 37°C overnight.

Identification

Test PPA positive strains for motility, indole production, fermention of mannitol and inositol, citrate utilization, H_2S production, gelatin liquefaction and urease. See Table 29.1.

Proteus species

Proteus vulgaris
This is indole positive, produces H_2S (lead acetate paper test) but does not ferment mannitol or inositol. Many strains swarm on ordinary media (see above).

It is found in the intestines of humans and animals and is an opportunist pathogen in the urinary tract, wounds and other lesions. It is a food spoilage organism, widely distributed in soil and sewage-polluted water.

Proteus mirabilis
Differs mainly from *P. vulgaris* in not producing indole. It is a doubtful agent of gastroenteritis, found in urinary tract and hospital-acquired infections and septic lesions. It is associated with putrefying animal and vegetable matter.

Providencia species

Providencia alcalifaciens
Differs from *Proteus* in not possessing urease. It is found in clinical material where it may be an opportunist pathogen.

Providencia rettgeri
Formerly *Proteus inconstans*, this ferments mannitol and inositol, produces H_2S (lead acetate paper) and is urease positive. It is associated with 'fowl typhoid', and urinary tract and opportunist infections in humans.

Providencia stuartii
Ferments inositol but not mannitol, may produce H_2S and is urease negative. It has been reported in hospital infections.

Morganella

Morganella morganii
This is the only species so far described. It does not ferment mannitol or inositol but is H_2S and urease positive. It is a commensal in the intestines of humans and other animals, found in hospital infections and may be associated with 'summer diarrhoea' in children.

Chapter

30

Key to miscellaneous aerobic non-sporing Gram-negative bacilli of medical importance

The 16 genera considered in this chapter may be divided into five groups, four on the basis of common characteristics and a fifth with no common factors.

(1) *Brucella, Bordetella* and *Moraxella* are non-motile, can grow anaerobically, are catalase and oxidase positive, do not change Hugh and Leifson medium and have no special growth requirements (Chapter 31).
(2) *Haemophilus, Gardnerella* and *Streptobacillus* are non-motile, can grow anaerobically, are fermentative in Hugh and Leifson medium and have special growth requirements (Chapter 32).
(3) *Campylobacter* and *Helicobacter* which are motile, can grow anaerobically, do not change Hugh and Leifson medium and have no special growth requirements, although they are microaerophilic (Chapter 33).
(4) *Actinobacillus, Pasteurella, Yersinia, Cardiobacterium* and *Francisella* are motile, can grow anaerobically, are fermentative in Hugh and Leifson medium and have no special growth requirements (Chapter 34).
(5) *Legionella, Bartonella* and *Mobiluncus* are included together for convenience (Chapter 35).

Table 30.1 shows the general characteristics and those that assist in identification within the groups are given in the appropriate chapters.

Table 30.1 Gram-negative non-sporing bacilli

Key to miscellaneous aerobic non-sporing Gram-negative bacilli of medical importance

	Motility	*Facultative anaerobe*	*Catalase*	*Oxidase*	*O/F*	*Growth factor requirement*
Brucella	−	−	+	+	−	−
Bordetella	−	−	+	v	−	−
Moraxella	−	−	−	+	−	−
Haemophilus	−	+	v	+	F	+
Gardnerella	−	+	−	+	F	+
Streptobacillus	−	+	−	−	F	+
Campylobacter	+	+	v	v	−	−
Helicobacter	+	+	v	v	−	−
Actinobacillus	−	+	+	+	F	−
Pasteurella	−	+	+	+	F	−
Yersinia	−	+	+	+	F	−
Cardiobacterium	−	+	−	+	F	−
Francisella	−	+	+	v	F	−
Legionella	+	−	+	v	F	+
Bartonella, *Mobiluncus*	See Ch. 35					

v, variable; F, fermentative

Chapter

31

Brucella, Bordetella and Moraxella

Brucella

The genus Brucella contains Hazard Group 3 pathogens. All manipulations that might produce aerosols should be done in microbiological safety cabinets in Biosafety/Containment Level 3 laboratories.

There are three important and several other species of small, Gram-negative bacilli in this genus. They are non-motile, reduce nitrates to nitrites, but carbohydrates are not metabolized when normal cultural methods are used. There are several biotypes within each species.

Isolation

Human disease
Do blood cultures. Any system may be used but the Casteñada double phase method (p.164) in which the solid and liquid media are in the same bottle lessens the risk of contamination during repeated subculture and also minimizes the hazards of infection of laboratory workers. Use Brucella agar plus growth and selective supplements.

Examine Castenada bottles weekly for growth on the solid phase. If none is seen, tilt the bottle to flood the agar with the liquid, restore it to a vertical position and reincubate. Do not discard until 6 weeks have elapsed.

Treat bone marrow and liver biopsies in the same way. Subculture on Brucella selective media and on serum glucose agar or chocolate blood agar.

Animal material
Plate uterine and cervical swabbings or homogenized fetal tissue on Brucella selective medium

Milk
Dip a throat swab in the gravity cream and inoculate Brucella selective medium.

Incubate all cultures at 37°C under 10% CO_2 and examine daily for up to 5 days.

Identification

Colonies on primary media are small, flat or slightly raised and translucent. Subculture on slopes of glucose tryptone agar with a moistened lead acetate paper in the upper part of the tube to test for hydrogen sulphide production.

Do dye inhibition tests. In reference laboratories several concentrations of thionin and basic fuchsin are used to identify biotypes. For diagnostic purposes inoculate tubes of Brucella medium or glucose tryptone serum agar containing (1) 1:50000 thionin, and (2) 1:50000 basic fuchsin (National Aniline Division, Allied Chemical and Dye Corp.).

Inoculate tubes or plates with a small loopful of a 24-h culture.

It is advisable to control *Brucella* identification with reference strains, obtainable from Type Culture Collections or CDCs: *B. abortus* 554, *B. melitensis* 16M and *B. suis* 1330 (see Table 31.1).

Table 31.1 *Brucella* species

	Growth in		*Needs CO_2*	*H_2S*
	Fuchsin	*Thionin*		
B. melitensis	+	+	−	−
B. abortus	+	−	+	+ (most strains)
B. suis (American)	−	+	−	+ +
B. suis (other biotypes)	−	+	−	−

Do slide agglutination tests with available commercial sera. Absorbed monospecific sera are not yet available commercially. Reference laboratories make their own and also use bacteriophage identification and typing methods, and metabolic tests (see Robertson *et al.*, 1980).

Detection in milk

The Ring test

Purchase the stained antigen from a veterinary or commercial laboratory. It is a suspension of *Brucella abortus* cells stained with haematoxylin or another dye. Store the milk samples overnight at 4°C before testing.

To 1 ml of well mixed raw milk in a narrow tube (75 × 9 mm), add one drop (0.03 ml) of the stained antigen. Mix immediately by inverting several times and allow to stand for 1 h at 37°C.

If the milk contains antibodies, these will agglutinate the antigen and the stained aggregates of bacilli will rise with the cream, giving a blue cream line above a white column of milk. Weak positives give a blue cream line and a blue colour in the milk. Absence of antibodies is shown by a white cream line above blue milk. It may be necessary to add known negative cream to a low-fat milk.

False positives may be obtained with milk collected at the beginning and end of lactation, probably due to leakage of serum antibodies into the milk.

The Ring test does not give satisfactory results with pasteurized milk or with goats' milk.

Whey agglutination test

This is a useful test when applied to milk from individual cows but is of questionable value for testing bulk milk.

Centrifuge quarter milk and remove the cream. Add a few drops of rennin to the skimmed milk and incubate at 37°C for about 6 h. When coagulated, centrifuge and set up doubling dilutions of the whey from 1:10 to 1:2560, using 1-ml amounts in 75 × 9 mm tubes. Add one drop of standard concentrated *B. abortus* suspension and place in a water-bath at 37°C for 24 h. Read the agglutination titre. More than 1:40 is evidence of udder infection unless the animal has been vaccinated recently.

Serological diagnosis
Apart from the standard agglutination test (above) there are others that are outside the scope of this book. For information on the mercaptoethanol test, antihuman globulin and complement fixation tests see Robertson *et al.* (1980).

Useful information about the laboratory diagnosis of brucellosis is also given in the WHO Monograph No. 5 (1981) and Corbell and Hendry (1985).

Species of Brucella

Brucella melitensis
Grows on blood agar and on glucose tryptone serum agar aerobically in 3–4 days. Does not need carbon dioxide to initiate growth. Does not produce hydrogen sulphide and is not inhibited by fuchsin or thionin in the concentrations used in the commercial media. This organism causes brucellosis in humans, the Mediterranean or Malta fever or undulant fever. The reservoirs are sheep and goats and infection occurs by drinking goats' milk.

B. abortus
There are eight biotypes. Most require 5–10% carbon dioxide to initiate growth, produce hydrogen sulphide and are inhibited by thionin but not by fuchsin. Some biotypes resemble *B. melitensis*. They cause contagious abortion in cattle. Drinking infected milk can result in undulant fever in humans. Veterinarians and stockmen are frequently infected from aerosols released during birth or abortion of infected animals. Laboratory infections are usually acquired from aerosols released by faulty techniques.

B. suis
The four biotypes of this species do not require carbon dioxide for primary growth. The American biotype produces abundant hydrogen sulphide: the others do not. Biotypes 1 and 2 are inhibited by fuchsin but none are inhibited by thionin. They cause contagious abortion in pigs and may infect humans, reindeer, hares and geese.

Bordetella

The bacilli in this genus are small, 1.5 μm × 0.3 μm and regular. Nitrates are not reduced and carbohydrates are not attacked.

Isolation and identification

Pernasal swabs are better than cough plates, but swabs are best conveyed in one of the commercial transport media. Plate on Bordetella, or Charcoal agar medium containing cephalexin 40 mg/l (Oxoid Bordetella supplement).

Incubate under conditions of high humidity and at 37°C for 3 days. Examine daily and identify by slide agglutination, using commercially available sera. FA reagents are also available.

Test for growth and pigment formation on nutrient agar, urease, nitratase and motility (see Table 31.2).

Table 31.2 *Bordetella* species

	Growth on nutrient agar	*Brown coloration*	*Urease*	*Nitrate reduction*	*Motility*
B. pertussis	−	−	−	−	−
B. parapertussis	+	+	+	−	−
B. bronchiseptica	+	−	+	+	+

Species of Bordetella

Bordetella pertussis
Growth on the above media has been described as looking like a 'streak of aluminium paint'. Colonies are small (about 1 mm in diameter) and pearly grey. This organism cannot grow in primary culture without heated blood, but may adapt to growth on nutrient agar on subculture. *B. pertussis* causes whooping cough. For detailed information about the bacteriological diagnosis of this disease (see Wardlaw, 1990).

B. parapertussis
Growth on blood agar may take 48 h and a brown pigment is formed under the colonies. It grows on nutrient and on MacConkey agar. On Bordetella or similar media, the pearly colonies ('aluminium paint') develop earlier than those of *B. pertussis*. It does not change Hugh and Leifson medium and does not metabolize carbohydrates. It is nitrate negative and urease positive.

This organism is one of the causative organisms of whooping cough.

B. bronchiseptica
This organism has been in the genera *Brucella* and *Haemophilus*. It forms small smooth colonies, occasionally haemolytic on blood agar, grows best as 37°C and is motile, urease positive and grows in citrate medium.

It causes bronchopneumonia in dogs, often associated with distemper, bronchopneumonia in rodents and snuffles in rabbits. It has been associated with whooping cough.

This genus is discussed in detail by Wardlaw (1990) and Pittman and Wardlaw (1991).

Moraxella

The bacilli are plump (2 μm × 1 μm), often in pairs end to end, and non-motile. They are oxidase positive and do not attack sugars. They are indole negative, do not produce hydrogen sulphide and are sensitive to penicillin. Some species require enriched media.

Plate exudate, conjunctival fluid, etc., on blood agar and incubate at 37°C overnight. Subculture colonies of plump, oxidase positive Gram-negative bacilli on blood agar and on nutrient agar (not enriched), on gelatin agar (use the plate method) and on Loeffler medium and incubate at 37°C. Do the nitrate reduction test (see Table 31.3).

Table 31.3 Species of *Moraxella*

	Growth on nutrient agar	*Gelatin liquefaction*	*Nitrate reduction*	*Catalase*	*Urease*	*PPA*
M. lacunata	v	+	+	+	−	−
M. nonliquefaciens	v	−	+	+	−	−
M. bovis	+	+	v	+	−	−
M. osloensis	+	−	v	+	−	−
M. kingii[a]	+	−	−	−	−	−
M. phenylpyruvica	+	−	+	+	v	+

PPA, phenylalanine test
[a] *Kingella kingii*

Species of Moraxella

Moraxella lacunata
Colonies on blood agar are small and may be haemolytic. There is no growth on non-enriched media. Colonies on Loeffler medium are not visible but are indicated by pits of liquefaction ('lacunae'). Gelatin is liquefied slowly. Nitrates are reduced.

This organism is associated with angular conjunctivitis and is known as the Morax-Axenfeld bacillus.

M. liquefaciens
This is similar to *M. lacunata* and may be a biotype of that species. It liquefies gelatin rapidly.

M. nonliquefaciens
This is also similar to *M. lacunata* but fails to liquefy gelatin.

M. bovis
Requires enriched medium, liquefies gelatin and may reduce nitrates. It causes pink eye in cattle but has not been reported in human disease.

M. osloensis (*M. duplex, Mima polymorpha var. oxidans*)
Enriched medium is not required. Gelatin is not liquefied; nitrates may be reduced. It is found on the skin and in the eyes and the respiratory tract of humans but its pathogenicity is uncertain.

M. kingii (*now Kingella kingii*)
Colonies are haemolytic and may be mistaken for haemolytic streptococci or haemolytic haemophilus. It does not require enriched media. This is the only member of the group which is catalase negative. It has been found in joint lesions and in the respiratory tract.

M. phenylpyruvica
This does not require enriched media. It is urease variable and reduces phenylalanine to phenylpyruvic acid (PPA positive) – the only species in this

group to do this. It has been isolated from assorted human material but its pathogenicity is uncertain.

Some workers include *Branhamella catarrhalis* in the genus *Moraxella*.

For more information see Borre and Hagen (1981) and *Cowan and Steel's Manual* (1993).

References

Borre, K. and Hagen, N. (1981) Neisseriaceae; rod-shaped species of the genera *Moraxella, Acinetobacter, Kingella* and *Neisseria*; and the *Branhamella* groups of cocci. In *The Prokaryotes: a Handbook of Habitats, Isolation and Identification of Bacteria* (eds M. P. Starr, H. Stolp, G. Truper, A. Balows and A. Schlegel), Springer-Verlag, New York, pp. 1506–1529

Corbell, M. J. and Hendry, D. McL. F. D. (1985) Brucellas. In *Isolation and Identification of Micro-organisms of Medical Importance* (eds C. H. Collins and J. M. Grange), Society for Applied Bacteriology Technical Series No.21, Academic Press, London, pp. 53–82

Cowan and Steel's Manual for the Identification of Medical Bacteria (1993) 3rd edn (eds G. I. Barrow and R. K. A. Feltham), Cambridge University Press, Cambridge

Pittman, M. and Wardlaw, A. C. (1981) The genus *Bordetella*. In *The Prokaryotes: a Handbook of Habitats, Isolation and Identification of Bacteria* (eds M. P. Starr, H. Stolp, G. Truper, A. Balows and A. Schlegel), Springer-Verlag, New York, pp. 1506–1529

Robertson I., Farrell I. D., Hinchcliffe, P. M. and Quaife, R. D. (1980) *Benchbook on Brucella*. Public Health Laboratory Service Monograph No.14, HMSO, London

Wardlaw, A. C. (1990) *Bordetella*. In *Topley and Wilson's Principles of Bacteriology, Virology and Immunity*, 8th edn, vol.2 (eds M. T. Parker and B. I. Duerden), Edward Arnold, London, pp. 321–338

WHO (1981) *A Guide to the Diagnosis, Treatment and Prevention of Human Brucellosis* (ed. S. S. Elberg), World Health Organization, Geneva

Chapter

32

Haemophilus, Gardnerella and Streptobacillus

Haemophilus

For *in vitro* culture these organisms require either or both of two factors that are present in blood: X factor, which is haematin, and V factor, which is diphosphopyridine nucleotide and can be replaced by co-enzymes I or II. The X factor is heat stable; the V factor is heat labile. V factor is synthesized by *Staphylococcus aureus*. The bacilli are small (1.5 μm × 0.3 μm), non-motile, usually regular, fail to grow on ordinary media and reduce nitrate to nitrite. There are several species.

Haemophilus antigen may be directly detected in cerebrospinal fluid by countercurrent immunoelectrophoresis (McIntyre, 1987)

Isolation and identification

Plate sputum, cerebrospinal fluid, eye swabs, etc., on chocolate agar, which is the best medium for primary isolation because V factor may be inactivated by enzymes present in fresh blood. Incubate at 37°C for 18–24 h. Colonies may be haemolytic or non-haemolytic, are 1 mm in diameter on blood, larger on chocolate agar, grey and translucent. Bacilli are usually small and regular but filamentous forms may be seen (especially in spinal fluids and cultures from them).

'Satellitism', i.e. large colonies around *S. aureus* colonies and around colonies of other organisms synthesizing V factor, may be observed on mixed, primary cultures on blood agar.

Test for X and V factor requirements as follows.

Pick several colonies and spread on nutrient agar to make a lawn. Place X, V and X + V factor discs on the medium and incubate at 37°C overnight. Observe growth around the discs, i.e. whether the organisms require both X and V or X or V or neither (see Table 32.1). To test sugar reactions, add Fildes' extract to glucose, maltose, lactose and mannitol fermentation broths. These reactions are not entirely reliable.

Agglutination and capsular swelling tests

Do slide agglutination tests using commercial sera (e.g. Wellcome) to confirm identity. These are not satisfactory with capsulated strains as agglutination can occur with more than one serum. To test for capsule swelling with homologous sera use a very light suspension of the organism in saline, coloured with filtered methylene blue solution. Mix a drop of this with a drop

Table 32.1 ***Haemophilus*** **species**

	Haemolysis	*Factors required*			*Growth enhanced by 10% CO_2*
		X + V	*V only*	*X only*	
H. influenzae[a]	−	+			−
H. parainfluenzae	−		+		−
H. haemolyticus	+	+			−
H. parahaemolyticus	+		+		−
H. haemoglobinophilus[b]	−			+	−
H. ducreyi	±			+	−
H. aphrophilus	−			V	+

[a] *H. aegyptius*, [b] *H. canis*

of serum on a slide and cover with a cover-slip. Examine with an oil immersion lens with reduced light. Swollen capsules should be obvious, compared with non-capsulated organisms. More information about these tests is given by Turk (1982).

Species of Haemophilus

Haemophilus influenzae

Requires both X and V factors, produces acid from glucose but not lactose or maltose and varies in indole production. It occurs naturally in the nasopharynx and is associated with upper respiratory tract infections, including sinusitis and life-threatening acute epiglottitis. It also causes pneumonia, particularly post-influenzal. In eye infections ('pink eye'), it is known as the Koch-Weekes bacillus. It is also found in the normal vagina and is one of the causative organisms of purulent meningitis and purulent otitis media. There are several serological types and growth 'phases'. *H. aegyptius* is probably this species.

H. parainfluenzae

This differs only in not requiring X factor. It produces porphyrins from 6-aminolaevulinic acid. Normally present in the throat, but may be pathogenic.

H. haemolyticus and parahaemolyticus

Haemolytic organisms with the same properties as *H. influenzae* and *H. parainfluenzae*, respectively.

H. haemoglobinophilus (*H. canis*)

Requires X but not V factor. It is found in preputial infections in dogs and in the respiratory tract of humans.

H. ducreyi

Ducrey's bacillus is associated with chancroid or soft sore. In direct films of clinical material the organisms appear in a 'school of fish' arrangement. They are difficult to grow but may be obtained in pure culture by withdrawing pus from a bubo and inoculating inspissated whole rabbit blood slopes or 30% rabbit blood nutrient agar or Mueller Hinton medium containing Isovitalex. Culture at 30–34°C but no higher, in 10% carbon dioxide. Colonies at 48 h are green, grey or brown, intact and easily pushed along the surface of the media. It requires X but not V factor.

H. aphrophilus
A dubious species. Colonies on chocolate agar are small (0.5 mm) at 24 h, smooth and translucent. Better growth is obtained in a 10% carbon dioxide atmosphere. Some strains require X factor. There is no growth on MacConkey agar. Acid is produced from glucose, maltose, lactose and sucrose but not from mannitol. Fermentation media should be enriched with Fildes' extract. It is oxidase and catalase negative.

Human infections, including endocarditis, have been reported.

This organism closely resembles *Actinobacillus actinomycetemcomitans* (p.344) but it does not produce acid from lactose and sucrose (see Table 34.1.)

For more information about the genus *Haemophilus* see Biberstein (1981), Kilian and Frederiksen (1981) and Slack (1990).

Gardnerella

The species *Gardnerella vaginalis* was previously known as *Haemophilus vaginalis* or *Corynebacterium vaginale.*

Direct examination
Giemsa-stained films made from vaginal secretions usually show many squamous epithelial cells covered with large numbers of organisms ('clue cells'). Gram-stained films show large numbers of Gram-indifferent bacilli rather than the usual Gram-positive lactobacilli.

Culture
Place specimens in transport medium (Stuart's or Amies'). Plate on Columbia blood agar plus supplement (Oxoid) and incubate under 5–10% carbon dioxide. Colonies at 48 h are very small (1 mm), glistening 'dew-drops' and β-haemolytic on human blood agar. Gram-stained films show thin, poorly stained bacilli (unlike the solidly stained diphtheroids). They are Gram positive when young but become Gram negative later.

Make lawns on two plates of Columbia blood agar. On one place a drop of 3% hydrogen peroxide. On the other place discs of trimethoprim (5 μg) and metronidazole (50 μg). Incubate for 48 h. Gardnerellas are inhibited by the peroxide and are sensitive to trimethoprim and metronidazole. Lactobacilli and diphtheroids are resistant. Neither X nor V factor is required. There is no growth on MacConkey agar and, apart from acid from glucose (add Fildes' extract), biochemical tests seem to be all negative.

Subculture on blood agar and apply discs of metronidazole (50 μg) and sulphonamide (100 μg) (Oxoid). *G. vaginalis* is sensitive to metronidazole but resistant to sulphonamide.

This organism is found in the human vagina and may be associated with non-specific vaginitis. See Easmon and Ison (1985).

Streptobacillus

There is only one species, *Streptobacillus moniliformis*, which may be filamentous with or without swelling. It is fastidious, requiring blood or serum for growth. It is a commensal of rodents, a pathogen of mice and is one of the agents of rat-bite fever in humans. Milk-borne disease (Haverhill fever) has been reported.

Isolation and identification

Blood culture is the usual method of isolation, but media containing Liquoid should be avoided as this inhibits the growth of the organism. Inoculate enriched broth medium with aspirated fluids from joints, etc.

The organism grows slowly in liquid media, forming fluffy balls or colonies that resemble breadcrumbs. Remove these with a pasteur pipette for Gram staining and subculture on blood agar. Incubate at 37°C for 3 days, when colonies will be raised, granular, and 1–4 mm in diameter. L-form colonies may be seen (Cowan and Steel, 1993).

S. moniliformis is non-motile, a facultative anaerobe, catalase and oxidase negative and fermentative in Hugh and Leifson medium. See Table 30.1.

References

Biberstein, E. L. (1981) *Haemophilus–Pasteurella–Actinobacillus*: their significance in veterinary medicine. In *Haemophilus, Pasteurella and Actinobacillus* (eds M. Kilian, W. Frederiksen and E. L. Biberstein), Academic Press, London, pp. 57–76

Cowan and Steel's Manual for the Identification of Medical Bacteria (1993) 3rd edn (eds G. I. Barrow and R. K. A. Feltham), Cambridge University Press, Cambridge

Easmon, C. S. F. and Ison, C. A. (1985) *Gardnerella vaginalis*. In *Isolation and Identification of Micro-organisms of Medical and Veterinary Importance* (eds C. H. Collins and J. M. Grange), Society for Applied Bacteriology Technical Series No. 21, Academic Press, London, pp. 115–122

Kilian, M. and Frederiksen, W. (1981) Ecology of *Haemophilus, Pasteurella* and *Actinobacillus*. In *Haemophilus, Pasteurella* and *Actinobacillus*. (eds M. Kilian, W. Frederiksen, and E. L. Biberstein), Academic Press, London, pp. 11–38

McIntyre, M. (1987) Counter-current immunoelectrophoresis for the rapid detection of bacterial polysaccharide antigen in body fluids. In *Immunological Techniques in Microbiology* (eds J. M. Grange, A. Fox and N. L. Morgan), Society for Applied Bacteriology Technical Series No. 24, Blackwell, Oxford, pp. 137–143

Slack, M. P. E. (1990) *Haemophilus*. In *Topley and Wilson's Principles of Bacteriology, Virology and Immunity*, 8th edn, Vol. 2 (eds M. T. Parker and B. I. Duerden), Edward Arnold, London, pp. 355–382

Turk, D. C. (1982) *Haemophilus influenzae*. Public Health Laboratory Service Monograph No. 17, HMSO, London

Chapter

33

Campylobacter and Helicobacter

These organisms were originally in the genus *Vibrio*, then in *Campylobacter*, from which *Helicobacter* was recently separated because of a number of taxonomic dissimilarities.

Campylobacter

These small, curved, actively motile rods are microaerophilic, reduce nitrates to nitrites but do not attack carbohydrates. They have become very important in recent years as a cause of food poisoning (Chapter 13).

Isolation

Plate emulsions of faeces on blood agar containing (commercially available) campylobacter growth and antibiotic selective supplements. Incubate duplicate cultures at 37 and 42°C for 24–48 h in an atmosphere of approximately 5% oxygen, 10% carbon dioxide and 85% nitrogen, preferably using a gas generating kit, otherwise a candle jar.

Foods
These may have been frozen, chilled or heated, all of which may cause sublethal damage to the organisms.

Milk
Filter 200 ml through a cotton-wool plug. Place the plug in a sterile jar containing 100 ml of campylobacter enrichment medium without antibiotics. Incubate at 37°C for 2 h. Then add selective antibiotics and incubate at 43°C for 36 h. Subculture from the surface layer to selective campylobacter agar and incubate plates at 43°C for 36–48 h under microaerophilic conditions.

Water
Filter 100 ml through a 0.45 μm membrane and place the membrane face down on selective campylobacter agar. Incubate at 43°C for 24 h under microaerophilic conditions. Remove the filter and reincubate plates at 43°C for 36–48 h.

Place the filter in 100 ml of enrichment broth and proceed as for milk.

Meat, poultry etc.
Add 10 g to 100 ml of campylobacter selective broth plus antibiotic supplements and incubate at 37°C for 2 h. Then continue incubation at 43°C and proceed as for milk.

ELISA techniques are now available for the detection of campylobacters in food (Fricker and Park, 1987).

Identification

Campylobacter colonies are about 1 mm diameter at 24 h, grey, watery and flat. Examine Gram-stained films and test for growth at 43°C, H_2S production in TSI medium, nitrate reduction, urease and growth in the presence of 1.5% NaCl. Use media that are known to support growth of the organisms and include controls with known species. See Table 33.1.

The CAMPY kit (BioMerieux) is useful.

C. jejuni
This is an important agent of acute enterocolitis in humans (see Campylobacter enteritis in food poisoning; Chapter 13). It is also a commensal in the intestines of many wild and domestic animals.

C. coli
Also responsible for acute gastroenteritis in humans, this is less common than *C. jejuni*.

C. lari
This is the least common cause of campylobacter enteritis.

C. fetus
The subspecies *fetus* is an agent of infectious abortion in cattle and sheep and occasionally infects humans.

C. hyointestinalis
This is a pathogen of pigs and cattle and is sometimes responsible for diarrhoea in humans.

C. sputorum
This species is a commensal in the mouth of humans.

Campylobacters have received much attention in the last few years. See Lander and Gill (1985), Griffiths and Park (1990), Skirrow (1990) and Bolton *et al.* (1992).

Table 33.1 ***Campylobacter*** **and** ***Helicobacter***

	Growth at 44°C	*Hippurate hydrolysis*	*H_2S*[a]	*Nitrate reduction*	*Urease*	*Growth in 1.5% NaCl*
C. jejuni	+	+	−	+	−	−
C. coli	+	−	−	+	−	−
C. lari	+	−	−	+	−	+
C. fetus	−	−	−	+	−	v
C. hyointestinalis	+	−	+	+	−	−
C. sputorum	v	−	+	+	−	+
H. pylori	−	−	−	v	+	−

[a] TSI medium; v, variable

The species that grow at 43°C are called the 'thermophilic campylobacters' and are those most often isolated from humans.

There are a number of subspecies and biotypes of campylobacters. For information about biotyping, etc. see Bolton *et al.* (1992).

Helicobacter

Although these organisms may apparently colonize the gastrointestinal tract without ill effects they are also associated with gastritis and duodenal ulcers.

Isolation and identification

Culture biopsy material on one of the campylobacter media plus supplements. Incubate for 3–5 days under microaerophilic conditions and high humidity as for campylobacters. Colonies are translucent.

Test for oxidase, catalase, H_2S and urease. See Table 33.1. There are at least two useful diagnostic kits – CAMPY and RAPIDEC Pylori (Biomerieux).

Helicobacter pylori

This is the only species. It is oxidase and catalase positive, does not produce H_2S and is urease positive.

For more information see Skirrow (1990).

References

Bolton, F. J., Waring, D. R. A., Skirrow, M. B. and Hutchinson, D. N. (1992) Identification and biotyping of campylobacters. In *Identification Methods in Applied and Environmental Microbiology* (eds R. G. Board, D. Jones and F. A. Skinner), Society for Applied Microbiology Technical Series No. 29, Blackwell, Oxford, pp. 151–162

Fricker, C. R. and Park, R. W. A. (1987). Competitive ELISA, co-agglutination and haemagglutination for the detection and serotyping of campylobacters. In *Immunological Techniques in Microbiology* (eds J. M. Grange, A. Fox and N. L. Morgan), Society for Applied Bacteriology Technical Series No. 24, Blackwells, Oxford, pp. 195–210

Griffiths, P. L. and Park, R. W. A. (1990) Campylobacters associated with human diarrhoeal disease. *Journal of Applied Bacteriology*, **69**, 281–301

Lander, K. P. and Gill, K. P. W. (1985) Campylobacters. In *Isolation and Identification of Micro-organisms of Medical and Veterinary Importance* (eds C. H. Collins and J. M. Grange), Society for Applied Bacteriology Technical Series No. 21, Academic Press, London, pp. 190–200

Skirrow, M. B. (1990) *Campylobacter*, *Helicobacter* and other motile, curved Gram-negative rods. In *Topley and Wilson's Principles of Bacteriology, Virology and Immunity*, 8th edn (eds M. Y. Parker and B. I. Duerden), Edward Arnold, London, pp. 531–549

Chapter

34

Actinobacillus, Pasteurella, Yersinia, Cardiobacterium and Francisella

The general properties of the organisms described in this chapter are shown in Table 30.1.

Actinobacillus

This genus contains non-motile Gram-negative rods that are fermentative but do not produce gas and vary in their oxidase and catalase reactions.

Isolation and identification

Plate pus, which may contain small white granules, or macerated tissue on blood agar and incubate at 37°C for 48–72 h under a 10% carbon dioxide atmosphere. Colonies are small, sticky and adherent. Subculture small flat colonies on to MacConkey agar, in Hugh and Leifson medium and test for motility, catalase, oxidase, acid production from lactose, mannitol and sucrose, and aesculin hydrolysis. See Tables 30.1 and 34.1.

Species of Actinobacillus

Actinobacillus lignieresii
This species grows on MacConkey agar, is fermentative and produces acid from lactose (late) mannitol and sucrose.

It is oxidase and catalase positive, urease and aesculin negative. It is associated with woody tongue in cattle and human infections have been reported.

A. equuli
This species resembles *A. lignieresii* but ferments lactose rapidly. The oxidase and catalase tests are positive; lactose, sucrose and mannitol are fermented, but aesculin and urease tests are negative. It is associated with joint ill and sleepy disease of foals, and is also known as *Shigella equirulis* or *S. equulis*.

A. actinomycetemcomitans
This organism requires carbon dioxide for primary isolation and some strains require the X factor. Use an enriched medium and place an X factor disc on the heavy part of the inoculum. There is no growth on MacConkey agar and the oxidase test is negative. Lactose and sucrose are not fermented but some strains may produce acid from mannitol. It is aesculin and urease negative.

This organism is difficult to distinguish from *H. aphrophilus* (see Table 34.1). It is sometimes associated with infections by *Actinomyces israelii* and it has been recovered from blood cultures of patients with endocarditis.

A. suis
This hydrolyses aesculin and does not ferment mannitol. It has been isolated from pigs.

For further information about the identification of these organism see *Cowan and Steel's Manual* (1993).

Yersinia

Yersinias cause plague, pseudotuberculosis and gastroenteritis in humans and animals. They are now classified among the enterobacteria but are retained here for pragmatic reasons. They are catalase positive, fermentative in Hugh and Leifson medium (slow reaction), reduce nitrates to nitrites and fail to liquefy gelatin.

The plague bacillus, Yersinia pestis, is in Hazard Group 3 and all work should be done in a microbiological safely cabinet in a Biosafety/Containment Level 3 laboratory. After isolation and presumptive identification cultures should be sent to a reference laboratory for any further tests.

Isolation and identification

Yersinia pestis
In suspected bubonic plague examine pus from buboes; in pneumonic plague, sputum; in septicaemic plague, blood; and in rats the heart blood, enlarged lymph nodes and spleen. Before examining rats immerse them in disinfectant for several hours to kill fleas which might be infected.

Make Gram- and methylene blue-stained films and look for small oval Gram-negative bacilli with capsules. Bipolar staining is often seen in films of fresh isolates stained with methylene blue. Plate on blood agar containing 0.025% sodium sulphite to reduce the oxygen tension, on 3% salt agar and on MacConkey agar. Incubate for 24 h.

Colonies of plague bacilli are flat or convex, greyish white and about 1 mm in diameter at 24 h. Subculture into nutrient broth and cover with liquid paraffin. Test for motility at 22°C, fermentation of salicin and sucrose, aesculin hydrolysis and urease.

Films of colonies on salt agar show pear-shaped and globular forms. *Yersinia pestis* grows on MacConkey agar and usually produces stalactite growth in broth covered with paraffin. It is non-motile at 37°C and 22°C. It produces acid from salicin but not sucrose, hydrolyses aesculin but is urease negative. See Table 34.2.

It is the causative organism of plague in rats, transmitted from rat to rat and from rat to humans by the rat flea. In humans plague is bubonic, pneumonic or septicaemic.

Table 34.1 Actinobacillus species

	Oxidase	*Growth on MacConkey*	*Acid from*			*Aesculin hydrolysis*	*Urease*
			Lactose	*Mannitol*	*Sucrose*		
A. lignieresii	+	+	+ w	+	+	−	−
A. equuli	+	+	+	+	+	−	−
A. suis	+	+	+	−	+	+	+
A. actinomycetemcomitans	−	−	−	v	−	−	−

+ w, weak; v, variable

Table 34.2 ***Yersinia*** **species**

	Motility at 22°C	*Acid from*		*Aesculin hydrolysis*	*Urease*
		Salicin	*Sucrose*		
Y. pestis	−	+	−	+	−
Y. pseudotuberculosis	+	+	−	+	+
Y. enterocolitica	+	−	+	−	−

Materials and cultures should be sent to the appropriate reference laboratory or Communicable Diseases Centre. Reference laboratories use FA methods, bacteriophage, specific agglutination tests, precipitin tests and animal inoculations.

Yersinia enterocolitica

Culture faeces on yersinia selective medium (Cefsulodin-Irgasan-Novobiocin agar, CIN). Incubate at 30°C for 24–48 h and look for 'bullseye' colonies. Enrichment is not usually recommended for isolating this organism from faeces. Some strains grow on DCA as small non-lactose fermenting colonies, even after incubation at 37°C.

Make a 10% suspension of food in peptone water diluent. Incubate some at 32°C for 24 h. Refrigerate the remaining suspension at 4°C for up to 21 days and subculture at intervals on yersinia selective agar. Media used for other enteric pathogens can be reincubated at 28–30°C for 24 h and examined for *Y. enterocolitica* but some strains of this organism are inhibited by bile salts and deoxycholate.

Gram-stained films show small, coccoid bacilli, unlike the longer bacilli of other enterobacteria.

Test for growth on nutrient and MacConkey agar, motility at 22°C, fermentation of salicin and sucrose, aesculin hydrolysis and urease. Kits may be used for all or most of these. Incubate at 30°C rather than at 37°C. See Table 34.2.

There are several serotypes and biotypes. Some are pathogenic for humans (O:3, O:5, O:8, O:9), others for animals. Serological tests for specific antibodies may be of value in diagnosis.

Y. enterocolitica causes enteritis in humans and animals. In humans there may also be lymphadenitis, septicaemia, arthritis and erythema nodosum. The reservoirs are pigs, cattle, poultry, rats, cats, dogs and chinchillas. The organism has been found in milk and milk products, water, oysters and mussels. It grows at 4°C and may therefore multiply in cold storage of food. Person to person spread occurs, mainly in families.

Y. fredericksenii, Y. kristensenii and *Y. intermedia* resemble *Y. enterocolitica* but their roles in disease are uncertain. For further information see Swaminathan *et al.* (1982), Mair and Fox (1986) Brewer and Corbel (1985) and Doyle and Cliver (1990).

Yersinia pseudotuberculosis

Homogenize lymph nodes, etc. in tryptone broth. Inoculate blood agar with some of the suspension. Retain the remainder at 4°C and subculture to blood agar every 2–3 days for up to 3 weeks.

Emulsify faeces in peptone water diluent and inoculate yersinia selective medium (CIN).

Colonies on blood agar are raised, sometimes umbonate, granular and about 1–2 mm in diameter. Rough variants occur. Growth on MacConkey agar is obvious but poor. *Y. pseudotuberculosis* is motile at 22°C, ferments salicin but not sucrose, hydrolyses aesculin and is urease positive (Table 34.2).

This organism causes pseudotuberculosis in rodents, especially in guinea pigs. Human infections occur and symptoms resemble those of infection with *Y. enterocolitica*. *Y. pseudotuberculosis* is rarely isolated from faeces, however, and serological tests for specific antibodies are useful.

Pasteurella

Pasteurella multocida

Bipolar staining and pleomorphism are less obvious than in *Yersinia* species. The bacilli are very small (about 1.5 μm × 0.3 μm) and non-motile. Colonies on nutrient agar are translucent, slightly raised, about 1 mm in diameter at 24 h; on blood agar they are slightly larger, more opaque and non-haemolytic. There is no growth on MacConkey agar. Acid is produced from mannitol; it is indole positive, does not hydrolyse aesculin and gives variable reactions in the urease test (Table 34.3).

Table 34.3 *Pasteurella* species

	Growth on MacConkey	*Acid from mannitol*	*Aesculin hydrolysis*	*Indole*	*Urease*	*β-haemolysis*
P. multocida	−	+	−	+	v	−
P. pneumoniae	−	−	−	+	+	−
P. ureae	−	+	−	−	+	−
P. haemolytica	+	+	+	−	−	+
P. gallinarum	−	−	−	−	−	−

v, variable

This organism causes haemorrhagic septicaemia in domestic and wild animals and birds, e.g. fowl cholera, swine plague, transit fever. Humans may be infected by animal bites or by inhaling droplets from animal sneezes. It has also been isolated from septic fingers.

It is often given specific names according to its animal host, e.g. *avicida, aviseptica, suilla, suiseptica, bovicida, boviseptica, ovicida, oviseptica, cuniculocida, lepiseptica, muricida, muriseptica*. There is some host interspecificity: Cattle strains are also pathogenic for mice but not for fowls; fowl cholera strains are pathogenic for cattle and mice; lamb septicaemia strains are not pathogenic for rodents.

P. pneumoniae, P. haemolytica and P. gallinarum

These species are not infrequently isolated from veterinary material.

Pasteurella ureae

P. uraea, a commensal found in the respiratory tract in humans, resembles *P. multocida* but is indole and ornithine decarboxylase negative. It gives a rapid positive urease test.

For further information about the genus *Pasteurella* see Carter (1984), Curtis (1985) and *Cowan and Steel's Manual* (1993).

Cardiobacterium

Cardiobacterium hominis
It is a facultative anaerobe and requires an enriched medium and high humidity. Growth is best under carbon dioxide. Colonies on blood agar are minute at 24 h and about 1 mm in diameter after 48 h and are convex, glossy and butyrous. There is no growth on MacConkey medium, the oxidase test is positive and glucose and sucrose, but not lactose, are fermented without gas production. The catalase, urea and nitrate reduction tests are negative. Hydrogen sulphide is produced but gelatin is not liquefied.

This organism has been isolated from blood cultures of patients with endocarditis but may also be found in the upper respiratory tract.

Francisella

Francisella tularensis is in Hazard Group 3. It is highly infectious and has caused many laboratory infections. It should be handled with great care in microbiological safety cabinets in Biosafety/Containment Level 3 laboratory.

Culture blood, exudate, pus or homogenized tissue on several slopes of blood agar (enriched), cystine glucose agar and inspissated egg yolk medium. Also inoculate tubes of plain nutrient agar. Incubate for 3–6 days and examine for very small drop-like colonies. If the material is heavily contaminated, add 100 μg/ml of cycloheximide or 200 units/ml of nystatin, and 2.5–5 μg/ml of neomycin.

Subculture colonies of small, swollen or pleomorphic Gram-negative bacilli on blood and nutrient agar, on MacConkey agar and in enriched nutrient broth. Test for acid production from glucose and maltose, motility at 22°C and urease activity.

The bacilli are very small, show bipolar staining and are non-motile. There is no growth on nutrient or MacConkey agar, poor growth on blood agar with very small, grey colonies; on inspissated egg, at 3 to 4 days, colonies are minute and drop-like. It is a strict aerobe. Acid is formed in glucose and maltose. Indole and urease tests are negative; there is no motility at 22°C.

The organism is responsible for a plague-like disease (tularaemia) in ground squirrels and other rodents in western USA and Scandinavia.

Cultures and materials in cases of suspected tularaemia should be sent to a reference laboratory. Fluorescent antibody procedures are used for rapid diagnosis.

The genus *Francisella* is described in detail by Eigelsbach and McGann (1984).

References

Brewer, R. A. and Corbel, M. J. (1985) *Yersinia enterocolitica* and related species. In *Isolation and Identification of Micro-organisms of Medical and Veterinary Importance* (ed C. H. Collins and J. M. Grange), Society for Applied Bacteriology Technical Series No.21, Academic Press, London, pp.83–104

Carter, G. R. (1984) *Pasteurella*. In *Bergey's Manual of Systematic Bacteriology*, Vol.1 (ed N. R. Krieg, and J. G. Holt), Baltimore, Williams and Wilkins, pp.552–558

Cowan and Steel's Manual for the Identification of Medical Bacteria (1993), 3rd edn (eds G. I. Barrow and R. K. A. Feltham), Cambridge University Press, Cambridge

Curtis, P. E. (1985) *Pasteurella multocida*. In *Isolation and Identification of Micro-organisms of Medical and Veterinary Importance* (ed C. H. Collins and J. M. Grange), Society for Applied Bacteriology Technical Series No.21, Academic Press, London, pp.43–52

Doyle, M. P. and Cliver, D.O. (1990) *Yersinia enterocolitica*. In *Foodborne Diseases* (ed D. O. Cliver), Academic Press, New York, pp.224–229

Eigelsbach, H. T. and McGann, V. G. (1984) Genus *Francisella*. In *Bergey's Manual of Systematic Bacteriology*, Vol.1 (eds N. R. Krieg, and J. G. Holt), Baltimore, Williams and Wilkins, p.394

Mair, N. S. and Fox, E (1986) *Yersiniosis: Laboratory Diagnosis, Clinical Features and Epidemiology*, Public Health Laboratory Service, London

Swaminathan, B., Harmon, M. C. and Mehlman, I. J. (1982) A review: *Yersinia enterocolitica*. *Journal of Applied Bacteriology*, **52**, 151–183

Chapter

35

Legionella, Bartonella and Mobiluncus

Legionella

Some 36 species and 57 serogroups have been described but only a few appear to be of public health significance, responsible for pneumonia (legionnaires' disease, legionellosis).

Serological diagnosis

For early diagnosis it is best to use one of the commercial kits for, e.g. the indirect fluorescent antibody test (IFAT) or the rapid micro-agglutination test (RMAT).

Direct (FA) microscopy

Examine endotracheal aspirates, bronchial secretions, etc. by fluorescence microscopy with commercial antisera. These ably identify L. *pneumophila* serogroup 1 which is the agent most commonly involved.

Isolation

Homogenize the material in peptone water diluent and plate on buffered charcoal yeast extract agar (BCYE) with and without commercial legionella growth and selective supplements. Note that legionellas have an absolute requirement for cysteine and (usually) for iron, so there will be no growth on ordinary blood agar.

Incubate under 5% CO_2 and high humidity and examine plates after 3, 5, 7 and 10 days before discarding them. Colonies of legionellas are small, round, up to 3 mm in diameter, grey or greenish brown and present a 'ground glass' appearance.

For the isolation of legionellas from water samples see Chapter 20.

Identification

Examine Gram-stained films of suspect colonies, test those that are Gram negative with FA sera and do agglutination tests with polyvalent antisera. Proceed with specific sera if available.

Subculture for catalase, oxidase and hippurate hydrolysis tests. Test also for gelatin liquefaction but ensure that the medium contains cysteine and iron salts. Legionellas are catalase positive. Other reactions vary with species (see below).

It is advisable to send cultures of legionellas to a reference laboratory for full identification and serogrouping.

Species of Legionella

Only the seven relatively common species are listed below. Biochemical properties are of limited use. Identification must rely on agglutination tests.

L. pneumophila
The species usually found. It is oxidase positive and the only one that hydrolyses hippurate. It liquefies gelatin.

L. bozemanii, L. dumoffii and L. gormanii
These are oxidase negative, do not hydrolyse hippurate but liquefy gelatin.

L. micdadei
Is oxidase positive, does not hydrolyse hippurate and does not liquefy gelatin.

L. longbeachii and L. jordanii
These are also oxidase positive, do not hydrolyse hippurate but do liquefy gelatin.

For further information about legionellas see Bartlett *et al.* (1988) and Harrison and Taylor (1988).

Bartonella bacilliformis

This small Gram-negative rod causes Oroya fever (Carrion fever; bartonellosis) and is transmitted by the sandfly (*Phlebotomus* spp.).

It is best observed in Giemsa-stained blood films and the bacilli are seen in and on red blood cells. They are pleomorphic, coccobacilli, often in chains or arranged like Chinese letters.

For blood culture use a leptospira medium and incubate at 25°C for up to 4 weeks. Subculture to lysed blood agar or semisolid leptospira medium. Incubate cultures for at least 10 days. It is best to send specimens or cultures to a reference laboratory.

Mobiluncus

An old species revived in 1980 and now of interest. It is found in vaginal secretions along with other agents of vaginitis but its significance is not fully documented.

It grows on the media used for vaginitis investigations and there appear to be two morphological variants: one long and Gram negative and the other short and Gram variable.

References

Bartlett, L. R., Macrae, A. D. and Macfarlane, J. T. (1988) *Legionella Infections*, Edward Arnold, London

Harrison, T. G. and Taylor, A. G. (1988) *A Laboratory Manual for Legionella*, Wiley, London

Chapter

36

Staphylococcus and Micrococcus

Staphylococci and micrococci are frequently isolated from pathological material and foods. Distinguishing between the two groups is important. Some staphylococci are pathogens; some are doubtful or opportunist pathogens; others, and micrococci, appear to be harmless but are useful indicators of pollution. Staphylococci are fermentative, capable of producing acid from glucose anaerobically; micrococci are oxidative and produce acid from glucose only in the presence of oxygen. Staphylococci divide in more than one plane, giving rise to irregular clusters, often resembling bunches of grapes

As with other groups of microorganisms recent taxonomic and nomenclatural changes have not exactly facilitated the identification of the Gram-positive cocci which are isolated in routine laboratories from clinical material and foods. The term 'micrococci' is still loosely but widely used to include many that cannot be easily identified (see also under *Aerococcus*, p.368 *Pediococcus*, p.369).

Isolation

Pathological material

Plate pus, urine, swabs, etc., on blood agar or blood agar containing 10 mg/l each of colistin and nalidixic acid to prevent the spread of proteus and inoculate salt meat broth. Incubate overnight at 37°C. Plate the salt meat broth on blood agar.

Foodstuffs

Prepare 10% suspensions in 0.1% peptone water in a Stomacher. If heat stress is suspected add 0.1 ml amounts to several tubes containing 10 ml of brain heart infusion broth. Incubate for 3–4 h and then subculture. Use one or more of the following media. Milk salt agar (MSA), phenolphthalein phosphate polymyxin agar (PPP), Baird-Parker medium (BP) or tellurite polymyxin egg yolk (TPEY). Incubate for 24–48 h. Test the PPP medium with ammonia for phosphatase-positive colonies (see below). There are also commercial tests for the detection of enterotoxins (Berry *et al.*, 1987).

MSA agar relies on the high salt content to select staphylococci. In PPP medium, the polymyxin suppresses many other organisms. BP medium is very selective but may be overgrown by proteus. This may be suppressed by adding sulphamethazine, 50 μg/ml to the media.

For enrichment or to assess the load of staphylococci, use salt meat broth or mannitol salt broth. Add 0.1 and 1.0 ml of the emulsion to 10 ml of broth and 10 ml to 50 ml. Incubate overnight and plate on one of the solid media. If staphylococci are present in large numbers in the food there will be a heavy growth from the 0.1-ml inoculum; very small numbers may give growth only from the 10-ml sample.

If counts are required, use the Miles and Misra method (p.154) and one of the solid media.

Staphylococcal food poisoning

There is no need to examine the stools of all the patients; a sample of 10–25% is sufficient. Inoculate Robertson's cooked meat medium containing 7–10% sodium chloride (salt-meat medium) with 2–5 g of faeces or vomit and incubate overnight and plate on blood agar or one of the special staphylococcal media. Examine all food handlers for lesions on exposed parts of the body and swab these as well as the noses, hands and fingernails of all kitchen staff. Swab chopping boards and utensils which have crevices and cracks likely to harbour staphylococci. Use ordinary pathological laboratory swabs: cotton wool wound on a thin wire or stick, sterilized in a test-tube (these swabs can be obtained from most laboratory suppliers). Plate on the solid medium and break off the cotton wool swab in cooked meat medium.

All strains of *S. aureus* isolated from patients, food, implements and food handlers in incidents of staphylococcal food poisoning should be sent to a reference laboratory. Not all strains can cause the disease and others are usually encountered during the investigation.

Identification

Colonies of staphylococci and micrococci on ordinary media are golden, brown, white, yellow or pink, opaque, domed 1–3 mm in diameter after 24 h on blood agar and are usually easily emulsified. There may be β-haemolysis on blood agar. Aerococci show α-haemolysis.

On Baird-Parker medium after 24 h, *Staphylococcus aureus* gives black, shiny, convex colonies, 1–1.5 mm in diameter; there is a narrow white margin and the colonies are surrounded by a zone of clearing 2–5 mm in diameter. This clearing may be evident only at 36 h.

Other staphylococci, micrococci, some enterococci, coryneforms and enterobacteria may grow and may produce black colonies but do not produce the clear zone. Some strains of *S. epidermidis* have a wide opaque zone surrounded by a narrow clear zone. Any grey or white colonies can be ignored. Most other organisms are inhibited (but not proteus: see above).

On TPEY medium, colonies of *S. aureus* are black or grey and give a zone of precipitation around and/or beneath the colonies.

Examine Gram-stained films. Do coagulase and DNase tests on Gram-positive cocci growing in clusters grown from clinical material. This is a short cut: strains positive by both tests are probably *S. aureus*.

Coagulase test

Possession of the enzyme coagulase which coagulates plasma is an almost exclusive property of *S. aureus*. There are two ways of performing this test:

(1) *Slide coagulase test* Emulsify one or two colonies in a drop of water on a

slide. If no clumping occurs in 10–20 s dip a straight wire into human or rabbit plasma (EDTA) and stir the bacterial suspension with it. *S. aureus* agglutinates, causing visible clumping in 10 s.

Use water instead of saline because some staphylococci are salt sensitive, particularly if they have been cultured in salt media. Avoid excess (e.g. a loopful) plasma as this may give false positives. Check the plasma with a known coagulase positive staphylococcus.

(2) *Tube test* Do this (a) to confirm the slide test, (b) if the slide test is negative.

Add 0.2 ml of plasma to 0.8 ml of nutrient (not glucose) broth in a small tube. Inoculate with the suspected staphylococcus and incubate at 37°C in a water-bath. Examine at 3 h and if negative leave overnight at room temperature and examine again. Include known positive and negative controls. In the tube test, citrated plasma may be clotted by any organism that can utilize citrate, e.g. by faecal streptococci (but these are catalase negative), *Pseudomonas* and *Serratia*. It is advisable therefore to use EDTA plasma (available commercially) or oxalate or heparin plasma. Check Gram films of all tube coagulase positive organisms.

S. aureus produces a clot, gelling either the whole contents of the tube or forming a loose web of fibrin. Longer incubation may result in disappearance of the clot due to digestion (fibrinolysis).

The slide test detects 'bound' coagulase ('clumping factor'), which acts on fibrinogen directly; the tube test detects 'free' coagulase, which acts on fibrinogen in conjunction with other factors in the plasma. Either or both coagulases may be present. There are several commercial kits for identifying *S. aureus* by latex agglutination and/or nuclease, e.g. Slidex (Biomerieux), Staphylex (Oxoid), Staphynuclease (API), as well as strip tests for this and other species, e.g RAPIDEC Staph (Biomerieux) and API Staph (Biomerieux).

Kits remove the potential hazard of using human plasma. Some combine tests for clumping factor and protein A. Others employ haemagglutination. Berke and Tilton (1981) compared kits with the standard coagulase technique.

DNase test

Inoculate DNase agar plates with a loop so that the growth is in plaques about 1 cm in diameter. Incubate at 37°C overnight. Flood the plate with 1 N hydrochloric acid. Clearing around the colonies indicates DNase activity. The hydrochloric acid reacts with unchanged deoxyribonucleic acid to give a cloudy precipitate. A few other bacteria, e.g. *Branhamella, S. pyogenes* and *Serratia*, may give a positive reaction.

If the organisms are coagulase and DNase negative they may be differentiated by the OF test using Baird-Parker's method and one or more of the following tests.

Schleifer and Kloos' test

Use commercial purple agar base containing 10 ml glycerol and 0.4 mg erythromycin/litre. Staphylococci change the colour of the indicator (bromocresol purple) to yellow: micrococci do not grow (Schleifer and Kloos, 1975).

Lysostaphin test

Prepare a solution of Lysostaphin (Sigma Chemicals) by dissolving 50 mg in 40 ml phosphate buffer (0.02 mol/l). Adjust to pH 7.4 and add NaCl to give 1% (w/v). Dispense in 1-ml amounts and store at − 60°C. To use, add 1 ml to 9 ml phosphate buffer and mix 5 drops of this with 5 drops of an overnight broth culture in a small test tube. Incubate at 35°C and examine for lysis at 30, 60 and 120 min. Include a control using phosphate buffer instead of Lysostaphin. Most staphylococci are lysed (Kloos and Schleifer, 1975).

Phosphatase test
Inoculate phenolphthalein phosphate agar and incubate overnight. Expose to ammonia vapour. Colonies of phosphatase positive staphylococci will turn pink.

S. aureus gives a positive test (but negative strains have been reported). Coagulase-negative staphylococci and micrococci are usually phosphatase negative (see Table 36.1 and 36.2).

Table 36.1 *Staphylococcus, Micrococcus* and *Aerococcus*

	OF	*SK*	*Lys*	*Coagulase*	*DNase*	*Phosphatase*	*Morphology*
S. aureus	F	+	+	+	+	+	Clusters
Staphylococcus spp.	F	+	+/−	−	−	−	Clusters
Micrococcus spp.	O	−	−	−	−	−	Clusters, tetrads or packets
Aerococcus viridans [a]	F	−	−	−	−	−	Clusters or pairs

OF, Hugh and Leifson, SK, Schleifer and Kloos (1975) Lys, lysostaphin sensitivity; F, fermentative; O, oxidative
[a] Includes *Gaffkya*

If identification to species is required (other than *S. aureus*) use the API Staph system or test for nitratase reduction, susceptibility to novobiocin (5 μg disc) and for fermentation of mannitol, trehalose and sucrose (Baird-Parker sugar medium, p.75). See under *Species* below and for further information Baird-Parker (1979), Kloos and Jorgensen (1988), Jones *et al.* (1990) and *Cowan and Steel's Manual* (1993).

Staphylococcus

There are now 29 species of *Staphylococcus* but some are of little interest. They may be divided into three groups: (1) coagulase positive; (2) coagulase negative and novobiocin susceptible; (3) coagulase negative and novobiocin resistant. Only known and opportunist pathogens are mentioned below.

Staphylococcus aureus
This species is coagulase, DNase and phosphatase positive, forms acid from mannitol, trehalose and sucrose and is sensitive to novobiocin.

Some strains are haemolytic on horse blood agar but the zone of haemolysis is relatively small compared with the diameter of the colony (differing from the haemolytic streptococcus).

Production of the golden yellow pigment is probably the most variable characteristic. Young cultures may show no pigment at all. The colour may develop if cultures are left for 1 or 2 days on the bench at room temperature.

Pigment production is enhanced by the presence in the medium of lactose or other carbohydrates and their breakdown products. It is best demonstrated on glycerol monoacetate agar.

Table 36.2 Properties of some *Staphylococcus* species

Species	*Pigment*	*Coagulase*	*DNAse*	*Phosphatase*	*Novobiocin*[a]	*Mannitol*	*Trehalose*	*Sucrose*
S. aureus	+	+	+	+	S	+	+	+
S. chromogenes	+	+	+	+	S	v	+	+
S. hyicus	−	+	+	+	S	−	+	+
S. intermedius	−	+	+	+	S	+	+	+
S. epidermidis	−	−	+	+	S	−	−	+
S. warneri	−	−	−	−	S	v	+	+
S. cohnii	−	−	−	+	R	+	+	−
S. saprophyticus	−	−	−	−	R	+	+	+

[a] 5-μg disc
S, susceptible; R, resistant; v, variable

In clinical laboratories, *S. aureus* is usually identified by *either* the coagulase *or* the DNase test. False-positive coagulase tests are possible with enterococci. *Pseudomonas* and *Serratia* if citrated plasma is used. *Serratia* may give a positive DNase test. Coagulase negative, DNase positive strains do occur. We believe, therefore, that both coagulase and DNase tests should be done.

S. aureus is a common cause of pyogenic infections and food poisoning (Chapter 13). Staphylococci are disseminated by common domestic and ward activities such as bedmaking, dressing or undressing. They are present in the nose, on the skin and in the hair of a large proportion of the population.

Methicillin-resistant staphylococci

Methicillin-resistant *S. aureus* (MRSA) and methicillin-resistant coagulase negative staphylococci (MRCNS) are involved in serious hospital infections and require early identification. Cultures are usually a mixture of resistant and sensitive organisms. Growth of resistant strains is encouraged by media containing 2% NaCl and incubation at 30°C. For a review of these organisms see Cookson and Phillips (1990).

Staphylococcus epidermidis (*S. albus*)

Is coagulase negative, DNase and phosphatase positive and may liquefy gelatin. Occasional strains ferment mannitol. It produces no pigment; the colonies are 'china white'. Antibiotic-resistant strains are not uncommon and are often isolated from clinical material, including urine, where they may be opportunist pathogens. (See Schleifer and Kloos, 1975.)

Staphylococcus epidermidis (sensu stricto) is associated with infections with implanted material, e.g. prosthetic heart valves and joints, ventricular shunts and cannulas. Strains resistant to many antibiotics occur. See Table 36.2.

S. saprophyticus

This causes urinary tract infections in young sexually active women. There is a diagnostic kit (Dermaci, Sweden).

Several other coagulase negative species occur. See Kloos and Jorgensen (1988).

Other species

S. chromogenes produces pigment like that of *S. aureus* and ferments sucrose. It is an opportunist pathogen and has been found in pigs and in cows' milk. *S. hyicus* is an opportunist pathogen and has been found in dermatitis of pigs, in poultry and cows' milk. *S. intermedius* is known to cause infections in dogs. *S. warneri* and *S. cohnii* are opportunist pathogens.

Staphylococcal toxins

There are kits for testing for the presence of these in foods and also for investigating toxic shock syndrome which is associated with the use of tampons (see Arbuthnot *et al.*, 1990).

Staphylococcal typing

Strains isolated from hospital infections and food poisoning incidents should be typed for epidemiological purposes. This is best done at a reference laboratory (see Richardson *et al.*, 1992). The genus is reviewed in a Society for Applied Bacteriology Symposium (Jones *et al.*, 1990).

Micrococcus

These are Gram-positive, oxidase negative, catalase positive cocci that differ from the staphylococci in that they utilize glucose oxidatively or do not produce enough acid to change the colour of the indicator in the medium. They are common saprophytes of air, water and soil and are often found in foods.

The classification is at present confused: the genus contains the organisms formerly called *Gaffkya* and *Sarcina*, tetrad- and packet-forming cocci. These morphological characteristics vary with cultural conditions and are not considered constant enough for taxonomic purposes. The genus *Sarcina* now includes only anaerobic cocci.

At present, it does not seem advisable to desçribe newly isolated strains by any of the specific names that abound in earlier textbooks.

References

Arbuthnott, J. P., Coleman, D. C. and De Azavedo, J. S. (1990) Staphylococcal toxins in human disease. In *Staphylococci* (eds D. Jones, R. G. Board and M. Sussman), Society for Applied Bacteriology Symposium Series No. 19, Blackwells, Oxford, pp.101S–108S

Baird-Parker, A. C. (1979) Methods for identifying staphylococci and micrococci. In *Identification Methods for Microbiologists*, 2nd edn (eds F. A. Skinner and D. W. Lovelock), Society for Applied Bacteriology Technical Series No. 14, Academic Press, London, pp.201–209

Berke, A. and Tilton, R. C. (1981) Evaluation of rapid coagulase methods for the identification of *Staphylococcus aureus*. *Journal of Clinical Microbiology*, **23**, 916–919

Berry, P. R., Weinecke, A. A., Rodhouse, J. C. and Gilbert, R. J. (1987) Use of commercial tests for the detection of *Clostridium perfringens* and *Staphylococcus aureus* enterotoxins. In *Immunological Techniques in Microbiology* (eds J. M. Grange, A. Fox and N. L. Morgan), Society for Applied Bacteriology Technical Series No. 24, Blackwells, London, pp.245–250

Cookson, B. and Phillips, I. (1990) Methicillin-resistant staphylococci. In *Staphylococci* (eds D. Jones, R. G. Board and M. Sussman), Society for Applied Bacteriology Symposium Series No. 19, Blackwells, Oxford, pp.55S–70S

Cowan and Steel's Manual for the Identification of Medical Bacteria (1993) 3rd edn (eds G. I. Barrow and R. K. A. Feltham), Cambridge University Press, Cambridge

Jones, D., Board, R. G. and Sussman, M. (eds) (1990) Staphylococci. *Journal of Applied Applied Bacteriology Symposium* No. 19, **69**, 1S–188S

Kloos, W. E. and Jorgensen, J. H. (1988) Staphylococci. In *Manual of Clinical Microbiology*, 4th edn (eds E. H. Lennette, A. Balows, W. J. Hauser and H. J. Shadomy), Association of American Microbiologists, Washington, pp.143–153

Kloos, W. E. and Schleifer, K. H. (1975) Simplified scheme for routine identification of human staphylococci. *Journal of Clinical Microbiology*, **1**, 82–88

Richardson, J. F., Noble, W. C. and Marples, R. M. (1992) Species identification and epidemiological typing of the staphylococci. In *Identification Methods in Applied and Environmental Microbiology* (eds R. G. Board, D. Jones and F. A. Skinner), Society for Applied Bacteriology Technical Series No. 29, Blackwells, Oxford, pp.193–220

Schleifer, K. H. and Kloos, W. E. (1975) Isolation and characterization of staphylococci from human skin. *International Journal of Systematic Bacteriology*, **25**, 50–61

Chapter

37

Streptococcus, Enterococcus, Lactococcus, Aerococcus, Leuconostoc and Pediococcus

This group includes organisms of medical, dental and veterinary importance as well as starters used in the food and dairy industries, spoilage agents and saprophytes. The cells divide in one plane or in two planes at right angles to one another.

The important characters that distinguish *Streptococcus, Enterococcus, Lactococcus* and *Aerococcus* are shown in Table 37.1.

Table 37.1 *Streptococcus, Enterococcus, Lactococcus* and *Aerococcus*

	Haemolysis	*Growth at 45°C*	*Growth in 6.5% NaCl*	*Aesculin hydrolysis*	*Lancefield group*
Streptococcus		−	−	v	ABCG
Enterococcus	v	+	+	+	D
Lactococcus	α	−	−	+	Others
Aerococcus	α	−	+	v	None

v, variable

Streptococcus

Gram-positive cocci that always divide in the same plane, forming pairs or chains; the individual cells may be oval or lanceolate. They are Gram-positive, non-sporing, non-motile and some are capsulated. Most strains are aerobic but it is best to culture clinical material anaerobically. Important streptococci will grow and many other organisms will be suppressed. The catalase and oxidase tests are negative.

Isolation

From clinical material

Blood agar is the usual primary medium. If the material is known to contain many other organisms place a 30 μg neomycin disc on the heavy part of the inoculum. Alternatively use blood agar containing 10 μg/ml colistin and

5 µg/ml oxolinic acid (COBA medium; Petts, 1984) which is very selective for streptococci. (Crystal violet blood agar, formerly recommended, is not very satisfactory because of batch variation in the dye affects colony size and amount of haemolysis.)

Islam's (1977) medium facilitates the recognition of Group B streptococci, e.g. in antenatal screening. Incubate anaerobically at 37°C.

Dental plaque
Plate on blood agar and trypticase yeast extract cystine agar or mitis salivarius agar.

Bovine mastitis
Use Edwards' medium. The crystal violet and thallous sulphate inhibit most saprophytic organisms; aesculin-fermenting saprophytic streptococci give black colonies and mastitis streptococci pale grey colonies.

Dairy products
Use yeast glucose agar for mesophiles and yeast lactose agar for thermophiles.

Air
For β-haemolytic streptococci, in hospital cross-infection investigations use crystal violet agar containing 1:500 000 crystal violet (satisfactory for this purpose) with slit samplers. For evidence of vitiation use mitis salivarius agar.

Identification of streptococci

Colonies on blood agar are usually small, 1–2 mm in diameter and convex with an entire edge. The whole colony can sometimes be pushed along the surface of the medium. Colonies may be 'glossy', 'matt' or 'mucoid'. Growth in broth is often granular, with a deposit at the bottom of the tube.

The primary classification is made on the basis of alteration of haemolysis on horse blood agar.

α-Haemolytic, often referred to as 'viridans' streptococci, produce a small, greenish zone around the colonies. This is best observed on chocolate blood agar.

β-Haemolytic streptococci give small colonies surrounded by a much larger, clear haemolysed zone in which all the red cells have been destroyed. Minute β-haemolytic colonies may be *S. anginosus*, formerly known as *S. milleri*.

Some streptococci show no haemolysis.

Haemolysis on blood agar is only a rough guide to pathogenicity. The β-haemolytic streptococci include those strains which are pathogenic for humans and animals but the type of haemolysis may depend on conditions of incubation and the medium used as a base for the blood agar. Some α-haemolytic streptococci show β-haemolysis on Columbia-based blood agar.

Some saprophytic streptococci are β-haemolytic; so are organisms in other genera having similar colonial morphology. Haemolytic *Haemophilus* spp. are often reported as haemolytic streptococci in throat swabs because films are not made. *Corynebacterium pyogenes* is also haemolytic.

Clinical strains
β-Haemolytic streptococci from clinical material are usually identified by rapid coagglutination methods (see below) or as *S. pyogenes* (group A) by their sensitivity to bacitracin (see below). *Streptococcus pneumoniae* (the pneumococcus) is identified by its sensitivity to optochin (see p.366). Other streptococci require further tests.

Streptococcal antigens

Species and strains of streptococci are usually identified by their serological group and type. There are 15 Lancefield groups (A-P, excluding I) characterized by a series of carbohydrate antigens contained in the cell wall. Rabbits immunized with known strains of each group produce serum which will react specifically *in vitro* by a precipitin reaction with an extract of the homologous organism. The carbohydrate is known as the C substance. The streptococci within Group A and a few other groups can be divided into serological Griffith types (1–30) by means of two classes of protein antigens, M and T.

M is a type-specific antigen near the surface of the organism which can be removed by trypsinization and is present in matt and mucoid types of colonies but not in glossy types. It is demonstrated by a precipitin test.

T is not type-specific, may be present with or independent of the M antigen, and is demonstrated by agglutination tests with appropriate antisera.

Grouping is usually carried out in the laboratory where the organisms are isolated, using sera and control extracts obtained commercially. Only sera for Groups A, B, C, D and G need be used for routine purposes.

Workers should be aware that the groups are not always species-specific. Some enterococci possess the G antigen in additon to their D antigen. *S. anginosus* may group as A, C, or G and, when haemolytic may be confused with pyogenic streptococci. Some 'viridans' streptococci may also group as A or C and under certain conditions may appear to be β-haemolytic.

Typing of haemolytic (Group A, etc.) strains is necessary for epidemiological purposes. This is best done by reference laboratories.

Serological grouping of streptococci

There are two approaches. Coagglutination which requires commercially available kits, is rapid and gives results within 1 h of obtaining a satisfactory growth on the primary culture. Precipitin tests take longer but allow a larger number of groups to be identified.

Coagglutination kit tests

Several kits are available. The streptococcal antibody is attached to staphylococci or latex particles. These antibody-coated particles are agglutinated when mixed on a slide with suspensions or extracts of streptococci of the same group. In the Phadebact (Pharmacia) method colonies from the primary plates are mixed with the reagents on a slide. In the Murex (Wellcome) and Oxoid systems the antigens are extracted before mixing with the latex reagent on a slide. Others include Pathodex (Diagnostic Products, USA) and Prolex (Pro-Lab, Canada). Groups A, B, C, D, F and G may be identified but it is recommended that biochemical tests are done on Group D streptococci to identify enterococci and 'viridans' streptococci.

Precipitin tube methods

These have largely been superseded by the slide (kit) methods but the following methods are still in use.

Centrifuge 50 ml of an overnight culture of the streptococcus in 0.1% glucose broth or use a suspension prepared by scraping the overnight growth from a heavily inoculated blood agar plate.

For Lancefield method suspend the deposit in 0.4 ml of 0.2 N hydrochloric acid and place in boiling water-bath for 10 min. Cool, add 1 drop of 0.02% phenol red and loopsful of 0.5 N sodium hydroxide solution until the colour changes to faint pink. Centrifuge; the supernatant is the extract.

For Fuller's formamide method suspend the deposit in 0.1 ml of formamide and place in an oil-bath at 160°C for 15 min. Cool and add 0.25 ml of acid alcohol (95.5 parts ethanol: 52 parts N hydrochloric acid). Mix and centrifuge. Remove the supernatant fluid and add to it 0.5 ml of acetone. Mix, centrifuge and discard the supernatant. To the deposit add 0.4 ml of saline, 1 drop of 0.02% phenol red and loopsful of 0.2 N sodium hydroxide solution until neutral. This is the extract. This is not a very good method for Group D streptococci.

In Maxted's method the C substance is extracted from streptococcal suspension by incubation with an enzyme prepared from a strain of *Streptomyces*.

To prepare this enzyme, obtain *Streptomyces* sp. No.787 from the National Collection of Type Cultures and grow it for several days at 37°C on buffered yeast extract agar. The medium is best sloped in flat bottles (120-ml 'medical flats'). When there is good growth, place the cultures in a bowl containing broken pieces of solid carbon dioxide. After 24 h allow the medium to thaw and remove the fluid; this contains the enzyme. Bottle it in small amounts and store in a refrigerator.

To prepare the streptococcal extract, scrape the growth from a heavily inoculated 24-h blood agar culture of the organisms into 0.5 ml of enzyme solution in a small tube and place in a water-bath at 37°C for 2 h. Centrifuge and use the extract to do precipitin tests as described below.

Prepare capillary tubes from pasteur pipettes. Dip the narrow end in the grouping serum so that a column a few millimetres long enters the tube. Place it in a block of Plasticine and, with a very fine pasteur pipette, layer extract on the serum so that the two do not mix but a clear interface is preserved. Some practice is necessary in controlling the pipette. If an air bubble develops between the two liquids, introduce a very fine wire, when an interface is usually produced.

A positive result is indicated by a white precipitate that develops at the interface. It is sometimes necessary to dilute the extract 1:2 or 1:5 with saline to obtain a good precipitate.

The bacitracin disc method

Inoculate a blood agar plate heavily and place a commercial bacitracin disc on the surface and incubate overnight. A zone of inhibition appears around the disc if the streptococci are Group A. This test is not wholly reliable and is declining in popularity except as a screening method.

Biochemical tests

There are now so many species that conventional tests become very time-consuming. It is best, therefore, to use the API20 Strep system and refer to the database (see Coleman *et al.*, 1992). Table 37.2 shows the characteristics of some of the commoner species.

Groups and species of streptococci

Group A

These are β-haemolytic, are the so-called haemolytic streptococci of scarlet

Table 37.2 *Streptococcus* species

	Haemolysis	*Hydrolysis of*		*Growth on bile agar*	*Arginine hydrolysis*	*Acid from*			*Lancefield group*
		Aesculin	*Hippurate*			*Mannitol*	*Sorbitol*	*Lactose*	
S. pyogenes	β	v	−	−	+	v	−	+	A
S. agalactiae	(β)	−	+	+	+	−	−	v	B
S. equi	β	−	−	−	+	−	−	−	C
S. zooepidemicus	β	v	−	−	+	−	+	+	C
S. equisimilis	β	v	−	−	+	−	−	v	C
S. dysgalactiae	α	−	−	−	+	−	v	+	C
S. bovis	(α/β)	+	v	+	−	v	−	+	D
S. equinus	α	+	−	+	−	−	−	−	D
S. canis	β	v	−	−	+	−	−	v	G
S. porcinus	β	+	−	−	+	+	+	v	*
S. suis	(β)	+	−	−	+	+	+	−	*
S. anguinosus	(β)	v	−	−	+	v	−	v	*
S. uberis	(α)	+	+	−	+	+	+	+	*
S. sanguis	α/β	+	−	v	+	−	−	+	*
S. acidominimus	−	+	−	−	−	+	+	+	*
S. mutans	(α)	+	−	v	−	+	+	+	*
S. oralis	α/β	v	v	+	−	−	−	v	*
S. salivarius	−	+	v	−	−	−	−	+	*

v, variable; *, not A, B, C, D, or G
α, alpha; β beta; (α), alpha or none
(β), beta or none; α/β, alpha or beta
(α/β), alpha, beta or none

fever, tonsillitis, puerperal sepsis and other infections of humans, and are known as *S. pyogenes*. Some strains are capsulated and form large (3-mm) colonies like water drops on the surface of the medium. These have been associated with milk-borne outbreaks. Capsule formation is, however, not uncommon when streptococci are grown in milk.

Group B

The β-haemolytic streptococci in this group correspond to *S. agalactiae*, the causative organism of chronic bovine mastitis. This is an important cause of neonatal meningitis and septicaemia. It is found in the female genital tract and gut. It has been reported as a cause of urinary tract infections. It is usually β-haemolytic but non-haemolytic variants occur. It grows on bile salt media.

Group C

The most important β-haemolytic member of this group is *S. equi*, which causes strangles in horses. Others include *S. equisimilis*, responsible for some human infections and *S. zooepidemicus* for outbreaks among animals. *S. dysgalactiae*, which is α-haemolytic, is associated with acute bovine mastitis (but is less common than *S. agalactiae*).

Some group C strains cause infections in humans.

Group D

This now includes *S. bovis* (a group rather than a species) and *S. equinus*.

S. bovis is present in the human bowel and animal intestines. In bowel disorders of humans it may enter the bloodstream, leading to endocarditis.

S. equinus inhabits animal intestines (especially in horses) and is found in milk. Human infections have been reported.

The D antigen is deeper in the cell than are the antigens of other streptococci. For this reason, extracts prepared by acid hydrolysates of the streptococci and the antisera prepared with them do not give good precipitation. The commercial antisera are satisfactory but Fuller's method should be used for grouping. Most, but not all, will group with coagglutination reagents.

Group D streptococci can be confused with aerococci but the latter do not grow at 45°C and do not hydrolyse arginine. Both cause greening of bacon.

Group G

S. canis, a pathogen of dogs, is also an agent of bovine mastitis.

Other streptococci found in human and animal material

S. porcinus is found in the respiratory tract and cervical abscesses of pigs.

S. suis is associated with meningitis in pigs and may infect humans.

S. anginosus, formerly *S. milleri*, probably consists of more than one species. It is found in the mouth, dental and other abscesses. Some strains, described as *S. intermedius*, have been found in brain and liver abscesses.

S. uberis is an animal species and one of the agents of bovine mastitis.

S. sanguis is probably not a single species. It is associated with endocarditis in humans.

S. acidominimus is a commensal in the genital tract of cows.

S. oralis, also known as *S. mitior*, is a commensal in the human mouth and a known cause of subacute endocarditis.

S. mutans, again, probably not a single species, is found in dental plaque and is associated with caries.

See Beighton (1985) for more information about oral streptococci.

S. thermophilus

Resembles Group D streptococci biochemically but has no D antigen. The optimum temperature for growth is 50°C. It is used (along with *Lactobacillus bulgaricus*) in the manufacture of yoghurt.

S. pneumoniae
The pneumococcus is a causative organism of lobar pneumonia, otitis media and meningitis in humans and of various other infections in humans and animals, including (rarely) mastitis. In pus and sputum, the organism appears as a capsulated diplococcus but usually grows on laboratory media in chains and then shows no capsule. There are several serological types, but serological identification is rarely attempted nowadays. Strains of serotype III usually produce highly mucoid colonies but most other strains give flat 'draughtsman'-type colonies 1–2 mm in diameter with a greenish haemolysis.

Pneumococci often resemble 'viridans' streptococci on culture. The following tests allow rapid differentiation.

(1) Pneumococci are sensitive to optochin (ethylhydrocupreine hydrochloride). Optochin discs are supplied by most media manufacturers.
   A disc placed on a plate inoculated with pneumococci will give a zone of inhibition of at least 10 mm when incubated aerobically (not in carbon dioxide).
(2) Pneumococci are lysed by bile; other streptococci are not. Add 0.2 ml of 10% sodium deoxycholate in saline to 5 ml of an overnight broth culture (do not use glucose broth). Incubate at 37°C. Clearing should be complete in 30 min.

Pneumococci are alone among the streptococci in fermenting inulin and are not heat resistant as are many 'viridans' streptococci. These two criteria should not be used alone for differentiation. For more details about species of streptococci see Coleman *et al.* (1992) and *Cowan and Steel's Manual* (1993).

Enterococcus

These aerobic Gram-positive cooci form pairs or short chains. Like streptococci, they are catalase negative, but grow at 45°C, in the presence of 40% bile and 6.5% NaCl and hydrolyse aesculin. They are in Lancefield Group D (but so are some streptococci — see Table 37.2).

Isolation

Clinical material
Plate on blood agar, MacConkey and Edwards medium. This contains aesculin, which when hydrolysed by enterococci, gives black colonies.

Identification

Test for growth at 10 and 45°C, aesculin hydrolysis, arginine dihydrolase, and acid production from mannitol, sorbitol, sucrose, arabinose and adonitol. See Table 37.3.

Only four of the relatively common species are described here. The API20 STREP system identifies these and the less common species.

Common Enterococcus species

The important biochemical reactions are shown in Table 37.3.

Table 37.3 Common *Enterococcus* species (Group D)

	Growth at		*Arginine dihydrolase*	*Acid from*				
	10°C	*45°C*		*Mannitol*	*Sorbitol*	*Sucrose*	*Arabinose*	*Adonitol*
E. faecalis	+	+	+	+	+	+	−	−
E. faecium	+	+	+	+	v	+	+	−
E. durans	+	+	+	−	−	−	−	−
E. avium	v	+	−	+	+	+	v	+

v, variable

E. faecalis
This is normally present in the human gut, hence its use as an indicator of faecal pollution. It is also found in the gut of chickens but less often in that of other animals. It is the only enterococcus that reduces 2,3,5 triphenyltetrazolium chloride (TTC; in Mead's medium). It grows in the presence of 0.03% potassium tellurite.

E. faecium
This is present in the intestines of pigs and other animals and has caused spoilage of 'commercially sterilized' canned ham. TTC is not reduced and it does not grow in media containing 0.03% potassium tellurite. It is more resistant to antibiotics than other enterococci.

E. durans
This is found in milk and dairy products but is uncommon in clinical material.

E. avium
Present in the gut of chickens and some other animals, this is not often found in clinical material.

For more information about enterococci see Facklam and Collins (1989), also Coleman *et al.* (1992) and *Cowan and Steel's Manual* (1993).

Lactococcus

Formerly in *Streptococcus*, there are currently five species in this genus. They are associated with milk, dairy products, vegetables and fish. They may be confused with *S. thermophilus* (see Table 37.4). Only three species are listed here.

L. lactis
This is present in souring milk, to which it imparts a distinctive flavour, and is used as a cheese starter.

L. cremoris
Now considered to be a subspecies of *L. lactis*, this is used as a starter in the dairy industry.

Table 37.4 *Lactococcus* species and *S. thermophilus*

	Growth at		Arginine hydrolysis	Acid from	
	10°C	45°C		Mannitol	Sorbitol
L. lactis	+	−	+	v	−
L. cremoris	+	−	−	−	−
L. plantarum	+	−	−	+	+
S. thermophilus	−	+	−	−	−

v, variable

L. plantarum
This has been implicated in ths spoilage of frozen vegetables.

Aerococcus

These usually occur in clusters, pairs or short chains.

Isolation

Use blood agar for food homogenates and for air sampling. Incubate at 30–35°C.

Identification

Aerococci give α-haemolysis on blood agar and resemble other 'viridans' streptococci. They differ from *S. oralis* in growing in the presence of 6.5% NaCl and are resistant to 60°C for 30 min. They grow on bile agar but unlike the enterococci do not grow at 45°C nor possess the D antigen.

Aerococcus species

Aerococcus viridans
Is the only named species. It is a common airborne contaminant and is also found in curing brines. It appears to be the same organism as *Gaffkya homari* which causes an infection in lobsters; they develop a pink discoloration on the ventral surface and the blood loses its characteristic bluish green colour, becoming pink. In humans it occurs on the skin, may infect wounds and be responsible for endocarditis.

Leuconostoc

This genus contains several species of microaerophilic streptococci of economic importance. Some produce a dextran slime on frozen vegetables. Leuconostocs are used as cheese starters and have been associated with human infections, when they were resistant to vancomycin. They grow at 10°C but not at 45°C.

Isolation

Culture material on semi-solid yeast glucose agar at pH 6.7–7.0. Incubate at 20–25°C under 5% carbon dioxide.

Identification

Inoculate MRS sugars: glucose, lactose, sucrose, mannose, arabinose and xylose and look for dextran slime on solid medium containing glucose (see Table 37.5)

Table 37.5 ***Leuconostoc*** **species**

	Acid from						*Slime*
	Arabinose	*Xylose*	*Glucose*	*Mannose*	*Lactose*	*Sucrose*	
L. cremoris	–	–	+	–	+	–	–
L. dextranicum	–	v	+	v	+	+	+
L. lactis	–	–	+	v	+	+	–
L. mesenteroides	+	v	+	+	v	+	+ +
L. paramesenteroides	v	v	+	+	v	+	–

Leuconostoc species

Leuconostoc mesenteroides
This is the most important species. It produces gas and a large amount of dextran slime on media and vegetable matter containing glucose. Colonies on agar media without the sugar are small and grey. Growth is very poor on media without yeast extract. It takes part in sauerkraut fermentation and silage production and is responsible for slime disease of pickles (unless *Lactobacillus plantarum* is encouraged by high salinity). It is also responsible for 'slimy sugar' or 'sugar sickness'. It spoils frozen peas and fruit juices.

L. paramesenteroides
Is similar to *L. mesenteroides* but does not produce a dextran slime. Widely distributed on vegetation and in milk and dairy products.

L. dextranicum
Produces rather less slime than *L. mesenteroides* and is less active biochemically. Widely distributed in fermenting vegetables and in milk and dairy products.

L. cremoris
Is rare in nature but extensively used as a starter in dairy products.

L. lactis
Not common. Found in milk and dairy products.

Pediococcus

Pediococci are non-capsulated microaerophilic cocci that occur singly and in

Table 37.6 *Pediococcus* species

	Growth at		*Acid from*							
	37°C	*45°C*	*Lactose*	*Sucrose*	*Maltose*	*Mannitol*	*Galactose*	*Salicin*	*Arabinose*	*Xylose*
P. damnosus	−	−	−	−	+	−	−	−	−	−
P. acidi-lactici	+	+	v	v	−	−	+	−	−	−
P. pentosaceus	+	+	+	v	+	−	+	+	+	+
P. urinae-equi	+	−	+	+	+	−	+	+	v	v
P. halophilus	+	−	v	+	+	+	+	+	v	v

pairs. They are nutritionally exacting and are of economic importance in the brewing, fermentation and food processing industries.

Isolation

Use enriched media and wort agar for beer spoilage pediococci. Tomato juice media at pH 5.5 is best. Incubate at temperatures according to species sought (see Table 37.6) under 5–10% carbon dioxide.

Pediococci differ from leuconostocs in hydrolysing arginine and in not producing slimy or mucoid colonies. They grow in acetate or acetic acid agar.

Identification

If it is necessary to proceed to species identification inoculate MRS sugar media: lactose, sucrose, galactose, salicin, maltose, mannitol, arabinose and xylose. Incubate at appropriate temperature. Gas is not produced by any species (see Table 37.6).

Pediococcus species

Pediococcus damnosus (*P. cerevisiae*)
Is found in yeasts and wort. Contamination of beer results in a cloudy, sour product with a peculiar odour — 'sarcina sickness'. *P. damnosus* is resistant to the antibacterial agents in hops.

P. acidi-lactici
Is found in sauerkraut, wort and fermented cereal mashes. Is sensitive to the antibacterial activity of hops.

P. pentosaceus
Is also found in sauerkraut as well as in pickles, silage and cereal mashes but not in hopped beer.

P. urinae-equi
Is a contaminant of brewers' yeasts. (The name indicates that it was originally isolated from urine of a horse.)

References

Beighton, D. (1985) *Streptococcus mutans* and other streptococci from the oral cavity. In *Isolation and Identification of Micro-organisms of Medical and Veterinary Importance* (eds C.H. Collins and J.M. Grange), Society for Applied Bacteriology Technical Series No.21, Academic Press, London, pp.177–190

Coleman, G., Efstatiou, A. and Morrison, D. (1992) Streptococci and related organisms. In *Identification Methods in Applied and Environmental Microbiology* (eds R. G. Board, D. Jones and F. A. Skinner), Society for Applied Bacteriology Technical Series No.29, Oxford, Blackwells, pp.221–250

Cowan and Steel's Manual for the Identification of Medical Bacteria (1993) 3rd edn (eds G. I. Barrow and R. K. A. Feltham), Cambridge University Press, Cambridge

Facklam, R. R. and Collins, M. D. (1989) Identification of *Enterococcus* species isolated from human infections by a conventional test scheme. *Journal of Clinical Microbiology*, **27**, 731–734

Islam, A. K. M. S. (1977) Rapid recognition of Group B streptococci. *Lancet*, **i**, 256–257

Petts, D. N. (1984) Colistin-oxolinic acid blood agar: a new selective medium for streptococci. *Journal of Clinical Microbiology*, **19**, 4–7

Chapter

38

Lactobacillus and Erysipelothrix

These two genera contain Gram-positive, non-motile rods that are microaerophilic and facultatively anaerobic. Tables 38.1 and 38.2 show the principal characteristics that distinguish them from one another and from *Listeria*.

Lactobacillus

These Gram-positive bacilli may form chains and may resemble corynebacteria. They are non-motile, catalase negative, fermentative and have specific growth requirements.

They are important in the food, dairy and brewing industries, are commensals of animals and are found in the human gut and vagina. Some species are found in plant material. Although generally considered to be non-pathogenic some strains are associated with caries and others have been recovered from blood cultures.

Isolation

Plate in duplicate on MRS, Rogosa and similar media at pH 5.0–5.8. LS Differential medium is particularly useful in the examination of yoghurt. Use Raka-Ray agar for beer. Many strains grow poorly, if at all, at pH 7.0, but plate human pathological material on blood agar (see below). Incubate one set of plates aerobically and the other in a 5:95 carbon dioxide–hydrogen atmosphere. Incubate cultures from food at 28–30°C and from human or animal material ar 35–37°C for 48–72 h. Examine for very small, white colonies. Few other organisms, apart from moulds, will grow in the acid media.

For provisional identification of *L. acidophilus*, which is the commonest species in human material, place on the blood agar plates with:

(1) a sulphonamide sensitivity disc which has been dipped in saturated sucrose solution, and
(2) a penicillin (5 unit) disc.

L. acidophilus gives a green haemolysis. Growth is stimulated by sucrose; it is resistant to sulphonamides and sensitive to penicillin.

Identification

Subculture if necessary to obtain pure growth and heavily inoculate the modified MRS broth containing the following sugars: lactose, salicin, mannitol (for arginine hydrolysis test), and sorbitol. Incubate 28–30°C for 3–4 days. Inoculate tubes of MRS broth and incubate at 15°C and at 45°C (see Table 38.1). Test for arginine dihydrolase and aesculin hydrolysis.

Table 38.1 Some *Lactobacillus* species

	Growth at		*Acid from*				*Arginine dihydrolase*	*Aesculin hydrolysis*
	15°C	*45°C*	*Lactose*	*Salicin*	*Mannitol*	*Sorbitol*		
L. casei	+	v	v	+	+	+	−	+
L. plantarum	+	−	+	+	+	−	−	+
L. brevis	+	−	v	−	−	−	+	v
L. leichmanhii	−	+	v	+	−	−	v	+
L. acidophilus	−	+	+	+	−	−	−	+
L. delbrueckii	−	+	v	v	−	−	v	−
L. salivarius	−	+	+	v	+	+	−	v

v, variable

Culture on solid medium low in carbohydrate to test for catalase (high-carbohydrate medium may give false positives). The catalase test is negative (corynebacteria and listeria, which may be confused with lactobacilli, are catalase positive).

Identification of lactobacilli is not easy; reactions vary according to technique and the taxonomic position of some species is still uncertain. Table 38.1 lists the properties of some of the more common lactobacilli but there are many more species. The API system is recommended for full identification of members of this genus.

Species of lactobacilli

Lactobacillus acidophilus and *L. casei*

These are widely distributed and found in milk and dairy products and as commensals in the alimentary tract of mammals. They are used, along with other organisms, to make yoghurt and are also associated with dental caries in humans as a secondary invader. *L. acidophilus* is probably the organism described as the Boas-Oppler bacillus, observed in cases of carcinoma of the stomach in humans, and also Doderlein's bacillus, found in the human vagina.

Colonies of this organism on agar media are described as either 'feathery' or 'crab-like'.

L. plantarum

Is widely distributed on plants, alive and dead, and is one of the organisms responsible for fermentation in pickles and is used in the manufacture of sauerkraut.

L. delbrueckii

Is found in vegetation and in grain, and is used as a 'starter' to initiate acid conditions in yeast fermentation of grains. This organism grows on agar medium as small, flat colonies with crenated edges. This species now includes the former *L. lactis* and *L. bulgaricus*.

L. leichmannii
Is found in most dairy products. It is used in the commercial production of lactic acid and grows on agar as small white colonies.

Other species
See Sharpe (1981) and Kandler and Weiss (1986). Some lactobacilli, isolated from meat, are now placed in the genus *Carnobacterium* (see Collins *et al.*, 1987).

Counting lactobacilli

Enumeration of lactobacilli is a useful guide to hygiene and sterilization procedures in dairy industries. Suitable dilutions of bottle or churn rinsings or of swabbings steeped in 0.1% peptone water are prepared (see Plate counts, p.152) in Rogosa medium or one of the other media recommended for lactobacilli. Plates are incubated at 37°C for 3 days or at 30°C for 5 days in an atmosphere of 5% carbon dioxide.

Dental surgeons and oral hygienists sometimes request lactobacillus counts on saliva. The patient chews paraffin wax to encourage salivation, then 10 ml of saliva are collected and diluted for plate counts. Usually dilutions of 1 : 10, 1 : 100 and 1 : 1000 are sufficient.

Counts of 10^5 lactobacilli/ml of saliva are thought to indicate the likelihood of caries. There is a simple colorimetric test in which 0.2 ml of saliva is added to medium, as a shake tube culture, and a colour change after incubation suggests a predisposition to caries.

For more information about this genus see Sharpe (1981) and Kandler and Weiss (1986).

Erysipelothrix

This genus contains small (2 × 0.3 μm) Gram-positive rods that are found in and on healthy animals and on the scales of fish. Humans may be infected through abrasions when handling such animals, particularly pigs and fish. The resulting skin lesions are called erysipeloid. Epizootic septicaemia of mice has also been reported. There is one species, *Erysipelothrix rhusiopathiae*. Table 38.2 shows the characteristics that distinguish *Erysipelothrix* from *Lactobacillus* and *Listeria*.

It is non-motile, produces acid from glucose but not from maltose and mannitol, does not hydrolyse aesculin and is catalase negative and is resistant to neomycin. it grows at 5°C. See Table 38.2.

Table 38.2 *Lactobacillus*, *Erysipelothrix* and *Listeria*[a]

	Motility	*Catalase*	*Growth at 5°C*	*Acid from salicin*	*H_2S*	*Growth requirement*
Lactobacillus	−	−	−	+	−	+
Erysipelothrix	−	−	+	−	+	−
Listeria	+	+	+	+	−	−

[a] Added for comparison

Isolation and identification

Culture tissue or fluid from the edge of a lesion. Emulsified biopsy material is better than swabs of lesions. Plate on blood agar and on crystal violet or azide blood agar and incubate at 37°C in 5–10% carbon dioxide for 24–48 h.

Colonies are either small (0.5–1.0 mm) and 'dew-drop' or larger and granular. Young cultures are Gram-variable; older cultures show filaments. Test for motility with a Craigie tube. Test for fermentation of glucose, maltose and mannitol using enriched medium, e.g. broth containing Fildes' extract. Inoculate aesculin broth. Inoculate two tubes of enriched agar and incubate one each at 5 and 37°C. Do a catalase test and test for neomycin sensitivity by the disc method.

For more information about the genus *Erysipelothrix* see Jones (1986).

References

Collins, M. D., Farrow, J. A., Phillips, B. A., Feresu, S. B. and Jones, D. (1987) Classification of *Lactobacillus divergens, Lactobacillus piscicola* and some catalase-negative, asporogenous, rod-shaped bacteria from poultry in a new genus *Carnobacterium. International Journal of Systematic Bacteriology*, **37**, 310

Jones, D. (1986) The genus *Erysipelothrix*. In *Bergey's Manual of Systematic Bacteriology*, Vol. 2 (eds P. H. A. Sneath *et al.*), Williams and Wilkins, Baltimore, p. 1245

Kandler, O. and Weiss, N. (1986) The genus *Lactobacillus*. In *Bergey's Manual of Systematic Bacteriology*, Vol. 2 (eds P. H. A. Sneath *et al.*), Williams and Wilkins, Baltimore, p. 1245

Sharpe, M. E. (1981) The genus *Lactobacillus*. In *The Prokaryotes: a Handbook on Habitats, Isolation and Identification of Bacteria* (eds M. Starr *et al.*), Springer, New York, p. 1653

Chapter

39

Corynebacteria

These organisms are frequently club shaped, thin in the middle with swollen ends that contain metachromatic granules. These are best seen when Albert's or Neisser's stain is used. The stained bacilli may also be barred, with a 'palisade' or 'Chinese letter' arrangement, due to a snapping action in cell division.

This section is restricted to corynebacteria of medical and veterinary importance. Other 'corynebacteria' or 'coryneforms' are discussed later.

The diphtheria group and diphtheroids

Diphtheria is now a very rare disease and may be clinically atypical. *If an organism is suspected of being a diphtheria bacillus because of its colonial and microscopical appearance, inform the physician at once and before proceeding to identify the bacillus*. It is in the best possible interests of the patient that this is done. No harm will result if the final result is negative, but a delay of 24 h or more to confirm a positive case may have serious consequences, especially in an unimmunized child.

A culture should be sent immediately to the nearest reference laboratory. A case of diphtheria will usually generate large numbers of swabs from contacts. For the logistics of mass swabbing and the examination of large numbers of swabs in the shortest possible time see Collins and Dulake (1983).

Isolation and identification of the diphtheria organism and related species

Plate throat, nasal swabs, etc., on Hoyle's or other tellurite media and on blood agar. Incubate at 37°C and examine at 24 and 48 h.

For colonial characteristics, see species descriptions.

Make Gram-stained films of suspicious colonies. *C. diphtheriae* tends to decolorize easily and appears Gram-negative compared with diphtheroids which stain solidly Gram positive. When staining, make a control film of staphylococci treated in the same way. This apparent reaction is a useful characteristic.

Subculture a colony from the tellurite to Loeffler serum as soon as possible. Incubate at 37°C for 4–6 h. The best results are obtained with Loeffler medium containing an appreciable amount of water of condensation and when the tube has a cotton wool plug or the cap is left loose. Examine Gram- and Albert-stained films for typical beaded bacilli which may show metachromic granules at their extremities. Inoculate Robinson's serum water sugars (glucose, sucrose and starch) and a urea slope. But do not wait for the result before doing the plate toxigenicity test.

Note: Robinson's sugar medium is buffered and contains horse serum. Unbuffered horse serum medium may give false-positive results because of fermentation of the small amount of glycogen it contains, which is metabolized by many corynebacteria. Similarly, media that contain unheated rabbit serum may give false-positive starch fermentation reactions. Natural amylases hydrolyse starch, forming glucose, which is fermented by the organism. Misidentification of *C. diphtheriae* by inexperienced workers using various other fermentation tests is not uncommon. The biochemical reactions are given in Table 39.1. The API CORYNE kit is useful.

Table 39.1 Corynebacteria from human sources

	Acid from				
	Glucose	*Sucrose*	*Starch*	*β-haemolysis*	*Urease*
C. diphtheriae					
gravis	+	−	+	−	−
mitis	+	−	−	+	−
intermedius	+	−	−	−	−
C. ulcerans	+	−	+	−	+
C. pseudodiphtheriticum [a]	−	−	−	−	−
C. xerosis	+	+	+	−	−
C. jeikanum	+	−	−	−	−
C. ammoniogenes [b]	+	−	−	−	+

v, variable; [a] *C. hoffmannii*; [b] formerly *Brevibacterium*

Plate toxigenicity test

Elek's method, modified (Davies, 1974): inoculate a moist slope of Loeffler's medium with a colony from the primary plate as early as possible in the morning. Also inoculate Loeffler slopes with the stock control cultures maintained on Dorset egg medium. NCTC 10648 is toxigenic and NCTC 10356 is non-toxigenic. Incubate at 37°C for 4–6 h.

Melt two 15-ml tubes of Elek's medium and cool to 50°C. Add 3 ml of sterile horse serum to each and pour into plastic or very clear glass petri dishes. Place immediately in the still liquid medium in each plate a strip of filter-paper (60 × 15 mm) which has been soaked in diphtheria antitoxin. The titres of available antisera vary. Laboratories should therefore titrate each new batch against stock cultures, using concentrations between 500 and 1000 units/ml. This should be done in anticipation, not when a diphtheria investigation is imminent. Place the strip along a diameter of the plate. Dry the plates. These may be stored in a refrigerator for several days.

On each plate, streak the 4–6 h Loeffler cultures at right-angles to the paper strip so that the unknown strain is between the toxigenic and non-toxigenic controls as in Figure 39.1, approximately 10-mm apart. Incubate at 37°C for 18–24 h. If no lines are visible re-incubate for a further 12 h. Examine by transmitted light against a black background. If the unknown strain is toxigenic its precipitation lines should join those of the toxigenic control, i.e. a reaction of identity. Strains that produce little toxin may not show lines but turn the lines of adjacent toxigenic strains.

An alternative method is that of Jamieson (1965). Prepare the Loeffler cultures as above. Pour the plates but do not add the antitoxin strip. Dry the plates and place along the diameter of each a filter-paper strip (75 × 5 mm) which has been soaked in diphtheria antitoxin (for concentration see above).

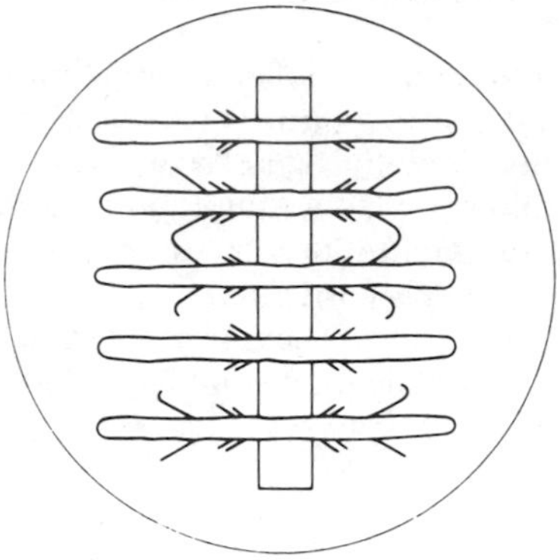

Figure 39.1 Elek plate toxigenicity test

Place the plates over a cardboard template marked out as in Figure 39.2 and heavily inoculate them from the Loeffler cultures in the shape of the arrow-heads. Place the unknown strain between the two controls. Incubate at 37°C for 18–24 h and examine for reactions of identify.

To be of value to the clinician, the plate toxigenicity test must give clear results in 18–24 h. If the toxigenic control does not show lines at 24 h, the medium or serum is unsatisfactory. The test requires a certain amount of experience and the medium and serum used must be tested frequently. For reasons not yet understood, certain batches of medium and serum give poor results. To test the medium and serum, use the control strains noted above and also NCTC 3894, which is weakly toxigenic. The strength of the antitoxin used is critical.

Report the organism as toxigenic or non-toxigenic *C. diphtheriae* cultural type *gravis, mitis* or *intermedius*.

Note that in any outbreak of diphtheria due to a toxigenic strain, a proportion of contacts may be carrying non-toxigenic *C. diphtheriae*.

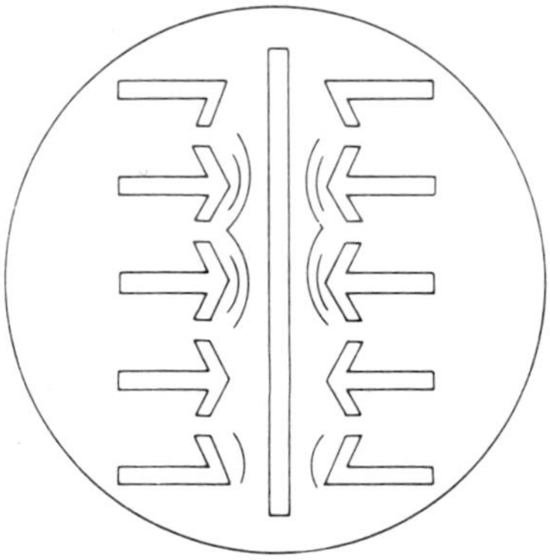

Figure 39.2 Jamieson's method for plate toxigenicity test

Species of corynebacteria from human sources

The following general descriptions of colony appearance and microscopic morphology of diphtheria bacilli and diphtheroids is a guide only. The appearance of the colony varies considerably with the medium and it is advisable to check this periodically with stock (NCTC and ATCC) organisms and, if possible, strains from recent cases of the disease. Even then, strains may be encountered that present new or unusual colony appearances.

C. diphtheriae
Three biotypes, *gravis*, *mitis* and *intermedius*, were originally given these names because of their respective association with the severe, mild and intermediate clinical manifestations of the disease. These names have become attached to the colonial forms and related to starch fermentation. These properties do not always agree and it is more important to carry out the toxigenicity test than to attempt to interpret the colonial and biochemical properties, which should be used only to establish that the organism belongs to the species. The following colony appearances relate to tellurite medium of the Hoyle type.

C. diphtheriae, gravis type
At 18–24 h, the colony is 1–2 mm in diameter, pearl grey with a darker centre; the edge is slightly crenated. It fractures easily when touched with a wire but is not buttery. At 48 h, it has enlarged to 3–4 mm and is much darker grey with the edge markedly crenated and with radial striations. This is the 'daisy head' type of the colony. The organism does not emulsify in saline. Films stained with methylene blue show short, often barred or beaded, bacilli arranged in irregular 'palisade' form or 'Chinese letters'. Well decolorized Gram films compared on the same slide with staphylococci or diphtheroids appear relatively Gram negative. Glucose and starch are fermented, but not sucrose. There is no haemolysis on blood agar.

C. diphtheriae, mitis type
At 18–24 h, colonies are 1–2 mm across and are dark grey, smooth and shining. The colonies fracture when touched but are much more buttery than the *gravis* type. At 48 h, they are larger (2-3 mm diameter), less shiny and darker in the centre, which may be raised, giving the 'poached egg' appearance. The organisms emulsify easily in saline. Films show long, thin bacilli, usually beaded and with granules, which may not be very Gram positive (see above). Glucose is fermented but not starch or sucrose. Colonies on blood agar are haemolytic.

C. diphtheriae, intermedius type
Colonies at 18–24 h are small (1 mm diameter), flat with sharp edges, black and not shining or glossy. At 48 h, there is little change. They are easily emulsified and films show beaded bacilli, intermediate in size between *gravis* and *mitis* types. Glucose is fermented but not starch or sucrose. Colonies on blood agar are non-haemolytic.

C. ulcerans
This organism resembles very closely the *gravis* type of diphtheria bacillus and is often mistaken for it. The colony tends to be more granular in the centre. Glucose and starch are fermented but not sucrose. *C. ulcerans* splits urea; *C. diphtheriae* does not. *C. ulcerans* also grow quite well at 25–30°C. It gives a reaction of incomplete identity with the plate toxigenicity test and *C. diphtheriae*

antitoxin. This organism does not cause diphtheria and is not readily communicable, but can cause ulcerated tonsils.

C. pseudodiphtheriticum (*C. hofmanii*)
Hoffman's bacillus at 18–24 h gives colonies 2–3 mm in diameter that are domed, shiny and dark grey or black. They are buttery (occasionally sticky), easily emulsified and films show intensely Gram-positive bacilli that are smaller and much more regular in shape, size and arrangement than are diphtheria bacilli. Carbohydrates are not fermented. These organisms are common and harmless commensals.

C. xerosis
At 18–24 h, colonies are 1–2 mm in diameter, flat, grey and rough, with serrated edges. At 48 h, they are 2–3 mm in diameter. Films show beaded bacilli, very Gram positive. This is a dubious species, a commensal of the conjunctivae and often found in the female genitalia. It is not a pathogen but resembles very closely the diphtheria bacillus in microscopical (but not colonial) morphology. It is, however, much more strongly Gram positive than *C. diphtheriae* and produces acid from glucose, usually sucrose but not starch.

C. jeikeinum
Formerly the 'JK diphtheroids', these have been isolated from blood and 'sterile' body fluids. They may colonize prostheses and are resistant to many antibiotics (see Jackman *et al.*, 1987).

C. ammoniogenes
Formerly *Brevibacterium*, this species is associated with 'nappy rash'.

Corynebacteria from animal sources

Isolation and identification

Culture, pus, etc., on blood agar and tellurite medium, incubate at 37°C and examine at 24 and 48 h for colonies of beaded Gram-positive bacilli. Some corynebacteria are haemolytic on blood agar. Small grey-black colonies are usually found on tellurite medium.

Subculture on Loeffler medium and examine Albert-stained films at 6–18 h for clubbed or beaded bacilli which may have metachromatic granules. From the Loeffler inoculate gelatin medium, nutrient agar, Robinson's serum water glucose, sucrose, mannitol, maltose, lactose and urea medium. Incubate at 37°C for 24–48 h (see Table 39.2). The API CORYNE kit is useful.

C. kutscheri (*C. murium*)
Colonies on blood agar are larger than *C. pyogenes*, flat, grey and non-haemolytic. The bacilli are slender, often filamentous. Loeffler medium is not liquefied. It is responsible for fatal septicaemia in mice and rats.

C. pseudotuberculosis (*C. ovis*)
Colonies are small, usually grey but may be yellowish and are haemolytic, especially in anerobic cultures. Loeffler serum may be liquefied. The bacilli are slender and clubbed, very like *C. diphtheriae*. It is responsible for epizootics in rodents and is a nuisance in laboratory animal houses.

Table 39.2 Corynebacteria from animal sources

	Gelatin liquefaction	Acid from					β-haemolysis	Urease
		Glucose	Sucrose	Mannitol	Maltose	Lactose		
C. kutscheri	–	+	+	–	+	–	–	v
C. pseudotuberculosis	v	+	v	–	+	v	v	+
C. renale[a]	–	+	–	–	v	–	–	+
C. bovis	–	+	–	v	+	v	–	–
C. pyogenes[b]	v	–	–	–	+	+	+	–
C. equi[c]	–		–	–	–	–	–	+

v, variable; [a] group; [b] now in *Actinomyces*; [c] now in *Rhodococcus*

This is the Preisz-Nocard bacillus which causes pseudotuberculosis in sheep, cattle and swine and lymphangitis in horses.

C. renale
This is now believed to be three separate species, *C. renale*, *C. pilosum* and *C. cystitidis*, all associated with urinary tract infections of cattle.

C. bovis
This is found in cows' udders and may cause mastitis.

C. pyogenes
Now considered to belong to *Actinomyces*. The colonies on blood agar are pinpoint and haemolytic and may be mistaken for haemolytic streptococci except for their slow rate of growth. They are often not apparent for 36–48 h. Films show very small bacilli that may not easily be recognized as corynebacteria. The catalase test is negative (other corynebacteria mentioned here are positive). It does not grow on nutrient agar but Loeffler's serum cultures are liquefied. This organism causes suppurative lesions including mastitis in domestic and wild animals.

C. equi
This is now regarded as *Rhodococcus equi* (see also p.433). It grows well on nutrient agar, forming large, moist, pink colonies. The bacilli are very pleomorphic, varying from coccoid to large clubbed forms with granules. It is non-haemolytic, does not liquefy Loeffler medium and does not form acid from carbohydrates. It is responsible for serious pyaemia in young horses.

'Coryneform bacilli'

Various bacilli that resemble corynebacteria in morphology are found in association with animals and dairy products. They are relatively heat resistant (70°C for 15 min). One such is *Microbacterium lacticum*, which is thermoresistant and a strict aerobe. It is non-motile, grows slowly at 30°C on milk agar, forms acid from glucose and lactose, is catalase positive, hydrolyses starch, is lipolytic and does not liquefy gelatin in 7 days. A pale yellow pigment may be formed. These organisms may cause spoilage in milk and indicate unsatisfactory dairy hygiene. They are also found in deep litter in hen houses.

References

Collins, C. H. and Dulake, C. (1983) Diphtheria: the logistics of mass swabbing. *Journal of Infection*, **6**, 227–230

Davies, J. R. (1974) Identification of diphtheria bacilli. In *Laboratory Methods 1* (eds C. H. Collins and A. T. Willis), Public Health Laboratory Service Monograph No. 4, HMSO, London

Jackman, P. J. H., Pitcher, D. G., Pelcznska, S. and Borman, P. (1987) Classification of corynebacteria associated with endocarditis (Group JK) as *Corynebacterium jeikeium* sp. nov. *Systematic and Applied Microbiology*, **9** 83–90

Jamieson, J. E. (1965) A modified Elek test for toxigenic *Corynebacterium diphtheriae*. *Monthly Bulletin of the Ministry of Health*, **24**, 55–58

Chapter

40

Listeria, Kurthia and Brochothrix

These are all aerobic Gram-positive, catalase-positive non-sporing rods. *Listeria* includes some human and animal pathogens, *Kurthia* is found in food products and occasionally in clinical material, and *Brochothrix* is a food-spoilage organism. Table 40.1 shows some distinguishing characteristics.

Table 40.1 *Listeria, Kurthia* and *Brochothrix*

	Growth			*Motile at 22°C*	*O/F*	*Acid from glucose*
	At 37°C	*Anaerobic*	*On STAA*			
Listeria	+	+	−	+	F	+
Kurthia	+	−	−	+	none	−
Brochothrix	−	+	+	−	F	+

STAA, Gardner's medium; F, fermentative

Listeria

These are aerobic, facultative anaerobic Gram-positive rods that are catalase positive and ferment carbohydrates. They do not grow on STAA medium.

They are found in grass, silage, soil, sewage and water. There are seven species, two of which, *L. monocytogenes* and *L. ivanovii*, are pathogenic for humans and animals. Listerias have become important food-borne pathogens (see Chapter 13). They survive in cold storage and may multiply at low temperatures.

Isolation

Clinical material

Plate cerebrospinal fluid (CSF), centrifuged deposits and blood cultures on blood agar and (Oxford) listeria selective agar.

Homogenize tissues in peptone water diluent and plate on blood agar and listeria selective agar. Treat the remainder of the homogenate as for faeces (below). Incubate primary plates at 37°C for 24–48 h.

Homogenize faeces in peptone water. Add 1 vol of homogenate to 9 vol of listeria selective broth. Subculture at 4 and 24 h to selective media. At 24 h add 0.1 ml to secondary listeria selective broth. Incubate at 37°C for 24–48 h and plate on selective media.

Food
Homogenize 25 g of the material in 225 ml of commercial listeria enrichment broth. Incubate at 30°C for 48 h. Subculture to listeria selective (Modified Oxford—MOX—or PALCAM) agar and incubate at 37°C for 48 h under microaerophilic conditions (e.g. with Campylobacter gas generating kit, Oxoid). Colonies on MOX are black, those on PALCAM are grey-green or black, with shrunken centres and a black halo. Neither medium is wholly selective for listerias.

Identification

Colonies on blood agar are small (1 mm diam.), smooth and surrounded by a narrow zone of β-haemolysis. On (Oxford) selective agar they are the same size but are black, with a black halo and a sunken centre. Colonies on translucent media, viewed by light reflected at 45° from below (Henry method, see Prentice and Neaves, 1992) are grey-blue in colour and finely textured. Colonies of most other organisms are opaque and/or otherwise coloured.

Examine for typical 'tumbling' motility at 20–22°C (no higher, as listerias are non-motile above 25°C). Test for acid production from glucose, rhamnose, xylose and mannitol (10% carbohydrate in purple broth) and do CAMP test.

CAMP (Christie, Atkins and Munch-Petersen) test
Use blood agar prepared from fresh, washed, defibrinated sheep blood. Make dense suspensions from colonies of the suspected listerias and of *S. aureus* (NCTC 1803) and *Rhodococcus equi* (NCTC 1621). Make streaks of the staphylococcus and the rhodococcus across the blood agar, about 2–3 cm apart. Streak the suspect listerias at right angles but do not touch the other streaks. Incubate at 37°C for 12–18 h (not longer).

L. monocytogenes and *L. seeligeri* give a narrow zone of β-haemolysis near the staphylococcus; *L. ivanovii* gives a larger, semicircular zone near the rhodococcus. Other species give no reactions. See Table 40.2.

Table 40.2 *Listeria* species

	β-haemolysis	*CAMP test*		*Acid from*		
		S. aureus	*R. equi*	*Rhamnose*	*Xylose*	*Mannitol*
L. monocytogenes	+	+	−	+	−	−
L. ivanovii	+	−	+	−	+	−
L. innocua	−	−	−	v	−	−
L. seeligeri	(+)	(+)	−	v	+	−
L. welshimeri	−	−	−	v	+	−
L. grayi	−	−	−	−	−	+
L. murrayi	−	−	−	v	−	+

v, variable; ( ), weak

Kit and commercial tests
In the Rosco Listeria ZYM (Lab M, UK), the reagents prepared from tablets are inoculated from colonies on selective media. The result is available in 4 h. *L. monocytogenes* is distinguished from *L. innocua* by β-haemolysis. The Mastalist

(Mast, UK) has 12 tests which are inoculated with a multipoint apparatus. The API listeria strip (BioMerieux) uses 10 tests to identify 7 species.

There is an ELISA kit which uses monoclonal antibodies (Listeria-TEC, Organon), and a gene probe—luminescence system (Laboratory Impex, UK).

Listeria species

L. monocytogenes
Responsible for human listeriosis, usually food-borne. Symptoms include meningitis (especially in neonates) and septicaemia.

L. ivanovii
This causes septicaemia, encephalitis and abortion in farm animals and also disease in humans.

The other species, *L. innocua, L. seeligeri, L. welshimeri, L. grayi* and *L. murrayi* are not clinically significant but are included in Table 40.2 because it is important to distinguish them from the pathogens.

For a review of listeriosis see Bahk and Marth (1990) and for more detailed information about isolation and identification see Prentice and Neaves (1992).

Kurthia

These are strictly aerobic Gram-positive rods that tend to occur in parallel chains when first isolated but assume coccoid forms in older cultures. Strains vary in motility. They are catalase positive, do not produce acid from carbohydrates but do produce acid from either ethanol or glycerol. There is no growth on STAA medium. There are two species, *K. zopfii* and *K. gibsonii.*

They are found in animal faeces, milk, water, cold-stored meat products and hence occasionally in clinical material.

Isolation

Culture on yeast extract agar. Incubate at 25-30°C for 4–5 days and observe colour of colonies. Culture on nutrient gelatine. Incubate at 20°C for 5 days and note 'bird-feather' nature of growth.

Identification

Test for growth at 4, 37 and 45°C, acid production from ethanol and glycerol, and DNase. See Table 40.3

This genus is reviewed by Keddie and Shaw (1986).

Table 40.3 *Kurthia* species

	Growth at		Acid from		*DNAse*	*Colony colour*
	37°C	*45°C*	*Ethanol*	*Glycerol*		
K. zopfii	v	–	+	–	–	none
K. gibsonii	+	+	–	+	+	cream or yellow

v, variable

Brochothrix

These are aerobic, facultative anaerobic Gram-positive, non-motile rods that occur singly or in short chains or filaments, often with pleomorphic forms. They are catalase positive, ferment carbohydrates, grow at 4°C and 25°C but not at 37°C. They grow poorly on MRS but well on STAA media.

There are two species: *B. thermosphactum* is a meat spoilage organism, found in souring ('British fresh') sausages at low temperatures, in film-packaged products, milk and cheese.

B. campestris is found in grass and soil.

Isolation

Culture on Gardner's STAA medium and incubate at 20–25°C.

Identification

The most important characteristic is growth on STAA medium. Test for growth in the presence of 8% NaC1 and for hippurate hydrolysis. *B. thermosphactum* grows in the NaC1 medium but is hippurate negative. *B. campestris* fails to grow in NaC1 medium but hydrolyses hippurate.

For more information see Gardner (1966) and Dodd and Dainty (1992).

References

Bahk, J. and Marth, E. H. (1990) Listeriosis and *Listeria monocytogenes*. In *Foodborne Diseases* (ed D.O. Cliver), Academic Press, New York, pp.248–259

Dodd, C. E. R. and Dainty, R. H. (1992) Identification of *Brochothrix* by intracellular and surface biochemical composition. In *Identification Methods in Applied and Environmental Microbiology* (eds R. G. Board, D. Jones and F. A. Skinner), Society for Applied Bacteriology Technical Series No. 29, Blackwells, Oxford, pp. 297–310

Gardner, G. A. (1966) A selective medium for the enumeration of *Microbacterium thermosphactum* in meat and meat products. *Journal of Applied Bacteriology*, **29**, 455–457

Keddie, R. M. and Shaw, S. (1986) *Kurthia*. In *Bergey's Manual of Systematic Bacteriology* (eds P. H. A. Sneath *et al.*), Williams and Wilkins, Baltimore, pp.1255–1257

Prentice, G. A. and Neaves, P. (1992) The identification of *Listeria* species. In *Identification Methods in Applied and Environmental Microbiology* (eds R. G. Board, D. Jones and F. A. Skinner), Society for Applied Bacteriology Technical Series No. 29, Blackwells, Oxford, pp.283–296

Chapter

41

Neisseria and Branhamella

The aerobic, Gram-negative cocci include the genera *Neisseria* and *Branhamella*. Both are oxidase and catalase negative.

Neisseria

These occur mostly as oval or kidney-shaped cocci arranged in pairs with their long axes parallel. They are non-motile, non-sporing, oxidase positive and reduce nitrates to nitrites. The two important pathogens are *N. gonorrhoeae* (the gonococcus), causative organism or gonorrhoea, and *N. meningitidis* (the meningococcus), one of the organisms causing meningitis. Both require media containing blood or serum. Other species, which are commensals or opportunist pathogens, will grow on nutrient agar.

Gonococcus

Examine Gram-stained films of exudates (urethra, cervix, conjunctivae, etc.) for intracellular Gram-negative diplococci. Exercise caution in reporting the presence of gonococci in vaginal and conjunctival material and in specimens from children.

Culture at once or use one of the transport media. Culture on deep plates (20–25 ml of medium) of New York City (NYC) or Thayer Martin (TM) media with commercial selective supplements. Incubate all cultures 35–36°C (better than 37°C) in 5–10% CO_2 and 70% humidity. Examine at 24 and 48 h. Colonies of *N. gonorrhoeae* are transparent discs about 1 mm in diameter, later increasing in size and opacity, when the edge becomes irregular. Test suspicious colonies by the Gram film and oxidase test. Typical morphology and a positive oxidase test are presumptive evidence of the gonococcus in material from the male urethra, but not from other sites.

For rapid results test colonies by FA method, coagglutination (Phadebact) or rapid biochemical tests (Gonocheck II, Rapid IDNH).

Subculture oxidase positive colonies on chocolate agar and incubate overnight in a 10% carbon dioxide atmosphere. Immediate subculture into carbohydrate test media may not be satisfactory as the primary medium contains antibiotics; other organisms may grow. Repeat the oxidase test on the subculture and emulsify positive colonies in about 1 ml of serum broth. Use this to test for acid production from glucose, maltose, sucrose and lactose in Flynn and Waitkin's sugar-free medium (p.77). Some serum media may contain a maltase that may give a false-positive reaction with maltose. The best results are obtained with a semisolid medium, as gonococci do not like liquid media (Table 41.1). API QUADFERM and API WH are useful.

Confirm fermentation results by agglutination or coagglutination tests (Phadebact).

Exercise caution in reporting these organisms without adequate experience, especially if they are from children, eye swabs, anal swabs or other sites. They may be meningococci or other *Neisseria* which are genital commensals that do not cause sexually transmitted disease. Sugar reactions and colonial morphology are not entirely reliable. Certain other organisms, *Moraxella*, (p.333) and *Acinetobacter calcoaceticus* (p.287) may resemble gonococci in direct films and on primary isolation. *Moraxella*, like *Neisseria*, is oxidase positive but *Acinetobacter* is oxidase negative.

For further information on gonococci: see Jephcott and Egglestone (1985), Jephcott (1987).

Meningococcus

Examine films of spinal fluid for intracellular Gram-negative diplococci. Inoculate blood agar and chocolate agar with cerebrospinal fluid (CSF), or its centrifuged deposit. Incubate, preferably in a 5% carbon dioxide atmosphere, at 37°C overnight.

Incubate the remaining CSF overnight and repeat culture on chocolate agar.

Meningococcal colonies are transparent, raised discs about 2 mm in diameter at 18–24 h. In CSF cultures, the growth will be pure, while in eye discharges and vaginal swabbings of young children, etc., other organisms will be present. Subculture oxidase positive colonies of Gram-negative diplococci in semisolid serum sugar media (glucose, maltose, sucrose). Meningococci give acid in glucose and maltose (see Table 41.1).

Agglutinating sera are available but should be used with caution. There are at least four serological groups. Slide agglutinations may be unreliable. It is best to prepare suspensions in formol saline for tube agglutination tests.

N. lactamica

This is important in that it might be confused with pathogenic species because it is oxidase positive, is ONPG positive, and does not grow at 22°C. It is fermentative and produces acid from lactose but this may be delayed. The pathogens do not change lactose and are oxidative.

Other species and similar organisms

These grow on nutrient agar. Colonies are variable: 1–3 mm, opaque, glossy or sticky, fragile or coherent; others may be rough and granular. They include *N. flavescens* which produces a yellow pigment, *N. mucosa*, *N. elongata* and *N. subflava*. For the properties of these see *Cowan and Steel's Manual* (1993).

Gemella haemolysans is sometimes confused with neisserias because although it is Gram positive it is easily decolorized. It differs from the neisserias in being β-haemolytic, oxidase and catalase negative.

Branhamella

This genus contains three species, but only one, *B. catarrhalis* (formerly known as *Neisseria catarrhalis*), is found in human material. Previously regarded as a commensal of the respiratory tract it is now considered to be an opportunist pathogen of the upper and lower respiratory system and is associated with lung abscesses. Middle ear infections are not uncommon. It grows well on blood agar, forming non-pigmented, non-haemolytic colonies. It is β-lactamase and DNase positive (Corkill and Makin, 1982; Johnson, 1983). Carbohydrates

Table 41.1 *Neisseria* and *Branhamella*

	Growth on		*Growth at 22°C*	*DNase*	*ONPG*	*Acid from*			
	Nutrient agar	*Thayer Martin*				*Glucose*	*Maltose*	*Sucrose*	*Lactose*
N. gonorrhoeae	−	+	−	−	−	+	−	−	−
N. meningitidis	−	+	−	−	−	+	+	−	−
N. lactamica	+	+	−	−	+	+	+	−	+
Other *Neisseria*	v	−	+	−	v	v	v	v	v
Branhamella	+	−	+	+	−	−	−	−	−

v, variable

are not fermented, oxidase and catalase tests are negative (see Table 41.1). It also degrades tributyrin (a commercial butylase strip test is available).

References

Corkill, J. E. and Makin, T. (1982) A selective medium for non-pathogenic aerobic Gram-negative cocci from the respiratory tract, with particular reference to *Branhamella catarrhalis*. *Medical Laboratory Sciences*, **39**, 3–10

Cowan and Steel's Manual for the Identification of Medical Bacteria (1993) 3rd edn (eds G. I. Barrow and R. K. A. Feltham) Cambridge University Press, Cambridge

Jephcott, A. E. (1987) Gonorrhoea. In *Sexually Transmitted Diseases*, Public Health Laboratory Service, London, pp.24–40

Jephcott, A. E. and Egglestone, S. I. (1985) *Neisseria gonorrhoeae*. In *Isolation and Identification of Micro-organisms of Medical and Veterinary Importance* (eds C. H. Collins and J. M. Grange), Society for Applied Bacteriology Technical Series No.21, Academic Press, London, pp.143–160

Johnson, A. P. (1983) The pathogenic potential of commensal species of *Neisseria*. *Journal of Clinical Pathology*, **36**, 213–223

Chapter

42

Bacillus

This is a large genus of Gram-positive spore-bearing bacilli that are aerobic, facultatively anaerobic and catalase positive. Many species are normally present in soil and in decaying animal and vegetable matter. One species, *Bacillus anthracis*, is responsible for anthrax in humans and animals. *B. cereus* causes food poisoning (Gilbert *et al.*, 1981; Kramer *et al.*, 1982). This and several other species are now known to cause disease in humans and animals (see Logan, 1988). Some are responsible for food spoilage.

The bacilli are large (up to 10 × 1 μm) and commonly adhere in chains. Most species are motile; some form motile colonies. Some species are capsulated, some produce a sticky levan on media that contain sucrose. Spores may be round or oval, central or subterminal.

Physiological characters vary. There are strict aerobes and species which are facultative anaerobes. A few are thermophiles. All are catalase positive. In old cultures, Gram-negative forms may be seen and some species are best described as 'Gram-indifferent'.

Anthrax

Isolation of *B. anthracis* from pathological material

Veterinarians often diagnose anthrax in animals on clinical grounds and by the examination of blood films stained with polychrome methylene blue. These show chains of large, square-ended bacilli with the remains of their capsules forming pink-coloured debris between the ends of adjacent organisms. This is M'Fadyean's reaction.

Culture animal blood, spleen substance or other tissue ground in a Griffith's tube or macerator. In humans, culture material from the cutaneous lesions, faeces, urine and sputum (pulmonary anthrax is now rare).

Plate on blood agar and on PLET medium which is brain heart infusion agar containing (per ml) polymyxin, 30 units; lysozyme, 40 μg; EDTA, 300 μg; thallous acetate, 40 μg and incubate overnight. Pick woolly or waxy 'medusa head' colonies of Gram-positive rods into Craigie tubes to test for motility and on chloral hydrate blood agar. It may be difficult to get pure growths without several subcultures. For identification, see below.

Isolation of *B. anthracis* from hairs, hides, feedingstuffs and fertilizers

The sample should be in a 200–300 ml screw-capped jar, and should occupy about 4 fluid ounces. Add sufficient warm 0.1% peptone solution to cover it. Shake, stand at 37°C for 2 h, decant and heat the fluid at 60°C for 30 min. Place 0.1, 1.0 and 2 ml of the uncentrifuged fluid and the deposit after centrifuging in petri dishes. Add to each plate either 0.5 ml of 5% egg albumen or 0.25 ml of 0.01% dibromopropamidine isethionate, pour on 15 ml of yeast extract agar at 50°C, mix well, allow to set and leave on the bench until 5 p.m. Incubate at 37°C until 9 a.m. the next day. Longer incubation produces atypical colonies. The lysozyme in the egg albumen and the dibromopropamidine compound reduce the numbers of other organisms. Polymyxin B (20 units/ml) in the agar is also useful for heavily contaminated material. Also plate 0.25 ml amounts of deposit on several plates of PLET medium.

Examine under a low-power binocular (plate) microscope. Ignore surface colonies and look for deep colonies that resemble dahlia tubers or Chinese artichokes. Pick into Craigie tubes and plate on chloral hydrate agar.

Identification of *B. anthracis*

Bacillus anthracis is non-motile. The Craigie tube is the best way of demonstrating this. *B. cereus*, which is commonly mistaken for *B. anthracis*, is motile.

The growth of *B. anthracis* on agar media has a characteristic woolly or waxy nature when touched with a wire (the 'tenacity test'). Inoculate ammonium salt sugar medium containing 1% salicin. Do not use media that contain peptones; sufficient ammonia may be produced by some species to mask acid production. Culture on nutrient agar containing 10 units/ml of penicillin, nutrient agar containing 0.3% 2-phenylethanol and in nutrient broth incubated at 45°C.

Test for the 'string of pearls' appearance: grow in nutrient broth containing 0.5 units of penicillin per ml for 3–6 h at 37°C and examine a wet preparation. Strings of spherical bodies suggest *B. anthracis*.

The biochemical properties are given in Table 42.1 The API method is useful but the best way to identify *B. anthracis* is by immunofluorescence. Sera are available commercially. If the anthrax γ-bacteriophage is available make a lawn of the suspect culture on blood agar, place one drop (0.02 ml) of the phage (titre *ca* 10^6 plaque-forming units (pfu)/ml) on the surface and incubate at 37°C overnight. *B. anthracis* gives a clear zone of lysis (so do some other *Bacillus* spp.).

It causes anthrax in animals, a fatal septicaemia. *B. anthracis* transmissible to humans, producing a localized cutaneous necrosis ('malignant pustule') or pneumonia (woolsorters' disease) due to inhalation of spores.

Imported bone meal, meat meal and other fertilizers and sometimes feeding-stuffs may be infected. These are often prepared from the remains of animals that died of anthrax. In most developed countries, an animal dying of anthrax must be buried in quicklime under the supervision of a veterinarian and below the depth at which earthworms are active.

There are regulations governing the importation of hairs and hides from foreign countries where anthrax is endemic.

For further information on the isolation of *B. anthracis* see Carman *et al.* (1985) and Turnbull (1990).

Table 42.1 Some *Bacillus* species

	Lecithinase	Acid[a] from				Growth		VP
		Mannitol	Xylose	Arabinose	Starch	Anaerobic	at 50°C	
B. anthracis	(+)	−	−	−	+	+	−	+
B. cereus	+	−	−	−	+	+	−	+
B. mycoides	+	−	−	−	+	+	−	+
B. laterosporus	+	v	−	−	−	+	−	+
B. subtilis	−	+	v	+	+	−	+	+
B. pumilus	−	+	+	+	−	−	v	+
B. coagulans	−	v	v	v	+	+	+	v
B. megaterium	−	v	+	v	+	−	−	−
B. polymyxa	−	v	+	v	+	−	−	+
B. macerans	−	v	+	+	+	+	+	v
B. circulans	−	+	+	v	+	+	v	v
B. brevis	−	−	−	−	−	−	+	v
B. licheniformis	−	+	+	+	+	+	+	v

( ), *weak; v, variable*
[a] In ammonium salt medium

Food poisoning

Bacillus cereus
May be associated with food poisoning (Chapter 13).

Make 10% suspensions of the food, faeces and vomit in 0.1% peptone water, using a Stomacher. Inoculate glucose tryptone agar and Bacillus Cereus Selective agar. Incubate at 30°C for 24–48 h.

Use a Stomacher to prepare a 10% suspension of the food in peptone water diluent. Prepare further dilutions, 10^{-1}–10^{-3} and plate 0.1 ml amounts on *B. cereus* selective agar. Incubate at 37°C for 24 h and then at 25–30°C for a further 24 h. This medium shows the egg yolk reaction, mannitol fermentation and encourages sporulation (see below). Count colonies of presumptive *B. cereus*; more than 10^5 cfu/g is suggestive.

Food spoilage

Examine Gram-stained films for spore-bearing Gram-positive bacilli. Make 10% suspensions of food in 0.1% peptone water, and pasteurize some of the suspension at 75–80°C for 10 min to kill vegetative forms.

Inoculate glucose tryptone agar and *B. cereus* selective media with both unheated and pasteurized material.

Incubate at 25–30°C overnight. If the food was canned, incubate replicate cultures at 60°C. Identify colonies of Gram-positive bacilli.

Cold-tolerant spore-formers are important spoilage organisms. Incubate at 5°C for 7 days.

Identification

Test colonies for lecithinase, acid production from xylose, arabinose and starch (ammonium salt sugar medium), anaerobic growth, growth at 50°C and Voges-Proskauer reaction (use 1% NaCl glucose broth). See Table 42.1. The API 20E and API50 CHB will identify more species.

Rapid staining method for B. cereus

Make films from colonies on *B. cereus* selective medium, which encourages sporulation and dry in air. Place the slide over boiling water and stain with 5% malachite green for 2 min. Wash and blot dry. Stain with 0.3% Sudan black in 70% alcohol for 15 min. Wash with xylol and blot dry. Counterstain with 0.5% safranin for 20 s. Wash and dry. *B. cereus* spores are oval, central or subterminal and do not swell the cells and are stained green. The vegetative cells stain red and contain black granules.

Bacillus species

Bacillus cereus

General properties are described above. Phospholipase activity produces large haloes around colonies on egg yolk agar. With most other species, activity is usually limited to beneath the colony. Methylene blue milk is rapidly decolorized. This organism is responsible for 'bitty cream', particularly in warm weather, and causes milk and ice-cream to fail the methylene blue test. It may cause food poisoning (see above). Bacteraemia and pneumonia associated with *B. cereus* have been reported. It is a common contaminant on shell eggs. It does not ferment mannitol, unlike *B. megaterium*, which it may otherwise resemble.

B. subtilis ('*B. mesentericus*')

Wrinkled or smooth and folded colonies, 4–5 mm across on glucose tryptone agar with yellow halo (acid production). Strongly alkaline reaction in Crossley medium but no gas or blackening. No zone around colonies on egg yolk agar. Responsible for spoilage in dried milk and in some fruit and vegetable products.

Causes 'ropy' bread. Used in industry to manufacture enzymes for biological washing products.

B. stearothermophilus

This and associated thermophiles form large colonies (4 mm in diameter) on glucose tryptone agar with a yellow halo due to acid production at 60°C but grow poorly at 37°C and not at 20°C. This organism is responsible for 'flat-sour' (i.e. acid but no gas) spoilage in canned foods. For more detailed investigation see Chapter 16.

B. mycoides

A variant of *B. cereus* with rhizoid colonial form.

B. licheniformis

May cause food poisoning. It does not produce lecithinase.

B. megaterium

A very large bacillus. Widely distributed. May cause *B. cereus*-like food poisoning.

B. pumilus

Found on plants. Common contaminant of culture media.

B. coagulans

Widely distributed. Found in canned foods.

B. anthracis
See p.391

B. polymyxa
Widely distributed in soil and decaying vegetables (possesses a pectinase). Source of the antibiotic polymyxin.

B. macerans
Common in soil. Found in rotting flax.

B. circulans
Motile colonies with non-motile variants. Found in soil.

B. laterosporus
Soil organism. Source of the antibiotics tyrothricin and gramicidin.

B. thuringensis
Insect pathogen, widely used in pest control.

B. sphaericus
Large spores, resembles *C. tetani* in appearance. Forms motile colonies. Soil organism.

B. rotans
Probably synonym of *B. sphaericus*. Motile colonies.

B. pasteuri
Found in decomposing urine. Grows only in media containing peptone and urea.

There are many other species. See Claus and Berkeley (1986).

References

Carman, J. A., Hambleton, P. and Melling, J. (1985) *Bacillus anthracis*. In *Isolation and Identification of Micro-organisms of Medical and Veterinary Importance* (eds C. H. Collins and J. M. Grange), Society for Applied Bacteriology Technical Series No. 21, Academic Press, London, pp.207–214

Claus, D. and Berkeley, R. C. W. (1986) The genus *Bacillus*. In *Bergey's Manual of Determinative Bacteriology* (eds P. H. A. Sneath, N. S. Mair and M. E. Sharpe), Williams and Wilkins, Baltimore, pp.1105–1139

Gilbert, R. J., Turnbull, P. C. B., Parry, J. M. and Kramer, J. M. (1981) *Bacillus cereus* and other *Bacillus* species: their part in food poisoning and other clinical infections. In *The Aerobic Endospore-forming Bacteria* (eds R. C. W. Berkeley and M. Goodfellow), Society for General Microbiology Special Publications Series No. 4, Academic Press, London, pp.297–314

Kramer, J. M., Turnbull, P. C. B., Munshi, G. and Gilbert, R. J. (1982) Identification and characterization of *Bacillus cereus* and other *Bacillus* species associated with foods and food poisoning. In *Isolation and Identification Methods for Food Poisoning Organisms* (eds J. E. L. Corry, D. Roberts and F.A. Skinner), Society for Applied Bacteriology Technical Series No. 17, Academic Press, London, pp.261–286

Logan, N. A. (1988) *Bacillus* species of medical and veterinary importance. *Journal of Medical Microbiology*, **25**, 157–165

Turnbull, P. C. B. (ed.) (1990) *Proceedings of an International Workshop on Anthrax, Winchester, 1989. Salisbury Medical Bulletin* No. 68 Special Supplement. PHLS Porton Down, Salisbury

Chapter

43

Gram-negative anaerobic bacilli and cocci

Gram-negative anaerobic bacilli

The Gram-negative anaerobic bacilli are found in the alimentary and genitourinary tracts of humans and other animals and in necrotic lesions, often in association with other organisms. They are pleomorphic, often difficult to grow and die easily in cultures. Identification is often difficult. Reference experts use gas-liquid chromatography. *Eikenella* are not strict anaerobes and will grow in 10% carbon dioxide. The organism known as *E. corrodens* may be confused with another corroder – *Bacteroides ureolyticus*.

In pathological material they are not uncommonly mixed with other anaerobes and capnophilic organisms. In general, primary infections above the diaphragm are due to *Fusobacterium*, but *Bacteroides* may occur anywhere and are frequently recovered from blood cultures.

There have been changes in the classification of these organisms. The genus *Bacteroides* now contains only the *B. fragilis* group. The original *B. melaninogenicus* group now contains two new genera: *Prevotella* (saccharolytic) and *Porphyromonas* (asaccharolytic). Other new genera include *Tisseriella, Sebaldella, Rokenella, Fibrobacter, Megamonas* and *Mitsuokella*. Another new genus, *Bilophila*, has been found in gangrenous appendices.

Motile anaerobic Gram-negative bacilli probably belong to newly described genera such as *Selenomonas* or *Wolinella* or to *Mobiluncus*. *Capnocytophaga* spp. are CO_2-dependent but grow anaerobically; they are found in the human and animal mouth.

Isolation

Examine Gram films of pathological material. Diagnosis of Vincent's angina can be made from films counterstained with dilute fuchsin, if fusiform bacilli can be seen in large numbers, associated with spirochaetes (*Borrelia vincenti*). In pus from abdominal, brain, lung or genital tract lesions, large numbers of Gram-negative bacilli of various shapes and sizes, with pointed or rounded ends suggest *Bacteroides* infection. Such material often has a foul odour. Infections involving *Prevotella melaniniogenica* and other members of the genera *Prevotella* and are rapidly diagnosed by viewing the pus under long wave UV (365 nm) light. *Porphyromonas* produces porphyrins which give a bright red fluorescence.

Culture pus and other material anaerobically, aerobically and under 5–10% carbon dioxide. For anaerobic cultures use both non-selective and enriched selective media, e.g. with haemin, menadione, sodium pyruvate, cysteine HCl. Or use commercial Fastidious Anaerobe Agar or Wilkins-Chalgren agar. The usual selective agents are neomycin 75 μg/ml, nalidixic acid 10 μg/ml and vancomycin 2.5 μg/ml. Various combinations may be used to restrict contaminating facultative flora from certain specimens, e.g. neomycin and vancomycin for pus from the upper respiratory tract.

A newly described selective medium containing josamycin, vancomycin and norfloxacin at 3, 4 and 1 μg/ml respectively, is promising (Brazier *et al.*, 1991).

Place a 5 μg metronidazole disc on the streaked-out inoculum for the rapid recognition of obligate anaerobes, all of which are sensitive. Incubate at 37°C for 48 h.

Compare growths on selective and non-selective media and the anaerobic with aerobic cultures.

Identification

Full identification of these organisms is difficult and rarely necessary. All that is required in a clinical laboratory is confirmation that anaerobes are involved. A simple report that '. . . mixed anaerobes are present – sensitive to metronidazole. . .' usually suffices.

Identification is more important when anaerobes are present in blood cultures and other normally sterile sites. These simple presumptive tests are useful.

Potassium hydroxide test
The Gram stain is important but may give equivocal results. Emulsify a few colonies in a drop of 3% KOH on a slide. Bacteroides tend to become 'stringy' and form long strands when the loop is moved away from the drop.

UV fluorescence
Examine under long wave UV light. Red fluorescence suggests *Prevotella* spp. or *Porphyromonas* spp. and develops much earlier than the characteristic black pigment. Yellow fluorescent colonies of Gram-negative rods are probably fusobacteria (Brazier, 1986).

Phosphomycin test
Place a disc soaked in phosphomycin, 300 μg/ml, on a seeded plate and incubate. Fusobacteria are usually sensitive and bacteroides resistant (Bennett and Duerden, 1985).

Carbohydrate fermentation
See Phillips (1976) and Holdeman *et al.* (1977).

Antibiogram and commercial identification kits
Antibiogram patterns give a low level of accuracy and offer only a limited range of identifications. Commercial kits are an improvement but results based on a generated number may be misleading. Much depends on the data base on which the scheme is based. The API 32 ATB kit is one of the most useful.

Anaerobic cocci

These may occur in cultures from a wide variety of human and animal material.

Peptostreptococci and peptococci

It is difficult to separate these genera. Cocci occur singly, in pairs, clumps or short chains. They grow on blood agar at 37°C and are non-haemolytic but grow better on Wilkins-Chalgren agar containing a supplement for non-sporing anaerobes.

These organisms are resistant to metronidazole (5 μg disc) and grow slowly. There is no need to distinguish between them.

Peptostreptococci may be pathogenic for humans and animals, associated with gangrenous lesions. They are commensals in the intestine and have been isolated from the vagina in health and in puerpural fever. They produce large amounts of hydrogen sulphide from high-protein media, e.g. blood broth and a putrid odour in ordinary media. The API system is useful for identifying these organisms.

The organism formerly known as *Gaffkya tetragenus* is now *Peptostreptococcus tetradius*.

This genus *Peptococcus* now contains only one species, *P. niger*.

Sarcina

Cocci in packets of eight. Obligate anaerobes, requiring incubation at 20–30°C and carbohydrate media. Found in air, water, soil and occasionally in clinical material.

Sarcina ventriculi is a large coccus arranged in clumps of four to eight. It is found in the human gut, particularly in vegetarians.

Veillonella

These are obligate anaerobes and occur as irregular masses of very small Gram-negative cocci, cultured from various parts of the respiratory and alimentary tracts and genitalia. They do not appear to cause disease.

Growth may be improved by adding 1% sodium pyruvate and 0.1% potassium nitrate to the basal medium.

Taxonomists now recognize seven species. *Veillonella atypica, V. dispar* and *V. parvula* are found in humans; *V. caviae, V. criceti, V. ratti* and *V. rodentium* in rodents. In clinical laboratories it is not necessary to identify species.

Fluorescence under UV light is generally weak, fades rapidly and is medium-dependent. It works only on brain infusion agars except for *V. criceti* which fluoresces on a range of media (Brazier and Riley, 1988). Phenotypic differentiation is based on catalase, fermentation of fructose and the origin of the strain. For further information see Mays *et al.* (1982).

References

Bennett, K. W. and Duerden, B. I. (1985) Identification of fusobacteria in a routine diagnostic laboratory. *Journal of Applied Bacteriology*, **59**, 171–181

Brazier, J. S. (1986) Yellow fluorescence of fusobacteria. *Letters in Applied Microbiology*, **2**, 125–126

Brazier, J. S. and Riley, T. V. (1988) UV red fluorescence of *Veillonella* spp. *Journal of Clinical Microbiology*, **26**, 383–384

Brazier, J. A., Citrol, D. M. and Goldstein, E. J. C. (1991) A selective medium for *Fusobacterium* spp. *Journal of Applied Bacteriology*, **71**, 343–346

Holdeman, L. V., Cato, E. P. and Moore, W. E. C. (eds) (1977) *Anaerobic Laboratory Manual*, 4th edn, Virginia State University, Blacksburg, VA

Mays, T. D., Holdeman, L. V., Moore, W. E. C. and Johnson, J. L. (1982) Taxonomy of the genus *Veillonella* Prevot. *International Journal of Systematic Bacteriology*, **32**, 28–36

Phillips, K. D. (1976) A simple and sensitive technique for determining the fermentation reactions of nonsporing anaerobes. *Journal of Applied Bacteriology*, **41**, 325–328

Chapter

44

Clostridium

This is a large genus of Gram-positive spore-bearing anaerobes (a few are aerotolerant) that are catalase negative.

They are normally present in soil; some are responsible for human and animal disease; others are associated with food spoilage

Clostridia are classified according to the shape and position of the spores (see Table 44.1) and by their physiological characteristics. They may be either predominantly saccharolytic or proteolytic in their energy-yielding activities.

Saccharolytic species decompose sugars to form butyric and acetic acids and alcohols. The meat in Robertson's medium is reddened and gas is produced.

Proteolytic species attack amino acids. Meat in Robertson's medium is blackened and decomposed, giving the culture a foul odour.

Most species are mesophiles; there are a few important thermophiles and some psychrophiles and psychrotrophs.

The clostridia are conveniently considered under four headings: food poisoning; tetanus and gas gangrene; pseudomembranous colitis; and food spoilage.

Food poisoning

Botulism

This is the least common but most often fatal kind of food poisoning (Chapter 13) and is caused by *Clostridium botulinum*. This is a strict anaerobe and requires a neutral pH and absence of competition to grow in food (e.g. canned, underprocessed). It is a difficult organism to isolate in pure culture.

Emulsify the suspected material in 0.1% peptone water and inoculate several tubes of Robertson's cooked meat medium which have been heated to drive off air and then cooled to room temperature. Heat some of these tubes at 75–80°C for 30 min and cool. Incubate both heated and unheated tubes at 35°C for 3–5 days. Plate them on pre-reduced blood agar and egg yolk agar and incubate under strict anaerobic conditions at 35°C for 3–5 days.

If the material is heavily contaminated, heat some of the emulsion as described above, make several serial dilutions of heated and unheated material and add 1-ml amounts to 15–20 ml of glucose nutrient agar melted and at 50°C in 152 × 16 mm tubes (Burri tubes, p. 408). Mix, cool and incubate. Cut the tube near the sites of suspected colonies, aspirate these with a pasteur pipette and plate on blood agar or on egg yolk agar containing (per ml): cycloserine, 250 μg; sulphamethoxazole, 76 μg; trimethoprim, 4 μg (Dezfullian *et al.*, 1981).

Clostridium botulinum gives a positive egg yolk and Nagler (half-antitoxin

Table 44.1 Morphology and colonial appearance of some species of *Clostridium*

	Spores	*Bacilli*	*Haemolysis*	*Colony appearance on blood agar*
C. botulinum	OC or S	normal	+	large, fimbriate, transparent
C. perfringens	[a]	large thick	+	flat, circular, regular
C. tetani	RT	normal	+	small, grey, fimbriate, translucent
C. novyi	OS	large	+	flat, spreading, transparent
C. septicum	OS	normal	+	irregular, transparent
C. fallax	OS	thick	−	large, irregular, opaque
C. sordellii	OC or S	large thick	+	small, crenated
C. bifermentans	OC or S	large thick	+	small, circular, transparent
C. histolyticum	OS	normal	−	small, regular, transparent
C. sporogenes	OS	thin	+	medusa head, fimbriate, opaque
C. tertium	OT	long thin	−	small, regular, transparent
C. cochlearium		thin	−	circular, transparent
C. butyricum	OC	normal	−	white, circular, irregular
C. nigrificans		normal		black
C. thermosaccharolyticum		normal		granular, feather edges
C. difficile	OS	normal	−	grey, translucent, irregular

Spores: O, oval; R, round; S, subterminal; C, central; T, terminal
[a] not usually seen

plate) reaction (see below). Types A, B and F liquefy gelatin, blacken and digest cooked meat medium and produce hydrogen sulphide. Types C, D and E do none of these. All strains produce acid from glucose, fructose and maltose and are indole negative. See p. 408 and Tables 44.1 and 44.2.

It is best to use fluorescent antibody methods and to send suspected cultures to a reference laboratory.

Type A *C. botulinum* is usually associated with meat. Type E is found in fish and fish products and estuarine mud and is psychrotrophic.

Perfringens (welchii) food poisoning

Clostridium perfringens is a common cause of food poisoning (Chapter 13).

Faeces

It is pointless to examine stools of food handlers and 'contacts' in outbreaks of *C. perfringens* food poisoning as the disease is not spread by human agency. Examine the stools from a proportion of cases (10–25% if the outbreak is large). Heat-sensitive *C. perfringens* are found in almost all stools from normal people but large numbers may be demonstrated if the individual is a victim of *C. perfringens* food poisoning. A quantitative or, more practically, a semiquantitative examination is therefore necessary. Make a 1:10 suspensions of faeces in nutrient broth and add 1 ml to each of two tubes of Robertson's cooked meat medium. Heat one tube at 80°C for 10 min and then cool. Incubate tubes at 37°C overnight and plate on blood agar with and without neomycin and on egg yolk agar. Place a metronidazole disc on the streaked-out inoculum. Incubate at 37°C anaerobically overnight.

Emulsify faeces in ethanol (industrial grade) to give a 50% suspension. Mix well and stand for 1 h. Inoculate media and incubate as above.

Food

Outbreaks involving both heat-resistant and heat-sensitive strains have been reported. See Chapter 13.

Use a Stomacher to prepare a 10% suspension of the food in peptone water diluent. Make dilutions, 10^{-1}–10^{-3} and plate both suspensions and dilutions on Willis and Hobbs medium, tryptose sulphite cycloserine agar and neomycin blood agar. Place a metronidazole disc on the inoculum. Incubate anaerobically at 37°C overnight.

To count the clostridia do pour plates with the dilutions in oleandomycin polymyxin sulphadiazine perfringens agar (OPSP). Or use the Miles and Misra method with neomycin blood agar. Incubate duplicate plates aerobically and anaerobically to distinguish between clostridia and other organisms.

Add some of the suspension to two tubes of Robertson's cooked meat medium. Heat one tube for 10 min at 80°C. Incubate overnight and plate out as indicated above.

Commercial kits are available for the detection of enterotoxins. See Berry *et al.* (1987).

Identification

Heat-resistant strains of *C. perfringens* yield very slightly haemolytic colonies (other strains usually show much more haemolysis) which are moist, raised and smooth, about 2 mm in diameter. Heat-sensitive strains do not grow from the pasteurized cultures. Spores are not usually seen in laboratory cultures of

Table 44.2 Cultural and biochemical properties of some species of clostridia

	RCM				*Purple milk*	*Acid from*				*Indole*	*Gelatin liquefaction*	*Lecithinase*
	Colour	*Digestion*	*Odour*	*Gas*		*Glucose*	*Sucrose*	*Lactose*	*Salicin*			
C. botulinum	black	+	–	+	D	+	v	–	–	–	+	–
C. perfringens	black	+	+	+	CD	+	+	+	v	–	+	+
C. tetani	black	+	+	–	C	–	–	–	–	+	+	–
C. novyi	red	–	–	+	GC	+	–	–	–	v	+	v
C. septicum	red	–	–	+	AC	+	–	+	+	–	+	–
C. fallax	red	–	–	+	AC	+	+	+	+	–	–	–
C. sordellii[a]	black	+	+	+	CD	+	–	–	–	+	+	+
C. bifermentans[a]	black	+	+	+	CD	+	–	–	v	+	+	+
C. histolyticum	black	+	+	–	D	–	–	–	–	–	+	–
C. sporogenes	black	+	+	+	D	+	–	–	v	–	+	–
C. tertium		–	–	+	AC	+	+	+	+	–	–	–
C. cochlearium	red	–	–	+		–	–	–	–	–	–	–
C. butyricum		–	–	+	ACG	+	+	+	+	–	–	–
C. nigrificans		–	–	–		–	–	–	–	–	–	–
C. thermo-saccharolyticum		–	–	+		+	+	+	+	–	–	–

RCM: Robertson's cooked meat medium
Purple milk: A, acid; C, dot; D, digestion; G, gas
[a]*C. sordellii* also splits urea; *C. bifermentans* does not.

this organism. Heat-sensitive strains may also cause food poisoning.

It is usually sufficient to identify *C. perfringens* by the Nagler half-antitoxin test (see below) but other properties are given in Tables 44.1 and 44.2. No great reliance should be placed on the 'stormy fermentation' of purple milk. General cultural methods are given on p.408.

Nagler half-antitoxin plate test

Dip a cotton wool swab in standard antitoxin and spread on one-half of a plate of Willis and Hobbs medium or on nutrient agar containing 10% egg yolk. Dry the plates and inoculate the organism across both halves of the plate. Incubate anaerobically overnight at 37°C.

The lecithinases of *C. perfringens* and *C. novyi* produce white haloes round colonies on the untreated half of the plate. This activity is inhibited by the half treated with antitoxin. The lecithinases of *C. bifermentans* and *C. sordellii* are partially inhibited by *C. perfringens* antitoxin, but they may be distinguished on Willis and Hobbs medium. A diffuse pink halo appears round colonies of *C. perfringens* due to fermentation of the lactose. *C. bifermentans* and *C. sordellii* give white haloes because they do not ferment lactose.

On Willis and Hobbs medium, some clostridia show a 'pearly layer' due to lipolysis. This is a useful differential criterion.

Tetanus and gas gangrene

Tetanus is caused by *C. tetani*. Clostridia associated with gas gangrene include *C. novyi (oedematiens)*, *C. perfringens, C. septicum* and *C. sordellii. C. histolyticum* is a possible pathogen and some other species, such as *C. sporogenes, C. tertium* and *C. bifermentans* may also be found in wounds but are not known to be pathogenic.

Examine Gram-stained and FA films of wound exudates. Bacilli with swollen terminal spores ('drumstick') suggest *C. tetani* but are not diagnostic. Thick, rectangular, box-like bacilli suggest *C. perfringens*. Very large bacilli might be *C. novyi* and thin, almost filamentous bacilli may be seen.

Inoculate two plates of blood agar, one for anaerobic and the other for aerobic culture, and also Willis and Hobbs or Lowbury and Lilly medium. One of the commercial media may be used as well as or instead of the latter. Inoculate two tubes of Robertson's cooked meat medium. Heat one at 80°C for 30 min and cool. Incubate blood agar and all the other media anaerobically at 37°C for 24 h. Plate out the Robertson's medium on blood agar and other clostridial media and incubate anaerobically with 10% carbon dioxide.

Identification

Test suspected colonies (see Table 44.1) with half-antitoxin plates (see above), inoculate Crossley milk medium, glucose, lactose, sucrose, maltose and salicin peptone waters and note nature of growth and appearance in the cooked meat medium (see Table 44.2).

Pseudomembranous colitis and antibiotic-associated diarrhoea

These are caused by *C. difficile*.

Treat faeces by the ethanol shock method (p.402). Plate on blood agar containing cefoxitin, 8 μg/ml, and cycloserine 250 μg/ml or on cycloserine cefoxitin fructose agar (Clostridium Difficile Selective Medium) and inoculate cooked meat medium. Or plate on Columbia agar containing these antibiotics and 5% egg yolk so that lecithinase-producing colonies may be ignored (*C. difficile* is lecithinase negative). Incubate anaerobically for 48 h.

Colonies of *C. difficile* are 2–5 mm in diameter, irregular and opaque. Confirm by latex agglutination (Mercia; Diagnostics Ltd). There may be cross reaction with *C. glycolicum*.

Examine blood agar cultures under long wave UV light. *C. difficile* colonies (and those of some other organisms) fluoresce yellow-green.

GLC of volatile fatty acids and *p*-cresol may be necessary for the final identification of *Cl. difficile*.

There are diagnostic kits for the toxin (Porton Cambridge, UK) and Premier *C. difficile* Toxin A (Meridian Diagnostics, USA).

Species of *Clostridium* of medical importance

C. botulinum

Strict anaerobe; requires also a neutral pH and absence of competition to grow in food (e.g. canned, underprocessed). Free spores may be seen. Cultural characteristics are variable within and between strains. There are six antigenic types (A-F). Types A, B and F are proteolytic, causing clearing on Willis and Hobbs medium; types C, D and E are not. Causative organism of botulism, one of the least common but most fatal kind of food poisoning. Type A is usually associated with meat and Type E (a psychrotroph) with fish products. Found in soil and marine mud.

C.perfringens (*C. welchii*)

This highly aerotolerant anaerobe may grow in broth, e.g. in MacConkey broth inoculated with water. Spores are rarely seen in culture (a diagnostic feature) but can be obtained on Ellner's medium and medium with added bile, bicarbonate and quinoline (Phillips, 1986). There are six antigenic types (A-F). Type A is associated with gas gangrene and with food poisoning and antibiotic-associated diarrhoea. Causes lamb dysentery, sheep 'struck' and pulpy kidney in lambs. Found in soil, water and animal intestines.

C. tetani

Strict anaerobe, easily dies on exposure to air. Swarms over the medium, but this may be difficult to see. It is, however, an important diagnostic characteristic. Found in soil, especially animal manured, and in the animal intestine. Causative organism of tetanus.

C. novyi (*C. oedematiens*)

A strict anaerobe which dies rapidly in air. There are four antigenic types (A-D). Types A and B cause gas gangrene, type D bacillary haemoglobinuria in cattle and 'black disease' of sheep.

C. septicum and chauvoei

Strict anaerobes. Once considered to be a single species, these are now known to be antigenically distinct. *C. septicum* causes gas gangrene in humans, braxy and blackleg in sheep. *C. septicum* is often isolated from blood cultures in patients with malignant disease of the colon. *C. chauvoei* is not pathogenic for humans but causes quarter evil, blackleg and 'symptomatic anthrax' of cattle and sheep. Both can be identified by fluorescent antibody methods.

C. bifermentans and C. sordellii
Separate species, distinguished biochemically but conveniently considered together because both have a toxin (lecithinase) similar to that of C. *perfringens* and give the same reaction as that organism on half-antitoxin plates with C. *perfringens* type A antiserum. C. *sordellii* is urease positive, C. *bifermentans* urease negative, C. *bifermentans* is not pathogenic but C. *sordellii* is associated with wound infections and gas gangrene. Both are found in soil.

C. difficile
May cause a wide spectrum of disease ranging from mild antibiotic-associated diarrhoea to full blown pseudomembraneous colitis which may be fatal.

Species of clostridia of doubtful or no significance in human pathological material

C. histolyticum
Not a strict anaerobe. Filamentous forms may grow as surface colonies on blood agar aerobically. Associated with gas gangrene but usually in mixed infection with other clostridia. Found in soil and intestines of humans and animals.

C. sporogenes
Not pathogenic but is a common contaminant. Widely distributed; found in soil and animal intestines.

C. tertium
Aerotolerant and non-pathogenic.

C. tetanomorphum
Of note because its sporing forms resemble the drumsticks of C. *tetani*. It is not pathogenic.

C. cochlearium
The only common clostridium that is biochemically inert. It is not pathogenic.

C. fallax
A strict anaerobe, associated with gas gangrene. Spore are not very resistant to heat and the organism may be killed in differential heating methods. Found in soil.

Food spoilage clostridia

Clostridium species are known to be responsible for spoilage of a wide variety of foods, including milk, meat products, fresh water fish and vegetables.

Prepare 10% emulsions of the product in 0.1% peptone water using a Stomacher or blender. Heat some of this emulsion at 75–80°C for 30 min (pasteurized sample). Inoculate several tubes of Robertson's cooked meat medium and/or liquid reinforced clostridium medium with pasteurized and unpasteurized emulsion and incubate pairs at various temperatures, e.g. 5–7°C for psychrophiles, 22 and 37°C for mesophiles and 55°C for thermophiles. Inoculate melted reinforced clostridial agar (RCM) in deep tubes with dilutions of the emulsions. Allow to set. Incubate at the desired temperature and look for black colonies, or blackening of the media which suggests clostridia. Inoculate several tubes of Crossley's Milk medium each with 2 ml of suspension and incubate at the required temperatures. This gives a useful guide to the identity of the clostridias.

Appearance in Crossley's milk medium

C. putrifaciens, C. sporogenes, C. oedematiens, C. histolyticum are indicated by slightly alkaline reaction, gas, soft curd subsequently digested leaving clear brown liquid, a black sediment and a foul smell.

C. sphenoides gives a slightly acid reaction, soft curd, whey and some gas.

C. butyricum produces acid, firm clot and gas.

C. perfringens usually gives a stormy clot.

C. tertium usually gives a stormy clot.

Note that aerobic spore bearers may give similar reactions, so plate out and incubate both aerobically and anaerobically.

Thermophiles associated with canned food spoilage

Two kinds of clostridial spoilage occur due to underprocessing: 'hard swell' and 'sulphur stinkers'.

Examine Gram-stained films for Gram-positive spore-bearers. Inoculate reinforced clostridial medium or glucose tryptone agar and iron sulphite medium in deep tube cultures by adding about 1-ml amounts of dilutions of a 10% emulsion of the food material in peptone water diluent to 152 × 16 mm tubes containing 15–20 ml of medium melted and at 50°C. Either solid or semisolid media may be used. Incubate duplicate sets of tubes for up to 3 days at 60 and 25°C.

C. thermosaccharolyticum
Produces white lenticular colonies in all three media and changes the colour of glucose tryptone agar from purple to yellow. This organism is responsible for 'hard swell'.

C. nigrificans
Produces black colonies, particularly in media that contain iron. This causes 'sulphur stinkers'.

Neither organism grows appreciably at 25°C.

Counting clostridia

Emulsify 10 g of the food in 90 ml of a 0.1% peptone water in a homogenizer. Divide into two portions and heat one at 75°C for 30 min.

Do MPN tests (10-, 1- and 0.1-ml amounts) using the five or three tube method, on each portion in liquid differential reinforced clostridial medium (DRCM). Clostridia turn this medium black ('black tube method'). The unheated portion gives the total count, the heated portion the spore count.

For the pour plate method add 0.1-ml amounts of serial dilutions of the emulsion to melted OPSP agar. Incubate anaerobically for 24 h and count the large black colonies.

It may be necessary to dilute the inoculum 1:10 or 1:100 if many clostridia are present. If the load of other organisms is heavy, add 75 units/ml of polymyxin to the medium used for the unheated count.

To recover individual colonies, Burri tubes may be used. These are open at both ends and are closed with rubber bungs. Use solid DRCM and, after incubation and counting, remove both stoppers and extrude the agar cylinder. Cut it with a sterile knife and aspirate colonies with a pasteur pipette. These tubes need not be incubated in an anaerobic jar. After inoculation, cover the medium with a layer of melted paraffin wax about 2 cm deep.

Miller-Prickett tubes (flattened test-tubes) are also useful. Pipette dilutions into the tubes and then add 15 ml of melted medium at 50°C to each tube. Seal with paraffin wax as above.

Membrane filters can be used for liquid samples. Roll up the filters and place in test-tubes; cover with melted medium.

DRCM is suitable for most counts but for *C. nigrificans* iron sulphite medium is good.

Select incubation temperatures according to the species to be counted.

General identification procedure

Subculture each kind of colony of anaerobic Gram-positive bacillus in the following media: Robertson's cooked medium; purple milk; peptone water for indole production; gelatin medium; glucose, lactose, sucrose and salicin peptone waters. These media should be in cotton wool-plugged test-tubes, not in screw-capped bottles. Omit the indicator from these as it may be decolorized during anaerobiosis. Test for acid production after 24–48 h with bromocresol purple. It may be necessary to enrich some media with Fildes' extract and an iron nail in each tube assists anaerobiosis. Gas production in sugar media is not very helpful and Durham's tubes can be omitted. The API 20A system for anaerobes is very helpful.

Do half-antitoxin (Nagler) plates as above using *C. perfringens* type A and *C. novyi* antitoxin (see Tables 44.1 and 44.2).

Food spoilage species

(1) Thermophiles	
C. nigrificans	Sulphur stinkers
C. thermosaccharolyticum	Hard swell
(2) Mesophiles	
C. butyricum	Cheese disorders. Butter and milk products
C. sporogenes *C. sphenoides** *C. novyi* *C. perfringens*	Meat and dairy products
(3) Psychrophiles	
*C. putrefaciens**	Bone taint, off-odours

* They are described briefly on p. 407.

For further information on clostridia see Willis (1977), Holdeman *et al.* (1977) and Phillips *et al.* (1985).

References

Berry, P. R., Weinecke, A. A., Rodhouse, J. C. and Gilbert, R. J. (1987) Use of commercial tests for the detection of *Clostridium perfringens* and *Staphylococcus aureus* enterotoxins. In *Immunological Techniques in Microbiology* (eds J. M. Grange, A. Fox and N. L. Morgan), Society for Applied Bacteriology Technical Series No. 24, Blackwells, London, pp. 245–250

Dezfullian, M., McCroskey, C. L., Hatheway, C. L. and Dowell, V. R. (1981) Isolation of *Clostridium botulinum* from human faeces. *Journal of Clinical Microbiology*, **13**, 526–531

Holdeman, L. V., Cato, E. P. and Moore, W. E. C. (eds) (1977) *Anaerobic Laboratory Manual*, 4th edn, Virginia State University, Blacksburg, VA

Phillips, K. D. (1986) A sporulation medium for *Clostridium perfringens. Letters in Applied Microbiology*, 3, 77–79

Phillips, K. D., Brazier, J. S., Levett, P. N. and Willis, A. T. (1985) Clostridia. *In Isolation and Identification of Micro-organisms of Medical and Veterinary Importance* (eds C. H. Collins and J. M. Grange), Society for Applied Microbiology Technical Series No. 21, Academic Press, London, pp. 215–236

Willis, A. T. (1977) *Anaerobic Bacteriology*, 3rd edn, Butterworths, London

Chapter

45

Mycobacterium

These organisms are acid-fast: if they are stained with a strong phenolic solution of an arylmethane dye, e.g. carbol fuchsin, they retain the stain when washed with dilute acid. Other organisms are decolorized. There are no grounds for the commonly held belief that some mycobacteria are acid- and alcohol-fast but that others are only acid-fast. These properties vary with the technique and the organism's physiological state.

The most important members of this genus are obligate parasites. They include the three 'species' of tubercle bacilli, *Mycobacterium tuberculosis, M. bovis* and *M. africanum; M. paratuberculosis* (Johne's bacillus) and the leprosy (Hansen's) bacillus *M. leprae*. Several species that appear to be free living are opportunist pathogens of humans and animals. Several species that normally live as saprophytes in the environment may also be opportunist pathogens of animals and humans.

The tubercle bacillus, and in some countries certain other species of mycobacteria (see national lists) are Risk/Hazard Group 3 pathogens and infectious by the airborne route. All manipulations that might produce aerosols should be done in microbiological safety cabinets in Biosafety/Containment Level 3 laboratories.

Tubercle bacilli

Direct microscopic examination.

Sputum

This is the most common material suspected of containing tubercle bacilli. It is often viscous and difficult to manipulate. To make direct films use disposable 10 μl plastic loops. These avoid the use of bunsens in safety cabinets. Discard them into disinfectant. Ordinary bacteriological loops are unsatisfactory and unsafe. Spread a small portion of the most purulent part of the material carefully on a slide, avoiding hard rubbing which releases infected airborne particles. Allow the slides to dry in the safety cabinet then remove them for fixing and staining. This is quite safe; aerosols are released during the spreading, not from the dried film, although these should not be left unstained for any length of time. Fixing may not kill the organisms and dried material is easily detached.

For direct microscopy of centrifuged deposits after homogenizing sputum for culture spread a 10-μl loopful over an area of about 1 cm^2. Dry slowly and handle with care. The material tends to float off the slide while it is being stained.

CSF
Make two parallel marks 2–3 mm apart and 10-mm long in the middle of a microscope slide. Place a loopful of the centrifuged deposit between the marks. Allow to dry and superimpose another loopful. Do this several times before drying, fixing and staining. Examine the whole area.

Urine
Direct microscopy of the centrifuged deposit is unreliable because urine frequently contains acid-fast bacilli which may resemble tubercle bacilli but which have entered the specimen from the skin or the environment.

Aspirated fluids, pus
Centrifuge if possible and make films from the deposit. Otherwise prepare thin films. Thick films float off during straining. Material containing much blood is difficult to examine and may give false positives.

Gastric lavage
Direct microscopy is unreliable because environmental mycobacteria are frequently present in food and hence in the stomach contents.

Laryngeal swabs
Direct microscopy is unrewarding.

Blood, faeces, milk
Direct microscopy is unreliable because acid-fast artefacts and saprophytic mycobacteria may be present. Microscopical examination of faeces and centrifugates of lysed blood may be useful in the provisional diagnosis of disseminated mycobacterial disease, usually caused by *M. avium-intracellulare*, in patients with acquired immune deficiency syndrome (AIDS; Kiehn *et al.*, 1985).

Tissue
Cut the tissue into small pieces. Work in a safety cabinet. Remove caseous material with a scalpel and scrape the area between caseation and soft tissue. Spread this on a slide. It is more likely to contain the bacilli than other material.

Staining

Stain direct films of sputum or other material by the ZN method. The bacilli may be difficult to find and it is often necessary to spend several minutes examining each film. Fluorescence microscopy, using the auramine phenol (AP) method, is popular in some clinical laboratories. A lower-power objective may be used and more material examined in less time. False positives are not uncommon with this method and the presence of acid-fast bacilli should be confirmed by overstaining the film with ZN stain. These staining methods are described in Chapter 7.

Reporting

Report 'acid-fast bacilli seen/not seen'. If they are present report the number per high power field. Table 45.1 gives a reporting system in common use in Europe for sputum films.

Table 45.1 Scale for reporting acid-fast bacilli in sputum smears examined microscopically

No. of bacilli observed	*Report*
0 per 300 fields	Negative for AFB
1–2 per 300 fields	(±) Repeat test
1–10 per 100 fields	+
1–10 per 10 fields	+ +
1–10 per field	+ + +
10 or more per field	+ + + +

Note that this is a logarithmic scale; this facilitates plotting.

False positives

Single acid-fast bacilli should be regarded with caution. They may be environmental contaminants, e.g. from water. They may be transferred from one slide to another during staining or if the same piece of blotting paper is used for more than one slide.

Isolation from pathological material

Mycobacteria are often scanty in pathological material and not uniformly distributed. Many other organisms may be present and the plating methods used in other bacteriological examinations are useless. Acid-fast bacilli usually grow at a very much slower rate than other bacteria and are overgrown on plate cultures. Some specimens are from normally sterile sites (blood, bone marrow, CSF, certain tissues) and may be inoculated directly on culture media. Others are likely to contain many bacteria and fungi and must first be treated with a reagent that is less lethal for mycobacteria than for other organisms and that reduces the viscosity of the preparation so that it may be centrifuged. No ideal reagent is known; there is a choice between 'hard' and 'soft' reagents.

Hard reagents are recommended for specimens that are heavily contaminated with other organisms. Exposure times are critical or the mycobacteria are also killed. Soft reagents are to be preferred when there are fewer other organisms, e.g. in freshly collected sputum. The reagent can be left in contact with the specimen for several hours without significantly reducing the numbers of mycobacteria. Soft reagents may permit the recovery of mycobacteria other than tubercle bacilli, which are often killed by hard reagents. Alternatively, specimens may be inoculated on media containing a 'cocktail' of antimicrobial agents that will kill virtually all microorganisms other than mycobacteria (p.415).

The methods described below are intended to reduce the number of manipulations, particularly in neutralization and centrifugation, so that the hazards to the operator are minimized.

Culture of the centrifuged deposit after treatment gives optimum results, but is avoided by some workers because of the dangers of centrifuging tuberculous liquids. Sealed centrifuge buckets (safety cups) overcome this problem (see Chapter 1). These reduce considerably the hazards of aerosol formation should a bottle leak or break in the centrifuge. They should be opened in a safety cabinet.

If material is not centrifuged, more tubes of culture media should be inoculated and a liquid medium should be included. This should contain an antibiotic mixture (p.415).

Sputum culture

Encourage the sputum from the specimen container into the preparation bottle with a pipette made from a piece of sterile glass tubing (200 × 5 mm) with one end left rough to cut through sputum strands and the other smoothed in a flame to accept a rubber teat.

Hard method
Add about 1 ml of sputum to 2 ml of 4% NaOH solution or a mixture of 2% NaOH and 1% *N*-acetyl-L-cysteine (a mucolytic agent) in a 25 ml screw-capped bottle. Stopper securely, place in a self-sealing plastic bag to prevent accidental dispersal of aerosols and shake mechanically, but not vigorously, for not less than 15 min but not more than 30 min. Thin specimens require the shorter time. Incubation does not help.

Remove the bottle from the plastic bag and add 3 ml of 14% (approximately 1 mol/l) dipotassium hydrogen orthophosphate (KH_2PO_4) solution containing enough phenol red to give a yellow colour. This neutralizing fluid should be dispensed ready for use and sterilized in small screw-capped bottles. It should not be pipetted from a stock bottle.

The colour change of the phenol red from yellow to orange pink indicates correct neutralization. Re-stopper, mix gently and centrifuge with the bottles in sealed centrifuge buckets ('safety cups') for 15 min at 3000 rev/min. Pour off the supernatant fluid carefully into disinfectant. Wipe the neck of the bottle with a piece of filter-paper, which is then discarded into disinfectant. Culture the deposit.

Soft method
To 2–4 ml of sputum in a screw-capped bottle, add an equal volume of 23% trisodium orthophosphate solution. Mix gently and stand at room temperature or in a refrigerator for 18 h. Add the contents of a 10-ml bottle of sterile distilled water to reduce the viscosity, centrifuge, decant and culture the deposit using the methods and precautions described above.

Laryngeal and other swabs
Place the swab in a tube containing 1 N sodium hydroxide solution for 5 min. Remove, drain and place in another tube containing 14% potassium dihydrogen orthophosphate (KH_2PO_4) solution for 5 min. Drain and inoculate culture media.

Urine culture
Use fresh, early morning mid-stream specimens. Do not bulk several specimens as this often leads to contamination of cultures. Centrifuge 25–50 ml at 3000 rev/min and assess the number of organisms in the deposit by examining a Gram-stained film. Treat the deposit with 2 ml of 4% sulphuric acid for 15–40 min depending on the load of other organisms. Neutralize with 15 ml of distilled water, centrifuge and culture the deposit.

Alternatively plate some of the urine on blood agar, incubate overnight and if this is sterile pass 50–100 ml of the urine through a membrane filter. Cut the filter into strips and place each strip on the surface of the culture media in screw-capped bottles.

Blood and bone marrow
Since the advent of AIDS it is often necessary to detect disseminated mycobacterial disease by examination of blood and bone marrow. Collect 8.5 ml of blood aseptically and add to a tube containing 1.5 ml of 0.35% of sodium

polyethanolium sulphate (SPS; this is the least toxic of the anticoagulants for mycobacteria and it lyses white cells), mix to avoid coagulation and inoculate 10 to 20 volumes of liquid media, e.g. Middlebrook 7H9 or 13A liquid medium containing 0.025% SPS and, if necessary, an antibiotic cocktail (p.415). The Isolator-10 lysis centrifugation system (Du Pont Co., Wilmington, DE, USA) is also suitable (Kiehn *et al.*, 1985). Collect 10 ml of blood into the tube which contains an anticoagulant and saponin which lyses all blood cells. Centrifuge at 3000 *g* for 30 min and inoculate deposit on Middlebrook 7H11 agar.

Alternatively, add blood or bone marrow directly to the medium. Check broth weekly for bacterial growth. Confirm acid fastness by Ziehl-Neelsen smear and inoculate solid media.

Radiometric techniques (Bactec) are ideal for blood culture and are used according to the manufacturer's instructions.

Faeces

Culture of faeces of AIDS patients with suspected intestinal disease caused by *M. avium-intracellulare* may be required (Kiehn *et al.*, 1985). Suspend 1 g of faeces in 5 ml of Middlebrook 7H9 broth, decontaminate as for sputum by the 'hard' method (p.412) and inoculate on Mitchison's antibiotic medium (p.415).

Culture of CSF, pleural fluids, pus, etc.

Plate original specimen or centrifuged deposit on blood agar and incubate overnight. Keep the remainder of specimen in a deep-freeze. If the blood agar culture is sterile, inoculate media for mycobacteria without further treatment. Use as much material and inoculate as many tubes as possible. If the blood agar culture shows the presence of other organisms, proceed as for sputum and/or culture directly into media containing antibiotics (see below). Half fill the original container with Kirchner medium and incubate. Some tubercle bacilli may adhere to the glass or plastic.

Automated culture

A commercial instrument (BACTEC: Becton Dickinson) has been developed for detecting the early growth of mycobacteria by a radiometric or infra-red method.

Sputum or other homogenates, decontaminated if necessary, are added to Middlebrook broth medium containing antibiotics to discourage the growth of other organisms, and ^{14}C-labelled palmitic acid. The medium is prepared commercially (Becton Dickinson) in rubber sealed bottles and is inoculated with a syringe and hypodermic needle. If growth occurs, $^{14}CO_2$ is evolved. The air space above the medium in each bottle is sampled automatically at fixed intervals and the amount of radioactive gas is estimated and recorded. Growth of mycobacteria may be detected in 2–12 days, but positive results require further tests to distinguish between tubercle bacilli and other mycobacteria. In the BACTEC, *p*-nitro-α-acetylamino-β-propiophenone (NAP) is used and this takes another 2–5 days. Mycobacteria that grow in media containing this substance are subcultured and identified by traditional methods (see Heifets, 1986 and p.126).

A system combining a broth medium and a slide culture system is commercially available (MB Check, Roche). Detection of mycobacterial growth is more rapid than by conventional culture on solid medium but not as rapid as radiometric methods.

Culture of tissue

Grind very small pieces of tissue, e.g. endometrial curettings, in Griffith's tubes or emulsify them in sterile water with glass beads on a Vortex mixer. Homogenize larger specimens with a stomacher or blender. Check the sterility and proceed as for CSF, etc. Keep some of the material in a deep freeze in case the cultures are contaminated and the tests need repeating.

Isolation from the environment

This may be done to trace the sources of opportunist and saprophytic mycobacteria which contaminate clinical material and laboratory reagents.

Water
Pass up to 2 litres of water from cold and hot taps through membrane filters (as for water bacteriology p.272). Drain the membranes and place them in 3% sulphuric or oxalic acid for 3 min and then in sterile water for 5 min. Cut them into strips and place each strip on the surface of the culture medium in screw-capped bottles. Use Lowenstein-Jensen medium and Middlebrook 7H11 medium containing antibiotics (see below).

Swab the insides of cold and hot water taps and treat them as laryngeal swabs.

Milk
Centrifuge at least 100 ml from each animal. Examination of bulk milk is useless. Treat cream and deposit separately by the NaOH method and culture deposit on several tubes of medium.

Dust and soil
Place 2–5 g samples in 1 litre of sterile distilled water containing 0.5% Tween 80. Mix gently to avoid too much froth. Pass through a coarse filter to remove large particles and allow to settle in a refrigerator for 24 h. Decant the supernatant and pass it through membrane filters as described above for waters. Centrifuge the sediment and treat it with sodium hydroxide as for sputum. Neutralize, centrifuge and inoculate several tubes of medium.

Culture media

Egg media are most commonly used and there seems little to choose between any of them. In the UK, Lowenstein-Jensen medium is most popular, but in the USA bacteriologists seem to favour ATS or Piezer media. Both glycerol medium (which encourages the growth of the human tubercle bacillus) and pyruvate medium (which encourages the bovine tubercle bacillus) should be used.

Pyruvate medium should not be used alone as some opportunist mycobacteria grow very poorly on it.

Some workers prefer agar-based media, e.g. Middlebrook 7H10 or 7H11. Kirchner liquid medium is also useful for fluid specimens that cannot be centrifuged or when it is desirable to culture a large amount of material.

The antibiotic media of Mitchison *et al.* (1987) are useful for contaminated specimens, even after treatment with NaOH, etc. They offer a safety net for non-repeatable specimens such as biopsies. These media should be used as well as, not in place of, egg media for tissues, fluids and urines. They are made as follows.

Add the following to complete Kirchner or Middlebrook 7H11 media: polymyxin, 200 units/ml: carbenicillin, 100 mg/l; trimethoprim, 10 mg/l; amphotericin, 10 mg/l. Dispense fluid medium in 10 ml amounts and make slopes of solid medium.

An alternative antibiotic mixture (PANTA) is used particularly in the (Bactec) system. This contains (final concentrations) polymyxin 50 units/ml; amphotericin 5 mg/l; nalidixic acid 20 mg/l; trimethoprim 5 mg/l; azlocillin 10 mg/l.

Inoculating culture media

Use an inoculum of at least 0.2 ml (not a loopful) for each tube of medium. Inoculate two tubes each of egg medium, one containing glycerol and the other pyruvic acid and other media as indicated above.

Incubation

Incubate all cultures at 35–37°C and cultures from superficial lesions also at 30–33°C for at least 8 weeks. Prolonging the incubation for a further 2–4 weeks may result in a small increment of positives, especially from tissues. Examine weekly.

Alternatives to culture

In view of the time taken for mycobacterial growth to be apparent, methods for the direct detection of mycobacterial components in clinical material have been developed. Antigens are detectable in 'clean' specimens such as CSF by antigen-capture enzyme-linked immunosorbent assay (ELISA; Ramkisson *et al.*, 1988) and specific mycobacterial lipids are detectable by mass spectrometry (French *et al.*, 1987). The use of the polymerase chain reaction (PCR) to detect very small numbers of specific DNA sequences holds great promise (Brisson-Noel *et al.*, 1991) although the sensitivity of the technique renders cross-contamination of specimens a serious problem.

Identification of tubercle bacilli

Most mycobacteria cultured from pathological material will be tubercle bacilli. No growth will be evident on egg media until 10–14 days. Colonies of the human tubercle bacillus are cream coloured, dry and look like breadcrumbs or cauliflowers. Those of the bovine tubercle bacillus are smaller, whiter and flat. Growth of this organism may be very poor; it grows much better on pyruvate medium than on glycerol medium.

On Middlebrook agar media, colonies of both types are flat and grey. In Kirchner fluid media, colonies are round, granular and may adhere to one another in strings or masses. They grow at the bottom of the tube and settle rapidly after the fluid has been shaken.

Make ZN films to check acid-fastness. Some yeasts and coryneform organisms grow on egg media and their colonies may resemble those of tubercle bacilli. Make the films in a drop of saturated mercuric chloride as a safety measure against dispersing live bacilli in aerosols, and spread the films gently. Note whether the organisms are difficult or easy to emulsify. Check the morphology microscopically. Tubercle bacilli may be arranged in serpentine cords, are usually uniformly stained and usually 3–4 μm, in length.

Make suspensions of the organisms as follows. Prepare small, glass screw-capped bottles containing two wire nails shorter than the diameter of the base of the bottles, a few glass beads 2–3 mm in diameter and 2 ml of phosphate buffer, pH 7.2–7.4 (the nails will rust if water is used; wash both nails and beads in dilute hydrochloric acid and then in water before use). Sterilize by autoclaving. Place several colonies of the organisms in a bottle and mix on a magnetic stirrer. Allow large particles to settle and use the supernatant.

Inoculate two tubes of egg medium and one tube of the same medium containing 500 μg/ml of *p*-nitrobenzoic acid (PNBA medium). To make the *p*-nitrobenzoic acid stock solution, dissolve 0.5 g of the compound in 50 ml of water to which a small volume of 1 N sodium hydroxide solution has been added. Carefully neutralize with hydrochloric acid and make up to 100 ml with water. This solution keeps for several months in a refrigerator. Incubate one egg slope at 25 ± 0.5°C, the other in an incubator with an internal light at 35–37°C. Incubate the *p*-nitrobenzoic acid egg slope at 35–37°C. If an internally illuminated incubator is not available, grow the organisms for 14 days and then expose to the light of a 25-W lamp at a distance of 2–3 ft for several hours and re-incubate for 1 week.

Tubercle bacilli have the following characteristics:

(1) Not easily emulsified in water (or concentrated $HgCl_2$ which is safer.)
(2) Regular morphology, 3–4 μm in length and usually showing serpentine cords.
(3) Relatively slow growing, taking 10 days or more to show visible growth.
(4) Produce no yellow, orange or red pigment.
(5) Fail to grow in the presence of 500 μg/ml of *p*-nitrobenzoic acid (the bovine organism may show a trace of growth).
(6) Fail to grow at 25°C (the bovine organism may show a trace of growth).

Acid-fast organisms that grow in 1 week or less, or yield yellow or red colonies, or emulsify easily, or are morphologically small, coccoid, or long, thin and beaded or irregularly stained, are unlikely to be tubercle bacilli.

Variants of tubercle bacilli and BCG

Although the identification of human, bovine and 'African' variants of *M. tuberculosis* is of no clinical value it may be necessary to recognize them for epidemiological work (Collins *et al.*, 1982). It is also useful to identify BCG, particularly if it may have been responsible for disseminated disease as a result of vaccination of neonates, HIV infection or the treatment of malignancy.

Use suspensions prepared as described above and inoculate media for the following tests and incubate at 37°C.

TCH susceptibility
Use Lowenstein-Jensen medium containing 5 μg/ml of thiophen-2-carboxylic acid hydrazide. Read at 18 days.

Pyruvate preference
Compare the growth on Lowenstein-Jensen medium containing glycerol with that containing pyruvate after 18 days' incubation.

Nitratase test
Use Middlebrook 7H10 broth and test after 18 days by method (3) described on p.115.

Oxygen preference
Use Kirchner medium made semisolid with 0.2% agar. Inoculate with about

0.02 ml of suspension. Mix gently to avoid air bubbles and incubate undisturbed for 18 days. Aerobic growth occurs at or near the surface; microaerophilic growth as a band 1–3 cm below the surface, sometimes extending upwards.

Pyrazinamide susceptibility
Use the method described on p.429.

Cycloserine susceptibility
Use Lowenstein-Jensen medium containing 20 µg/ml cycloserine. Read after 18 days.

See Table 45.2. Identification of a strain as 'Asian' or 'African' does not indicate origin of strain or ethnic group of the patient. The variants are widely distributed.

Table 45.2 Variants of the mammalian tubercle bacilli

	TCH	*Nitratase*	*Oxygen preference*	*Pyrazinamide resistance*
M. bovis	S	–	M	R
M. africanum type I	S	–	M	S
M. africanum type II	S	+	M	S
M. tuberculosis Asian	S	+	A	S
M. tuberculosis Classical	R	+	A	S
BCG[a]	S	–	A	R

S, sensitive; R, resistant, M, microaerophilic; A, aerobic
[a] Differs from the others in being resistant to cycloserine

The niacin test

It was originally claimed that this test distinguished between human tubercle bacilli, which synthesize niacin, and bovine tubercle bacilli, which do not. Unfortunately some human strains give negative niacin results and some other mycobacteria do synthesize niacin. It is an unreliable test and uses a hazardous chemical (cyanogen bromide). It has little to commend it.

Species of tubercle bacilli

M. tuberculosis
This species grows well (eugonic) on egg media containing glycerol or pyruvate. Colonies resemble breadcrumbs and are cream coloured. Films show clumping and cord formation – the bacilli are orientated in ropes or cords, especially on moist medium. There is no growth at 25 or 42°C or on PNBA medium. It is usually resistant to TCH (the Asian variants are sensitive), is nitratase positive, aerobic and usually susceptible to pyrazinamide. It causes tuberculosis in humans and may infect domestic and wild animals (usually directly or indirectly from humans). Simians are particularly likely to become infected.

M. bovis
Is also known as the bovine variant of *M. tuberculosis*. It grows poorly (dysgonic) on egg medium containing glycerol but growth is usually enhanced on pyruvate medium. Colonies are flat and grey or white; growth may be effuse. Cords are present in films. There is no or very poor growth at 25°C and

on PNBA medium compared with a control slope at 37°C. It is sensitive to TCH, nitratase negative, microaerophilic and resistant to pyrazinamide. This is the classic bovine tubercle bacillus, common in milch cows before eradication schemes were introduced. It is still occasionally associated with human disease both pulmonary and extrapulmonary, but such infections are now rarely associated with the consumption of infected milk. Most disease represents reactivation of infections acquired much earlier.

M. africanum
This name is given to the African group of strains with properties intermediate between those of the human and bovine tubercle bacillus. (See Collins *et al.*, 1982.) Growth is usually dysgonic, enhanced by pyruvate. It is susceptible to TCH, microaerophilic and susceptible to pyrazinamide. The nitratase test may be negative (most West African strains), or positive (most East African strains). It causes tuberculosis, clinically indistinguishable from that caused by the other tubercle bacilli.

BCG
This is the bacillus of Calmette and Guerin, an organism of attenuated virulence used in immunization against tuberculosis. It is eugonic; growth is not enhanced by pyruvate. It is susceptible to TCH, nitratase negative, aerobic, resistant to pyrazinamide and (unlike other strains in this group) is resistant to cycloserine.

Animal inoculation tests

Cultural methods have now largely superseded animal inoculation. They are generally more reliable and much less expensive. Tubercle bacilli that are resistant to isoniazid usually show limited or no virulence to guinea pigs, and so do some Asian or South Indian strains. Reasons for ending the routine guinea pig test were given by Marks (1972).

Opportunist mycobacteria

Some mycobacteria, normally present in the environment, may, under some circumstances, cause or be associated with human disease. Such diseases are correctly called mycobacterioses, not tuberculosis. Before specific names were given to these organisms and their taxonomic position was clarified, they were known as 'atypical' or 'anonymous' mycobacteria. Various temporary systems of classification were proposed and were used until species were delineated.

A number of other mycobacteria, not so far known to be associated with human disease, are frequently cultured in clinical laboratories. These are usually contaminants of the specimen or are introduced during collection and/or laboratory processing. Mostly they come from water but they may also be present in dust, soil and food. Occasionally they are seen in direct films but fail to grow on cultures incubated at 35–37°C.

It is also difficult to assess the significance of opportunists. Single isolations may be regarded with suspicion. Whenever possible, attempts should be made to recover the organisms from subsequent specimens.

Most of these mycobacteria may be identified with the aid of the following tests and with Table 45.3.

Table 45.3 Usual properties of some opportunist mycobacteria and others that may be encountered in clinical material

Species	*Pigment*	*TZ*	*Nitratase*	*Tween hydrolysis*	*Growth at*					*Sulphatase*[a]		*Catalase*[b]	*Tellurite*[c] *reduced*	*Growth on N-medium*	*Rapid growth*
					20	25	33 (°C)	42	44	3 days	21 days				
M. kansasii	P	S	+	+	−	+	+	v	−	−	+	+ + +		−	−
M. marinum	P	R	−	+ (late)	+	+	+	−	−	−	+ +	+ +		v	+
M. xenopi	− /S	R	−	−	−	−	−	+	+	−	+ + +	−	−	−	−
M. avium-intracellulare	− /S	R	−	−	−	+	+	+	v	−	v	−	+	−	−
M. scrofulaceum	S	v	−	−	v	+	+	−	−	−	v	−	−	−	−
M. malmoense	−	R	+	+ (late)	−	+	+	−	−	−	−	−	−	−	−
M. simiae	P	R	−	−	−	+	+	−	−	−	+	+ +	−	−	−
M. szulgai	S/P[d]	R	+	+	−	+	+	−	−	−	+ + +	+	−	−	−
M. fortuitum	−	R	+	−	+	+	+	v	−	+ + +	+ + +	v	+	+	+
M. chelonei	−	R	−	−	+	+	+	−	−	+ + +	+ + +	v	+	+	+
M. gordonae	S	R	−	+	+	+	+	−	−	−	+	+ +		−	−
M. flavescens	S	R	+	+	+	+	+	−	−	−	+	+		+	+
M. gastri	−	R	−	+	+	+	+	−	−	−	+	+	−	−	−
M. terrae	−	R	+	+	v	+	+	−	−	−	+	+ + +	−	−	−
M. triviale	−	R	−	+	−	+	+	−	−	−	+	+	−	−	−
M. nonchromogenicum	−	R	−	+	v	+	+	−	−	−	+	+ + +	−	−	−
M. smegmatis	−	R	+	+	+	+	+	+	+	−	+	+ +	+	+	+
M. phlei	S	R	+	+	+	+	+	+	+	−	+	+ +	+	+	+
M. ulcerans	− /S	R	−	−	−	−	+	−	−	−	−	−	−	−	−

TZ, thiacetazone; P, photochromogen; S, scotochromogen; +, usually positive; −, usually negative or none; v, variable

[a] Sulphatase: + + +, deep pink; + +, pink; +, pale pink

[b] Catalase: Amount of foam, + + +, more than 20 mm; + +, 10–20 mm; +, 5–10 mm; −, less than 5 mm

[c] Test not done on pigmented strains

[d] *M. szulgai:* At 25°C = photochromogen; at 37°C = scotochromogen

Inoculation
Unless otherwise stated, inoculate the media with one loopful or a 10-μl drop of a suspension prepared as described on p.427.

Pigment production
Inoculate two egg medium slopes. Incubate both at 35–37°C, one exposed to light and the other in a light-proof box. Examine at 14 days. Photochromogens show a yellow pigment only when exposed to light. Scotochromogens are pigmented in light and dark, but the culture exposed to light is usually deeper in colour. False photochromogenicity may arise if the inoculum is too heavy and pigment precursor is carried over. Continue to incubate the slope exposed to light and look for orange-coloured crystals of carotene pigment which may form in the growth.

Temperature tests
Inoculate egg medium slopes with a single streak down the centre and incubate at the following (exact) temperatures: 20, 25, 42 and 44°C. Examine for growth at 3 and 7 days and thereafter weekly for 3 weeks.

Thiacetazone susceptibility
Inoculate egg medium slopes containing 20 μg/ml of thiacetazone (TZ). Make a stock 0.1% solution of this drug in formdimethylamide (*caution*). This solution keeps well in a refrigerator. Incubate slopes for 18–21 days and observe the growth in comparison with that on an egg slope without the drug.

Nitratase test
Use method (3) on p.115.

Sulphatase test
Use the method described on p.117. If the organism is a rapid grower, add the ammonia after 3 days.

Catalase test
Prepare the egg medium in butts in screw-capped tubes (20 × 150 mm). Inoculate and incubate at 35–37°C for 14 days with the cap loosened and then add 1 ml of a mixture of equal parts of 30% hydrogen peroxide (*caution*) and 10% Tween 80. Allow to stand for a few minutes and measure the height in millimetres of the column of bubbles. More than 40 mm of froth is a strong positive result (+ + +) (Wayne's test).

Tween hydrolysis
Add 0.5 ml of Tween 80 and 2.0 ml of a 0.1% aqueous solution of neutral red to 100 ml of M/15 phosphate buffer. Dispense this in 2-ml lots and autoclave at 115°C for 10 min. This solution keeps for about 2 weeks in the dark and should be a pale straw or amber colour. Add a large loopful of culture from solid medium and incubate at 35–37°C. Examine at 5 and 10 days for a colour change from amber to deep pink.

Tellurite reduction
Grow the organisms in Middlebrook 7H10 medium for 14 days or until there is a heavy growth. Add 4 drops of a sterile 0.02% aqueous solution of potassium tellurite (*caution*) and incubate for a further 7 days. If tellurite is reduced there will be a black deposit of metallic tellurium. Ignore grey precipitates. This test is of no value for highly pigmented organisms.

Growth in N medium
Inoculate the medium with a straight wire and incubate. Rapid growers give a turbidity with or without a pellicle in 3 days. 'Frosting', i.e. adherence of a film of growth to the walls of the tube above the surface of the medium, is often seen. Ignore a very faint turbidity.

Resistance to antibacterial agents
Methods of doing these tests are described below. Some mycobacteria exhibit consistent patterns which may be useful for identification.

Aerial hyphae
Some *Nocardia* spp. are partially acid-fast and resemble rapidly growing mycobacteria. Do a slide culture (p.139) and look for aerial hyphae. In very young cultures of mycobacteria, mycelium may be seen, but it fragments early into bacilli. Aerial hyphae are not formed except by *M. xenopi*.

Emulsifiability and morphology
When making films for microscopic examination, note if the organisms emulsify easily. Note the morphology, whether very short, coccobacilli, long, poorly stained filaments or beaded forms.

Other available techniques include lipid chromatography (Jenkins, 1981) pyrolysis mass spectrometry (Sisson *et al.*, 1991) and DNA probes.

Species of opportunist and other mycobacteria

M. kansasii
This is a photochromogen (but see below). Optimum pigment production is observed if the inoculum is not too heavy, if there is a plentiful supply of air (loosened cap) and if exposure to light is continuous. Young cultures grown in the dark and then exposed to light for 1 h will develop pigment, but old cultures that have reached the stationary phase may not show any pigment after exposure. Continuous incubation in the light results in the formation of orange-coloured crystals of carotene. The morphology is distinctive; the bacilli are long (5–6 μm) and are beaded. *M. kansasii* is nitratase positive, sensitive to 20 μg/ml of thiacetazone, hydrolyses Tween 80 rapidly and grows at 25°C but not at 20°C; some strains grow poorly at 42°C but not at 44°C. The sulphatase test is weakly positive and the catalase strongly positive (more than 40 mm in the Wayne test). It is resistant to streptomycin, isoniazid and PAS but susceptible to ethionamide, ethambutol and rifampicin.

Some strains are susceptible to amikacin and erythromycin by the disc technique.

Occasional non-chromogenic or scotochromogenic strains have been reported but these may be recognized by their biochemical reactions and morphology.

This organism is an opportunist pathogen, associated with pulmonary infection. It is rarely significant when isolated from other sites. It has been found in water supplies.

M. xenopi
Pigment, if any, is more obvious on cultures incubated at 42–44°C, which become pale yellow. Morphologically it is easily recognized; it stains poorly and the bacilli are long (5–6 μm) and filamentous. It is nitratase negative, resistant to thiacetazone and does not hydrolyse Tween 80. It is a thermophile, growing well at 44°C, slowly at 35°C, when growth is effuse, and not at 25°C.

(This is the only opportunist mycobacterium which fails to grow at 25°C.) The sulphatase test is strongly positive, the catalase test negative. It does not reduce tellurite. It is more susceptible to isoniazid than other opportunist mycobacteria, usually giving a resistance ratio of 4. It is susceptible to ethionamide, but its susceptibility to other antituberculosis drugs varies between strains. Some strains are susceptible to amikacin and erythromycin by the disc method.

It is an opportunist pathogen in human lung disease and is rarely significant in other sites. It is a frequent contaminant of pathological material, especially urines, and has been found in hospital hot water supplies. 'Outbreaks' of laboratory contamination are not uncommon. This organism is very common in south-east England and north-west France.

M. avium-intracellulare

The two species in this group may be separated by agglutination serology but as this is not done by most reference laboratories they are usually grouped together as the MAI bacilli, after the initial letters of its components. A feeble yellow pigment, not influenced by light, is produced by some species. The bacilli are small, almost coccoid. Some strains are susceptible to thiacetazone. All are nitratase negative, catalase negative or weakly positive, reduce tellurite and do not hydrolyse Tween 80. The sulphatase reaction varies from strongly positive to negative. There is growth on egg medium at 25°C, growth at 20°C and 42°C is variable. Resistance to antituberculosis drugs is usual, but some strains are susceptible to ethionamide.

MAI bacilli are opportunist pathogens of humans, associated with cervical adenitis (scrofula), especially in young children, but pulmonary infections also occur. They frequently cause opportunist disease in patients with the acquired immune deficiency syndrome (AIDS). Such disease is often disseminated and the organisms may be isolated from many sites including blood, bone marrow and faeces (Kiehn *et al.*, 1985). They are also opportunist pathogens of pigs and birds (*M. avium* was once described as the avian tubercle bacillus). They have also been found in soil and water.

M. scrofulaceum

This, the 'scrofula scotochromogen', is phenetically similar to the MAI bacilli but is scotochromogenic and fails to reduce tellurite to tellurium. It may also be differentiated by agglutination serology.

M. marinum (formerly M. balnei)

This is missed in clinical laboratories that do not culture material from superficial lesions at 30–33°C. Primary cultures do not grow at 35–37°C. It is a photochromogen. Beading or banding similar to that of *M. kansasii* may be evident. It is nitratase negative, resistant to thiacetazone and hydrolyses Tween 80. After laboratory subculture its temperature range is modified. It will grow at 25°C and at 37°C but not at 44°C. The catalase and sulphatase tests are weakly positive and growth may occur in N medium. Resistance to streptomycin and isoniazid is usual; resistance to other drugs is variable. Some strains are susceptible to cotrimoxazole, erythromycin and amikacin by the disc method. This is an opportunist pathogen responsible for superficial infections, known as swimming pool granuloma, fish tank granuloma or fish-fanciers' finger. It is found in sea-bathing pools and in tanks where tropical fish are kept, and is a pathogen of some fish. (See the review by Collins *et al.*, 1985a.)

M. gordonae

This is also known as the tap water scotochromogen, although other scotochromogens are found in water supplies. Organisms in this group grow slowly, produce a deep orange pigment in light and usually a yellow pigment in the dark. No crystals of carotene are formed. If such crystals are seen, the organism may be a scotochromogenic strain of *M. kansasii*. Morphology is not distinctive. They are nitratase negative, resistant to thiosemicarbazone and hydrolyse Tween 80. Growth occurs at 20°C but not at 44°C. Some strains are psychrophilic. The sulphatase reaction is weak and the catalase test is usually strongly positive. Growth does not occur in N medium. They are usually resistant to isoniazid and PAS but susceptible to streptomycin and other antituberculosis drugs.

The organisms in this group are very rarely associated with human disease. They are not infrequent contaminants of pathological material but usually appear as single colonies on egg medium. They may be found in tap water, dust and soil.

M. szulgai

This scotochromogen differs from *M. gordonae* in being nitratase positive, strongly sulphatase positive and giving a weak catalase reaction. It is a rare human opportunist pathogen.

Rapidly-growing scotochromogens

There are many species in this group, e.g. *M. flavescens, M. gilvum, M. duvalii* and *M. vaccae*. With very rare exceptions they are not known to cause human disease but occur in the environment and sometimes contaminate pathological material.

M. fortuitum and M. chelonei

These two non-pigmented rapid growers are conveniently considered together. There have been taxonomic problems: *M. fortuitum* has also been called *M. ranae, M. giae* and *M. peregrinum; M. chelonei* was formerly named *M. abcessus, M. runyonii* and *M. borstelense*. On subculture, growth occurs within 3 days on most media but on primary isolation from clinical or environmental specimens growth may, paradoxically, not be apparent for several weeks or months. The bacilli tend to be rather fat and solidly stained. The nitratase test is positive (*M. fortuitum*) or negative (*M. chelonei*). Tween 80 is not hydrolysed. There is growth at 20°C and in some strains at 42°C. Psychrophilic strains, which fail to grow at 37°C, are not uncommon. The sulphatase test is positive at 3 days and the catalase reaction is variable. Tellurite is reduced rapidly. There is growth in N medium in 3 days. There is a general resistance to all antituberculosis drugs, except that *M. fortuitum* is usually susceptible to ethionamide and the quinolones but *M. chelonei* is resistant.

These are opportunist pathogens usually occurring in superficial infections (e.g. injection abcesses) and occasionally as secondary agents in pulmonary disease. They are common in the environment and frequently appear as laboratory contaminants.

M. smegmatis and M. phlei

These two rapidly growing saprophytes are more common in textbooks than in clinical laboratories. Growth occurs in 3 days. They are nitratase positive and hydrolyse Tween 80 in 10 days. They grow at 20°C and at 44°C, *M. phlei* grows at 52°C. The 3-day sulphatase test is negative, but longer incubation gives positive reactions. The catalase test is strongly positive. Tellurite is reduced. There is growth in N medium.

They are not associated with humans or animal disease.

They have contributed to the myth of acid-fast versus acid- and alcohol-fast bacilli (p.410).

Slowly-growing nonchromogens
There are at least four species, *M. terrae, M. nonchromogenicum, M. triviale* and *M. gastri*. All grow at 25°C but poorly or not at all at 42°C. All hydrolyse Tween 80 and none reduces tellurite, which permits differentiation from MAI bacilli. *M. terrae* and *M. triviale* give a positive nitratase test but *M. nonchromogenicum* and *M. gastri* are negative. *M. gastri* is occasionally photochromogenic and is usually susceptible to thiacetazone and may therefore be confused with *M. kansasii* to which, though non-pathogenic, it is genetically closely related. Susceptibility to anti-tuberculosis drugs is variable.

M. ulcerans
This is likely to be missed by clinical laboratories because it does not grow at 37°C. It grows only between 31 and 34°C. Growth is very slow, taking 10–12 weeks to give small colonies resembling those of *M. bovis*. It is biochemically inert and undistinguished. It causes Buruli ulcer, a serious skin infection in certain regions within the tropics and in Australia.

M. simiae
Originally isolated from monkeys, this is also known as *M. habana*. It is a photochromogen, is nitratase negative and does not hydrolyse Tween 80. Pulmonary and disseminated infections have been reported.

M. malmoense
This resembles the MAI organisms, grows very slowly and is sometimes confused with *M. bovis* by inexperienced workers. This suggests that it is commoner than might be expected. It is difficult to identify except by lipid chromatography. It causes pulmonary disease in adults and cervical adenopathy in children.

M. haemophilum
Another rare species, possibly missed because it grows poorly or not at all on media that do not contain iron or haemin. It grows at 30°C on LJ medium containing 2% ferric ammonium citrate and on Middlebrook 7H11 medium plus 60 μg/ml of haemin. All the usual tests are negative. It has been isolated from the skin and subcutaneous tissues of immunologically compromised patients.

Drug susceptibility tests

Most cases of tuberculosis yield tubercle bacilli that are sensitive to the commonly used antituberculosis drugs. Susceptibility tests are, however, frequently required by physicians at the commencement of and during treatment, particularly if the patient's condition does not improve. Susceptibility tests for opportunist mycobacteria are of questionable value as patients often respond to drug regimens despite *in vitro* resistance.

It is customary for physicians to treat patients with combinations of at least three drugs. Treatment with one drug only usually results in the emergence of resistant organisms.

The drugs in general use are isoniazid, rifampicin, ethambutol and pyrazinamide. Streptomycin, being an injectable drug, is now usually reserved for special purposes.

Susceptibility tests with mycobacteria are complex and are usually done in specialist and reference laboratories. There are four principal methods:

(1) the absolute concentration method which is popular in Europe,
(2) the proportion method, also used in Europe and popular in America,
(3) the resistance ratio method used in the UK,
(4) the radiometric method.

Different techniques are used for pyrazinamide susceptibility tests.

The absolute concentration method

Carefully measured amounts of standardized inocula are placed on control media and on media containing varying amounts of drugs and the lowest concentration that will inhibit all, or nearly all, of the growth is reported. It is very difficult however, to standardize the *active* concentration of drugs in the media and the method gives different results in different laboratories. It is not possible to use a medium that must be heated after the addition of the drug (e.g. egg medium) because some drugs are partially heat labile and heating times are rarely constant even in the same laboratory. Middlebrook 7H10 or 7H11 media are used but this method has not found much favour in the UK.

The proportion method

Several dilutions of the inoculum are made and media containing no drug and standard concentration of drugs are inoculated. The number of colonies growing on the control from suitable dilutions of the inoculum is counted and also the number growing on drug-containing medium receiving the same inoculum. Comparison of the two shows the proportion of organisms that are resistant. This is usually expressed as a percentage.

This method is popular in the USA and in Europe but it is technically very difficult and as it is usually done in petri dishes we regard it as highly hazardous. There are also more risks attached to standardizing the inocula than with the resistance ratio method.

The resistance ratio method

The minimal inhibitory concentrations of the drug that inhibit test strains are divided by those which inhibit control strains to give the resistance ratios. Ratios of 1 and 2 are considered susceptible those of 4 or more resistant (see below). These results usually correlate with the clinical findings. This method is used extensively in the UK and is described below.

Each of the above methods may be used for 'direct' tests on sputum homogenates that contain enough tubercle bacilli to give positive direct films as well as for 'indirect' tests on cultures. The latter are more reliable and reproducible.

For background information about these tests, see Canetti *et al.* (1969) and Vestal (1975). For a discussion on the design of resistance ratio tests, see Marks (1961) and Collins *et al.* (1985b).

Technique of resistance ratio method

Dilutions are incorporated in Lowenstein-Jensen medium, which is then inspissated. As some of these drugs are affected by heat, it is essential that the inspissation procedure is standardized. The inspissator must have a large circulating fan so that all tubes are raised to the same temperature in the same time. The load (i.e. number of tubes) and the time of exposure must be constant. To ensure that the drug-containing medium is not overheated, the machine should be raised to its correct temperature *before* it is loaded and racks can be devised that allow loading in a few seconds. A period of 45 min at 80°C is usually sufficient to coagulate this medium and no further heat is necessary if it is prepared with a reasonably aseptic technique.

Control strains

The resistance ratio method compares the MIC of the unknown strain with that of control strains on the same batch of medium.

Some workers use the H37Rv strain of *M. tuberculosis* as the control strain, but the susceptibility of this to some drugs does not parallel that of wild tubercle bacilli and may give misleading ratios. It is better to use the modal resistance method of Leat and Marks (1970). The unknown strains are compared with the modal resistance (i.e. that which occurs most often) of a number of known susceptible strains of recent origins.

Drug concentrations

Each laboratory must determine its own ranges, as these will vary slightly according to local conditions. Initially use those suggested in Table 45.4 and inoculate at least 12 sets with known susceptible organisms to arrive at a baseline for future work. A drug-free control slope must be included.

Table 45.4 TB susceptibility tests: suggested concentrations of drugs in Lowenstein-Jensen medium

	Final concentration (μg/ml)						
Isoniazid	0.007	0.015	0.03	0.06	0.125	0.25	0.5
Ethambutol	0.07	0.15	0.31	0.62	1.25	2.5	0.5
Rifampicin	0.53	1.06	3.12	6.25	12.5	25	50
Streptomycin	0.53	1.06	3.12	6.25	12.5	25	50

These are for the preliminary titration. Choose six that give confluent or near confluent growth in the two lowest and no growth in the next four.

Stock solutions of the drugs are conveniently prepared as 1% solutions in water (except for rifampicin which should be dissolved in formdimethylamide (*caution*)). All stock solutions keep well at −4°C.

Bacterial suspension

Smooth suspensions must be used. Large clumps or rafts of bacilli give irregular results and make readings difficult.

Sterilize 7 ml (bijou) screw-capped bottles containing a wire nail (shorter than the diameter of the bottle), a few glass beads and 2 ml of phosphate buffer (13.3 g of anhydrous Na_2HPO_4 and 3.5 g of KH_2PO_4 in 2 litres of water; pH 7.4). Into each bottle place a scrape of growth equal to about three or four large colonies and place the bottle on a magnetic stirrer for 3–4 min. Allow to stand for a further 5 min for any lumps to settle and use the supernatant to inoculate culture media.

Methods for inoculation

Place the bottle containing the suspension on a block of Plasticine at a suitable angle. Inoculate each tube with a 3-mm loopful of the suspension, withdrawn edgewise. Plastic disposable 10 μl loops are best.

A better and quicker method uses an automatic micropipette (the Jencon Micro-Repette is ideal) fitted with a plastic pipette tip. Deposit 10 μl of the suspension near to the top of each slope of medium. As it runs down it spreads out to give a suitable inoculum. The plastic tips are sterilized individually in small, capped test-tubes and after use are deposited in a jar containing glutaraldehyde, which is subsequently autoclaved. The tips may be re-used.

Incubation

Incubate at 37°C for 18–21 days.

Reading results

Examine tubes with a hand lens and record as follows: confluent growth, CG; innumerable discrete colonies, IC; between 20 and 100 colonies, +; less than 20 colonies, 0.

The drug-free control slope must give growth equal to CG or IC. Less growth invalidates the test. The *modal* resistance, i.e. the MIC occurring most frequently in the susceptible control strains, should give readings of CG or IC on the first two tubes (Table 45.5). The range must therefore be adjusted to give this result during the preliminary exercises. Once the range is set it is seldom necessary to vary it by more than one dilution.

Table 45.5 'Modal resistance'

Tube no.		*Drug concentration*					
		1	*2*	*3*	*4*	*5*	*6*
Strain	A	CG	CG	0	0	0	0
	B	CG	IC	0	0	0	0
	C	CG	CG	+	0	0	0
	D	IC	IC	0	0	0	0
	E	CG	CG	+	0	0	0
Mode		CG	CG	0	0	0	0

Interpretation

The resistance ratio is found by dividing the MIC of the test strain by the modal MIC of the control strains. When the readings are all CG or IC, this is easy, but when there are tubes showing +, care and experience may be required in interpretation. In general, a resistance ratio of two or less can be reported as *susceptible*, 4 as *resistant* and 8 as *highly resistant*. A ratio of 3 is *borderline* except for ethambutol, when it probably indicates resistance. Mixed *susceptible* and *resistant* strains occur. Examples of these findings are shown in Table 45.6.

The radiometric method

The BACTEC instrument (p.126) may be used to obtain rapid results, e.g. within 7 days. Appropriate amounts of each drug are added to the vials followed by the inoculum. For more information see Heifets (1991).

Table 45.6 Interpretation of TB drug susceptibility

Tube no.		*Drug concentration*					*Resistance ratio*	*Interpretation*
	1	*2*	*3*	*4*	*5*	*6*		
Mode	CG	IC	0	0	0	0		
Strain A	CG	0	0	0	0	0	0.5	Susceptible
B	CG	CG	0	0	0	0	1	Susceptible
C	CG	IC	+	0	0	0	1	Susceptible
D	CG	CG	CG	0	0	0	2	Susceptible
E	CG	CG	CG	+	0	0	3	Borderline[a]
F	CG	CG	CG	IC	0	0	4	Resistant
G	CG	CG	CG	CG	IC	0	8	Highly resistant
H	CG	CG	CG	CG	CG	CG	16	Highly resistant
I	CG	CG	CG	+	+	+	–	Mixed susceptible and resistant

[a] Probably resistant with ethambutol

Pyrazinamide sensitivity tests

Pyrazinamide acts upon tubercle bacilli in the lysosomes, which have a pH of about 5.2. For reliable susceptibility tests the medium should therefore be at this pH. Tubercle bacilli do not grow well on egg medium at pH 5.2 and agar medium (e.g. Middlebrook 7H11) are frequently used. We have had good and consistent results with Yates' (1984) modification of Marks (1964) stepped pH method. This is a Kirchner semisolid medium layered on to butts of Lowenstein-Jensen medium.

Pyrazinamide stock solution

Dissolve 0.22 g of dry powder in 100 ml of distilled water and sterilize by filtration or steaming.

Solid medium

Use Lowenstein-Jensen medium containing only one half the usual concentration of malachite green (at an acid pH less dye is bound to the egg and the higher concentration of 'free' dye is inhibitory to tubercle bacilli). Adjust 600 ml of medium to pH 5.2 with N HCl. To 300 ml add 9 ml of water (control); to the other 300 ml add 9 ml of the stock pyrazinamide solution (test). Tube each batch in 1-ml amounts and inspissate upright, to make butts, at 87°C for 1 h.

Semisolid medium

Add 1 g of agar and 3 g of sodium pyruvate to 1 litre of Kirchner medium. Adjust pH to 5.2 with 5N HCl. To 500 ml add 15 ml of water (control); to the other 500 ml add 15 ml of the stock pyrazinamide solution. Steam both bottles to dissolve the agar. Cool to 40°C and add to each bottle 40 ml of OADC supplement.

Final medium

Layer 2 ml of the control semisolid medium on butts of the control Lowenstein-Jensen medium. Do the same with the test media. Store in a refrigerator and use within 3 weeks. (It may keep longer; we have never tried it.)

Bacterial suspensions

Two inocula are required. Use one as described above for other susceptibility tests and also a 1 : 10 dilution of it in sterile water.

Inoculation

With a suitable pipette (e.g. the Jencons MicroRepette and plastic tip) add 20 μl (approximately) of (a) the undiluted, and (b) the diluted suspension to pairs of test and control media.

Interpretation

Two concentrations of inoculum are used because

(1) some strains require a heavy inoculum to give growth at an acid pH, even in the control medium, and
(2) a heavy inoculum of some other strains may overcome the inhibitory action of pyrazinamide.

Colonies of tubercle bacilli should be distributed throughout the semisolid medium in one or both controls. If there is no growth in the test bottle from either the heavy or light inoculum report the strain as *susceptible*. If there is growth in the test bottle that received the heavy inoculum but not in that which received the diluted suspension, report the strain as *susceptible*. If there is growth in both test bottles, comparable with that in their respective controls, report the strain as *resistant*.

NB. Human strains of tubercle bacilli from untreated patients are invariably susceptible. 'Classic bovine' strains and all other species of mycobacteria found in clinical material are naturally resistant.

References

Brisson-Noel, A., Aznar, C., Chureau, C. *et al.* (1991) Diagnosis of tuberculosis by DNA amplification in clinical practice evaluation. *Lancet*, **338**, 364–366

Canetti, G. and seven others (1969) Advances in techniques of testing mycobacterial drug sensitivity and the use of sensitivity tests in tuberculosis control programmes. *Bulletin of the World Health Organisation*, **41**, 21–43

Collins, C. H., Yates, M. D. and Grange, J. M. (1982) Subdivision of *Mycobacterium tuberculosis* into five variants for epidemiological purposes: methods and nomenclature. *Journal of Hygiene (Cambridge)*, **89**, 235–242

Collins, C. H., Grange, J. M. and Yates, M. D. (1985a) *Mycobacterium marinum* infections in man. *Journal of Hygiene (Cambridge)*, **94**, 135–149

Collins, C. H., Grange, J. M. and Yates, M. D. (1985b) *Organization and Practice in Tuberculosis Bacteriology*. Butterworths, London, pp.90–97

French, G. A., Teoh, R., Chan, C. Y., Humphries, M. J., Cheung, S. W. and O'Mahoney, G. (1987) Diagnosis of tuberculous meningitis by detection of tuberculostearic acid in cerebrospinal fluid. *Lancet*, **ii**, 117–119

Heifets, L. (1986) Rapid automated methods (BACTEC system) in clinical mycobacteriology. *Seminars in Respiratory Infections*, **1**, 242–249

Heifets, L. (ed.) (1991) Drug susceptibility in the management of chemotherapy of tuberculosis. In *Drug Susceptibility in the Chemotherapy of Mycobacterial Infections*, CRC Press, Boca Raton, pp.89–121

Jenkins, P. A. (1981) Lipid analysis for the identification of mycobacteria: an appraisal. *Reviews of Infectious Diseases*, **3**, 862–866

Kiehn, T. E., Edwards, F. F., Brannon, P. *et al.* (1985) Infections caused by *Mycobacterium avium* complex in immunocompromised patients: diagnosis by blood culture and fecal examination, antimicrobial susceptibility tests, and morphological and seroagglutination characteristics. *Journal of Clinical Microbiology*, **21**, 168–173

Leat, J. L. and Marks, J. (1970) Improvements in drug sensitivity tests on tubercle bacilli. *Tubercle*, **51**, 68–73

Marks, J. (1961) The design of sensitivity tests on mycobacteria. *Tubercle*, **42**, 314–316

Marks, J. (1964) A 'stepped pH' technique for the estimation of pyrazinamide sensitivity. *Tubercle*, **45**, 47–50

Marks, J. (1972) Ending the routine guinea pig test. *Tubercle*, **53**, 31–34

Mitchison, D. A., Allen, B. J. and Manickavasagar, D. (1987) Selective Kirchner medium for the culture of specimens other than sputum for mycobacteria. *Journal of Clinical Pathology*, **36**, 1357–1361

Ramkisson, A., Coovadia, Y. M. and Coovadia, H. M. (1988) A competition ELISA for the detection of mycobacterial antigen in tuberculous exudates. *Tubercle*, **69**, 209–212

Sisson, P. R., Freeman, R., Magee, J. G. and Lightfoot, N. F. (1991) Differentiation between mycobacteria of the *Mycobacterium tuberculosis* complex by pyrolysis mass spectrometry. *Tubercle*, **72**, 206–209

Vestal, A. L. (1975) *Procedures for the Isolation and Identification of Mycobacteria*, DHEW (CDC) 75–8230, Government Printing Office, Washington DC

Yates, M. D. (1984) The differentiation and epidemiology of the tubercle bacilli and a study on the identification of other mycobacteria. *MPhil Thesis*, University of London

Chapter

46

Nocardia, Actinomadura, Streptomyces and Rhodococcus

This group contains Gram-positive aerobes that may form filaments, branches and aerial mycelium. Some are partly acid fast.

Isolation

Culture pus directly and wash some of it to recover granules as described on p.435. Treat sputum by the 'soft' method used for culturing mycobacteria (p.412) or use one of the sputum digesting agents. Culture on blood agar and on duplicate Lowenstein-Jensen (LJ) slopes. Incubate at 37°C for several days. Incubate one of the LJ slopes at 45°C at which nocardias may grow while other organisms are discouraged.

Identification

Subculture aerobic growth on blood agar and incubate at 37°C for 3–10 days. Colonies vary in size and may be flat or wrinkled, sometimes 'star shaped'. They may be pigmented (pink, red, green, brown or yellow) and often covered with a whitish, downy or chalky aerial mycelium. Colonies of nocardias are usually on the surface; those of streptomyces may be embedded in the medium.

Examine Gram- and ZN-stained films (minimum decolorization with the latter). Do slide culture (see below) and inoculate lysozyme broth: dissolve 20 mg lysozyme in 2 ml of 50% ethanol in water and add 50 μl to 2 ml of nutrient broth at pH 6.8. Test for the hydrolysis of casein, xanthine, hypoxanthine and tyrosine (see Table 46.1). Test also for acid production from arabinose and xylose. The API ZYM system may be used to distinguish between nocardias, streptomyces and related organisms.

Slide cultures

Melt a tube of malt agar, dilute with an equal volume of distilled water and cool to 45°C. Draw about 1 ml of this mixture into a pasteur pipette, followed by a drop of a thin suspension of the organisms. Run this over the surface of a slide in a moist chamber. No cover-glass is needed. Examine daily with a low-power microscope until a mycelium is seen. Spores may be observed as powdery spots on the surface of the medium.

Table 46.1 ***Nocardia, Rhodococcus, Actinomadura*** **and** ***Streptomyces***

	Acid fast	*Aerial mycelium*	*Lysozyme*	*Hydrolysis of*		
				Casein	*Xanthine*	*Tyrosine*
N. asteroides	+ [a]	+	R	–	–	–
N. brasiliensis	+ [a]	+	R	+	–	+
N. otitidiscaviarum	+ [a]	+	R	–	+	–
Actinomadura	–	+	S	+	v	+
Streptomyces	–	+	S	v	v	v
Rhodococcus	–	–	S	–	–	–

R, resistant; S, sensitive; v, variable
[a] Rarely complete. Acid-fast elements may be few in number or absent

Partially acid-fast, Gram-positive mycelium which does not branch and which fragments early in slide cultures may be tentatively identified as nocardia. Non-acid fast Gram-positive mycelium which branches with aerial mycelium and conidiophores abstricted in chains suggest streptomyces.

Nocardia species

Two (possibly three) species are of medical importance (Grange, 1992). Acid-fast elements may be rare or occur in cultures at different times and are best observed on Middlebrook media. Nocardias are resistant to lysozyme and may be differentiated by hydrolysis of casein, xanthine and tyrosine (Table 46.1).

N. asteroides
The most common; causes severe pulmonary infections, brain abscesses and occasionally cutaneous infections.

N. brasiliensis
Mostly confined to North America and the southern hemisphere and usually causes cutaneous infections but may be associated with systemic disease.

N. otitidiscaviarum (*N. caviae*)
Rare cause of human infections.

Actinomadura **and** ***Streptomyces*** **species**

A. *madurae* causes madura foot, A. *pelleteri* and *S. somaliensis* cause mycetomas.
They are difficult to distinguish bacteriologically although *A. madurae* is said to hydrolyse aesculin while *S. somaliensis* does not. Unlike the saprophytic streptomyces they do not produce acid from lactose and xylose. Cell wall analysis is necessary to sort out these and associated species.

Rhodococcus **species**

Several species are found in plant material but none are pathogenic for humans although one, *R. equi*, usually known as *Corynebacterium equi*, is a pathogen of horses. Rhodococci are important only in that they may be confused with nocardias, actinomaduras and *S. somaliensis*. See also p.382.

For more information on the organisms in this chapter see Goodfellow (1986); Goodfellow and Lechevalier (1986); Gordon (1988) and Collins *et al.* (1988).

References

Collins, C. H., Uttley, A. H. C. and Yates, M. D. (1988) Presumptive identification of nocardias in a clinical laboratory. *Journal of Applied Bacteriology*, **65**, 55–59

Goodfellow, M. (1986) Genus *Rhodococcus* Zopf 1891. In *Bergey's Manual of Determinative Bacteriology*, Vol. 2 (eds P. H. A. Sneath, N. S. Mair, M. E. Sharpe and J. G. Holt), William and Wilkins, Baltimore, pp. 1472–1475

Goodfellow, M. and Lechevalier, M. P. (1986) Genus *Nocardia* Trevisan 1889. In *Bergey's Manual of Determinative Bacteriology*, Vol. 2 (eds P. H. A. Sneath, N. S. Mair, M. E. Sharpe and J. G. Holt), William and Wilkins, Baltimore, pp. 1460–1514

Gordon, M. A. (1988) Aerobic pathogenic Actinomycetaceae. In *Manual of Clinical Microbiology*, 4th edn (eds. E. H. Lennette, A. Balows, W. J. Hausler and H. J. Shadomy), American Society for Microbiology, Washington, pp. 249–262

Grange, J. M. (1992) Actinomyces and Nocardia. In *Medical Microbiology*, 14th edn, (eds D. Greenwood, R. Slack and J. Peutherer), Churchill Livingstone, Edinburgh, pp. 265–269

Chapter

47

Actinomyces, Propionibacterium and Bifidobacterium

This group contains Gram-positive, anaerobic or microaerophilic 'diphtheroid-like' organisms that tend to branch. *Actinomyces* require CO_2 for growth and are catalase negative; *Propionibacterium* does not require CO_2 and is catalase positive; *Bifidobacterium* does not require CO_2 and is catalase negative.

Species of medical and veterinary importance occur in pus and discharges as colonies or granules.

Actinomyces

Isolation

Wash pus gently with saline in a sterile bottle or petri dish. Look for 'sulphur granules', about the size of a pinhead or smaller. Aspirate granules with a pasteur pipette and crush one between two slides and stain by the Gram method. Look for Gram-positive mycelia in an amorphous matrix surrounded by a zone of large Gram-negative, club-like structures. The club forms are weakly acid fast on Ziehl-Neelsen staining.

Crush granules with a sterile glass rod in a small sterile tube containing a drop of broth. Inoculate blood agar plates, brain heart infusion agar and enriched thioglycollate medium. These may be made selective for Gram-positive non spore-bearers by adding 30 μg/ml nalidixic acid and 10 μg/ml metronidazole. Do cultures in triplicate and incubate:

(1) anaerobically plus 5% carbon dioxide,
(2) in a 5% carbon dioxide atmosphere, and
(3) aerobically at 37°C for 2–7 days.

Subculture any growth in the broth media to solid media and incubate under the same conditions. Look for colonies resembling 'spiders' or 'molar teeth'.

Identification

Do catalase tests on colonies from anaerobic or microaerophilic cultures that show colonies of Gram-positive coryneform or filamentous organisms.

Actinomyces are catalase negative; corynebacteria and bifidobacteria are usually catalase positive. It is often difficult to distinguish actinomyces from bifidobacteria and propionibacteria by simple tests. Gas-liquid chromatography (GLC) may be necessary. The spot indole test applied to colonies is useful; *Actinomyces* spp. are negative, unlike *Propionibacterium*.

Subculture for nitratase test and test for acid production from mannitol, xylose, raffinose, and aesculin hydrolysis (see Table 47.1). The API ZYM system may be useful in distinguishing between actinomyces and related genera.

Table 47.1 *Actinomyces* and *Bifidobacterium*

	Nitratase	*Acid from*			*Hydrolysis of*	
		Mannitol	*Xylose*	*Raffinose*	*Starch*	*Aesculin*
A. bovis	−	−	−	−	+	−
A. israelii	v	v	+	+	−	+
A. naeslundii	+	−	v	+	v	+
A. odontolyticus	+	−	v	−	v	v
A. meyeri	−	−	+	+	−	−
A. pyogenes	v	−	−	−	v	−
B. eriksonii	−	+	+	+	−	v

Species of Actinomyces

Actinomyces bovis

Colonies at 48 h are pinpoint and smooth, later becoming white and shining with an entire edge. Spider and molar tooth colonies are rare. In thioglycollate broth, growth is usually diffuse but occasionally there are colonies that resemble breadcrumbs. Microscopically the organisms are coryneform, rarely branching.

Nitrates are not reduced, acid is not produced from mannitol, xylose or raffinose. Starch is hydrolysed; aesculin is not hydrolysed.

This is a pathogen of animals, usually of bovines, causing lumpy jaw.

A. israelii

Colonies at 48 h are microscopic, with a spider appearance, later becoming white and lobulated with the appearance of molar teeth. In thioglycollate broth there are distinct colonies with a diffuse surface growth and a clear medium. The colonies do not break when the medium is shaken. Microscopically the organisms are coryneform, with branching and filamentous forms.

Nitrates may be reduced, acid is produced from xylose and raffinose; aesculin is hydrolysed but not starch.

This species causes human actinomycosis and may be responsible for intrauterine infections in women fitted with IUCDs.

A. naeslundii

Colonies at 48 h are similar to those of *A. israelii*; spider forms are common, but not molar tooth forms. In thioglycollate broth, growth is diffuse and the medium is turbid. Microscopically the organisms are irregular and branched, with mycelial and diphtheroid forms.

Nitrates are reduced, acid is produced from raffinose; aesculin is hydrolysed; starch hydrolysis is variable. It is a facultative aerobe.

It is not known to be a human pathogen but has been found in human material.

A. odontolyticus
Colonies at 48 h are 1–2 mm in diameter and grey but they may develop a deep reddish colour after further incubation. CO_2 is required for growth. The organisms are coryneform, rarely branched. Nitrate is reduced; acid is rarely produced (except occasionally from xylose); starch and aesculin hydrolysis is variable.

The normal habitat is the human mouth but the organisms have been isolated from the tear ducts.

A. meyeri
Colonies at 48 h are pinpoint, greyish white and rough. CO_2 is required for both aerobic and anaerobic growth. The organisms are short and coryneform but branching is rare. Nitrate is not usually reduced; acid is produced from xylose and raffinose; starch and aesculin are not hydrolysed.

The normal habitat is the human mouth but the organism has been isolated from abscesses.

A. pyogenes
This is involved in suppurative lesions and mastitis in domestic and wild animals.

Produces pin-point α-haemolytic colonies at 36–48 h.

For more information about actinomyces see Schaal (1984 and 1986).

Propionibacterium

These are pleomorphic coryneforms varying from coccoid to branched forms. They are aerobic but may be aerotolerant, non-motile and grow poorly in the absence of carbohydrates. They are catalase positive and some strains may liquefy gelatin. The optimum temperature is 30°C. There are several species of interest to food microbiologists. They occur naturally in the bovine stomach and hence in rennet and are responsible for flavours and 'eyes' in Swiss cheese. A method for presumptive identification is given on p.258. It is difficult to identify species.

There is one species of medical interest: *P. acnes*.

Propionibacterium acnes
The bacilli are small, almost coccoid, or about 2.0 × 0.5 μm, and may show unstained bands. The best growth is obtained anaerobically. The colonies on blood agar are either small, flat, grey-white and buttery, or are larger, heaped up and more granular. Both colony forms are β-haemolytic. Acid is not usually produced. The spot test for indole is positive. It is nitratase negative, catalase positive and liquefies gelatin. Slide agglutination is useful.

It is a commensal on human skin, in hair follicles and in sweat glands. It appears to be involved in the pathogenesis of acne. Oral and systemic infections have been suspected.

This genus is described by Cummings and Johnson (1986).

Bifidobacterium

Bifidobacteria are small Gram-positive rods resembling coryneforms which predominate in the faeces of breast-fed infants, but they are also commensal in the adult bowels, mouth and vagina. They are anaerobic and may need to be differentiated from anaerobic corynebacteria. Bifidobacteria are catalase negative, nitratase negative and do not produce gas from glucose. CO_2 is not required but it does improve growth.

There is one pathogen, *B. eriksonii*, formerly *Actinomyces eriksonii*, associated with mixed infections in the upper respiratory tract.

For information about bifidobacteria see Scardovi (1986).

References

Cummings, C. S. and Johnson, K. L. (1986) The genus *Propionibacterium*. In *The Prokaryotes: A Handbook on Habitats, Isolation and Identification of Bacteria*, Vol.2 (eds M. P. Starr *et al.*), Springer, New York, p.1864

Scardovi, V. (1986) The genus *Bifidobacterium*. In *Bergey's Manual of Systematic Bacteriology*, Vol.2 (eds P. H. A. Sneath *et al.*), Williams and Wilkins, Baltimore, p.1418

Schaal, K. P. (1984) Laboratory diagnosis of actinomycete diseases. In *The Biology of the Actinomycetes* (eds M. Goodfellow *et al.*), Academic Press, London, p.425

Schaal, K. P. (1986) The genus *Actinomyces*. In *Bergey's Manual of Systematic Bacteriology* Vol. 2 (eds P. H. A. Sneath *et al*), Williams and Wilkins, Baltimore, p.1332

Chapter

48

Spirochaetes

Three genera that are of medical importance are considered briefly here: *Borrelia, Treponema, Leptospira.*

Borrelia

There are at least 19 species, classified by their arthropod vectors.

Borrelia duttonii and *B. recurrentis* cause relapsing fever, transmitted by lice and ticks.

Lyme disease (inflammatory arthropathy), first described in Lyme, Connecticut, USA and now known in Europe, is caused by *B. burgdorferi* (Burgdorfer, 1984, 1985). It is a tick-borne zoonosis and small woodland rodents and wild deer are reservoirs. Infection is also transmitted to domesticated and farm animals. Unlike the relapsing fever spirochaetes, this species is antigenically stable.

Identification

Take blood during a febrile period. For the relapsing fever species examine wet preparations by dark field using a high power dry objective. Stain thick and thin films with Giemsa stain.

Method
Fix smear in 10% methyl alcohol for 30 s. Stain with 1 part Giemsa stock stain and 49 parts Sorensen's buffer at pH 7 in a Coplin jar for 45 min. Wash off with Sorensen's buffer. Dry in air and examine.

Inoculate two laboratory rats with 1–2 ml of fresh or refrigerated defibrinated blood and examine the animal's blood daily between the second and seventh days.

Morphology
They form shallow, coarse, irregular, highly motile coils 0.25–0.5 × 8–16 μm.

Lyme disease
Diagnosis is made by fluorescence antibody methods (Technicon kit), ELISA or silver staining of biopsies.

Treponema

There are at least 14 species. In medicine the most important are *T. pallidum, T. pertenue, T. carateum and T. vincentii* [*Borrelia vincentii*].

They are found in the oral cavity, intestinal and genital tracts.

Identification

In suspected syphilis
Examine exudate (free from blood and antiseptics) from lesion by dark field or fluorescence microscopy. For detailed methods see Sequira (1987).

T. pallidum
Is the causative organism of syphilis, transmitted by direct contact.

Morphology
It forms tightly wound slender coils 0.1–0.2 × 6–20 μm with pointed ends each having three axial fibrils. They are sluggishly motile with drifting flexuous movements.

In suspected Vincent's angina
Stain smears with dilute carbol fuchsin. Large numbers of spirochaetes are seen together with fusobacteria.

T. vincentii
Is a causative organism of Vincent's angina (ulcerative lesions of the mouth or genitals) and pulmonary infections. It can be spread by direct contact.

Culture
Inoculate peptone yeast extract medium with added serum or ascitic fluid under anaerobic conditions at 37°C. Small, white colonies 1.2–1.5 mm in diameter having the appearance of a slight haze are visible after 2 weeks.

Morphology
It is a loosely wound single contoured spirochaete 0.2–0.6 μm × 7–18 μm.

Other treponemes

T. pertenue
This organism causes yaws and is principally transmitted by direct contact. It is morphologically indistinguishable from *T. pallidum*. Laboratory tests are unhelpful.

T. carateum
Causes pinta. The mode of spread is, in common with yaws, by direct contact. It is morphologically indistinguishable from *T. pallidum*. Diagnosis is by silver impregnation of tissue.

Leptospira

Identification is important only for epidemiological purposes. *Leptospira interrogans* (19 subgroups) is primarily an animal parasite but in humans causes an acute febrile illness with or without jaundice, conjunctivitis and meningitis (Weil's disease). *L. biflexa* is non-pathogenic.

Isolation and identification

Examine blood during the first week of illness and urine thereafter. Add 9 parts of blood to 1 part of 1% sodium oxalate in phosphate buffer at pH 8.1 and centrifuge 15 min at 1500 rev/min. Examine clear plasma under dark-field microscopy using low power magnification. If this is negative, centrifuge the remainder of the plasma at 10 000 rev/min for 20 min and examine the sediment. Direct films rarely show leptospira either in dark field or Giemsa-stained preparations. Centrifuge urine and examine deposit by dark field within 15 min.

Morphology

Leptospiras are short, fine, closely wound spirals, 0.25 × 6–20 μm, resembling a string of beads. The ends are bent at right angles to the main body to form hooks.

Culture

The methods given here are adapted from those of Waitkins (1985).

Add 2 drops of fresh blood, CSF or urine (at pH 8) to 5 ml of EMJH/5FU medium (p.78) and make five serial dilutions in the same medium to dilute out antibody. Incubate at 30°C, examine daily for 1 week, then weekly for several weeks. Use dark-field, phase contrast or fluorescence microscopy.

Identification

If leptospiras are seen subculture to EMJH medium and incubate at 30°C and 13°C. Subculture also to EMJH medium containing 225 mg azoguanine per litre.

L. interrogans grows at 30°C, not at 13°C and not in azoguanine medium.

L. biflexa grows at 13 and 30°C and also in azoguanine medium.

Serological diagnosis

Macroslide agglutinations may be done with patients' sera and (commercial) genus-specific antigen but further agglutination tests are best done in reference laboratories.

There is a rapid test kit (Leptese: Bradsure Biologicals).

For more information about leptospires, culture, serology and epidemiology see Waitkins (1985).

References

Burgdorfer, W. (1984) Discovery of the Lyme spirochaete and its relationship to tick vectors. *Yale Journal of Biology and Medicine*, **57**, 165–168

Burgdorfer, W. (1985) *Borrelia*. In *Manual of Clinical Microbiology*, 4th edn (eds E. H. Lennette, A. Balows, W. J. Hauser and H. J. Shadomy), Association of American Microbiologists, Washington, pp.154–175

Sequira, P. J. L. (1987) Syphilis. In *Sexually Transmitted Diseases* (ed. A. E. Jephcott), Public Health Laboratory Service, London, pp.6–22

Waitkins, S. A. (1985) Leptospiras and leptospirosis. In *Isolation and Identification of Micro-organisms of Medical and Veterinary Importance* (eds C. H. Collins and J. M. Grange), Society for Applied Bacteriology Technical Series No.21, Academic Press, London, pp.251–296

Chapter

49

Mycoplasmas

Mycoplasmas are the smallest free-living bacteria and are usually 0.2–0.3 μm in diameter. They lack cell walls and are therefore resistant to penicillin. They can pass through bacteria-excluding filters. Their morphology is variable: cells may be coccoid, filamentous or star-shaped. Individual cells cannot be stained by Gram's method. Mycoplasmas are usually recognized by their characteristic colonial morphology as observed by low-power microscopy.

Isolation

Specimens

These include sputum, throat swabs, bronchial lavage fluid, lung biopsies, cerebrospinal fluid, urine, urethral, vaginal and cervical swabs (Table 49.1).

Homogenize sputum by shaking it with an equal volume of mycoplasma broth in a bottle containing glass beads. Homogenize tissue in mycoplasma broth. Transport swabs to the laboratory in mycoplasma broth; squeeze the swabs into the broth to express material. Culture other specimens (e.g. cerebrospinal fluid, urine) without pretreatment or centrifuge them and use the deposit.

Culture

Basal mycoplasma media are prepared as agar, broth or in diphasic form (p. 82). They include thallous acetate and penicillin to inhibit the growth of bacteria and amphotericin B to suppress fungi. They may be supplemented with glucose, urea or arginine as indicated in Table 49.2.

Inoculate mycoplasma agar plates, mycoplasma broth and either SP4 broth or selective media in duplicate, with 0.1 ml of the specimen. Incubate at 37°C in anaerobic-type jars under the atmospheric conditions shown in Table 49.2. Examine agar media for colonies daily for up to 20 days. Colonies that appear between 1 and 4 days are likely to be *M. hominis* or *Ureaplasma urealyticum*, and between 5 and 20 days *M. pneumoniae* or *M. genitalium*. Subculture broth media to solid media when the appropriate colour changes have occurred.

Colonies of some species have a dark central zone and a lighter peripheral zone, giving them a 'fried egg' appearance; others, notably *Mycoplasma pneumoniae*, lack this zone and have a mulberry-like appearance.

Identification

Mycoplasmas cannot be removed from agar media for microscopy and subcultured in the usual way with inoculating wires and loops. With a sterile scalpel

Table 49.1 The isolation of mycoplasmas causing human disease

Species	*Clinical specimens*[a]	*Culture media*	*Substrate*	*pH*	*Atmosphere of culture*[b]	*Colony characteristic*	*Growth rate*
M. pneumoniae	Bronchoalveolar lavage fluid, throat swabs, sputum	Diphasic medium; SP4 medium; horse serum agar and broth	Glucose	7.8	1 + 2	Mulberry	Slow (over 5 days)
M. genitalium	Urethral swabs, throat swabs, sputum	SP4 broth or agar	Glucose	7.8	1 + 2	Mulberry	Slow (over 5 days)
U. urealyticum	Vaginal swabs, cervical swabs, urethral swabs, urine	Urea broth or agar supplemented with $MnCl_2$ which will stain colonies	Urea	6.0	3	Small fried egg (10–15 μm)	Medium (over 2 days)
(neonates)	Endotracheal secretions						
M. hominis	Vaginal swabs, cervical swabs, urethral swabs	Horse serum broth or agar	Arginine	7.0	1 + 2	Fried egg (50–500 μm)	Rapid (occasionally by 24 h)

[a] Swabs should be transported to the laboratory in a suitable medium (see text).
[b] Atmosphere 1 = 5% CO_2/95% N_2; 2 = air; 3 = 10–20% CO_2/80–90% N_2.

cut out a block of agar about 5 mm square and containing one or more colonies. Proceed as indicated below.

Microscopy
Place an agar block, colonies uppermost, on a slide. Place a drop of Diene's stain on a cover glass and invert it over the agar block. Examine under a low power microscope.

Imprints of mycoplasma colonies are stained dark blue and those of ureaplasmas greenish blue. Bacteria are also stained blue but they lose the colour after 30 min, whereas the mycoplasmas retain it for several hours. This stain differentiates mycoplasmas from artefacts such as crystals which may have a similar appearance.

Subculture
Place an agar block, colonies down, on the surface of mycoplasma agar and move it around. New colonies will develop along the track of the block.

Inhibition by hyperimmune serum
Cut out agar blocks containing colonies and place them in mycoplasma broth. Incubate for 4–10 days and then flood mycoplasma agar plates with the culture. Allow to dry and apply filter paper discs impregnated with hyperimmune sera (UK source: Division of Laboratory Reagents, PHLS Colindale, London). Incubate until colonies are visible. Zones of growth inhibition will be seen around discs that are impregnated with the homologous antibody to the isolate.

Urease activity
This is most useful for identifying *U. urealyticum*, of which there are 15 serotypes. Pour a solution containing 1% urea and 0.8% $MnCl_2$ over the colonies on mycoplasma agar. Urea-positive colonies appear light brown in colour.

Haemadsorption test
Pour a 0.5% suspension of washed guinea pig erythrocytes in phosphate buffered saline (PBS) over an agar plate bearing about 100 colonies. Leave at room temperature for 30 min, remove the cell suspension, wash the agar surface with PBS and examine under a low power microscope for adsorption of erythrocytes which occurs with colonies of *M. pneumoniae*.

Haemolysis test
Prepare an agar overlay containing 1 ml of a 20% suspension of washed guinea pig or sheep erythrocytes in PBS and 3 ml of mycoplasma agar base (at 45°C). Pour it gently over a plate on which there are colonies of mycoplasmas, allow to solidify and incubate aerobically at 37°C. After 24 h and again at 48 h refrigerate the plates at 4°C for 30 min and examine for haemolysis. Colonies of *M. pneumoniae* are surrounded by a zone of β-haemolysis. Other mycoplasmas are surrounded by zones of α-haemolysis, are smaller and take longer to develop the zone.

Immunofluorescent staining

This is a reliable method for identifying mycoplasma colonies. Transfer colonies from agar media by gently pressing them with the bottom of a previously

Table 49.2 Micoplasmas causing disease in humans and animals

Host	*Mycoplasma*	*Primary isolation site*	*Disease association*
Human	*M. pneumoniae*	Respiratory tract	Atypical pneumonia
	M. genitalium	Genital and respiratory tract	Non-specific urethritis
	M. hominis	Genital; occasionally respiratory	Septic wound infections
	M. fermentans	Genital tract	Genital (and other?) infections
	U. urealyticum	Genital tract and oropharynx	Non-gonococcal urethritis and chronic lung disease
Cattle	*M. mycoides* subsp. *mycoides*	Respiratory tract	Contagious bovine pleuropneumonia
	M. bovis	Milk, respiratory tract	Mastitis and respiratory disease
	M. spp. (bovine group 7)	Joints, milk, genital tract	Septic arthritis and mastitis
Sheep/goats	*M. agalactiae*	Joints, milk, blood	Contagious agalactia
	M. mycoides subsp. apr 1	Respiratory tract	Contagious caprine pleuropneumonia
	M. capricolum	Joints, udder, blood	Septicaemic arthritis
Poultry	*M. gallisepticum*	Respiratory tract	Air sacculitis, sinusitis and arthritis
	M. synoviae	Joints, respiratory tract	Air sacculitis and arthritis
Pigs	*M. hypopneumoniae*	Respiratory tract	Enzootic pneumonia
Rats/mice	*M. arthritidis*	Joints	Arthritis
	M. pulmonis	Respiratory tract	Murine respiratory mycoplasmosis

boiled and dried rubber or neoprene bung. Press the bung on several microscope slides in turn, increasing the pressure on each successive slide.

Dry the slides in air and fix in acetone. Stain by standard indirect immunofluorescence method with specific antibodies raised in rabbits (currently not commercially available).

Species of mycoplasmas and associated human disease

Table 49.1 lists the mycoplasmas responsible for human and animal disease. Those associated with human disease are described briefly below. For details of the clinical aspects of diseases caused by mycoplasmas see Taylor-Robinson (1990, 1992).

M. pneumoniae

This forms mulberry-like colonies in 4–5 days in media containing glucose, incubated in an atmosphere of 5% CO_2 and 95% N_2 as well as aerobically.

It is a major cause of primary atypical pneumonia, accounting for up to 15% of hospital-admitted pneumonias. Epidemics occur over a 4-year cycle. During periods of peak infection all available laboratory tests should be used to ensure adequate diagnosis and prompt and appropriate chemotherapy, usually with erythromycin or a tetracycline.

M. genitalium

Colonies and growth conditions are the same as for *M. pneumoniae*. Colonies may take 5 days to develop.

M. genitalium was originally isolated from the urethras of men with non-gonococcal urethritis. Subsequently, it has been isolated from the respiratory tract in association with *M. pneumoniae* with which it has several properties in common, including the ability to metabolize glucose and some antigenic cross-reactions. Its frequency in the urogenital and respiratory tracts has yet to be established.

M. hominis

This forms 'fried egg' colonies on media containing arginine under the same atmospheric conditions as *M. pneumoniae*. Growth is rapid, occasionally visible in 24 h.

It is present on the genital tract of about 20% of women and may be regarded as an opportunist pathogen. It is sometimes isolated from blood cultures of women with mild postpartum fever and may also cause pelvic inflammation and infertility.

U. urealyticum

Small 'fried egg' colonies appear on media containing urea and incubated in an atmosphere containing 10–20% CO_2 and 80–90% N_2. Colonies may take 2 or more days to develop.

About 60% of healthy women carry this organism in their genital tracts. Like *M. hominis* it may be regarded as an opportunist pathogen as it is also occasionally isolated from blood cultures of women with mild postpartum fever and may also cause pelvic inflammation and infertility. It is associated with non-specific urethritis in men and chronic lung disease in infants with low birth weights.

Serological diagnosis

Serological tests have been described for a number of mycoplasmal infections, notably those caused by *M. pneumoniae*. Culture, however, remains the most useful diagnostic method at present.

The complement fixation test is the one most often used to detect antibody to *M. pneumoniae*. More sensitive tests include indirect haemagglutination or agglutination of antigen-coated gelatin particles and μ-capture ELISA methods for specific IgM antibodies (Shearman *et al.*, 1993). Historical methods, such as the cold agglutination test are relatively insensitive and should be reserved for rapid diagnosis during epidemics.

Immunofluorescence and ELISA tests have been described for the serodiagnosis of *U. urealyticum* infections but at present the regents are not commercially available.

For further information about serological tests see Taylor and Gaya (1985).

References

Shearman, M. J., Cubie, H. A. and Inglis, J. M. (1993) *Mycoplasma pneumoniae* infection: early diagnosis of specific IgM by immunofluorescence. *British Journal of Biomedical Science*, **50**, 305–308

Taylor, P. and Gaya, H. (1985) *Mycoplasma pneumoniae*. In *Isolation and Identification of Micro-organisms of Medical and Veterinary Importance* (eds C. H. Collins and J. M. Grange), Society for Applied Bacteriology Technical Series No. 21, Academic Press, London, pp.313–327

Taylor-Robinson, D. (1990) The Mycoplasmatales: *Mycoplasma, Ureaplasma Acholeplasma, Spiroplasma* and *Aneroplasma*. In *Topley and Wilson's Principles of Bacteriology, Virology and Immunity*, 8th edn, Vol. 2, (eds M.T. Parker and B.I. Duerden), Edward Arnold, London, pp.663–682

Taylor-Robinson, D. (1992) Mycoplasmas. In *Medical Microbiology*, 14th edn, (eds R. C. B. Slack, D, Greenwood and J. E. Peutherer), Churchill Livingstone, Edinburgh, pp.459–470

Chapter

50

Yeasts

Yeasts are fungi whose main growth form is unicellular and which usually replicate by budding. Many can also grow in the hyphal form and the distinction between a yeast and a mould is one of convention only. Some yeast-like fungi such as *Acremonium, Geotrichum* and the 'black yeasts' *Exophiala, Aureobasidium* and *Phialophora* are traditionally excluded from texts on yeasts.

The yeasts are not a natural group. Many are Ascomycotina, a few are Basidiomycotina and others are asexual forms of these groups (Deuteromycotina or Fungi Imperfecti). Some genera are even 'convenience groups': for example both *Candida* and *Torulopsis* contain asexual forms of Ascomycotina and Basidiomycotina. In addition, as these two genera are distinguished only by their ability to produce pseudohyphae some workers group them together (as *Candida*).

Identification

The methods given in Chapter 9 will yield information about morphology, sugar fermentation and assimilation of carbon and nitrogen sources. The morphological key (Table 50.1) will allow most yeasts isolated in clinical and food laboratories to be identified to genus level. Specific identification may then be made by reference to Table 50.2 and Figure 50.1 (see also Campbell *et al.*, 1985).

Candida

Candidas are commensals in the human gut. The principal species of medical importance is *Candida albicans*, which causes infections of skin and mucous membranes, particularly in immunosuppressed patients.

C. parapsilosis is a normal skin commensal and frequently infects nails. It is also a common cause of candida endocarditis.

Other species rarely causing human disease include *C. tropicalis, C. kefyr* (*C. pseudotropicalis*), *C. krusei, C. guillermondii, C. lusitaniae* and *C. zeylanoides*.

All produce creamy, white, smooth colonies on malt or peptone agar, except *C. krusei* which has a ground glass appearance (see Table 50.2).

Cryptococcus

Like other encapsulated yeasts (*Rhodotorula, Trichosporon*), these are the asexual forms of Basidiomycetes although the sexual forms are not normally encountered.

One species, *C. neoformans*, is pathogenic for humans, causing meningitis (which may be fulminant or slowly progressive and is common in AIDS patients) and subcutaneous or deep granulomata.

Table 50.1 Key to identification of genera of yeasts

Cells reproducing by fission, not budding	*Schizosaccharomyces*
Budding cells only (corn meal and coverslip preparation); no ascospores	
(1) Cells small (2-3 μm), bottle-shaped, with broad base to daughter bud	*Pityrosporum*
(2) Cells oval with prominent lateral 'spur'; colony pinkish, depositing mirror image of colony on lid of inverted petri dish	*Sporobolomyces*
(3) Cells not as in 1 or 2	
(a) Urease negative	*Torulopsis*
(b) Urease positive; colonies pink/red; cells large, encapsulated	*Rhodotorula*
(c) Urease positive; colonies white or cream; cells large; encapsulated	*Cryptococcus*
Budding cells, some with ascospores	
(1) Ascospores liberated from parent cells, kidney-shaped or elongate	*Kluyveromyces*
(2) Ascospores liberated, round with or without disc-like flange; nitrate assimilated	*Hansenula*
(3) Ascospores as in 2; nitrate not assimilated	*Pichia*
(4) Ascospores retained in parent cells, round, smooth	*Saccharomyces*
(5) Ascospores retained, round, warty and ridged	*Debaromyces*
Budding cells borne on pseudomycelium; no ascospores	*Candida* (most spp.)
Budding cells born on pseudomycelium which grows out to true mycelium after 3–4 days	
(1) Chlamydospores present	*C. albicans*
(2) No chlamydospores	*C. tropicalis*
Extensive true mycelium, fragmenting to form arthrospores	
(1) Budding on short side branches	*Trichosporum*
(2) No budding	*Geotrichum*

In advanced infections agglutination tests may be negative because the capsular material is poorly metabolized by the body, and the accumulation of it causes a high dose immunological tolerance. This antigen can be detected in serum or in CSF by the agglutination of latex sensitized with γ-globulin from an immune rabbit serum as described on p.143. For details, see MacKenzie *et al*. (1980).

Torulopsis

This is a large group of asexual yeasts, placed in the genus *Candida* by many authorities. The commonest species in clinical mycology is *T. glabrata*, a small-celled species commensal on the skin. It occasionally causes peritonitis and urinary tract infections in debilitated patients. *T. candida* is a large-celled type, occasionally found in similar sites.

Rhodotorula

These are frequent skin contaminants. They form red or orange colonies on malt or glucose peptone agar. Morphologically they resemble cryptococci and may be capsulated. They are of no clinical significance.

Table 50.2 Biochemical differentiation of some common yeasts

	Fermentation of				*Assimilation of*								
	glu	*ma*	*su*	*la*	*glu*	*ma*	*su*	*la*	*mn*	*ra*	*ce*	*er*	NO_3
Candida albicans	+	+	−	−	+	+	+	−	+	−	−	−	−
C. tropicalis	+	+	+	−	+	+	+	−	+	−	+	−	−
C. kefyr	+	−	+	+	+	−	+	+	+	+	+	−	−
C. parapsilosis	+	−	−	−	+	+	+	−	+	−	−	−	−
C. guillermondii	+	−	+	−	+	+	+	−	+	+	+	−	−
C. krusei	+	−	−	−	+	−	−	−	−	−	−	−	−
Torulopsis glabrata	+	−	−	−	+	−	−	−	−	−	−	−	−
T. candida	±	−	±	−	+	+	+	+	+	+	+	±	−
Cryptococcus neoformans	−	−	−	−	+	+	+	−	+	+w	+w	±	−
C. albidus	−	−	−	−	+	+	+	±	+	+w	+	−	+
C. laurentii	−	−	−	−	+	+	+	+	−	−	−	−	−
Trichosporum cutaneum	−	−	−	−	+	+	+	+	±	±	±	±	−
T. capitatum	−	−	−	−	+	−	−	−	−	−	−	−	−
Saccharomyces cerevisiae	+	+	+	−	+	+	+	−	±	±	−	−	−

glu, glucose
ma, maltose
su, sucrose
la, lactose
mn, mannitol

ra, raffinose
ce, cellobiose
er, erythritol
NO_3, nitrate

+, gas produced or growth occurs
−, no gas produced or no growth occurs
w, weak

Sporobolomyces

These yeasts live on leaf surfaces and reproduce by ballistospores which are ejected into the air. The salmon pink colonies of *S. roseus* produce mirror images on the lids of petri dishes if left undisturbed. They are of no clinical significance.

Trichosporum

T. beigelli is the one most frequently isolated in clinical work. It is a skin contaminant but occasionally causes deep infections in immunocompromised patients. *T. capitatum* causes 'white piedra' in which masses of yeasts grow on the outer surfaces of axillary and other hair which is kept permanently moist.

The colonies of these species are whitish, wrinkled, tough, and often slightly 'hairy' on the surface, due to the presence of aerial hyphae.

Sexual yeasts

Saccharomyces

This genus contains the brewing ('pitching') and baking yeasts. *S. cerevisiae* is the 'top yeast' used in making beer, and *S. carlsbergensis* the 'bottom yeast' used in making lager. There are many special strains and variants used for particular purposes.

S. pastorianus, S. rouxii and S. mellis

A 'bottom yeast' with long sausage-shaped cells, which produces unpleasant flavours in beer. *S. rouxii* and *S. mellis*, which are now placed in the genus *Zygosaccharomyces*, are 'osmophilic yeasts' which will grow in concentrated sugar solution (*cf.* aspergilli of the glaucus group) and spoil honey and jam.

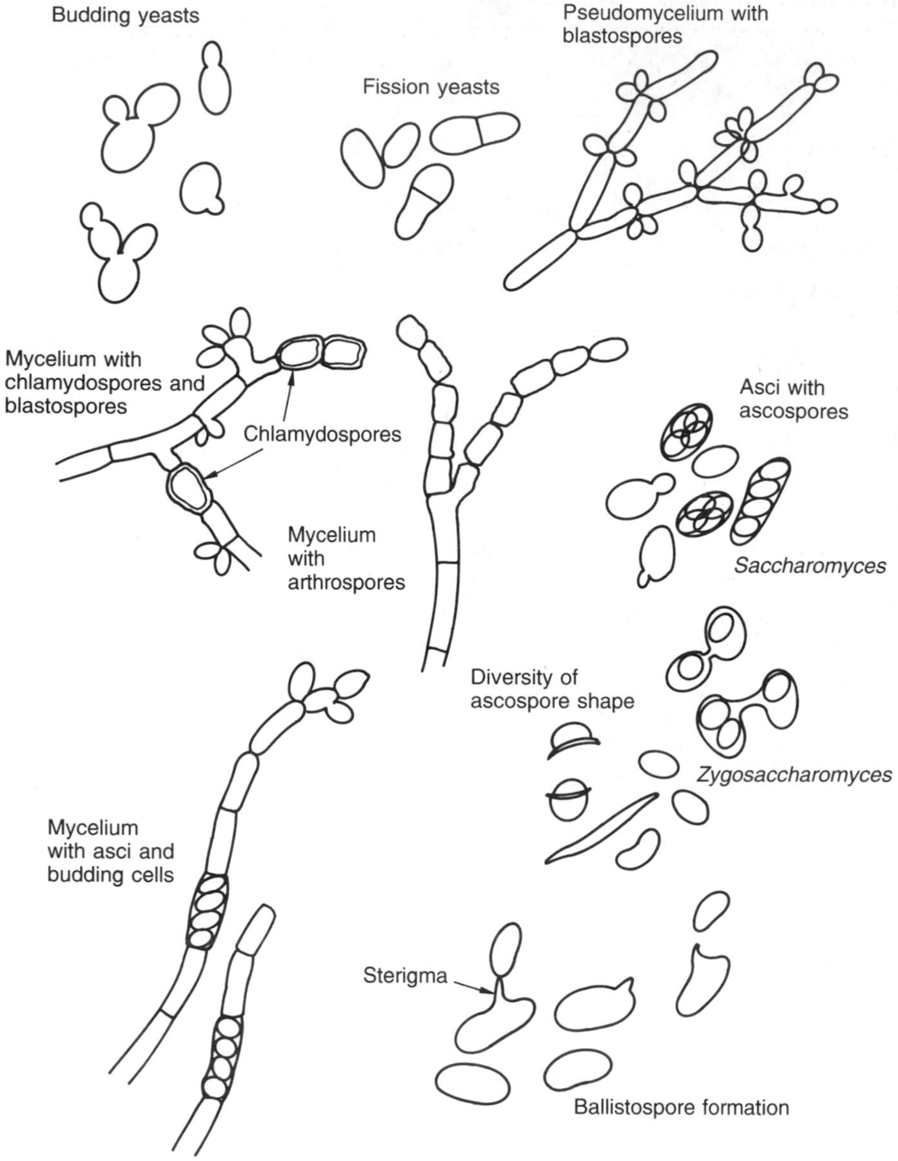

Figure 50.1 Yeasts

Zygosaccharomyces baillii
This can grow at low pH in the presence of some preservatives and has become a major problem in some areas of the food and beverage industry.

Pichia and Hansenula
Occur as contaminants in alcoholic liquors. They can use alcohol as a source of carbon, and are a nuisance in the fermentation industry. They grow as a dry pellicle on the surface of the substrate. Some species occur as the 'flor' in sherry and in certain French wines, to which they give a distinctive flavour. They also cause pickle spoilage, particularly in low-salt (10–15%) brines.

Debaromyces
These are also 'pellicle forming' yeasts. They are found in animal products such as cheese, glue and rennet, and are also responsible for spoilage in high-salt (20–25%) pickle brines.

Apiculate yeasts
These lemon-shaped yeasts are common contaminants in food. Perfect forms are placed in the genus *Hanseniaspora* and imperfect forms in the genus *Kloeckera*.

Yeasts are considered in detail by Davenport (1981) and Kreger-Van Rij (1984).

References

Campbell, C. K., Davis, C. and Mackenzie, D. W. R. (1985) Detection and isolation of pathogenic fungi. In *Isolation and Identification of Micro-organisms of Medical and Veterinary Importance* (eds C. H. Collins and J. M. Grange), Society for Applied Bacteriology Technical Series No. 21, Academic Press, London, pp. 329–343

Davenport, R. R. (1981) Yeasts and yeast-like organisms. In *Smith's Introduction to Industrial Mycology* (eds A. H. S. Onions, D. Allsop and M. O. W. Eggins), Edward Arnold, London, pp.65–92

Kreger-Van Rij, N. J. W. (ed.) (1984) *The Yeasts, A Taxonomic Study*, 3rd edn, Elsevier, Amsterdam

Mackenzie, D. W. R., Philpot, C. M. and Proctor, A. G. J. (1980) *Basic Serodiagnostic Methods for Diseases Caused by Fungi*, Public Health Laboratory Service Monograph No. 12, HMSO, London

Chapter

51

Common moulds

Many common moulds occur as contaminants on ordinary culture media. Most of them are saprophytes, causing spoilage of food and other commodities; others are plant pathogens. Some moulds (e.g. *Aspergillus*) are described in here although they are also recognized causes of human disease, especially in immunosuppressed patients or are of public health importance because they produce mycotoxins (e.g. *Penicillium*). Each isolation of a mould from human material must therefore be assessed individually and some reliance must be placed on the observation of hyphae by direct microscopy. The preliminary identification guide (Table 51.1) therefore includes some of the pathogens described in Chapter 52.

Alternaria and Ulocladium

Large genera of plant parasites and saprophytes. Without knowledge of the host plant speciation is almost impossible. Rare cases of subcutaneous granulomata in humans have been reported.

Colonies form a high, densely fluffy mat, white at first, becoming dark grey, deepest in colour in the centre. On some media the whole colony is black, with less aerial growth. Spores are large, brown pigmented, club shaped, multicellular with some septa longitudinal or oblique. In *Alternaria* (Figure 51.1) the spores are in chains: in *Ulocladium* they are single or only occasionally joined together. A tape impression mount will demonstrate this difference. *Ulocladium chartarum* may be responsible for discoloration of paintwork and wall paper on damp walls.

Arthrinium

Common contaminants of no medical importance. Most species are from plant material, on which they look quite unlike the *in vitro* growth.

The growth on laboratory media is dense, vigorous and at first pure white in colour. The centres of older cultures become grey because of the abundance of spores which are unicellular, rounded and black.

Chaetomium

Saprophytes on plant material. Some are troublesome pests of paper and cellulosic materials. There are no medical implications although some species are known to produce toxic metabolites.

Colonies are flat with low, often sparse aerial growth, with black or greenish discrete bodies standing on the agar surface. Low power examination shows that these are covered with long black spines (Figure 51.1). Under the high power these are seen to be straight, spiral or irregularly branched according to species. Large numbers of globose asci are produced inside the perithecium, each containing dark brown, ovoid ascospores. In *Chaetonium* the ascus walls

Table 51.1 Colour guide to mould cultures

Black or dark brown mycelium spores or both	
Alternaria	*Cladosporium*
Arthrinium	*Exophiala*
Aureobasidium	*Fonsceae*
Botrytis	*Madurella*
Chaetomium	*Phialophora*
Aspergillus (*niger*)	*Sporothrix*
Some shade of green predominant	
Trichoderma	*Aspergillus* (*flavus/parasiticus*)
Chaetomium	*Penicillium*
Some shade of red predominant	
Monascus	*Trichophyton*
Paecilomyces (*lilacinus*)	*Acremonium*
Aureobasidium	*Fusarium*
Colonies white or cream	
Chrysosporium	*Coccidioides*
Geomyces	*Blastomyces*
Dermatophytes	*Mucor*
Histoplasma	
Colonies sandy-brown to khaki	
Paecilomyces (*varioti*)	*Epidermophyton floccosum*
Scopulariopsis	*Scopulariopsis*
Aspergillus terreus	

rapidly break down, releasing the spores and giving the impression that the perithecium itself is full of loose spores.

Aureobasidium

Occurs on damp cellulosic materials and grows on painted surfaces.

Colonies are at first yeast-like, often mucoid, with pink, white and black areas. In time the black areas increase in size and the colony becomes drier and rough surfaced. There may be aerial mycelium. They are often referred to as 'black yeasts'.

Microscopy shows a bizarre mixture of oval to long budding yeasts and pigmented, irregularly-shaped hyphal cells. Under low power the edge of the colony shows yeasts clustered in round masses along the submerged hyphae. This is best seen in a slide culture or needle mount (Figure 51.1).

A. pullulans is the commonest of the saprophytic species in the group.

Botrytis

A group of plant pathogens, of which the common *B. cinerea* has a very wide host range and is also a successful saprophyte. It is well known as the grey mould of fruits such as strawberries.

Growth is vigorous, producing a high, grey turf which develops discrete black bodies (sclerotia) at the agar surface, especially at the edge of the culture

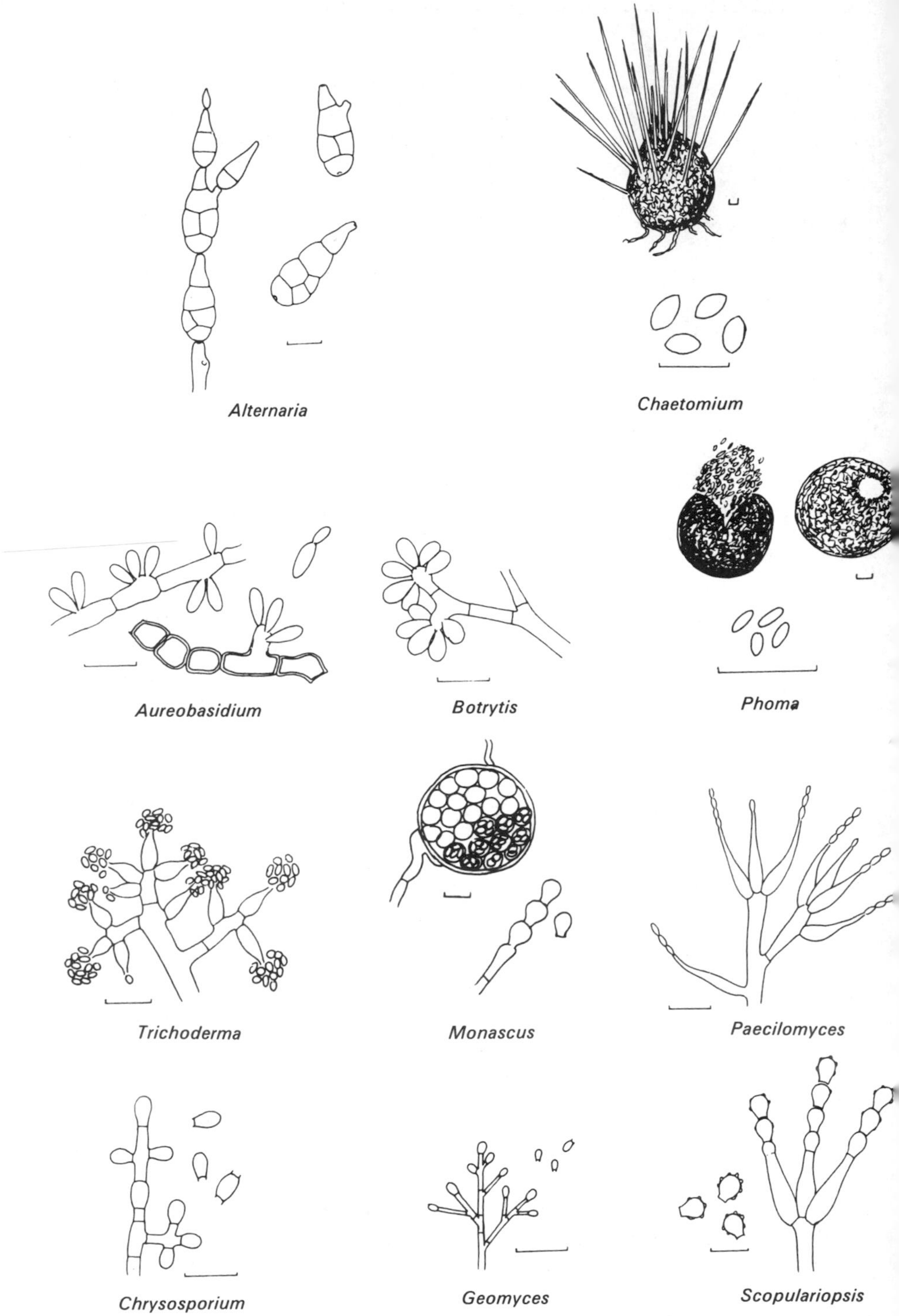

Figure 51.1 Common moulds (bar = 10 μm)

vessel. The large, colourless unicellular spores are produced in grape-like clusters on long, robust black conidiophores (Figure 51.1).

Phoma

This is a representative of a large number of pycnidial genera, all plant parasites, any of which may be encountered as contaminants. It is included here because it is probably one of the commonest and many others were once classified with it.

Some species produce low, flat colonies with areas of black mixed with pink (spore masses). Others may give dense, fluffy growths with black or grey shading. Microscopically they all show pycnidia — spherical structures containing large numbers of small conidia, often released in a wet, worm-like mass through a round ostiole. These are best seen by examining the growing colony *in situ* under low power (Figure 51.1).

Trichoderma

Growth is white, rapidly spreading and cobweb-like, with irregular patches of dark green spores. These are abundant small, ovoid, arising in ball-like groups on short right-angled branches (Figure 51.1). This fungus may be mistaken for a *Penicillium* but the rapidly spreading growth form, often covering the agar surface and even the sides of the petri dish, would be very unusual for *Penicillium*.

Monascus

A saprophyte. Colonies are flat, rather granular with a deep red pigment. The characteristic ascospores are produced inside spherical ascocarps. These are unusual in that each develops from a single (or a very few) hyphae and they have a thin wall (Figure 51.1). Chains of conidia superficially resembling *Scopulariopsis* may also be present.

Paecilomyces

Usually seen as a spoilage fungus but one species, *P. variotti*, has been found in rare, deep infections in immunosuppressed patients.

It resembles *Penicillium*, producing either pale purple or yellowish olive colonies depending on species. This feature, the distinctly elongated spores, and the long, tapering tips to the spore-producing cells of the 'penicillus' distinguish it from *Penicillium* (Figure 51.1).

Chrysosporium

Soil dwellers, many of which specialize in utilizing keratinaceous substrates.

This genus is close to the dermatophytes and like them is unaffected by cyclohexamide in the medium. Colonies of most species are white or cream coloured and flat with felt-like surfaces. Spores are similar to those of dermatophyte microconidia but mostly longer than 5 μm (Figure 51.1).

Geomyces

Saprophytes. One species, *G. pannorus*, a soil organism, is often mistaken for a dermatophyte and it has been reported on the surfaces of meat carcases in cold storage at temperatures as low as $-6°C$ (usually under the name *Sporotrichum carnis*).

Colonies are restricted, often heaped up, with a very thin white crust of aerial spores. These resemble dermatophyte microconidia but are smaller than most (< 1.5 μm long) and produced on minute, acutely-branched Christmas tree-like structures (Figure 51.1).

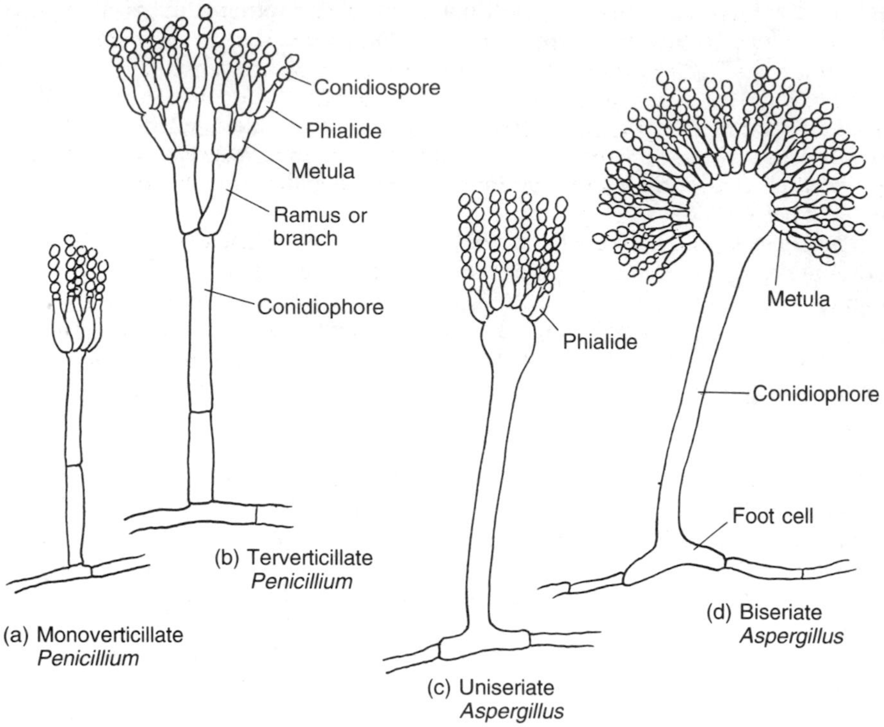

Figure 51.2 (a) and (b) *Penicillium*; (c) and (d) *Aspergillus*

Scopulariopsis
One species, *S. brevicaulis*, is a saprophyte, living in soil but does seem to be adapted to utilize keratin and is one of the commoner invaders of human nail tissue. It is also notable for releasing gaseous arsines from the arsenical pigment Paris green.

Colonies are often folded and heaped up, but they may be flat. When sporing they are cinnamon-brown in colour and have a powdery surface. The spores are large, rounded, with a flat detachment scar and are produced in long chains from a penicillium-like branching structure. In most strains the spore surfaces are distinctly roughened (Figure 51.1).

Aspergillus and Penicillium

These genera are closely related but although there are several species that appear intermediate in morphology, there is generally no difficulty in placing isolates in one genus or the other. Both produce long, dry chains of conidia by repeated budding through the end of a bottle-shaped cell, the phialide.

In *Penicillium* the phialides are clustered at the tips of tree-like conidiophores and develop unevenly.

In *Aspergillus* they are clustered on a club-shaped or spherical vesicle and develop synchronously (Figure 51.2).

In the subgenus *Aspergilloides* of *Penicillium* the phialides are attached directly to the tip of the conidiophore (monoverticillate), which may be distinctly swollen, giving the impression of a very simple *Aspergillus* fruiting body. However, the conidiophore of *Penicillium* is septate, and very similar in appearance to the mycelium from which it has branched, whereas that of *Aspergillus* is a more specialized structure, usually non-septate and developing from a distinct foot cell differentiated clearly from the supporting mycelium.

In both groups the mycelium is usually colourless, but may be coloured by the production of pigments retained in the cytoplasm or secreted into the medium; any colour on the obverse is due to the coloured spores and on the reverse due to secreted pigments. *Penicillium* spores are usually some shade of green or blue.

Aspergillus spores may be green, brown or black. In both genera there are species with white spores.

Some *Penicillium* spp. are plant parasites but the majority of both genera occur on decaying vegetation. Because of their wide range of secondary metabolites several are of importance in industrial fermentation and synthetic processes. They are also important in that they may produce mycotoxins in foods.

A. glaucus group

Members of this group flourish in conditions of physiological dryness, e.g. on the surface of jam, on textiles and on tobacco. They are very distinctive in appearance, with bluish green or grey-green conidia and large yellow cleistothecia.

A. restrictus group

Like *A. glaucus*, these will grow on dry materials, such as slightly dry textiles. As their name suggests, they are slow-growing organisms.

A. fumigatus

So-called from its smoky green colour. It is a common mould in compost, and can grow at temperatures above 40°C and hence it is pathogenic for birds. It spores more readily at 37°C than at 26°C. It is now notorious as a human pathogen, underlying pulmonary eosinophilia. In this condition, the fungus may be found in the mucous plugs that are expectorated. It also invades old tuberculous cavities and gives rise to aspergilloma of the lung. When isolated from such cases, the organism often does not produce spores; incubation at 42°C may encourage it to do so. Patients who harbour this organism develop precipitating antibody, which is rather weak in pulmonary eosinophilia but strong in cases of aspergilloma. For methods, see p. 141. If *A. fumigatus* is examined with a lens, the heads are seen to be columnar. It looks like a test-tube brush.

A. niger

On the other hand, this organism has round heads which are large enough to appear discrete to the naked eye. This, together with their black colour, makes them easy to recognize.

The flavus-oryzae group

Members of this group are notorious for the production of aflatoxin in groundnuts and other foods such as millet when badly stored (see Moss *et al.* 1989). They are also of value commercially, as a source of diastatic enzymes such as takadiastase. The specific name, which means yellow, is an error; the colony colour in this series is green. The heads are round, but smaller than those of *A. niger*.

Table 51.2 shows the morphological properties of some *Aspergillus* species/groups.

Table 51.2 Morphology of *Aspergillus* species

	Aspergillus						
	fumigatus	*flavus*	*niger*	*terreus*	*versicolor*	*nidulans*	*glaucus*
Finely roughened stalks	−	+	−	−	−	−	−
Stalks pale brown	−	−	−	−	−	+	−
Green colony	+	+	−	−	+	+	+
Black colony	−	−	+	−	−	−	−
Sand-brown colony	−	−	−	+	−	−	−
Metulae present	−	+/−	+	+	+	+	−

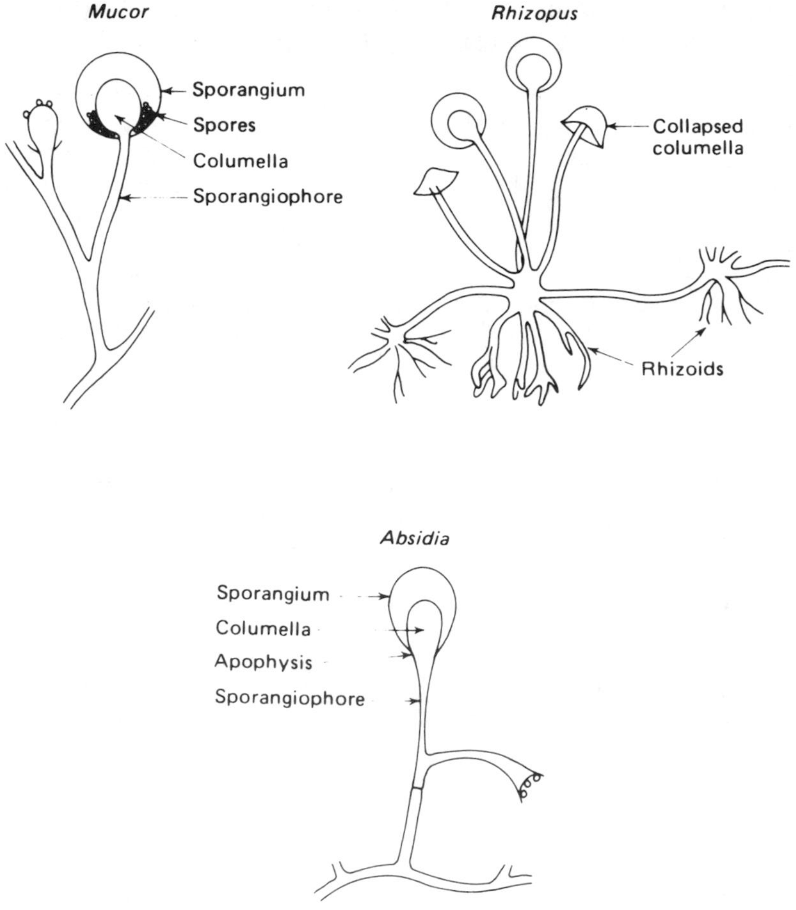

Figure 51.3 Zygomycetes

Zygomycotina

Fungi with non-septate, wide hyphae in which sexual fusion results in a thick-walled resting zygospore. These are seldom seen, however, and distinction of genera and species is largely based on the asexual structures. The group contains two main orders. Many of them grow rapidly and produce aerial mycelium resembling sheep wool.

Mucorales

These reproduce by many-spored sporangia. The majority belong to three genera: *Mucor, Rhizopus* and *Absidia.*

Rhizopus
Columella enlarges greatly after rupture of the sporangium and collapses to form a characteristic mushroom shape. Spores are angular and delicately striated. Root-like rhizoids may occur at the base of sporangium stalk (Figure 51.3).

Absidia
Stalk trumpet-shaped and sporangium more or less continuous (Figure 51.3).
The absence of features characteristic of the latter two genera suggests *Mucor* spp.
The common pathogenic species, all causing mucormycosis, are *M. pusillus, M. corymbifera* and *R. oryzae.* They are thermophilic: check their identity by culturing at 37°C.

Reference

Moss, M. O., Jarvis, B. and Skinner, F. A. (1989) Filamentous fungi in foods and feeds. *Journal of Applied Bacteriology* (Symposium Suppl. No. 18), **67**, 1S–144S

Chapter

52

Pathogenic moulds

The following fungi are considered in this chapter: Dermatophytes, causing infections of epidermal tissues; opportunist deep-tissue pathogens; agents of subcutaneous infections; agents of coccidiomycosis, histoplasmosis and blastomycosis.

Dermatophytes

There are three genera of common dermatophytes, classified according to the type of macroconidia they produce:

Microsporum—spindle-shaped and roughened
Trichophyton—cylindrical and smooth
Epidermophyton—club-shaped and smooth

Microsporum and *Epidermophyton* are easily recognized by their macroconidia but *Trichophyton* rarely produces them.

The nature of the disease assists in identification as many of these fungi have characteristic infection patterns (Table 52.1). This is particularly helpful in identifying strains with abnormal morphology.

Table 52.1 Characteristic infection patterns of dermatophytes

Patient	*Body site*	*Geographical origin*	*Organism*
Adult	Feet, hands, groin	Any	*T. rubrum* *T. interdigitale* *E. floccosum*
Adult or child	Scalp, face, arms, chest, legs	Any	*M. canis* *M. gypseum* *T. verrucosum* *T. mentagrophytes*
Adult or child	Scalp, face, arms, chest, legs	N Africa, E and SE Asia	*T. violaceum*
Child	Scalp	Any	*T. tonsurans*
Child	Scalp	W and Central Africa, Caribbean	*M. audouinii*
Child	Scalp	N and E Africa	*T. soudanaense*
Child	Scalp	N and E Africa, Middle East	*T. schoenleinii*

Appearance in clinical material

In nail and skin cleared with caustic soda fungal hyphae may be seen, some of them breaking into arthrospores, but specific identification cannot be made from their arrangement.

Some dermatophyte-infected hairs fluoresce under a Wood's lamp (examine in a darkened room).

Table 52.2 indicates the appearance of infected hairs and Figure 52.1 shows various types of infected hairs, skin and nails.

Table 52.2 Appearance of infected hairs

Spore arrangement	*Fluorescence*	*Organism*
Small spore ectothrix	+	*M. canis*
	+	*M. audouinii*
	−	*T. mentagrophytes*
Large spore ectothrix	−	*M. gypseum*
	−	*T. verrucosum*
Endothrix	−	*T. soudanense*
	−	*T. violaceum*
'Favic' hairs	weak	*T. schoenleinii*

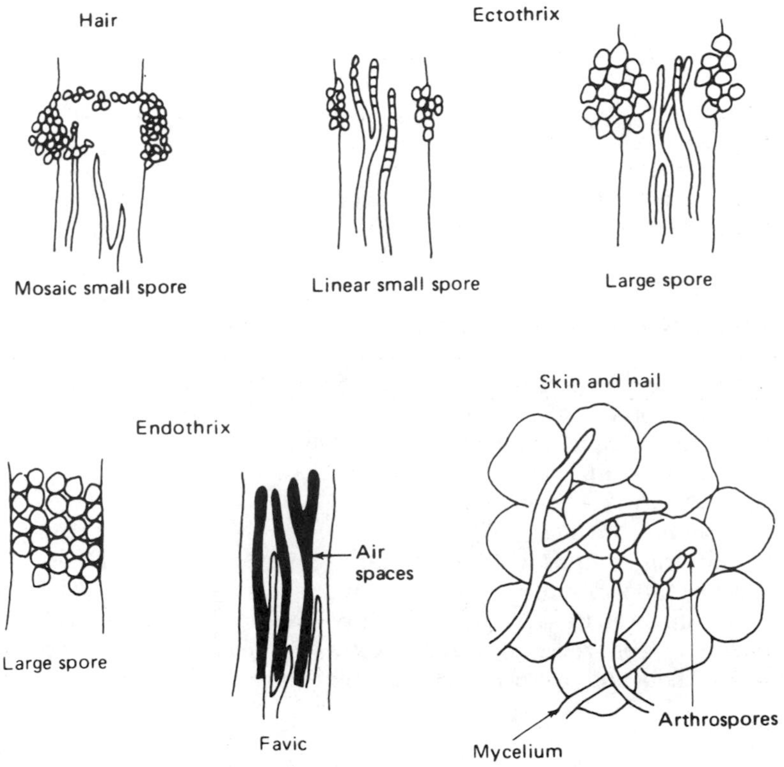

Figure 52.1 Infected hair, skin and nail

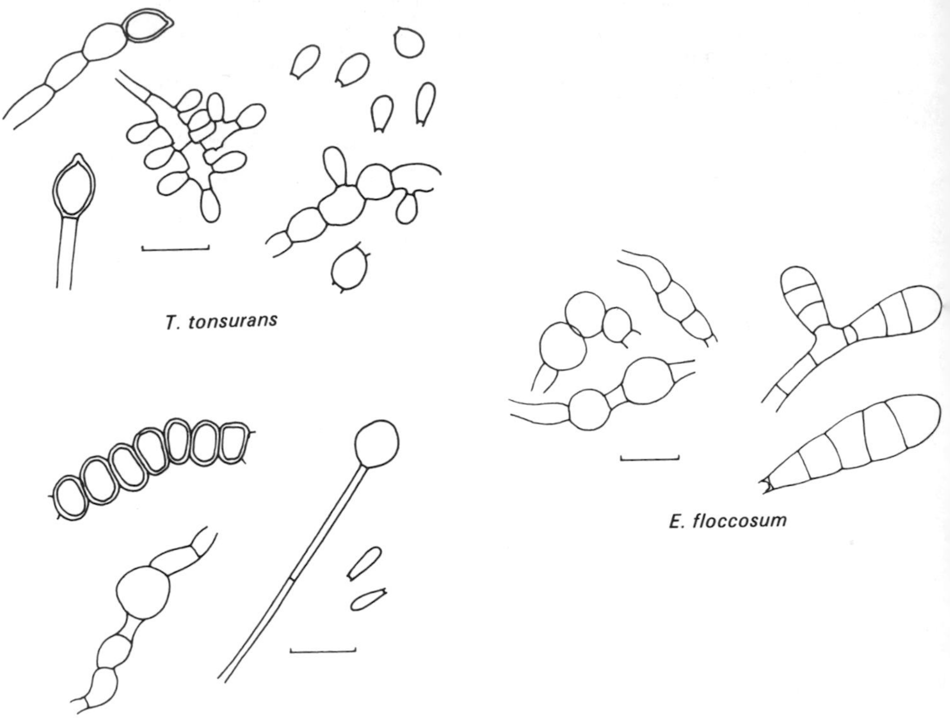

Figure 52.2 Microscopical appearances of some dermatophytes

Identification

Identification to species level depends on colonial and microscopical appearances, both of which are influenced by the growth medium.

Use a good brand of Sabouraud dextrose agar which has been shown to give adequate pigmentation and spore production.

Microsporum audouinii

This is a scalp fungus with no animal or soil reservoir. It spreads from child to child in schools, and infection usually resolves at puberty. The organism is of medium growth rate, with a thinly-developed, white to pale buff, aerial mycelium. The reverse of the colony is typically a pale apricot colour. Macroconidia and microconidia are rare (Figure 52.2).

M. canis

As its name suggests, this is a common parasite of dogs and cats, where the infection is often difficult to detect. Children are commonly infected from pets.

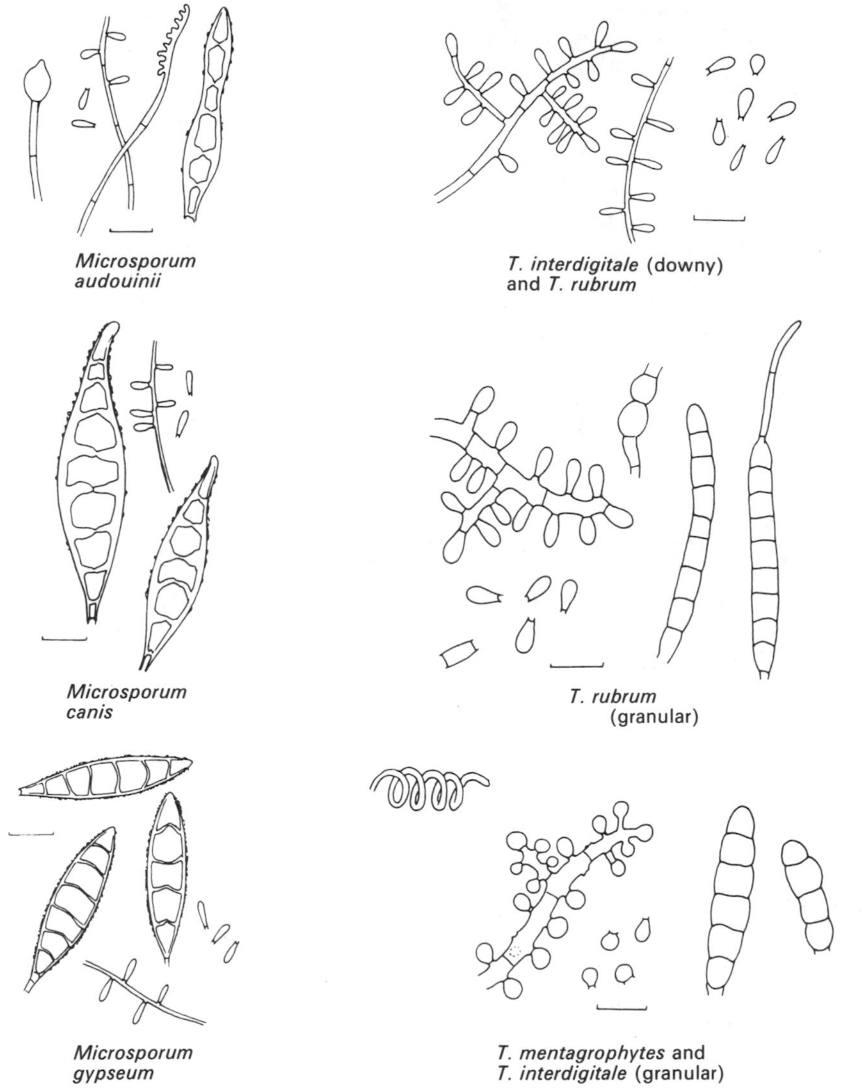

Figure 52.2 Microscopical appearances of some dermatophytes

This organism, which is fast growing, usually produces a vivid yellow pigment and fairly abundant macroconidia. Occasionally, strains are non-pigmented and slow growing. To differentiate them from *M. audouinii*, grow on rice (5 g of rice and 20 ml of water in a 100-ml flask, autoclaved at 115°C for 20 min and cooled). *M. canis* produces aerial mycelium; *M. audouinii* does not (Figure 52.2).

M. gypseum
This occurs as a saprophyte in soil and has a low virulence for humans. The occasional case which occurs usually shows a solitary lesion, related to contact with the soil. The fungus grows fast, and is soon covered with a buff-coloured granular coating of spores. These tend to be arranged in radial strands on the surface, rather like a spider's web. Macroconidia are abundant (Figure 52.2).

Trichophyton rubrum
Commonest dermatophyte seen in dermatology clinics in developed countries, but the less common *T. interdigitale* is probably more widespread in the general population.

There are many varieties of *T. rubrum*; that most often seen gives a downy, white colony with a red-brown pigment and a narrow white edge on reverse. Microscopy shows only club-shaped microconidia (Figure 52.2).

Other varieties are 'melanoid' in which a dark brown pigment diffuses into the medium; 'yellow', which is a non-sporing pale yellow form; and 'granular' which has the usual dark red reverse but powdery aerial growth with patches of pink. This variety has larger microconidia and usually produces macroconidia (Figure 52.2).

Non-pigmented strains occur which resemble the downy variety of *T. interdigitale* (Figure 52.2). To distinguish these, subculture on urea agar. *T. rubrum* does not usually produce the colour change by 7 days but most other dermatophytes do.

Trichophyton mentagrophytes complex
The commonest form in this group is the anthropophilic *T. interdigitale*, which causes athlete's foot.

Colonies resemble those of the zoophilic *T. mentagrophytes* (see below), with a uniformly powdery, cream-coloured surface made up of an abundance of nearly spherical microconidia. Spiral hyphae and macroconidia may be present (Figure 52.2). The downy variety forms a pure white cottony colony and has elongated microconidia resembling those of *T. rubrum*.

The zoophilic (usually from rodents) *T. mentagrophytes* has coarsely granular colonies with areas of agar surface showing between radiating zones of white or cream sporing growth. Reverse pigmentation often shows dark brown 'veins'.

T. tonsurans
A cause of scalp ringworm. Its colony is slow growing and velvety, often with a folded centre. Most strains show a mixture of microconidia and swollen hyphal cells (chlamydospores). Microconidia are large, oval and arranged along wide, often empty hyphae (Figure 52.2).

T. verrucosum
The organism of cattle ringworm. It causes scalp, beard or nail infections in farm workers, and other people such as slaughtermen and veterinary surgeons who come into contact with cattle. In culture it grows very poorly; suspected cultures should be examined carefully with a lens after 14 days' incubation, as colonies are often submerged and minute. This, together with the large, thin walled, balloon-like chlamydospores at the end of a straight hypha, and chains of thick-walled cells (Figure 52.2) is sufficient to identify the organism.

T. schoenleinii
The cause of favus. The colony is slow growing, hard and leathery, with a surface like white suede. The only microscopic features are thick, knobbly hyphae, like arthritic fingers—the so-called 'favic chandeliers'—and chlamydospores.

T. violaceum
Another cause of scalp ringworm, microscopically similar to *T. verrucosum*, although the hyphae are usually thinner. The characteristic feature is a deep

red-purple pigment, which appears as a spot in the centre of a young colony, and gradually spreads to the edge.

Epidermophyton
There is only one species, *E. floccosum*. This is one cause of tinea cruris ('dhobi itch') and 'jungle rot'. It also occasionally infects feet. Its growth on malt agar is khaki in colour with a powdery surface due to masses of macronidia whose shape is very characteristic (Figure 52.2)

For further information about dermatophytes see Mackenzie and Philpot (1981).

Entomophthorales

Insect pathogens, rarely seen in clinical material. The asexual spore form is a single, relatively large 'conidium' which is forcefully discharged.

Colonies are wrinkled, cream coloured and waxy. A haze of discharged spores forms a mirror image of the colony on the lid of the petri dish.

Basidiobolus
Cause of tropical subcutaneous zygomycosis.

Conidiobolus coronatus (*Entomophthora coronata*)
Causes tropical rhinoentomophthoromycosis of the nasal mucosa.

Miscellaneous pathogenic moulds

Madurella
M. mycetomatis and *M. grisea* cause black grain mycetoma. The former produces flat or wrinkled, yellowish colonies, often releasing melanoid pigments into the medium. The latter grows as domed, densely fluffy, mouse-grey to black colonies. Neither forms conidia.

Exophiala
Colonies are dark grey to black, intensely fluffy or wet and yeast-like depending on degree of conidial budding, strain and growth conditions. All produce conidia from small, tapering nipples which grow in length as more conidia are cut off (Figure 52.3).

E. jeanselmei is a cause of mycetoma, *E. werneckii* of tinea nigra, *E. dermatitidis* of chromomycosis and *E. spinifera* of subcutaneous cysts.

Phialophora
Black moulds in which the spore apex has a definite collarette or funnel-shaped cup in which the conidia are formed (Figure 52.3).

P. verrucosa is one cause of chromomycosis; *P. parasitica* is occasionally found in subcutaneous cysts.

Fonsecaea
Black moulds in which the conidia are formed in short chains by apical budding of the cell below. Differs from *Cladosporium* only in that the chains consist of no more than four spores (Figure 52.3).

F. pedrosi is the common cause of chromoblastomycosis; *F. compacta* is less common.

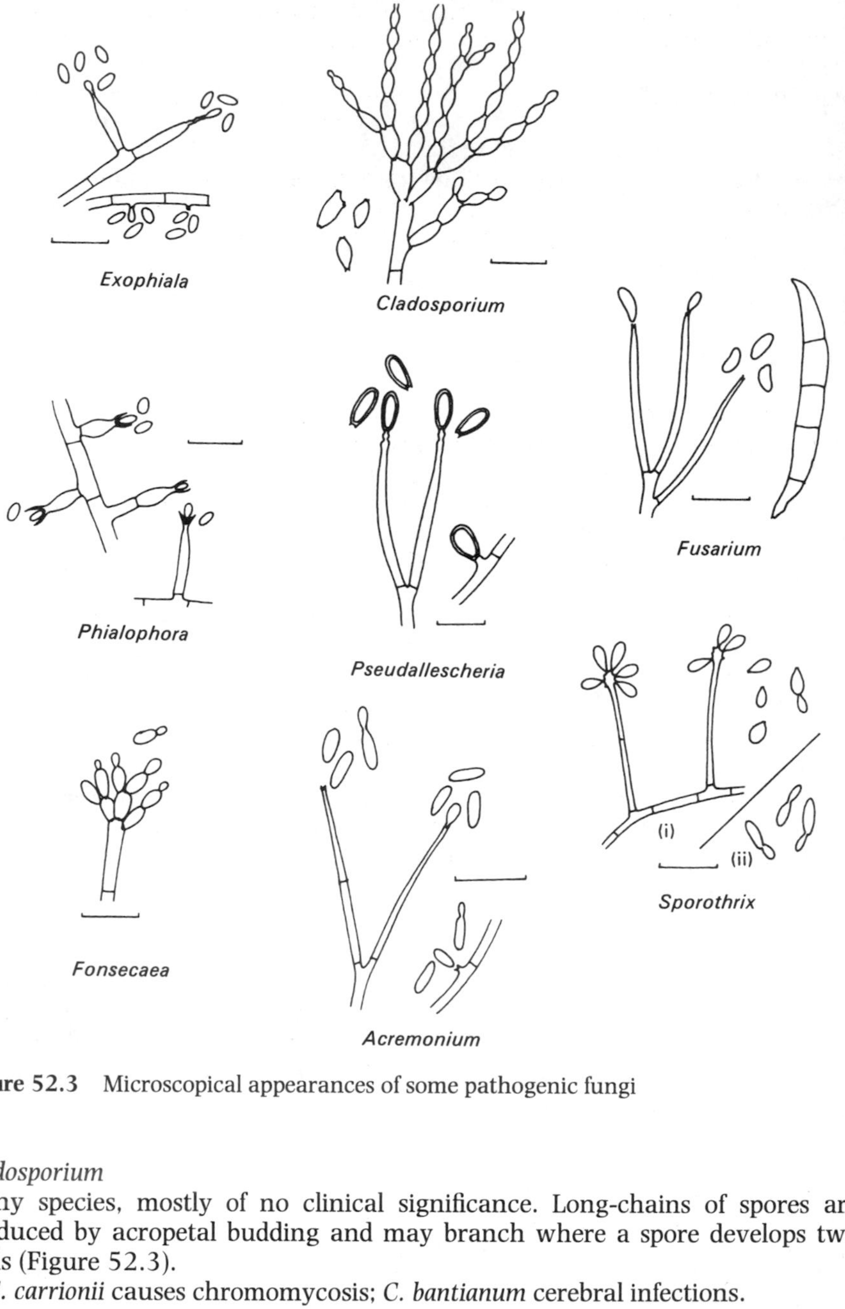

Figure 52.3 Microscopical appearances of some pathogenic fungi

Cladosporium
Many species, mostly of no clinical significance. Long-chains of spores are produced by acropetal budding and may branch where a spore develops two buds (Figure 52.3).

C. carrionii causes chromomycosis; *C. bantianum* cerebral infections.

Pseudallescheria boydii
Colonies fluffy, white to light grey, producing oval, slightly yellowish spores in succession at the tips of long, straight hyphae. Each spore has a basal scar. Occasional strains may produce black ascomata beneath the agar (Figure 52.3).

Variously named *Allescheria*, *Petriellidium* or *Monosporium*. Causes pale grain mycetoma and miscellaneous other infections (otitis externa in the UK).

Acremonium (*Cephalosporium*)
A very large group of white or pink moulds with wet heads of conidia produced one by one from the tips of straight hyphae or lateral nipples. In some, the conidia bud secondary spores and the colony becomes yeast-like (Figure 52.3).

A. kilense is a cause of pale grain mycetoma; other species invade nail tissue.

Fusarium
Colonies often loose and fluffy with a pink or purple pigment. Characterized by curved, multiseptate macroconidia, but these may be absent. Many strains produce microconidia similar to those of *Acremonium* but usually somewhat curved (Figure 52.3).

F. solani and *F. oxysporum* cause a variety of infections including mycetomas and keratitis.

Sporothrix
At low temperatures (25–30°C) colonies are grey to black, dry, membranous and wrinkled. Hyaline or pigmented spores are produced initially as rosettes on minute, teeth-like points on expanded apical knobs (Figure 52.3).

Differentiation from non-pathogenic species requires conversion to yeast form by culture at 37°C.

Causes sporotrichosis in which it exists as a budding, spindle-shaped yeast.

Blastomyces, Histoplasma and Coccidioides

These cause systemic or generalized infections resulting from the inhalation of airborne conidia. Heavily sporing mould cultures are more hazardous than those in the yeast-like phase. All manipulations that might release conidia should be done in microbiological safety cabinets in Containment Level 3 laboratories.

A useful cultural feature of these organisms is that they are not inhibited by cycloheximide.

Blastomyces dermatitidis
Causes North American blastomycosis. Infection is almost always by inhalation and skin lesions are usually secondary to a pulmonary infection, even a mild one. The mycelial form, at 26°C, is a moist colony at first, developing a fluffy aerial mycelium with microconidia. At 37°C or in tissue, large oval multinucleate yeasts are seen (Figure 52.4).

Histoplasma capsulatum
Causes histoplasmosis, which is very common in the southern half of the USA although it is also found in other countries. Most people recover rapidly: only an occasional patient develops progressive disease. The route of infection is inhalation, and the pathology similar to tuberculosis (Figure 52.4). See caution above.

At 26°C, the organism grows with a white aerial mycelium, which becomes light brown. The spores are striking and characteristic. At 37°C, small, oval budding yeasts are found. In tissue, these yeasts are intracellular, except in necrotic material.

Coccidioides immitis
Causes coccidiomycosis. This disease is limited to a few special areas in the south west of the USA and in Mexico and Venezuela. It is inhaled in desert

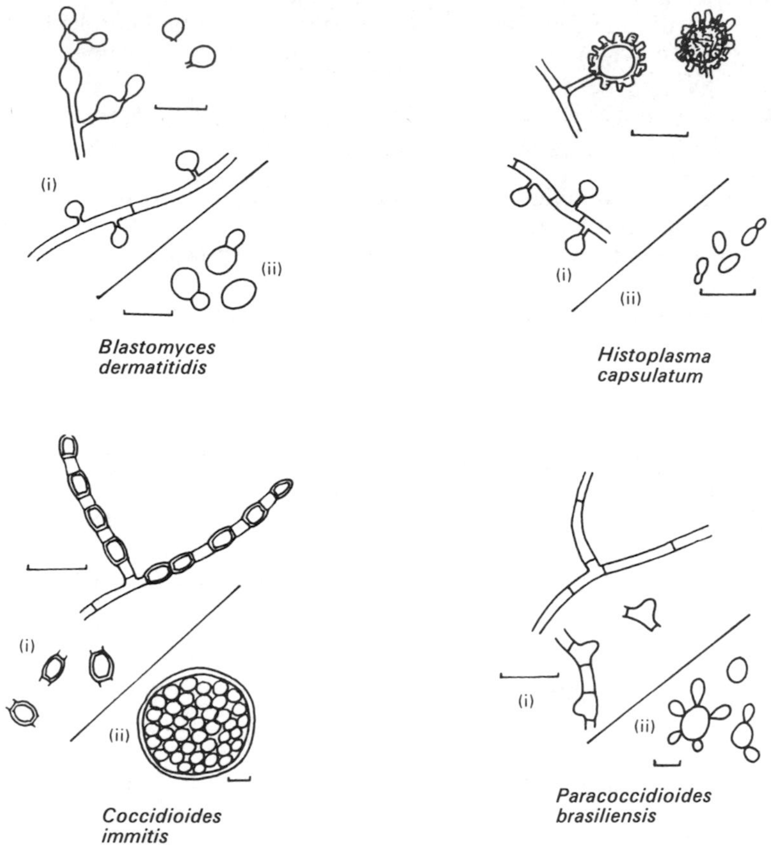

Figure 52.4 Microscopical appearances of *Blastomyces dermatitidis; Histoplasma capsulatum; Coccidiodes immitis; Paracoccidioides brasiliensis;* (i) mycelial form; (ii) tissue form

dust. Again, most people suffer only a mild, transient, 'flu-like' illness. Occasionally, cases progress to generalized destructive lesions. See caution above.

At 26°C, the colony is at first moist, then develops an aerial mycelium, becoming brown with age and breaking up into arthrospores. There is no second form in artificial culture. In the body, however, large, thick-walled cells are seen. When mature, these are full of spores (Figure 52.4).

Paracoccidioides braziliensis

Causes South American blastomycosis. The primary lesion is almost always in the oral mucosa, possibly because of the habit of cleaning the teeth with wood splinters and chewing bark in the places where the disease is found. The spread is through the lymphatics. The organism is very slow growing, white and featureless at 26°C. At 37°C on rich media, characteristic yeast cells are formed (Figure 52.4).

For more information about pathogenic fungi see Campbell and Stewart (1980), MacKenzie and Philpot (1981) and Clayton and Midgeley (1985).

References

Campbell, M. C. and Stewart, J. C. (1980) *The Medical Mycology Handbook*, J. Wiley & Sons, New York and Chichester

Clayton, Y. and Midgeley, G. (1985) *Medical Mycology*, Gower Medical Publishing, London and New York

Mackenzie, D. W. R. and Philpot, C. M. (1981) *Isolation and Identification of Ringworm Fungi*, Public Health Laboratory Service Monograph No.15, HMSO, London

Index